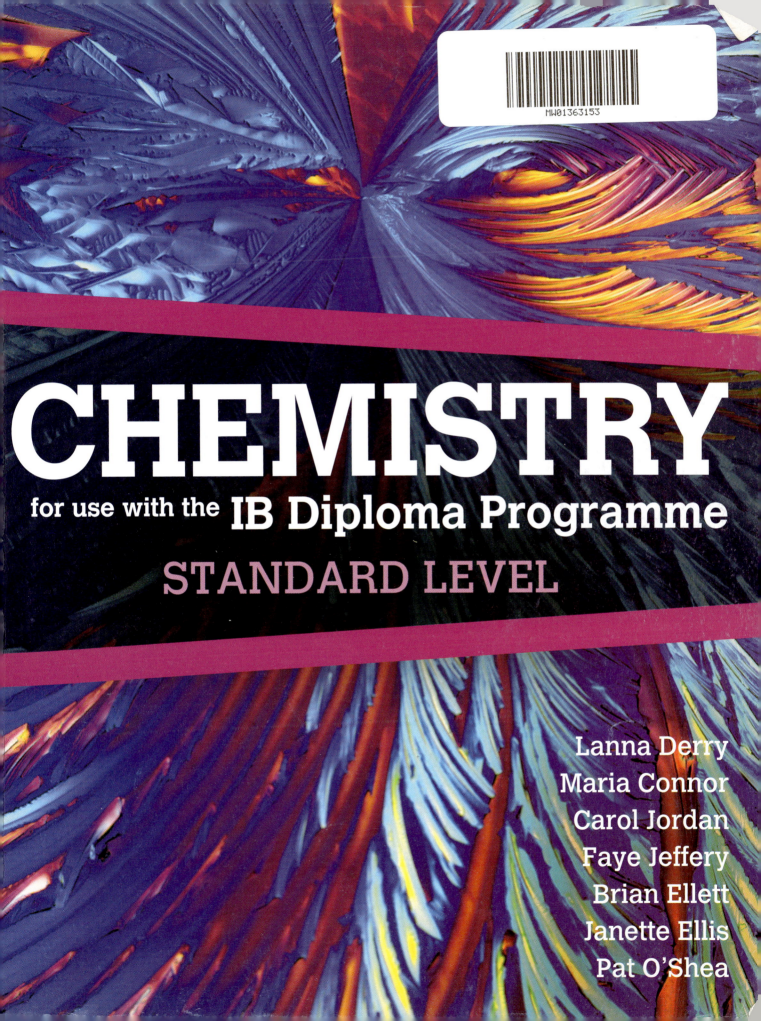

**Pearson Australia**
(a division of Pearson Australia Group Pty Ltd)
707 Collins Street, Melbourne, Victoria 3008
PO Box 23360, Melbourne, Victoria 8012
www.pearson.com.au

Copyright © Pearson Australia 2008
(a division of Pearson Australia Group Pty Ltd)
First published 2008 by Pearson Australia
Reprinted 2009, 2010 (twice), 2011, 2012

**Reproduction and Communication for educational purposes**
The Australian *Copyright Act 1968* (the Act) allows a maximum of one chapter or 10 per cent of the pages of this work, whichever is the greater, to be reproduced and/or communicated by any educational institution for its educational purposes provided that the educational institution (or the body that administers it) has given remuneration notice(s) to Copyright Agency Limited (CAL) under the Act. For details of the CAL licence for educational institutions contact Copyright Agency Limited (www.copyright.com.au).

**Reproduction and Communication for other purposes**
Except as permitted under the Act (for example a fair dealing for the purposes of study, research, criticism or review) no part of this book may be reproduced, stored in a retrieval system, communicated or transmitted in any form or by any means without prior written permission. All inquiries should be made to the publisher at the address above.

Edited by Marta Veroni
Designed by Meaghan Barbuto
Typeset by Nikki M Group Pty Ltd
Cover design by Glen McClay
Cover image description: Macro-photograph obtained with polarized light of crystals of cholesterol. A lipid found in vertebrates and in foods from animal sources, largely located in the brain, spinal cord and the liver; the major site of cholesterol synthesis.
Cover image by Science Photo Library
Produced by Pearson Australia
Printed in China (SWTC/06)

| | |
|---|---|
| Author: | Derry, Lanna. |
| Title: | CHEMISTRY: For use with the IB Diploma Programme Standard Level / authors, Lanna Derry, Maria Connor, Carol Jordan. |
| Edition: | 1st ed. |
| Publisher: | Melbourne: Pearson Australia, 2008. |
| ISBN: | 9780733993756 (pbk.) |
| Target Audience: | For secondary school age. |
| Subjects: | Chemistry--Textbooks. |
| Other Authors/Contributors: | Connor, Maria |
| | Jordan, Carol. |
| Dewey Number: | 540 |

Every effort has been made to trace and acknowledge copyright. However, should any infringement have occurred, the publishers tender their apologies and invite copyright holders to contact them.

# CONTENTS

## 1 Atomic structure — 1
- 1.1 The atom — 4
- 1.2 The mass spectrometer — 9
- 1.3 Electron arrangement — 14
- Chapter 1 Summary — 20
- Chapter 1 Review questions — 21
- Chapter 1 Test — 23

## 2 Bonding — 26
- 2.1 Ionic bonding — 28
- 2.2 Metallic bonding — 34
- 2.3 Covalent bonding — 36
- 2.4 Covalent bonding in network lattices — 51
- 2.5 Intermolecular forces — 57
- 2.6 Physical properties — 62
- Chapter 2 Summary — 66
- Chapter 2 Review questions — 70
- Chapter 2 Test — 72

## 3 Periodicity — 75
- 3.1 The periodic table — 77
- 3.2 Physical properties of the elements — 81
- 3.3 Chemical properties of elements and their oxides — 87
- Chapter 3 Summary — 91
- Chapter 3 Review questions — 92
- Chapter 3 Test — 94

## 4 Quantitative chemistry — 97
- 4.1 The mole concept and Avogadro's constant — 98
- 4.2 Calculations of mass and number of mole — 103
- 4.3 Empirical and molecular formulas — 108
- 4.4 Chemical equations — 114
- 4.5 Mass relationships in chemical reactions — 118
- 4.6 Factors affecting amounts of gases — 128
- 4.7 Gaseous volume relationships in chemical reactions — 131
- 4.8 Solutions — 142
- Chapter 4 Summary — 154
- Chapter 4 Review questions — 156
- Chapter 4 Test — 158

# 5 Measurement and data processing — 160

- 5.1 Uncertainty and error in measurement — 161
- 5.2 Uncertainty in calculated results — 171
- 5.3 Graphical techniques — 175
- Chapter 5 Summary — 182
- Chapter 5 Review questions — 183
- Chapter 5 Test — 185

# 6 Energetics — 188

- 6.1 Exothermic and endothermic reactions — 189
- 6.2 Calculation of enthalpy changes — 196
- 6.3 Hess's law — 204
- 6.4 Bond enthalpies — 207
- Chapter 6 Summary — 210
- Chapter 6 Review questions — 211
- Chapter 6 Test — 213

# 7 Kinetics — 216

- 7.1 Rates of reaction — 217
- 7.2 Collision theory — 225
- Chapter 7 Summary — 237
- Chapter 7 Review questions — 238
- Chapter 7 Test — 239

# 8 Equilibrium — 243

- 8.1 Dynamic equilibrium — 244
- 8.2 The position of equilibrium — 246
- 8.3 Industrial processes — 261
- Chapter 8 Summary — 268
- Chapter 8 Review questions — 269
- Chapter 8 Test — 271

# 9 Acids and bases — 274

- 9.1 Theories of acids and bases — 275
- 9.2 Properties of acids and bases — 282
- 9.3 Strong and weak acids and bases — 286
- 9.4 The pH scale — 292
- Chapter 9 Summary — 296
- Chapter 9 Review questions — 297
- Chapter 9 Test — 299

# 10 Oxidation and reduction — 301

10.1 Oxidation and reduction — 302
10.2 Redox equations — 307
10.3 Voltaic cells — 312
10.4 Reactivity — 320
10.5 Electrolytic cells — 327
Chapter 10 Summary — 333
Chapter 10 Review questions — 336
Chapter 10 Test — 339

# 11 Organic chemistry — 341

11.1 Introduction to organic chemistry — 342
11.2 Introducing functional groups — 356
11.3 Reactions of alkanes — 371
11.4 Reactions of alkenes — 376
11.5 Reactions of alcohols — 383
11.6 Reactions of halogenoalkanes — 387
11.7 Reaction pathways — 390
Chapter 11 Summary — 394
Chapter 11 Review questions — 397
Chapter 11 Test — 399

# Periodic Table — 402

# Appendix 1 Table of relative atomic masses — 403

# Appendix 2 Physical constants, symbols and units — 404

# Solutions — 406

# Glossary — 417

# Index — 425

# CHEMISTRY
## for use with the IB Diploma Programme
### STANDARD LEVEL

*The complete chemistry package*

**CHEMISTRY: For use with the IB Diploma Programme Standard Level** is the most comprehensive chemistry text specifically written for the IB Diploma Programme Chemistry course, Standard Level (Core).

The content is easy to follow and provides regular opportunities for revision and consolidation. All assessment statements of the IB Diploma Programme Chemistry syllabus are covered in highly structured and meaningful ways.

## Coursebook includes Student CD

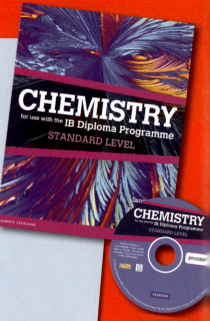

Each chapter in the coursebook includes:
- focus on the IB Standard Level (Core) Diploma Programme Chemistry syllabus, topics 1 to 11
- Syllabus Assessment Statements given beside the relevant theory
- stimulating photos and full colour illustrations to support learning of chemical concepts
- theory broken into manageable chunks for ease of learning
- comprehensive exercises for ongoing review and consolidation
- **Chem Complement** boxes, which engage students with interesting extension material and applications to Aims 8 and 9 for Experimental Sciences
- **Theory of Knowledge** boxes, which allow easy integration of this requirement of the syllabus
- ICT activities, which address Aim 7 for Experimental Sciences and are available on the *Companion Website*

- **Chapter summary**, which includes chapter glossary and key points
- **Review questions** to revise all chapter content
- **Comprehensive topic test** of examination questions.

*Student CD* contains:
- an electronic version of the coursebook
- fully worked solutions to all coursebook questions
- a link to *Pearson Places*.

# Teacher's Resource CD

The *Teacher's Resource CD* provides a wealth of teacher support material, including:

- **fully worked solutions** to coursebook questions
- **worksheets** for practising skills and consolidating theory; answers are also included
- **teacher demonstrations** to engage students and enhance understanding of concepts
- **practical investigations** to enhance the learning of chemical concepts and for use in meeting the mandated time allocation for practical work
- **practical notes** for the teacher/lab technician
- **risk assessments** for practical activities.

This time-saving resource contains documents available as:

- Microsoft Word documents that can be edited, allowing you to modify and adapt any resources to meet your needs
- PDFs to make printing easy.

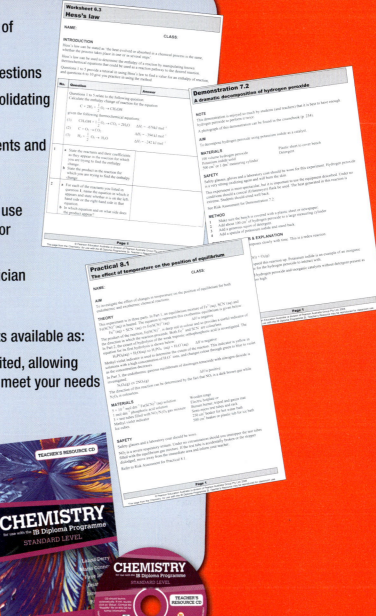

## Pearson Places

 www.pearsonplaces.com.au.

*Pearson Places* addresses Aim 7 for Experimental Sciences by providing easy integration of technology into the classroom. It contains a wealth of support material for students and teachers to enhance teaching and learning in chemistry.

The interactive material on the *Pearson Places* website allows students to review their work and revise fundamental concepts, as well as providing an opportunity for accelerated learning.

*Pearson Places* contains:

- **Review Questions**—auto-correcting multiple-choice questions for exam revision
- **Interactive Animations**—to engage students in exploring concepts
- **QuickTime Videos**—to explore chemical concepts in a visually stimulating way
- **3D Molecules Gallery**—for interactive viewing and manipulating of molecular structures
- **Web Destinations**—a list of reviewed websites that support further investigation and revision.

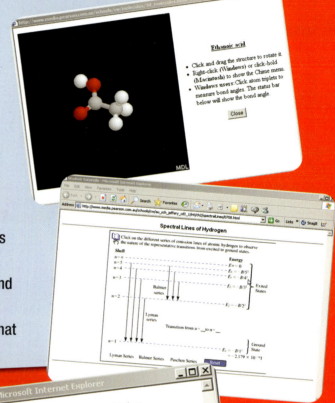

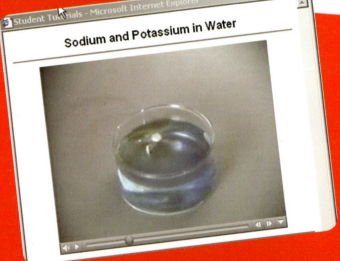

**For more information on *CHEMISTRY: For use with the IB Diploma Programme*
visit www.pearsonplaces.com.au**

# MEET THE AUTHORS

## Lanna Derry

Lanna Derry, the lead author of the *CHEMISTRY: For use with the IB Diploma Programme* series, is a highly experienced teacher of IB Chemistry. She has taught senior Chemistry in independent schools for more than twenty years and has authored revision guides for Year 12 Chemistry. Lanna is currently teaching IB Chemistry at Tintern Girls Grammar School, Ringwood East, Victoria, Australia.

## Maria Connor

Maria Connor is an experienced IB Chemistry examiner. She has taught IB and senior Chemistry for many years and is currently teaching IB Chemistry at Tintern Girls Grammar School, Ringwood East, Victoria, Australia.

## Carol Jordan

Carol is currently teaching at the Shanghai American School, Shanghai, China. She is an experienced teacher of IB Chemistry, IB Environmental Systems and Theory of Knowledge. She has been an assistant examiner and senior moderator for internal assessment for IB Chemistry. Carol is a workshop leader and was part of the team responsible for developing the new IB Diploma Programme Chemistry Guide.

**Faye Jeffery** is currently teaching at Melbourne Centre for Adult Education. She has taught Chemistry and Biology for more than twenty years. Faye has written a number of texts for Chemistry and Science.

**Brian Ellett** has taught senior Chemistry for more than twenty years and has written a number of texts for Chemistry. He is currently Head of Science at Salesian College, Chadstone, Victoria, Australia.

**Janette Ellis** has experience teaching both IB Chemistry and senior Chemistry. After teaching in Victoria for many years, she is now at Kambala, Rose Bay, New South Wales, Australia.

**Pat O'Shea** is a highly experienced teacher of Chemistry. He is currently Deputy Principal at Loreto College, Ballarat, Victoria, Australia. Pat has presented at many workshops for senior Chemistry teachers.

# HOW TO USE THIS BOOK

Our aim has been to present chemistry as exciting, accessible and relevant. The content is carefully structured with regular opportunities for revision and consolidation to prepare students for the IB Diploma Programme Standard Level Chemistry examinations.

## Major features

- *Chapter opening pages* that include a stimulating photo and a simple, student-friendly syllabus-related list of what students should be able to do by the end of the chapter
- **Chem Complement** boxes that engage students with interesting extensions of the Chemistry theory and applications to Aims 8 and 9 for Experimental Sciences
- **Theory of Knowledge** boxes that address the links between the syllabus and aspects of the scientific way of knowing as required by the syllabus
- *ICT activities* that address Aim 7 for Experimental Sciences and are available on the Companion Website
- *Comprehensive exercises* that encourage students to consolidate their learning in a systematic manner while familiarising students with the IB command terms
- *Glossary of terms* and a *summary of concepts* at the end of each chapter
- *Review questions* that draw together all aspects of the topic
- *End-of-chapter tests* that allow students to test their knowledge of the topic thoroughly using questions from past IBO examinations

# Icons in the coursebook

- **Assessment Statement** icons denote Assessment Statements from the IB Diploma Programme Standard Level (Core) Chemistry syllabus.

- **Worksheet** icons denote when a worksheet giving extra practice on a key part of the topic is available. These can be found on the **Teacher's Resource CD**.

- **Prac** icons denote when a practical investigation is available. These can be found on the **Teacher's Resource CD**.

- **Demo** icons denote when a teacher demonstration is available. These can be found in the **Teacher's Resource CD**.

- **Companion Website—Interactive Animation** icons denote when links to an animation are available to support a topic in the coursebook. These can be accessed on Pearson Places.

- **Companion Website—QuickTime Video** icons denote when links to a QuickTime video are available to support a topic in the coursebook. These can be accessed on Pearson Places.

- **Companion Website—Web Destinations** icons denote when Web links and Review Questions are available to support a topic in the coursebook. These can be accessed on Pearson Places.

# Other features

- **Worked examples** of calculations and chemical structures to aid mastery of difficult concepts
- **Glossary** at the end of the text as well as at the end of each chapter
- **Periodic table** with relative atomic masses included on the inside front cover to provide a quick and easy reference

# Student CD

This interactive resource contains:
- an electronic version of the coursebook
- fully worked solutions (including diagrams) to all coursebook questions
- a link to the live Companion Website (Internet access required) to provide access to course-related Web links.

# Other components

- Pearson Places  www.pearsonplaces.com.au.
- Teacher's Resource CD

## Other books in the series

CHEMISTRY: *For use with the IB Diploma Programme Higher Level*

CHEMISTRY: *For use with the IB Diploma Programme Options: Standard and Higher Level*

# ACKNOWLEDGEMENTS

We would like to thank the following for permission to reproduce photographs, texts and illustrations. The following abbreviations are used in this list: t = top, b = bottom, c = centre, l = left, r = right.

**AAP**: Simone Crepaldi: p. 230.

**Alamy Limited**: p. 384.

**Brith-Marie Warn**: p. 372 (petrol).

**Corbis Australia Pty Ltd**: pp. 100t, 192, 201, 203, 216, 233t; Dave G Houser: p. 120; James L Amos: p. 121; Stuart Westmorland: p. 57.

**Debbie Irwin, Ross Farrelly, Deborah Vitlin, Patrick Garnett, Chemistry Contexts 2 2ed, Pearson Education Australia, 2006**: p. 251.

**D R Stranks, M L Heffernan, K C Lee Dow, P T McTigue & G R A Withers, Chemistry a Structural View, Melbourne University Press, 1970**: p. 61.

**DK Images**: p. 362 (ant).

**Fairfax**: p. 263cr.

**Fundamental Photographs**: Richard Megma: p. 99c.

**Getty Images Australia Pty Ltd**: Bob Elsdale: p. 243; Rischgitz: p. 64; Stu Forster: p. 190r.

**Greenpeace**: p. 290br.

**JupiterImages Corporation © 2008**: pp. 14, 17, 217 (car), (217 people), 352, 366.

**Lanna Derry**: pp. 77, 198, 217 (test tube), 234, 302br, 373l.

**NASA**: p. 343.

**Newspix**: AFP Photo/William West: p. 52.

**Oregon State University adapted from Linus Pauling and The Nature of the Chemical Bond: A Documentary History, Special Collections**: p. 47.

**Otto Schott**: p. 166.

**Pearson Australia**: Katherine Wynne: p. 378; Michelle Jellett: p. 100b; Natalie Book: pp. 54(pencils), 309, 382; Peter Saffin: pp. 208, 233bl, 253b, 253t, 274, 275b, 275t, 285, 379.

**Photolibrary Pty Ltd**: pp. Cover, 1, 3, 7, 11, 15, 27, 29, 33, 53(graphite), 53(pencil), 53(tennis), 54b, 55b, 69, 75, 77b, 77t, 80, 115, 119, 131, 133t, 147, 160, 161t, 189b, 190l, 217(flask), 249, 266, 276, 291, 301, 302l, 303b, 312, 313t, 321, 322l, 329, 362b, 372(bunsen), 372b, 375; Andrew Lambert: pp. 88b, 89r, 161cr, 162, 220, 341, 383b; David Taylow: p. 307; Irene Windridge: p. 188; Mark J Winter: p. 368b; Martin Dohrn: p. 196; Martyn F Chillmaid: pp. 161br, 189t, 223, 373br, 373tr, 383cr; Russel Kightley: p. 26.

**Prentice Hall, Inc**: p. 133b.

**The Picture Source**: pp. 88l, 88t.

**Theodore Gray © 2008**: p. 77br.

**Uwe H. Friese**: p. 362(nettle).

Thanks to the **International Baccalaureate Organization** (IB Organization) for permission to reproduce IB intellectual property. This material has been developed independently of the International Baccalaureate Organization (IB Organization) which in no way endorses it.

Every effort has been made to trace and acknowledge copyright. However, should any infringement have occurred, the publishers tender their apologies and invite copyright owners to contact them.

The Publishers wish to thank
    Maria Connor
    Carol Jordan
    Michael McCann
for reviewing the text.

# 1 ATOMIC STRUCTURE

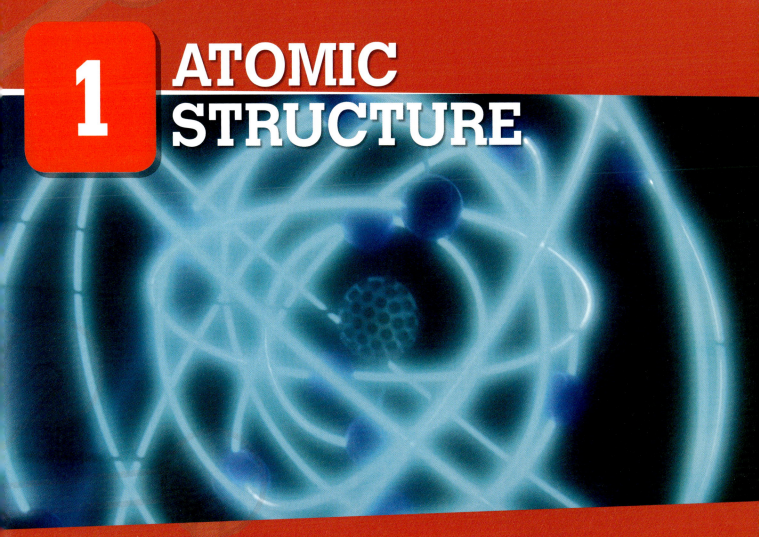

## Chapter overview

This chapter covers the IB Chemistry syllabus Topic 2: Atomic Structure.

**By the end of this chapter, you should be able to:**

- describe atomic structure in terms of number of protons, neutrons and electrons in the atom
- state the relative masses and relative charges of protons, neutrons and electrons
- represent isotopes using atomic numbers and mass numbers, and use these together with ionic charges to calculate the number of protons, neutrons and electrons in atoms and ions
- give definitions for atomic number ($Z$), mass number ($A$) and isotopes of an element
- compare isotopes of an element in terms of their properties, and discuss the uses of radioisotopes
- explain the function of each major component of the mass spectrometer and use data from the mass spectrometer to calculate relative atomic masses and abundance of isotopes
- describe and identify parts of the electromagnetic spectrum such as the ultraviolet, visible and infrared regions
- describe emission spectra as line spectra in contrast to a continuous spectrum, and explain how the emission spectrum of hydrogen is formed
- use Bohr's model of the atom to write electron arrangements for atoms and ions with an atomic number $\leq 20$.

IBO Assessment statements 2.1.1 to 2.3.4

The Greek philosopher Democritus (460–370 BCE) was the first to use the word *atom* to describe the small, indivisible particles from which he hypothesized substances must be made. His ideas were lost, however, when Aristotle (384–322 BCE) concluded that the world consisted of earth, air, fire and water. This idea was pursued by the alchemists of Europe and Asia for centuries until John Dalton (1766–1844), an English chemist, described elements and chemical reactions in terms of atomic theory. Dalton proposed that each element was made from a unique atom. It was Dalton's idea to give a symbol to each of these elements, although the symbols he chose are not used today.

Electrons were the first subatomic particle to be identified by charge by J.J. Thomson in 1899. Thomson devised a model of the atom, referred to as the 'plum pudding' model, in which negative particles were dotted throughout a mass of positive charge. This model was disproved by Rutherford's famous gold leaf experiment and Rutherford, in turn, formulated his own model of the atom. In 1919 Rutherford's experiments led to the discovery of protons, but neutrons were still invisible to those who sought to find the nature of the atom. It was not until 20 years after the invention of the mass spectrometer that the true nature of isotopes was able to be explained. In 1932 James Chadwick finally identified the elusive neutron, and the model of the atom as we know it was complete.

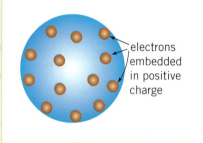

**Figure 1.0.1** Thomson's 'plum pudding' model of the atom.

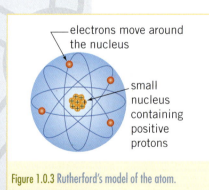

**Figure 1.0.3** Rutherford's model of the atom.

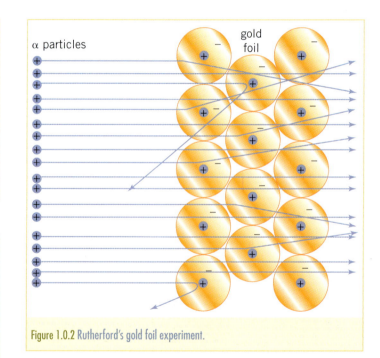

**Figure 1.0.2** Rutherford's gold foil experiment.

## THEORY OF KNOWLEDGE

To scientists, the phrase 'the *theory* of …' is an explanation for a hypothesis that has been thoroughly tested and verified. Theories are often represented by visual or mathematical models.

Stephen Hawking in *A Brief History of Time* said:

> a theory is a good theory if it satisfies two requirements: It must accurately describe a large class of observations on the basis of a model that contains only a few arbitrary elements, and it must make definite predictions about the results of future observations.

He goes on to state:

> Any physical theory is always provisional, in the sense that it is only a hypothesis; you can never prove it. No matter how many times the results of experiments agree with some theory, you can never be sure that the next time the result will not contradict the theory. On the other hand, you can disprove a theory by finding even a single repeatable observation that disagrees with the predictions of the theory.

- Comment on the statement that 'without Thomson, Rutherford's theory of the atom would not exist'.
- Is it possible to propose a theory that has not been experimentally tested and verified?
- How accurately do the theories and models scientists create describe and make predictions about the natural world?

## CHEM COMPLEMENT

### The difference one person can make

One New Zealander, Ernest Rutherford, played an important part in the discovery of subatomic particles. Rutherford's story is a compelling one. New Zealand had only 14 postgraduate students in 1893, Rutherford being one of them. He could not find a job in New Zealand, not even as a teacher. He applied for a scholarship to study in England but came second. As luck would have it, the winner did not accept the prize and Rutherford was on his way to the Cavendish Laboratory at Cambridge University, England, to work under J.J. Thomson.

Rutherford's contribution to the development of the atom was not isolated to his gold leaf experiment and later discovery of protons. Many of his students also made important contributions. James Chadwick, who discovered the neutron, worked with Rutherford at the Cavendish Laboratory in Cambridge. Henry Moseley was a member of Rutherford's team at Manchester, England, as was Niels Bohr. Moseley used X-ray experiments to determine the atomic number of the elements and so rearranged Mendeleev's periodic table. Sadly Henry Moseley was one of many soldiers killed in the Gallipoli landings in Turkey during World War I.

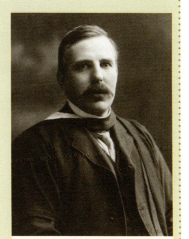

Figure 1.0.4 The New Zealander Ernest Rutherford contributed in many ways to the development of atomic structure in the early 20th century.

Rutherford's experiment

> **THEORY OF KNOWLEDGE**
> The language of Chemistry is constructed from the words of many different languages. For example, *atom* originates from the Greek word *atomos* meaning 'cannot be cut'. Can you think of any other examples of Chemistry vocabulary that have their origins in a language other than English? What are the origins of these words?

> **THEORY OF KNOWLEDGE**
> A paradigm is a set of shared beliefs that guides research and understanding. Thomas Kuhn, an influential American philosopher, caused a revolution in the way scientific thought was believed to change over time. In *The Structure of Scientific Revolutions*, published in 1962, Kuhn challenged the existing view that scientists work on problems associated with proving existing models with little change in thinking. He proposed that these periods of normal 'science' were characterized by periods of radical changes in thinking. These 'paradigm shifts', as he called them, occurred when new information provided better ways of thinking, leading to existing models being rejected.
>
> Think about the changes in the model of the atom from Democritus to Chadwick.
>
> - Outline the dominant paradigms during this period.
> - Do you think our current model of the atom was the result of changes that accumulated slowly or the result of one or more major paradigm shifts?
> - What role did the development of new technologies play in the growth of our knowledge of the atom?

## 1.1 THE ATOM

**AS 2.1.1**
State the position of protons, neutrons and electrons in the atom. © IBO 2007

**AS 2.1.2**
State the relative masses and relative charges of protons, neutrons and electrons. © IBO 2007

When Niels Bohr was working with Rutherford in 1913, he proposed a model of the atom that explained emission spectra that had been observed. Bohr's model proposed that electrons moved around the nucleus in shells, which were regions of space with fixed energies. Nearly 100 years later, Bohr's model is still acceptable as a model that explains the atom to our satisfaction, although more complex and more correct models do exist.

### Comparing subatomic particles

The nuclear atom can be summarized as follows:

**TABLE 1.1.1 PROPERTIES AND POSITIONS OF THE MAJOR SUBATOMIC PARTICLES**

| Subatomic particle | Symbol | Mass (kg) | Relative mass | Charge (C) | Relative charge | Position in the atom |
| --- | --- | --- | --- | --- | --- | --- |
| Proton | p | $1.6726 \times 10^{-27}$ | 1 | $+1.6022 \times 10^{-19}$ | +1 | In the nucleus |
| Neutron | n | $1.6749 \times 10^{-27}$ | 1 | 0 | 0 | In the nucleus |
| Electron | $e^-$ | $9.1094 \times 10^{-31}$ | $5 \times 10^{-4}$ or $\frac{1}{1836}$ | $-1.6022 \times 10^{-19}$ | −1 | Orbiting the nucleus |

**Protons** and **neutrons** have approximately the same mass, with the neutron being very slightly heavier than the proton. Given that their actual mass is so small, it is simpler to talk about their *relative masses*. Protons and neutrons have approximately the same mass, and so they are each assigned a relative mass of 1. The **electron** is much smaller, and has a relative mass roughly one two-thousandth that of either the proton or neutron. Electrons do not contribute significantly to the overall mass of the atom.

Neutrons have no charge. The electron is negatively charged and has a charge equal in magnitude but opposite in sign to that of the proton, which is positively charged. Thus, the electron is assigned a relative charge of –1, while the proton has a relative charge of +1. This means that in a neutral (uncharged) atom, the number of protons is equal to the number of electrons.

Together, protons and neutrons make up the **nucleus** of an atom. This is where most of the mass of the atom is found. The electrons orbit the nucleus in regions of space called **electron shells** (see section 1.3).

Discovering the charge on the electron

A neutral atom has an equal number of protons and electrons. It usually has at least as many neutrons as protons and often more neutrons than protons.

Protons, neutrons and electrons are the major subatomic particles, but the subatomic world is also populated by leptons, gluons, quarks (with such intriguing names as strange and charm) and so on. In this course, however, we will restrict our discussion to the three major particles.

> **THEORY OF KNOWLEDGE**
> Both direct and indirect evidence is used by chemists to explain the nature of matter. Direct evidence comes from one's own observations—what we see, hear and touch—indirect evidence comes from interpreting the work of others or using the evidence provided by technology tools. For example, subatomic particles cannot be observed directly but we know of their existence indirectly.
> - What indirect evidence was provided by Rutherford's gold foil experiment and what conclusions did he make?
> - Describe an investigation in chemistry in which you acquired knowledge by using indirect evidence.

## Atomic number, mass number and isotopes

The **atomic number**, symbol $Z$, is the number of protons in the nucleus. Hence $Z$ is sometimes called the proton number. It is this number that distinguishes one element from another. For example, atoms of carbon ($Z = 6$) all have 6 protons in the nucleus. Even if the numbers of electrons and neutrons in these atoms were to change, they would still be carbon atoms. All atoms of sulfur ($Z = 16$) will have 16 protons in the nucleus. All atoms of neon ($Z = 10$) will have 10 protons in the nucleus.

The **mass number** (which like the atomic number must be an integer), symbol $A$, is the sum of the number of protons and neutrons in the nucleus.

mass number = number of protons + number of neutrons

The invention of the **mass spectrometer** in 1919 by Francis Aston (for which he won the 1922 Nobel Prize) allowed very accurate measurements of mass to be made. These accurate masses suggested that sometimes atoms of the same element had more than one mass. To explain this puzzle, Frederick Soddy had suggested earlier (in 1913) that many atoms come in more than one form. These different forms have the same number of protons and similar properties but different masses. He called these different forms of an element **isotopes**. The term *isotope* comes from the Greek meaning 'at the same place'. The name was suggested to Frederick Soddy by Margaret Todd, a Scottish doctor, when Soddy explained that it appeared from his investigations as if several elements occupied each position in the periodic table. Soddy won the Nobel Prize in 1921 for his work.

Isotopes are atoms of the same element with the same number of protons, but different numbers of neutrons; that is, they have the same atomic number, but a different mass number.

 2.1.3
**Define the terms *mass number* (*A*), *atomic number* (*Z*) and *isotopes* of an element.**
© IBO 2007

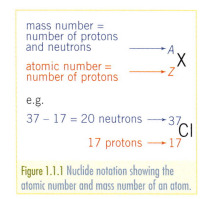
Figure 1.1.1 Nuclide notation showing the atomic number and mass number of an atom.

Isotopes
Atomic notation

**AS 2.1.4**
Deduce the symbol for an isotope given its mass number and atomic number. © IBO 2007

A simple way to represent a particular isotope of an element is by using nuclide notation. This combines mass number, atomic number and the symbol for the element as shown in figure 1.1.1. For example, there are two isotopes of silver: silver-107 and silver-109. In nuclide notation these would be written as $^{107}_{47}Ag$ and $^{109}_{47}Ag$.

This notation may sometimes be simplified by omitting the atomic number. The symbol for silver-107 would be $^{107}Ag$. This simplified form still describes the atom accurately, as the symbol for the atom is synonymic with the atomic number.

Both these isotopes would have 47 protons because the atomic number is 47, and they have 47 electrons because they are neutral atoms. However, the first isotope has 107 – 47 = 60 neutrons and the second has 109 – 47 = 62 neutrons.

**Ions** are atoms that have lost or gained electrons and so have a charge. A positive ion, or **cation**, has fewer electrons than the corresponding neutral atom, and a negative ion, or **anion**, has more electrons than the neutral atom. The number of protons and the number of neutrons for an ion are exactly the same as the neutral atom. For example, a particular magnesium ion may be represented as $^{24}_{12}Mg^{2+}$. Because this is a magnesium ion, it has 12 protons. The mass number is 24, so the ion has 24 – 12 = 12 neutrons, and because the ion has a 2+ charge, it has 12 – 2 = 10 electrons.

**AS 2.1.5**
Calculate the number of protons, neutrons and electrons in atoms and ions from the mass number, atomic number and charge. © IBO 2007

> ### CHEM COMPLEMENT
> 
> #### The origin of Z for atomic number
> 
> We often ponder over the use of $Z$ for atomic number and $A$ for mass number. To English speakers it seems quite strange. However, the German for atomic number is *Atomzahl*, so it is possible that the symbol $Z$ for atomic number came from $Z$ for *Zahl* (number). The *Encyclopaedia of Symbols* has a more poetic interpretation:
> 
> > The letter Z is one of the signs for the highest god in Greek mythology, Zeus. In modern physics $Z$ represents the greatest energy, nuclear power, in its potential form, nuclear charge.
> 
> Why the symbol $A$ was used for mass number is still a mystery, but with the use of M as a unit of concentration in Chemistry (molar = mol dm$^{-3}$), as well as being used to represent molar mass, it is most likely that $A$ was used rather than $M$ to avoid confusion.

**WORKSHEET 1.1**
Using nuclide symbol notation

The actual mass of an atom is of course incredibly small, of the order of $1 \times 10^{-26}$ kg for a carbon atom. Chemists have devised a relative atomic mass scale for convenience. We will examine this scale in section 1.2.

### Properties of isotopes

The chemical properties of atoms are determined by their electronic structure; however, their physical properties depend largely on their nuclei. This means that although the chemical properties are the same for two isotopes of the same element, their physical properties can vary. The most obvious example of this is the differing masses of isotopes, which allow the mass spectrometer to be used to separate the isotopes of an element.

**AS 2.1.6**
Compare the properties of the isotopes of an element. © IBO 2007

Density may also vary between isotopes of an element. For example, heavy water ($^2H_2O$) is denser and takes up about 11% less volume than ordinary water ($^1H_2O$). Other physical properties that can vary between isotopes are boiling point, melting point and rate of diffusion.

Other differences in physical properties are more sophisticated. For example several forms of spectroscopy rely on the unique nuclear properties of specific isotopes. Nuclear magnetic resonance (NMR) spectroscopy can be used only for isotopes with a non-zero nuclear spin. Carbon has three isotopes: $^{12}_{6}C$, $^{13}_{6}C$ and $^{14}_{6}C$. Similarly, hydrogen also has three isotopes: $^{1}_{1}H$, $^{2}_{1}H$ and $^{3}_{1}H$. Unlike the other isotopes of hydrogen and carbon, the isotopes $^{1}_{1}H$ and $^{13}_{6}C$ have a non-zero nuclear spin and are able to be used in NMR spectroscopy. NMR spectroscopy will be discussed in Option A: Modern Analytical Chemistry.

DEMO 1.1
Vacuum tubes

Many isotopes are radioactive and a number of these radioactive isotopes, or **radioisotopes**, have proved useful. For example, in living things, the isotope of carbon, $^{14}_{6}C$ exists in a set ratio with $^{12}_{6}C$. When the organism dies, $^{14}_{6}C$ decays, but $^{12}_{6}C$ does not. The percentage of $^{14}_{6}C$ decreases as the age of the dead organism increases. This percentage is used to estimate the age of the organism. This process is called radiocarbon dating. Some useful isotopes are listed in table 1.1.2.

**AS** 2.1.7
**Discuss the uses of radioisotopes** © IBO 2007

## CHEM COMPLEMENT

### Radiation: a useful but dangerous tool

The discovery of radiation is attributed to German scientist Wilhelm Roentgen in 1895. Roentgen was using a cathode tube covered in black paper when he noticed a screen on the other side of a darkened room fluorescing. Some invisible rays must have been passing from the tube to the screen. Roentgen named the rays X-rays; he even thought to X-ray his wife's hand. Medical science ran with this new idea in a big way. A year later, Henri Becquerel, a French scientist, found that materials such as uranium emit X-rays. Marie Curie and her husband Pierre found that the ore pitchblende is even more radioactive than uranium. Curie isolated the elements polonium and radium from this ore. Marie Curie died of leukaemia, believed to have been caused by prolonged exposure to radiation during her research work.

Figure 1.1.4 The age of ancient papyrus scrolls found in the ruins of Herculaneum, Italy, has been confirmed by radiocarbon dating.

We now know that a radioactive element decays. This means that the nucleus is unstable and it ejects small particles. The particles ejected have been labelled alpha, beta and gamma particles. Alpha particles are helium nuclei, beta particles are electrons and gamma particles are a stream of photons. When alpha particles are ejected from the nucleus a new element is formed.

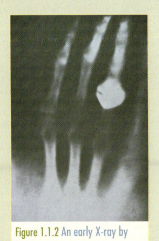

Figure 1.1.2 An early X-ray by Roentgen.

Figure 1.1.3 Types of radiation.

**TABLE 1.1.2 RADIOISOTOPES AND THEIR USES**

| Radioisotope | Symbol | Use |
|---|---|---|
| Carbon-14 | $^{14}_{6}C$ | Radiocarbon dating. The ratio of carbon-12 to carbon-14 is calculated to determine the age of an object. |
| Iodine-131 | $^{131}_{53}I$ | As a medical tracer in the treatment of thyroid disorders. The radioactive iodine is taken up by the thyroid gland and then the radiation kills part of it. |
| Iodine-125 | $^{125}_{53}I$ | As a medical tracer in the treatment of prostate cancer and brain tumours. It is also taken up by the thyroid gland. |
| Cobalt-60 | $^{60}_{27}Co$ | Radiotherapy, levelling devices and to sterilize foods and spices. |
| Americium-241 | $^{241}_{95}Am$ | Smoke detectors. Emits a beam of alpha particles which, if interrupted by smoke, will set the device off. |
| Technetium-99 | $^{99}_{43}Tc$ | Radiotherapy for cancer and for studying metabolic processes. Emits low energy radiation, so small doses can be administered. |

Medical tracers, radioactive forms of atoms, can be attached to molecules that target specific tissues in the body, such as cancerous tumours or organs such as the liver, lungs, heart or kidneys that are not functioning normally. The isotopes $^{131}_{53}I$ and $^{125}_{53}I$ are examples of tracers that target the thyroid gland in particular. The radioisotope allows the location of the tumour to be determined.

The life-saving use of these medical tracers is in strange contrast to the usually dangerous nature of radioisotopes to living things. *Radiation poisoning* is the term that is generally used to refer to acute problems caused by a large dosage of radiation from radioisotopes in a short period. Large amounts of radiation interfere with cell division, and this results in many of the symptoms of radiation poisoning.

> **THEORY OF KNOWLEDGE**
> Knowledge of isotopes has provided many social benefits to society. Can you outline ways in which you, your family or your friends have benefited from advances in scientific knowledge? Can you think of any uses of science that are not beneficial?

### Section 1.1 Exercises

1. State the missing words which complete the following paragraph.
   The major subatomic particles are _____, _____ and _____.
   The _____ and _____ are found in the _____, while the _____ move at great speed around the _____.

2. Compare the mass of a proton with that of a neutron.

3. Compare the mass and charge of an electron and a proton.

4. Define the term *isotopes of an element*. Include an example in your answer.

5. a State the chemical names for the quantities represented by the numbers in $^{15}_{7}N$.
   b Explain how you can use the information represented to make an electron shell diagram of a nitrogen atom.

6. Determine the number of protons, neutrons and electrons in each of the following.
   a $^{62}_{30}Zn$
   b $^{81}_{36}Kr$
   c $^{24}_{12}Mg^{2+}$
   d $^{81}_{35}Br^{-}$

7 Identify isotopes of the same element from the following list. Explain your choice.

$^{90}_{45}X, ^{92}_{43}X, ^{90}_{43}X, ^{95}_{45}X, ^{99}_{44}X$

8 Describe two ways in which isotopes of the same element may differ from each other.

9 Explain how radioisotopes can be used in modern medicine.

10 Describe how radiocarbon dating is used to determine the age of a dead organism.

## 1.2 THE MASS SPECTROMETER

### Separating atoms by mass

A mass spectrometer is a complex instrument that can be considered as a number of separate components, each of which performs a particular function. The underlying principle of its operation is that the movement of charged particles will be affected as they pass through a magnetic field. The degree to which these particles are deflected from their original path will depend on their mass and their charge—their mass/charge ($m/z$) ratio.

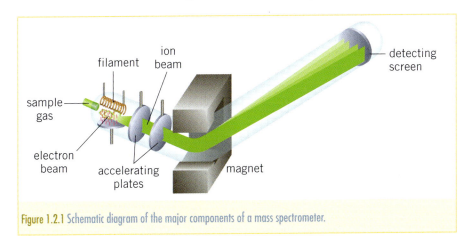

Figure 1.2.1 Schematic diagram of the major components of a mass spectrometer.

The operation of a mass spectrometer can be regarded as a series of stages as the particles move through the instrument.

1 *Vaporization*: The sample to be analysed is heated and vaporized and passed into an evacuated tube. This results in particles that are separate from one another.

2 *Ionization*: The atoms or molecules are then bombarded by a stream of high energy electrons and one or more electrons are knocked off each atom or molecule. This results in ions with, most commonly, a 1+ charge, but sometimes with a 2+ charge.

3 *Acceleration*: The positively charged ions are accelerated along the tube by attraction to negatively charged plates and the ions pass through slits that control the direction and velocity of their motion.

4 *Deflection*: The stream of ions is passed into a very strong magnetic field, which deflects the ions through a curved path. If the size of the magnetic field is fixed, a light ion will be deflected more than a heavy ion and a 2+ ion will be deflected more than a 1+ ion of the same mass.

### THEORY OF KNOWLEDGE

- Describe how the discovery of the mass spectrometer changed our understanding of mass, and explain the significance of this in the development of our knowledge of the structure of an atom.

- A new atomic theory was developed in the early 20th century based on mathematical models. Considering that there is no direct observable evidence that subatomic particles exist, this theory has the potential to develop when technology becomes more advanced. Explain what mathematical models are and the role they play in the development of new knowledge in science.

**AS 2.2.1**
**Describe and explain the operation of a mass spectrometer.** © IBO 2007

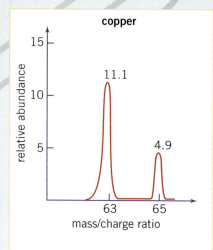

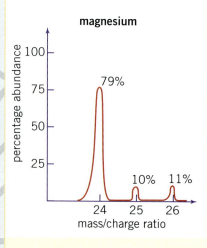

Figure 1.2.2 Mass spectra of copper and magnesium.

DEMO 1.2
A model mass spectrometer

 2.2.2
**Describe how the mass spectrometer may be used to determine relative atomic mass using the $^{12}$C scale.** © IBO 2007

PRAC 1.1
Interpretation of the mass spectrum of air

**The deflection of the ions depends on the mass/charge (*m/z*) ratio.**

In modern mass spectrometers the strength of the field is variable. If the ions are to be deflected to the same point, a stronger magnetic field is required to deflect a heavy ion than a lighter ion. Similarly, a stronger magnetic field is required to deflect an ion with a 1+ charge than a 2+ charge.

5 *Detection*: The ions are detected electronically by a device that measures both the location and the number of particles that collide with it.

6 *Recording*: The percentage abundance
($\frac{\text{number of isotopes of a particular type}}{\text{total number of particles in sample}} \times \frac{100}{1}$) of the different isotopes is
recorded as a graph called a mass spectrum. A peak is produced in the mass spectrum for each isotope (ion with a particular mass and charge). The position of the peaks along the horizontal axis indicates the ratio of $\frac{\text{mass of ion}}{\text{charge on ion}}$.

In simple elemental mass spectra (in which the ions generated carry only single charges) the number of peaks recorded indicates the number of isotopes of the element present and their isotopic masses. The height of each peak is a measure of the relative abundance of the isotope; the higher the peak the more of that isotope is present in the sample. These peak heights can be converted easily to an abundance fraction or *percentage abundance* to allow for the calculation of relative atomic mass. The abundance fraction for a particular isotope is the height of the peak for that isotope divided by the sum of the heights of all peaks in the spectrum.

From figure 1.2.2 it can be seen that copper has two isotopes and magnesium three. Given that the peak heights of the two copper isotopes are 11.1 and 4.9 units respectively, the abundance fractions of the isotopes can be determined:

$$\frac{11.1}{11.1 + 4.9} = \frac{11.1}{16.0} \text{ and } \frac{4.9}{11.1 + 4.9} = \frac{4.9}{16.0}$$

To convert these abundance fractions to percentage abundances we simply multiply by 100:

$$\frac{11.1}{16.0} \times \frac{100}{1} = 69.4\% \text{ and } \frac{4.9}{16.0} \times \frac{100}{1} = 30.6\%$$

To generate the relative scale of atomic masses, chemists chose the most abundant isotope of the element carbon, the carbon-12 isotope ($^{12}_{6}$C), and assigned it a relative mass of exactly 12 units. The element carbon was chosen as the reference for a number of important reasons:

- Carbon is very cheap and is widely available.
- It is relatively easy to isolate and purify this isotope.
- Carbon is not toxic in any way.

It was decided to assign carbon a mass of 12 units, rather than 1 as may have been expected, as this number mirrored the mass number of the isotope. As protons and neutrons are the basic building blocks of atoms (in addition to the very light electrons), the relative atomic mass will closely parallel the number of these fundamental particles in the nucleus of the element. Using a mass spectrometer, the lightest of all the elements was found to be deflected 12 times further than the standard carbon-12 isotope, and the most common isotope of magnesium, $^{24}_{12}Mg$, was deflected half as far as $^{12}_{6}C$. Thus hydrogen, the lightest of all elements, has a relative mass of close to 1 and $^{24}_{12}Mg$ has a relative mass of approximately 24. An element well known for its high density is lead. Lead atoms have a relative mass of approximately 207 and are, on average, about 17 times more massive than carbon-12 atoms.

Mass spectrometers are now primarily used for analysis of substances, often in conjunction with other specialized instruments such as the NMR (nuclear magnetic resonance) or IR (infrared) spectrometers. The relative isotopic masses of all isotopes have already been determined and are readily available.

Figure 1.2.3 A technician uses a mass spectrometer to analyse the surface molecules on a macromolecule such as a polymer or protein.

## CHEM COMPLEMENT

### Mass spectra help in identifying elephant poachers

Over the past 20 years the number of elephants in Africa has declined by more than 50%, from approximately 1.3 million to 610 000. In 1989 an international treaty was signed to prohibit the sale of ivory, although unfortunately large quantities are still sold on the black market. To help track the source of ivory obtained from the tusks of the animals, scientists have developed an extensive database of the isotopic composition of ivory from elephants living across Africa. The relative amounts of isotopes such as $^{12}_{6}C$, $^{13}_{6}C$, $^{14}_{7}N$, $^{15}_{7}N$, $^{86}_{38}Sr$ and $^{87}_{38}Sr$ allow scientists to locate the habitat of the elephant from which the ivory was taken to within a range of 150 km. The variations in the isotopic abundances arise from the diet of the animal (whether mainly grasses or trees) and the local ecology. Unfortunately, the destruction of its habitat and the demand for its meat and ivory will continue to endanger this magnificent animal.

Figure 1.2.4 Mass spectroscopy may be used to track the source of illegally obtained ivory from elephant tusks.

## Calculating relative atomic mass

The formal definition of relative atomic mass is useful in helping to recall how to mathematically determine its value for a particular element.

The **relative atomic mass** (RAM) of an element is defined as the weighted mean of the masses of its naturally occurring isotopes on a scale in which the mass of an atom of the carbon-12 isotope, $^{12}_{6}C$, is 12 units exactly. The symbol for RAM is $A_r$.

To determine the RAM of any element X, we multiply the **relative isotopic mass** (RIM, symbol $I_r$) of each naturally occurring isotope by its abundance fraction and add these values. For each naturally occurring isotope this may be written in mathematical terms as:

$$A_r(X) = \Sigma(I_r \times \text{abundance fraction})$$

If the abundance fraction is expressed as a percentage, the formula becomes:

$$A_r(X) = \frac{\Sigma(I_r \times \text{abundance fraction})}{100}$$

**AS 2.2.3**
Calculate non-integer relative atomic masses and abundance of isotopes from given data.
© IBO 2007

Mass spectrometer

**WORKSHEET 1.2**
Calculation of relative masses

---

**THEORY OF KNOWLEDGE**
Explain why symbols are used in certain aspects of Chemistry. Use examples to support your answer.

---

## Worked example 1

Use the data provided to determine the relative atomic mass of magnesium.

| Isotope | Relative isotopic mass | Percentage abundance |
|---|---|---|
| $^{24}Mg$ | 23.99 | 78.70 |
| $^{25}Mg$ | 24.99 | 10.13 |
| $^{26}Mg$ | 25.98 | 11.17 |

### Solution

$$A_r(Mg) = \frac{\Sigma(I_r \times \% \text{ abundance})}{100}$$
$$= \frac{23.99 \times 78.70}{100} + \frac{24.99 \times 10.13}{100} + \frac{25.98 \times 11.17}{100}$$
$$= 18.88 + 2.53 + 2.90$$
$$= 24.31$$

The relative atomic mass of magnesium is 24.31.

## Worked example 2

Gallium has two naturally occurring isotopes: $^{69}Ga$ with a relative isotopic mass of 68.93 and $^{71}Ga$ with a relative isotopic mass of 70.92. Given that the relative atomic mass of gallium is 69.72, determine the percentage abundances of each isotope.

### Solution

Let the percentage abundance of the lighter isotope be $x\%$. The abundance of the other isotope must be $(100 - x)\%$, so:

$$A_r(Ga) = \frac{\Sigma(I_r \times \% \text{ abundance})}{100} = 69.72$$
$$\frac{68.93 \times x}{100} + \frac{70.92 \times (100 - x)}{100} = 69.72$$
$$6972 = 68.93x + 70.92(100 - x)$$
$$6972 = 68.93x + 7092 - 70.92x$$
$$6972 - 7092 = 68.93x - 70.92x$$
$$-120 = -1.99x$$
$$x = 60.30$$

The percentage abundance of $^{69}Ga$ is 60.30% and of $^{71}Ga$ 39.70%.

Note: Relative atomic masses for all elements are provided in the periodic table inside the front cover of this book. Students are not expected to commit relative atomic mass data to memory, but you will most likely find that the values of some of the more common elements will be memorized as you solve the problems associated with this section of the course.

## Section 1.2 Exercises

1. Draw a flowchart to summarize the major parts of a mass spectrometer. Annotate the flowchart to explain the function of each part of the mass spectrometer.

2. Draw a mass spectrum for chlorine, which has 75% of the chlorine-35 isotope and 25% of the chlorine-37 isotope. Use labels to show the part of the spectrum that indicates the isotopic mass and the part that shows the abundance of each of the isotopes.

3. Define the term *relative atomic mass*.

4. An isotope of an element is deflected twice as much as an atom of carbon-12. What can be deduced about the mass of that isotope?

5. Carbon has two stable natural isotopes, carbon-12 and carbon-13. (The radioactive isotope carbon-14 is widely used to determine the approximate age of fossilized material.) Calculate the relative atomic mass of carbon, given that the relative isotopic masses and percentage abundances are 12.00 (98.89%) and 13.00 (1.11%) respectively.

6. The element thallium has two isotopes, $^{203}$Tl and $^{205}$Tl. The relative isotopic masses and relative abundances are 202.97 (11.4) and 204.97 (26.6) respectively. Determine the abundance fraction of each isotope and the relative atomic mass of thallium.

7. Calculate the relative atomic mass of silicon from the following data.

| Isotope | Abundance (%) | Isotopic mass |
|---|---|---|
| $^{28}$Si | 92.2 | 28.0 |
| $^{29}$Si | 4.7 | 29.0 |
| $^{30}$Si | 3.1 | 30.0 |

8. Boron has a relative atomic mass of 10.81. It has two isotopes, $^{10}$B of RIM 10.01 and $^{11}$B of RIM 11.01. Determine the percentage abundances of each isotope.

9. The mass spectrum of copper is shown in figure 1.2.2. Use the information given in that graph to calculate the relative atomic mass of copper.

10. It is very difficult to separate $^{72}_{32}$Ge and $^{74}_{32}$Ge but relatively easy to separate $^{74}_{32}$Ge and $^{74}_{34}$Se, although the first two have different mass numbers and the second two have the same mass numbers. Explain why this difference occurs.

11. A pure sample of $^{40}$Ca is passed through a mass spectrometer. Explain why the mass spectrum is found to have a small peak at $m/z = 20$ and a much larger peak at $m/z = 40$.

12. For each of the following pairs select the ion that will require the greater magnetic field to deflect it by a fixed amount.
    a. $^{39}$K$^+$ and $^{28}$Si$^+$
    b. $^{10}$B$^+$ and $^{10}$B$^{2+}$
    c. $^{35}$Cl$^{2+}$ and $^{37}$Cl$^+$

**AS** 2.3.1
Describe the electromagnetic spectrum. © IBO 2007

Electromagnetic spectrum

## 1.3 ELECTRON ARRANGEMENT

### The electromagnetic spectrum

To understand the production of a spectrum we need to review our basic understanding of light and the **electromagnetic spectrum**. Light consists of electromagnetic waves. The **wavelength** of light (the distance between two successive crests) ranges from around $8 \times 10^{-7}$ m for red light to around $4 \times 10^{-7}$ for violet light. The **frequency** of light (the number of waves passing a given point each second) ranges from about $4 \times 10^{-14}$ for red light to about $8 \times 10^{-14}$ for violet light. The wavelength and energy of light are related by the equation

$$E = \frac{hc}{\lambda}$$

where $E$ is the energy
$\lambda$ is the wavelength
$h$ is a constant (Planck's constant)
$c$ is the speed of light.

It is clear from this relationship that the energy of light increases as the wavelength decreases.

Similarly the frequency and energy of light are related by the equation

$$E = h\upsilon$$

where $\upsilon$ is the frequency.

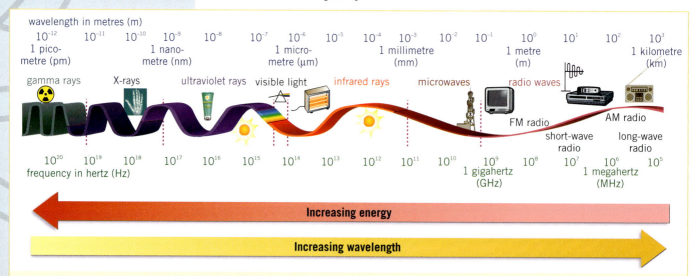

Figure 1.3.1 The electromagnetic spectrum.

Figure 1.3.2 Light passing through raindrops produces a colourful continuous spectrum.

As energy increases, the frequency of the light also increases. Thus red light is of lower energy than violet light. One way to remember this is that it is ultraviolet light that can damage our skin (high energy and short wavelength), whereas infrared ('ray') lamps (low energy and long wavelengths) are often used to help heal sporting injuries such as muscle strains. High energy is more dangerous than low energy. Visible light is part of the broader electromagnetic spectrum that includes gamma rays, X-rays, ultraviolet rays, infrared waves, microwaves and radio waves. The ultraviolet region of the electromagnetic spectrum contains waves with shorter wavelengths than visible light and the infrared region of the electromagnetic spectrum contains waves with longer wavelengths than visible light (see figure 1.3.1).

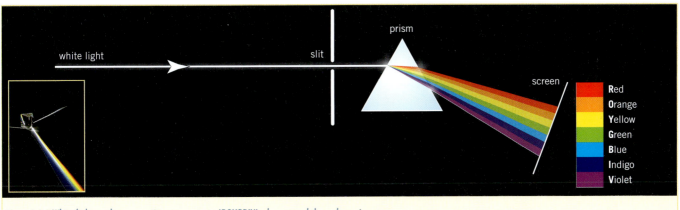

Figure 1.3.3 White light produces a continuous spectrum (ROYGBIV) when passed through a prism.

When sunlight (which contains all wavelengths of visible light) passes through a prism, the different wavelengths are bent (or refracted) through different angles so that the light is broken into its components, producing a **continuous spectrum** of colours. A similar effect is seen when sunlight passes through raindrops to produce a rainbow.

Finding the visible spectrum

## The Bohr model of the atom

In 1913, Danish physicist Niels Bohr proposed that the electrons move around the nucleus in fixed energy levels called shells. He proposed that each atom has a series of these shells. The shells close to the nucleus are of low energy and those further out are of higher energy. Shells are numbered outwards from the nucleus (1, 2, 3 …). They are also identified by letters (K, L, M …). Electrons move around the nucleus in these shells in pathways called orbits.

According to Bohr's model of the atom, different shells (or energy levels) hold different numbers of electrons. There is a maximum number of electrons that can fit in any shell. This maximum number is $2n^2$, where $n$ = the shell number.

For example, sodium ($Z = 11$) has 11 electrons. Two electrons are in the first shell, eight are in the second shell and one electron is in the third shell. This is called the **electron arrangement** or electron configuration of sodium and is written as 2,8,1. The electrons in the outer shell can also be called **valence electrons**. As all of the electrons are as close to the nucleus as possible, this is the lowest energy state of a sodium atom. The lowest energy state of an atom is known as the **ground state**.

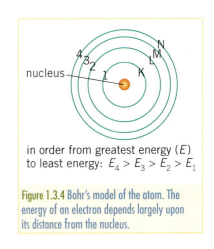

in order from greatest energy ($E$) to least energy: $E_4 > E_3 > E_2 > E_1$

Figure 1.3.4 Bohr's model of the atom. The energy of an electron depends largely upon its distance from the nucleus.

### TABLE 1.3.1 ELECTRONS IN SHELLS

| Shell number ($n$) | Maximum number of electrons in this shell ($2n^2$) |
|---|---|
| 1 | 2 |
| 2 | 8 |
| 3 | 18 |
| 4 | 32 |

### TABLE 1.3.2 ELECTRON ARRANGEMENTS OF SOME ELEMENTS

| Element name | Atomic number | Electron arrangement |
|---|---|---|
| Nitrogen | 7 | 2,5 |
| Oxygen | 8 | 2,6 |
| Neon | 10 | 2,8 |
| Chlorine | 17 | 2,8,7 |

## CHEM COMPLEMENT

### Why K, L, M, and not A, B, C?

Charles G. Barkla was a spectroscopist who studied the X-rays emitted by atoms and found that there appeared to be two types, which he originally named A and B. Later, he renamed them K and L, to leave room for the possibility that the K type was not the highest energy X-ray an atom can emit. We now know that this is the highest energy X-ray, produced when an electron in the innermost shell is knocked out and then recaptured. The innermost shell is therefore called the K shell. Barkla won the 1971 Nobel Prize for Physics.

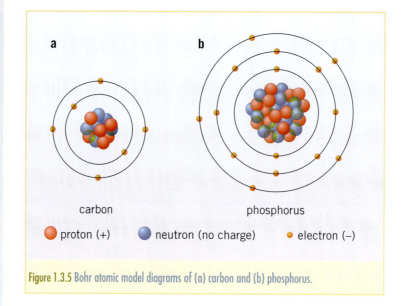

Figure 1.3.5 Bohr atomic model diagrams of (a) carbon and (b) phosphorus.

Atoms may also be represented diagrammatically. The Bohr model of the atom can be shown in full detail with numbers of protons, neutrons and electrons fully labelled.

The electron arrangement of an ion will be different from that of the atom from which it was formed, because an ion is an atom that has lost or gained electrons. Positive ions are atoms that have lost electrons and negative ions are atoms that have gained electrons.

**TABLE 1.3.3 ELECTRON ARRANGEMENTS OF SOME ELEMENTS AND THEIR IONS**

| Element and ion name | Symbol of ion | Atomic number | Charge on ion | Electron arrangement |
|---|---|---|---|---|
| Nitride ion | $N^{3-}$ | 7 | 3− | 2,8 |
| Oxide ion | $O^{2-}$ | 8 | 2− | 2,8 |
| Sodium ion | $Na^+$ | 11 | 1+ | 2,8 |
| Calcium ion | $Ca^{2+}$ | 20 | 2+ | 2,8,8 |

**AS 2.3.2** Distinguish between a continuous spectrum and a line spectrum. © IBO 2007

**AS 2.3.3** Explain how the lines in the emission spectrum of hydrogen are related to electron energy levels. © IBO 2007

PRAC 1.2 Flame tests and emission spectra

### Evidence for the Bohr model: line spectra

Experimental evidence for Bohr's model came from studies of the **emission spectra** of atoms. These spectra are the emissions of light from atoms that have been provided with energy such as heat, light or electricity. The bright colours of fireworks are the result of such emissions.

Bohr explained emission spectra by suggesting that if atoms are subjected to large amounts of energy from heat, light or electricity, the electrons can change energy levels. The electrons jump to energy levels further from the nucleus than they would usually occupy. The atom is said to be in an excited state when this happens. When the electrons return to the ground state this extra energy is released in the form of light. The electrons make specific jumps, depending on the energy levels involved, therefore the light released has a specific wavelength. The emitted light, a **line** (or emission) **spectrum**, looks like a series of coloured lines on a black background. Some of the emissions may be radiation of a wavelength that is not visible to the naked eye. The study of this light emitted from the atom is called emission spectroscopy.

Figure 1.3.6 Metal atoms in fireworks emit coloured light.

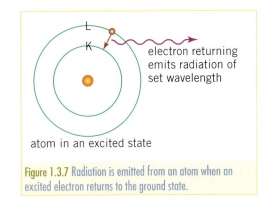

Figure 1.3.7 Radiation is emitted from an atom when an excited electron returns to the ground state.

Bohr's model worked well for the simplest element of all, hydrogen. His model enabled him to predict correctly an emission line that had previously not been detected. The electrons in larger atoms are more complex, however, and Bohr's model was unable to correctly predict the energy changes involved or the intensity of the spectral lines.

Flame tests for metals

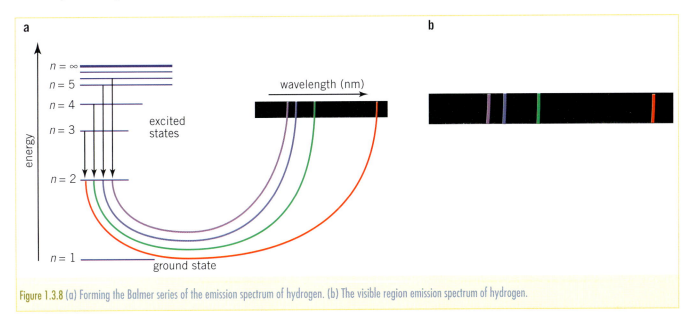

Figure 1.3.8 (a) Forming the Balmer series of the emission spectrum of hydrogen. (b) The visible region emission spectrum of hydrogen.

Spectra of light have been studied extensively since Isaac Newton first produced a rainbow by allowing sunlight to fall onto a prism in 1666. Many scientists contributed to the growing body of knowledge about spectra and the mysterious black lines that were found in the continuous spectrum of sunlight by Joseph von Fraunhofer in 1814. In 1885 Johann Balmer was able to calculate the wavelengths of the four lines in the hydrogen emission spectrum. The energy of these lines corresponds to the difference in energies between outer electron shells and the second electron shell of hydrogen. This group of lines became known as the Balmer series. Similar work by Theodore Lyman in 1906 identified a set of lines in the ultraviolet region of the spectrum as corresponding to the transitions from higher energy levels to the 1st shell, and in 1908 Friedrich Paschen identified a set of lines in the infrared region of the spectrum as the transitions from higher energy levels to the third electron shell.

Spectral lines of hydrogen

### THEORY OF KNOWLEDGE

The visible spectrum is only a small part of the total electromagnetic spectrum and the only part we can observe directly. However, most of what scientists know about the structure of atoms comes from studying how they interact with electromagnetic radiation in the infrared and ultraviolet parts of the spectrum, knowledge that is dependent entirely on technology.

- Could there be knowledge about the structure of the atom that is currently not known, because the technology needed to reveal this knowledge does not yet exist?
- What are the knowledge implications of this?

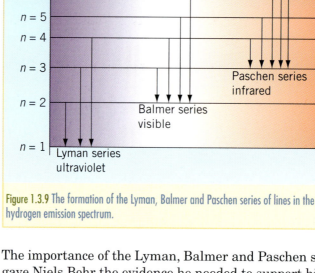

**Figure 1.3.9** The formation of the Lyman, Balmer and Paschen series of lines in the hydrogen emission spectrum.

The importance of the Lyman, Balmer and Paschen sets of lines is that they gave Niels Bohr the evidence he needed to support his theory that electrons existed in shells which had a specific energy. Each line in the hydrogen emission spectrum corresponds to a transition between two energy levels of the hydrogen atom. Within each set, the lines become closer to each other (converge) as the wavelength decreases.

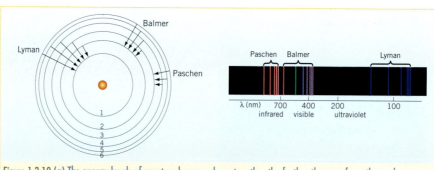

**Figure 1.3.10** (a) The energy levels of an atom become closer together the further they are from the nucleus. (b) The lines in each series of the emission spectrum become closer together as the energy increases (wavelength decreases).

Bohr's model of the atom explained the increasing closeness of the emission lines in terms of the decreasing difference between the energies of shells as their distance from the nucleus increased. The lines become closer together as their energy increases because the energy of the shells is increasing by diminishing amounts. Shell 4 is closer in energy to shell 3 than shell 3 is to shell 2 and shell 2 is to shell 1.

The energy of the line produced by the transition from shell 3 to shell 1 is larger than that from shell 2 to shell 1 because the difference in energy between shells 3 and 1 is greater than that between shells 2 and 1. At the outermost edge of the atom, the energies of the electron shells are so close that they are indistinguishable from each other, so it follows that at the highest energy of each series of lines in the emission spectrum, they merge into a **continuum**. This is called **convergence**.

Formation of line spectra

**WORKSHEET 1.3**
Emission spectra and electron configurations

## Section 1.3 Exercises

1. Outline the model of electron movement around the nucleus proposed by Bohr.

2. Identify the electron that will have the greater energy: an electron in shell 1 or one in shell 2. Explain your answer.

3. Draw a Bohr diagram for a magnesium atom, indicating the number and position of each subatomic particle.

4. **a** Determine the electron arrangement for each of the following elements.
   - **i** $^{31}_{15}P$
   - **ii** $^{19}_{9}F$
   - **iii** $^{40}_{18}Ar$
   - **iv** $^{39}_{19}K$

   **b** Determine the electron arrangement for each of the following ions.
   - **i** $^{24}_{12}Mg^{2+}$
   - **ii** $^{19}_{9}F^{-}$
   - **iii** $^{32}_{16}S^{2-}$
   - **iv** $^{39}_{19}K^{+}$

5. **a** State how many electrons there are in the valence (outermost) shell of each of the following atoms.
   - **i** $^{12}_{6}C$
   - **ii** $^{27}_{13}Al$
   - **iii** $^{19}_{9}F$

   **b** State how many electrons are in the valence shell of each of the following ions.
   - **i** $^{27}_{13}Al^{3+}$
   - **ii** $^{31}_{15}P^{3-}$
   - **iii** $^{35}_{17}Cl^{-}$

6. Compared to the visible region of the electromagnetic spectrum, state where you would find:
   **a** the ultraviolet region
   **b** the infrared region.

7. Consider the emission spectrum of hydrogen. Identify the electron shell to which electrons are falling for the following series.
   **a** the Balmer series
   **b** the Lyman series
   **c** the Paschen series of spectral lines.

8. Draw a labelled flowchart to describe how an emission spectrum is produced for an element such as hydrogen.

9. Explain how each of the four lines in the visible region of the hydrogen emission spectrum is related to an energy level in hydrogen.

10. Predict which is larger: the energy released by an electron transition between shell 6 and shell 5 or the energy released by an electron transition between shell 4 and shell 3. Explain your answer.

11. The term *convergence* describes the decreasing distance between the lines in an emission spectrum as the energy of a set of spectral lines increases. Explain why this occurs.

12. Draw labelled diagrams to distinguish between a continuous spectrum and a line spectrum.

---

**THEORY OF KNOWLEDGE**

Bohr's theory was controversial at the time, but it led to a better, more developed model of the atom. It showed, for example, how the electrons are arranged around the nucleus of the atom in energy levels.

- Explain why Bohr's theory was controversial.
- Explain why Bohr's model is still relevant 100 years later.
- Models exist that are more complex and more correct than Bohr's model. What are these models? Why are they more correct? What is their relevance?

# Chapter 1 Summary

## Terms and definitions

**Anion** A negatively charged ion.

**Atomic number** The number of protons in the nucleus of an atom.

**Cation** A positively charged ion.

**Continuous spectrum** A spectrum of light in which there are no gaps, so that each region blends directly into the next.

**Continuum** A series of lines becomes so close together that they merge.

**Convergence** The decreasing of the distance between lines in an emission spectrum as the energy of a set of spectral lines increases.

**Electromagnetic spectrum** The range of all possible electromagnetic radiation.

**Electron** A negatively charged subatomic particle that orbits the nucleus of the atom.

**Electron arrangement** The pattern of electrons around a nucleus, written as a series of numbers each of which represents the number of electrons in an electron shell, starting from the shell closest to the nucleus and proceeding outwards.

**Electron shell** The region of space surrounding the nucleus in which electrons may be found.

**Emission spectrum** A line spectrum generated when an element is excited and then releases energy as light.

**Frequency** The number of waves passing a given point each second.

**Ground state** The lowest energy state of an atom.

**Ions** Atoms that have lost or gained electrons and so have a charge.

**Isotopes** Atoms that have the same atomic number but different mass numbers.

**Line spectrum** Discrete lines that represent light of discrete energies on a black background.

**Mass number** The sum of the numbers of protons and neutrons in the nucleus of an atom.

**Mass spectrometer** An instrument that enables the relative masses of atoms to be determined.

**Neutron** An uncharged subatomic particle found in the nucleus of the atom.

**Nucleus** The small dense central part of the atom.

**Proton** A positively charged subatomic particle found in the nucleus of the atom.

**Radioisotope** An isotope that is radioactive.

**Relative atomic mass ($A_r$)** The weighted mean of the relative isotopic masses of the isotopes of an element.

**Relative isotopic mass ($I_r$)** The mass of a particular isotope measured relative to carbon-12.

**Valence electrons** Electrons in the outer shell (the highest main energy level) of an atom.

**Wavelength** The distance between successive crests of a light wave.

## Concepts

- Atoms contain a range of subatomic particles including protons, neutrons and electrons.

- Positively charged protons and neutrons with no charge are in the central dense nucleus of the atom, while negatively charged electrons move around the nucleus.

- Protons and neutrons have a relative mass of 1 and electrons have a relative mass of $5 \times 10^{-4}$.

- Some elements have isotopes. Isotopes of an element have the same number of protons but different numbers of neutrons.

- Each element can be represented in a nuclide notation in terms of its mass number, atomic number and charge.

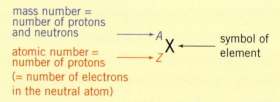

- The actual masses of atoms are very small. Chemists use a relative mass scale to compare atomic masses.

- Relative isotopic mass (RIM) of an atom is defined as the mass of the atom on the scale on which the mass of an atom of the carbon-12 isotope ($^{12}$C) is 12 units exactly. The symbol for RIM is $I_r$.

- Relative atomic mass (RAM) of an element is defined as the weighted mean of the masses of the naturally occurring isotopes on the scale on which the mass of an atom of the carbon-12 isotope ($^{12}$C) is 12 units exactly. The symbol for RAM is $A_r$.

- Relative isotopic masses (RIM) and abundance fractions are determined using a mass spectrometer, the main components of which are shown below.

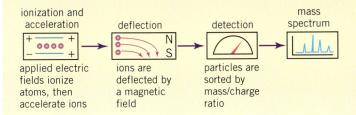

- The electromagnetic spectrum is the range of all possible electromagnetic radiation and includes ultraviolet, visible and infrared light.
- Emission spectra provide evidence for the electron arrangements of atoms.
- The emission spectrum of hydrogen is made up of coloured lines on a black background
- Convergence in an emission spectrum describes the increasing closeness of the lines in the spectrum due to the decreasing differences in energy levels as the distance from the nucleus increases.

- The electron arrangement for an atom or ion is written showing the electrons in order from closest to the nucleus outwards. Each electron shell can hold a maximum of $2n^2$ electrons where $n$ is the shell number.

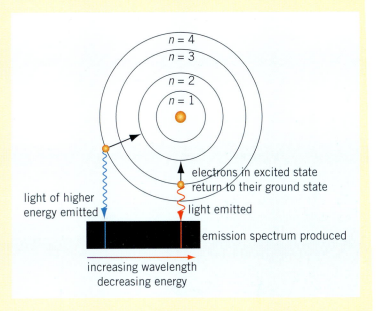

## Chapter 1 Review questions

1 For each of the following atoms, state:
   i the number of protons
   ii the number of neutrons
   iii the name of the element.
   a $^{54}_{25}Mn$
   b $^{83}_{36}Kr$
   c $^{6}_{3}Li$
   d $^{256}_{98}Cf$
   e $^{86}_{37}Rb$

2 The following is a table of the atomic structure of some of the elements which occur naturally in only one detected isotopic form. Determine the missing values in the table.

|   | Element | Atomic number | Mass number | Number of protons | Number of neutrons | Number of electrons |
|---|---------|---------------|-------------|-------------------|--------------------|---------------------|
| a | Beryllium | 4 | 9 | | | |
| b | Fluorine | 9 | 19 | | | |
| c | Scandium | 21 | | | 24 | |
| d | Arsenic | | | | 75 | 33 |

3 Carbon has three isotopes: $^{12}_{6}C$, $^{13}_{6}C$ and $^{14}_{6}C$.
   a Explain the term *isotopes* using carbon as an example.
   b $^{14}_{6}C$ is radioactive. Describe how $^{14}_{6}C$ is commonly used.
   c The mass of $^{14}_{6}C$ is very similar to the mass of $^{14}_{7}N$. Explain why $^{14}_{6}C$ is not considered an isotope of nitrogen.

4 For a particular element $Z = 13$ and $A = 27$.
   a Identify the element.
   b Determine the number of neutrons it has.
   c State its electron arrangement.

5 a i State the ground state electron arrangement of $^{40}_{20}Ca$.
     ii Determine the electron arrangement of $Ca^{2+}$.
   b i State the mass number of calcium.
     ii State the atomic number of calcium.
   c Determine the number of electrons, protons and neutrons in the ion $^{39}_{20}Ca^{2+}$.

6  Identify the elements with the following electron arrangements.
   a  2,7     b  2,4     c  2,8,2
   d  2,5     e  2,8,6

7  a  Explain one way in which protons and electrons are similar.
   b  Explain two ways in which protons and neutrons are similar.
   c  Explain two ways in which neutrons and electrons are different.

8  Explain the purpose of each of the following stages of mass spectrometry:
   a  Vaporization    b  Ionization
   c  Acceleration   d  Deflection
   e  Detection

9  Geologists believe that they have discovered a new element. When a small sample of this element, dubbed newium (symbol Nw), is analysed in a mass spectrometer it is found that it has only one isotope which is deflected only 0.218 times as far as carbon-12.
   a  Determine whether newium is heavier or lighter than carbon-12.
   b  Estimate the relative isotopic mass of newium.
   c  Discuss whether it is possible that newium is a new element, or whether the geologists are mistaken.

10  List the following particles in order of how much they will be deflected in a mass spectrometer with a fixed magnetic field: $^{70}Ga^+$, $^{40}Ca^+$, $^{56}Fe^{2+}$, $^{56}Fe^+$, $^{16}O^+$, $^{16}O^{2+}$.

11  Sketch, approximately to scale, the mass spectrum of lead, using the information in the table.

| Mass number of isotope | Abundance (%) |
|---|---|
| 204 | 1.48 |
| 206 | 23.6 |
| 207 | 22.6 |
| 208 | 52.3 |

12  The two isotopes of antimony have masses 120.90 and 122.90, with a relative abundance of 57.25% and 42.75% respectively. Using these figures, distinguish clearly between atomic mass, isotopic mass and mass number.

13  Lithium has 2 isotopes. $^6Li$ with a relative isotopic mass of 6.02 and a relative abundance of 7.42% and $^7Li$ with a relative isotopic mass of 7.02 and a relative abundance of 92.58%. Calculate the relative atomic mass of lithium.

14  Iridium is a rare and valuable metal that finds significant use as an alloying agent with platinum. Iridium has two isotopes: $^{191}Ir$ with a relative isotopic mass of 190.97 and a relative abundance of 37.3% and $^{193}Ir$ with a relative isotopic mass of 192.97 and a relative abundance of 62.7%. Calculate the relative atomic mass of iridium.

15  Chlorine has a relative atomic mass of 35.45. It has two isotopes: $^{35}Cl$ with relative isotopic mass 34.97 and $^{37}Cl$ of relative isotopic mass 36.97. Determine the percentage abundance of each isotope.

16  Gallium has an atomic mass of 69.72. It has only two isotopes. One of the isotopes has a relative abundance of 64% and a relative isotopic mass of 69.0. Calculate the relative isotopic mass of the other isotope.

17  Explain the difference between a continuous spectrum and a line spectrum.

18  Explain how lines in the visible region of the emission spectrum of hydrogen gas are formed.

19  The emission spectrum of hydrogen has a series of lines in the ultraviolet region of the electromagnetic spectrum. These have been identified as corresponding to transitions from higher energy levels to the first electron shell. Sketch a diagram showing the relationship between the electron transitions and the formation of these lines in the hydrogen emission spectrum.

20  Convergence describes the decreasing distance between lines in an emission spectrum until a continuum is reached. Explain why this occurs at the higher energy 'end' of each series of lines in the hydrogen emission spectrum.

21  Describe how technological advancement has helped to develop the model of the atom. Given that protons, neutrons and electrons cannot be seen directly, describe how technological advancements have made it possible to know of their existence.

www.pearsoned.com.au/schools

Weblinks are available on the Chemistry: For use with the IB Diploma Programme Standard Level Companion Website to support learning and research related to this chapter.

# Chapter 1 Test

## Part A: Multiple-choice questions

1. Which statement is correct about the isotopes of an element?
   A They have the same mass number.
   B They have the same electron arrangement.
   C They have more protons than neutrons.
   D They have the same numbers of protons and neutrons.
   © IBO 2006, Nov P1 Q5

2. What are *valence* electrons?
   A Electrons in the energy level closest to the nucleus
   B Electrons in the highest main energy level
   C The number of electrons required to complete the highest main energy level
   D The total number of electrons in the atom
   © IBO 2006, Nov P1 Q6

3. Which statement is correct about a line emission spectrum?
   A Electrons absorb energy as they move from low to high energy levels.
   B Electrons absorb energy as they move from high to low energy levels.
   C Electrons release energy as they move from low to high energy levels.
   D Electrons release energy as they move from high to low energy levels.
   © IBO 2005, Nov P1 Q6

4. Information is given about four different atoms:

   | Atom | Neutrons | Protons |
   |---|---|---|
   | W | 22 | 18 |
   | X | 18 | 20 |
   | Y | 22 | 16 |
   | Z | 20 | 18 |

   Which **two** atoms are isotopes?
   A W and Y
   B W and Z
   C X and Z
   D X and Y
   © IBO 2005, Nov P1 Q5

5. A certain sample of element Z contains 60% of $^{69}Z$ and 40% of $^{71}Z$. What is the relative atomic mass of element Z in this sample?
   A 69.2  B 69.8  C 70.0  D 70.2
   © IBO 2004, Nov P1 Q5

6. What is the difference between two neutral atoms represented by the symbols $^{59}_{27}Co$ and $^{59}_{28}Ni$?
   A The number of neutrons only
   B The number of protons and electrons only
   C The number of protons and neutrons only
   D The number of protons, neutrons and electrons
   © IBO 2004, Nov P1 Q6

7. What is the correct number of each particle in a fluoride ion, $^{19}F^-$?

   |   | Protons | Neutrons | Electrons |
   |---|---|---|---|
   | A | 9 | 10 | 8 |
   | B | 9 | 10 | 9 |
   | C | 9 | 10 | 10 |
   | D | 9 | 19 | 10 |

   © IBO 2003, Nov P1 Q5

8. Which statement is correct for the emission spectrum of the hydrogen atom?
   A The lines converge at lower energies.
   B The lines are produced when electrons move from lower to higher energy levels.
   C The lines in the visible region involve electron transitions into the energy level closest to the nucleus.
   D The line corresponding to the greatest emission of energy is in the ultraviolet region.
   © IBO 2003, Nov P1 Q6

9. Consider the composition of the species W, X, Y and Z below. Which species is an anion?

   | Species | Number of protons | Number of neutrons | Number of electrons |
   |---|---|---|---|
   | W | 9 | 10 | 10 |
   | X | 11 | 12 | 11 |
   | Y | 12 | 12 | 12 |
   | Z | 13 | 14 | 10 |

   A W  B X  C Y  D Z
   © IBO 2003, May P1 Q5

**10** Which species have electronic configurations 2,8,8; 2,8 and 2,8,1 respectively?

  A  Ne, F, Na
  B  $K^+$, $F^-$, $Mg^{2+}$
  C  $Ca^{2+}$, F, $Na^+$
  D  $Cl^-$, $F^-$, Na

© IBO 2001, May P1 Q7

**11** Which electron transition in the hydrogen atom releases the most energy?

  A  $n = 2 \to n = 1$
  B  $n = 3 \to n = 2$
  C  $n = 4 \to n = 3$
  D  $n = 5 \to n = 4$

© IBO 2001, NovP1 Q7

**12** Copper consists of the isotopes $^{63}$Cu and $^{65}$Cu and has a relative atomic mass of 63.55. What is the most likely composition?

| | $^{63}$Cu | $^{65}$Cu |
|---|---|---|
| A | 30% | 70% |
| B | 50% | 50% |
| C | 55% | 45% |
| D | 70% | 30% |

© IBO 2002, May P1 Q6

(12 marks)

www.pearsoned.com.au/schools

For more multiple-choice test questions, connect to the Chemistry: For use with the IB Diploma Programme Standard Level Companion Website and select Chapter 1 Review Questions.

## Part B: Short-answer questions

**1 a** State the electron arrangements of the following species.

  **i** Si
  **ii** $P^{3-}$

(2 marks)

**b** Identify the numbers of protons, neutrons and electrons in the species $^{33}S^{2-}$.

(1 mark)

© IBO 2006, Nov P2 Q3b, c

**2 a** State a physical property that is different for isotopes of an element.

(1 mark)

**b** Chlorine exists as two isotopes, $^{35}$Cl and $^{37}$Cl. The relative atomic mass of chlorine is 35.45. Calculate the percentage abundance of each isotope.

(2 marks)

© IBO 2003, Nov P2 Q3

**3 a** Define the terms *atomic number* and *mass number*.

(2 marks)

**b** For each of the species shown in the table, state the number of each subatomic particle present.

| Species | Protons | Neutrons | Electrons |
|---|---|---|---|
| $^{14}_{6}C$ | | | |
| $^{19}_{9}F^-$ | | | |
| $^{40}_{20}Ca^{2+}$ | | | |

(3 marks)

© IBO 2001, May P2 Q3

**4** Describe the difference between a continuous spectrum and a line spectrum.

(2 marks)

**5** A sample of germanium is analysed in a mass spectrometer. The first and last processes in mass spectrometry are vaporization and detection.

**a i** State the names of the other three processes in the order in which they occur in a mass spectrometer.

(2 marks)

**ii** For each of the processes named in part **a i**, outline how the process occurs.

(3 marks)

**b** The sample of germanium is found to have the following composition:

| Isotope | $^{70}$Ge | $^{72}$Ge | $^{74}$Ge | $^{76}$Ge |
|---|---|---|---|---|
| Relative abundance (%) | 22.60 | 25.45 | 36.73 | 15.22 |

**i** Define the term *relative atomic mass*.

(2 marks)

**ii** Calculate the relative atomic mass of this sample of germanium, giving your answer to two decimal places.

(2 marks)

© IBO 2005, HL Nov P2 Q2

**6** Some vaporized magnesium is introduced into a mass spectrometer. One of the ions that reaches the detector is $^{25}Mg^+$.

The $^{25}Mg^{2+}$ ion is also detected in this mass spectrometer by changing the magnetic field. Deduce and explain, by reference to the *m/z* values of these two ions of magnesium, which of the ions $^{25}Mg^{2+}$ and $^{25}Mg^+$ is detected using a stronger magnetic field.

(2 marks)

© IBO 2006, HL NovP2 Q5(c)

## Part C: Data-based questions

1. A sample of the element nickel was analysed by mass spectrometry. The spectrum obtained is shown below.

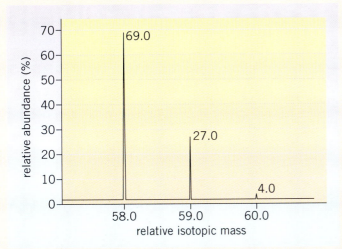

Calculate the relative atomic mass of nickel. (Full credit will only be gained when all working is shown, together with the correct answer.)

(3 marks)

2. Five of the lines in the Balmer series line spectrum of hydrogen are described in the table below.

| Line number (for reference) | Colour of line | Wavelength of line (nm) |
|---|---|---|
| 1 | Red | 656 |
| 2 | Green | 486 |
| 3 | Violet | 434 |
| 4 | Violet | 410 |
| 5 | Ultraviolet | 397 |

a Calculate the difference in wavelength between each adjacent pair of lines (e.g. lines 1 and 2, lines 2 and 3 etc.).

(2 marks)

b Describe how the difference in wavelength of adjacent lines changes as the wavelength decreases.

(1 mark)

c Explain why this change occurs.

(2 marks)

## Part D: Extended-response question

a Describe how the lines in the emission spectrum of an element are produced.

(4 marks)

b Draw a fictitious (made up) emission spectrum in the visible region of light, clearly showing the pattern of the lines.

(2 marks)

Total marks = 50

# 2 BONDING

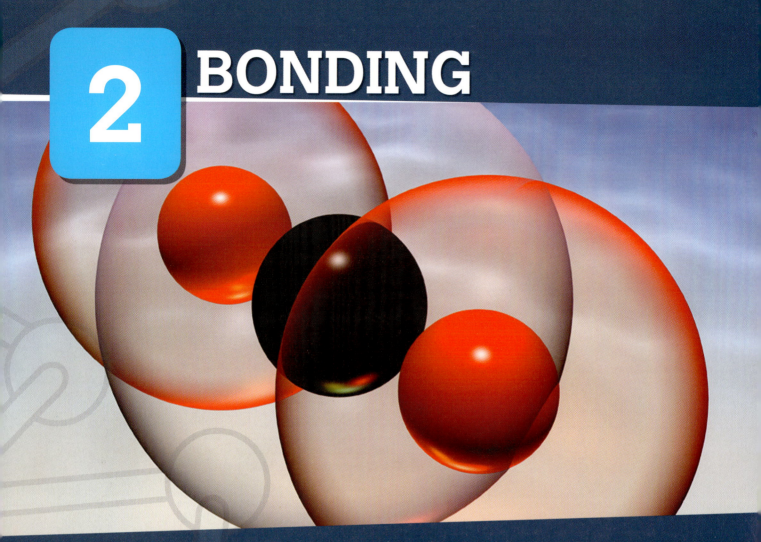

## Chapter overview

This chapter covers the IB Chemistry syllabus Topic 4: Bonding.

**By the end of this chapter, you should be able to:**

- recognize that all forms of bonding between particles are electrostatic in nature
- describe the formation of ions as the result of the transfer of one or more electrons from a metal atom to a non-metal atom
- describe the regular lattice arrangement of ions in a solid ionic compound
- write the names and formulas of ionic compounds containing one or more common polyatomic ions and transition metal ions
- describe the type of bonding and structure found in metals using the 'electron sea' model
- describe the covalent bond as the sharing of electrons between two atoms
- use Lewis structures to explain how simple molecules and ions form and use VSEPR theory to predict the shape and bond angles of simple molecules
- use electronegativities, shape and bond polarities to predict molecule polarity
- describe the relationship between the strength of intermolecular bonding forces and the melting and boiling points of molecular substances
- explain the structure and bonding of covalently bonded substances such as diamond, graphite, $C_{60}$ fullerene, silicon and silicon dioxide
- explain how the properties of ionic, metallic and covalent substances result from the bonding present.

**IBO Assessment statements 4.1.1 to 4.5.1.**

A chemical bond forms when (generally) outer-shell electrons of different atoms come close enough to each other to interact and re-arrange themselves into a more stable arrangement—one with a lower overall chemical potential energy. All chemical bonds are based on the **electrostatic attraction** between positive and negative particles. When two or more atoms approach one another to form a bond, it is their outer-shell electrons that generally interact. Outer-shell electrons are so important in the generation of chemical bonds that they are given a special name—**valence electrons**.

The most stable arrangement of electrons, either within a single atom or when grouped in compounds, is to have an outer-shell of eight electrons (except in the case of hydrogen, where two electrons meets this criterion), such as is found in the gases of group 0 (VIII) of the periodic table. Indeed, the group 0 gases—helium, neon, argon, krypton, xenon and radon—were originally called the inert gases because it was believed that they would never react with any other element. In 1962, Neil Bartlett and co-workers at the University of California in Berkley found that xenon would react with fluorine at very high voltages; however, these elements are still regarded as unreactive under normal conditions. It seems that a full outer-shell of electrons confers this stability, and other elements behave in such a way as to attain this outer-shell configuration.

This concept of eight outer-shell electrons being the most stable and preferable arrangement was first postulated by Gilbert Lewis (1875–1946), an American chemist who coined the phrase *octet rule* to summarize his findings. 'Octet' means 'a group of eight'. Application of this important principle allows us to recognize why atoms chemically bond in the ratios they do to form compounds.

Figure 2.0.1 Gilbert Lewis (1875–1946) developed models of electron-bonding behaviour. He also made important contributions in the fields of spectroscopy, acid–base theory and thermodynamics.

There are three different groups into which **strong chemical bonds** can be classified, each according to the types of atoms involved:

- Metal atoms combine with non-metal atoms to form ionic bonds (see section 2.1).
- Metal atoms combine with metal atoms to form metallic bonds (see section 2.2).
- Non-metal atoms combine with non-metal atoms to form covalent bonds (see section 2.3).

The rearrangement of valence electrons to form chemical bonds tends to result in the formation of very strong bonds, as the following table of bond dissociation enthalpies shows. **Bond dissociation enthalpy** can be defined as the enthalpy required to break the bonds between 1 mole of bonded atoms. The larger the value, the stronger the bond.

| TABLE 2.0.1 BOND DISSOCIATION ENTHALPIES | | | | |
|---|---|---|---|---|
| Compound or element | Formula | Atom types | Bonding type | Dissociation enthalpy (kJ mol$^{-1}$) |
| Sodium chloride | NaCl | Metal/non-metal | Ionic | 411 |
| Magnesium oxide | MgO | Metal/non-metal | Ionic | 601 |
| Aluminium fluoride | AlF$_3$ | Metal/non-metal | Ionic | 1504 |
| Hydrogen | H$_2$ | Non-metal | Covalent | 436 |
| Nitrogen | N$_2$ | Non-metal | Covalent | 945 |
| Water | H$_2$O | Non-metal | Covalent | 285 |
| Sodium | Na | Metal | Metallic | 109 |
| Iron | Fe | Metal | Metallic | 414 |
| Tungsten | W | Metal | Metallic | 845 |

## 2.1 IONIC BONDING

An **element** is a substance made up of atoms that are all of the same type. The majority of the 92 naturally occurring elements are metals. They are found on the left-hand side and the middle and lower sections of the periodic table. Metals tend to be shiny, have high melting and boiling points, and are generally good conductors of heat and electricity. Metal atoms tend to have low numbers of electrons (usually one or two) in their valence shells and, for this reason, have a tendency to lose these electrons so as to gain an outer-shell octet of electrons. Non-metal elements, on the other hand, are found on the right-hand side of the periodic table. Non-metals are generally poor conductors of electricity and heat. They have low melting points, and several are gases at room temperature. They have high numbers of electrons in their valence shells and will readily accept further electrons in order to reach the desired outcome of eight outer-shell electrons. This will be discussed further in chapter 3.

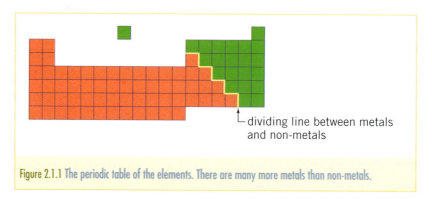

Figure 2.1.1 The periodic table of the elements. There are many more metals than non-metals.

### The nature of the ionic bond

**Ionic bonding** occurs as the result of a metal atom donating its valence electron(s) to a non-metal atom. As the metal atom loses electrons, it will gain an overall positive charge, becoming a positively charged ion (a cation). Similarly, the non-metal atom accepts electrons, becoming a negatively charged ion (an anion). The positive and negative ions are attracted to each other by electrostatic attraction, producing a neutral compound. An ionic bond can be defined as the electrostatic attraction between positive and negative ions. The ionic compound formed is hard and brittle and is often described as a 'salt'.

**AS 4.1.1**
Describe the ionic bond as the electrostatic attraction between oppositely charged ions.
© IBO 2007

Figure 2.1.2 (a) Sodium, a soft, silvery metal, and chlorine, a greenish gas, react vigorously. (b) White crystalline table salt is formed.

Sodium reacts explosively with chlorine gas to produce the white crystalline solid sodium chloride. Sodium is a highly reactive metallic element in group 1 of the periodic table. It has an electron arrangement of 2,8,1. The one valence electron is lost when sodium forms a bond and the sodium ion, $Na^+$, is formed. Chlorine has 17 electrons and an electron arrangement of 2,8,7. Chlorine has a strong affinity (or attracting power) for electrons and will readily accept one additional electron to fill its valence shell, resulting in the chloride ion, $Cl^-$.

The formation of the two ions can be usefully represented diagrammatically. Although such a diagram clearly shows the transfer of one electron from the metal to the non-metal, it is important to note that ionic bonding involves the formation of many ions that are then held together by electrostatic attraction in a continuous lattice of positive and negative ions.

## CHEM COMPLEMENT

### The role of salt in human history

Salt has been used from prehistoric times in foods for flavouring, pickling, preserving, curing meat and fish, for tanning animal hides, and in trade and politics, making it an important part of the history of human culture and civilization. Salt has also become symbolized in our language as well. For example in Roman times, soldiers were paid a part of their wages in salt, an indication of the value attached to it. This payment was known as *salarium*, from which the modern term *salary* derives.

Formation of sodium chloride
The sizes of anions and cations

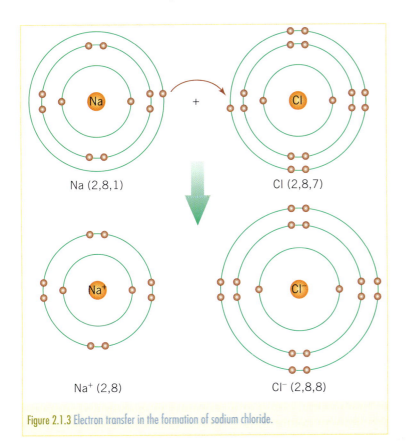

Figure 2.1.3 Electron transfer in the formation of sodium chloride.

**AS 4.1.2**
Describe how ions can be formed as a result of electron transfer. © IBO 2007

**AS 4.1.3**
Deduce which ions will be formed when elements in groups 1, 2 and 3 lose electrons. © IBO 2007

**AS 4.1.4**
Deduce which ions will be formed when elements in groups 5, 6 and 7 gain electrons. © IBO 2007

**AS 4.1.5**
State that transition elements can form more than one ion. © IBO 2007

**WORKSHEET 2.1**
Formation of ionic compounds

**AS 4.1.7**
State the formula of common polyatomic ions formed by non-metals in periods 2 and 3. © IBO 2007

Notice in figure 2.1.3 that when a sodium atom becomes a sodium ion its size decreases. The positive ion is smaller than its parent atom, as one electron shell has been lost. Less obvious perhaps is the increase in the size of the chlorine atom as it becomes a chloride ion. The presence of an extra electron in the outer shell leads to greater repulsion, so that the anion is larger than its parent atom. This will be discussed further in chapter 3.

Ionic compounds can be formed by any combination of positive and negative ions. In an ionic lattice a positive ion will always be surrounded by negative ions, and a negative ion will always be surrounded by positive ions. The exact configuration varies depending on the relative sizes of the ions involved.

In the formation of an ionic compound, metal atoms in groups 1, 2 and 3 will lose electrons to form ions with a 1+, 2+ and 3+ charge respectively, and elements in groups 5, 6 and 7 will gain electrons to form ions with a 3–, 2– and 1– charge respectively.

Transition metals also lose electrons when forming ionic compounds; however their more complex electron arrangement means that they can generally form more than one type of ion. For example, iron can form ions with either a 2+ or a 3+ charge. The following table gives some examples of ions formed by a range of metals and non-metals.

**TABLE 2.1.1 FORMATION OF SOME SIMPLE IONS**

| Element | Group | Name of ion | Formula of ion |
|---|---|---|---|
| Sodium | 1 | Sodium | $Na^+$ |
| Magnesium | 2 | Magnesium | $Mg^{2+}$ |
| Aluminium | 3 | Aluminium | $Al^{3+}$ |
| Chlorine | 7 | Chloride | $Cl^-$ |
| Oxygen | 6 | Oxide | $O^{2-}$ |
| Nitrogen | 5 | Nitride | $N^{3-}$ |
| Iron | Transition metals | Iron(II) | $Fe^{2+}$ |
| Iron | Transition metals | Iron(III) | $Fe^{3+}$ |
| Copper | Transition metals | Copper(I) | $Cu^+$ |
| Copper | Transition metals | Copper(II) | $Cu^{2+}$ |

### Naming and writing formulas for ionic compounds

The ability to write chemical formulas accurately is vital in the study of chemistry. Ionic formulas form an important part of your chemical vocabulary and, as such, should be committed to memory so that they can be used automatically. Just as words are needed to write sentences, in Chemistry, formulas are needed to write chemical equations.

As we have seen in the previous section, metals form positive ions and non-metals form negative ions. Simple ions are made up of single atoms that have lost or gained valence electrons to achieve a full valence shell. Some ions are more complex; they are a combination of several atoms with an overall charge. These ions are called **polyatomic ions**. Polyatomic ions are usually made up of non-metal atoms, although a few include metal atoms. Examples of some polyatomic atoms are $SO_4^{2-}$, $CO_3^{2-}$, $PO_4^{3-}$, $OH^-$, $NO_3^-$, $NH_4^+$ and $H_3O^+$.

Tables 2.1.2 and 2.1.3 show the formulas of a number of positive and negative ions that you may encounter in your studies of Chemistry.

**TABLE 2.1.2 FORMULAS OF A RANGE OF COMMON POSITIVE IONS INCLUDING POLYATOMIC IONS**

| +1 | +2 | +3 | +4 |
|---|---|---|---|
| Lithium $Li^+$ | Magnesium $Mg^{2+}$ | Aluminium $Al^{3+}$ | Tin(IV) $Sn^{4+}$ |
| Sodium $Na^+$ | Calcium $Ca^{2+}$ | Chromium(III) $Cr^{3+}$ | Lead(IV) $Pb^{4+}$ |
| Potassium $K^+$ | Strontium $Sr^{2+}$ | Iron(III) $Fe^{3+}$ | |
| Hydrogen $H^+$ | Barium $Ba^{2+}$ | | |
| Copper(I) $Cu^+$ | Iron(II) $Fe^{2+}$ | | |
| Silver $Ag^+$ | Copper(II) $Cu^{2+}$ | | |
| Ammonium $NH_4^+$ | Zinc $Zn^{2+}$ | | |
| Hydronium $H_3O^+$ | Lead(II) $Pb^{2+}$ | | |

**TABLE 2.1.3 FORMULAS OF A RANGE OF COMMON NEGATIVE IONS, INCLUDING POLYATOMIC IONS**

| −1 | −2 | −3 |
|---|---|---|
| Fluoride $F^-$ | Oxide $O^{2-}$ | Nitride $N^{3-}$ |
| Chloride $Cl^-$ | Sulfate $SO_4^{2-}$ | Phosphate $PO_4^{3-}$ |
| Bromide $Br^-$ | Sulfide $S^{2-}$ | |
| Iodide $I^-$ | Carbonate $CO_3^{2-}$ | |
| Hydroxide $OH^-$ | Chromate $CrO_4^{2-}$ | |
| Nitrate $NO_3^-$ | Dichromate $Cr_2O_7^{2-}$ | |
| Permanganate $MnO_4^-$ | Thiosulfate $S_2O_3^{2-}$ | |
| Cyanide $CN^-$ | | |
| Hydrogen sulfate $HSO_4^-$ | | |
| Hydrogen carbonate $HCO_3^-$ | | |
| Ethanoate ion $CH_3COO^-$ | | |

All ionic formulas are made up of positive and negative ions. The total positive charge will always equal the total negative charge because, overall, a compound is neutral. Sometimes multiple ions will be needed to achieve this neutrality. Ionic formulas are written as empirical formulas. An **empirical formula** is the lowest whole number ratio of the atoms in a compound.

Naming ionic compounds
Writing ionic formulas

All areas of knowledge have conventions or rules, sets of common understandings to make communication easier. For example, every language has grammar conventions. To name and write a balanced formula for a particular ionic compound, the following conventions are used.

- When naming an ionic compound, the positive ion is generally written first, followed by the negative ion. For example, NaCl is called sodium chloride. (The main exceptions here are salts of organic acids, such as sodium ethanoate, $CH_3COONa$ (see chapters 9 and 11).)
- As compounds do not carry an overall charge, it is necessary to balance the charges of the anion and cation components, for example NaCl, MgO, KOH, $CaCO_3$, $HNO_3$.
- If more than one of each ion is required to balance the overall charge, a subscript is used to indicate the number of each species required, for example $H_2SO_4$, $AlCl_3$, $K_2S$, $PbCl_2$, $K_2Cr_2O_7$.

**WORKSHEET 2.2**
Charges on ions

- In the case of some polyatomic ions, it may be necessary to use brackets to ensure no ambiguity is present. For example, $CuOH_2$ is incorrect; it should be $Cu(OH)_2$. Similarly $Al_2SO_{43}$ is incorrect; it should be $Al_2(SO_4)_3$.
- For metals that are able to form ions of different charges, Roman numerals are used to indicate the relevant charge on the ion; for example, $FeCl_2$ would be named as iron(II) chloride and $FeCl_3$ would be named as iron(III) chloride.
- The name of the ion depends on its composition. For example, the ending -ate for a polyatomic ion indicates the presence of oxygen; the ending -ide indicates that the ion is made up of a single atom with a negative charge such as sulfide, $S^{2-}$, chloride, $Cl^-$.
- The name of the ion often gives a clear indication of the atoms present. The hydroxide ion is made up of hydrogen and oxygen, $OH^-$, while hydrogencarbonate ions are made up of hydrogen, carbon and oxygen, $HCO_3^-$.

### Worked example 1

Write balanced chemical formulas for the following compounds.

a Potassium bromide
b Sodium nitrate
c Magnesium sulfide
d Sodium oxide
e Calcium chloride
f Iron(III) sulfate

### Solution

| Compound | Cation | Anion | Formula |
|---|---|---|---|
| a | $K^+$ | $Br^-$ | KBr |
| b | $Na^+$ | $NO_3^-$ | $NaNO_3$ |
| c | $Mg^{2+}$ | $S^{2-}$ | MgS |
| d | $Na^+$ (× 2) | $O^{2-}$ | $Na_2O$ |
| e | $Ca^{2+}$ | $Cl^-$ (× 2) | $CaCl_2$ |
| f | $Fe^{3+}$ (× 2) | $SO_4^{2-}$ (× 3) | $Fe_2(SO_4)_3$ |

### Worked example 2

State the name of each of the following compounds, given their formulas.

a NaOH
b $KMnO_4$
c $Cu_2SO_4$
d $Ca(NO_3)_2$
e $FePO_4$
f $(NH_4)_2S$

### Solution

a Sodium hydroxide
b Potassium permanganate
c Copper(I) sulfate
d Calcium nitrate
e Iron(III) phosphate
f Ammonium sulfide

## The structure of ionic compounds

**4.1.8**
Describe the lattice structure of ionic compounds. © IBO 2007

So far we have seen that the positive charge on the cation must be balanced by the negative charge on the anion to produce a neutral compound. Ionic compounds do not exist in nature as separate units of, for example, one sodium ion and one chloride ion (NaCl). In fact, the ions that make up ionic compounds arrange themselves into a regular pattern, a lattice structure (an **ionic lattice**), containing many millions of ions that extend in all three dimensions. No fixed number of ions is involved, but the ratio of cations to anions is constant for a given compound and is shown in the empirical formula.

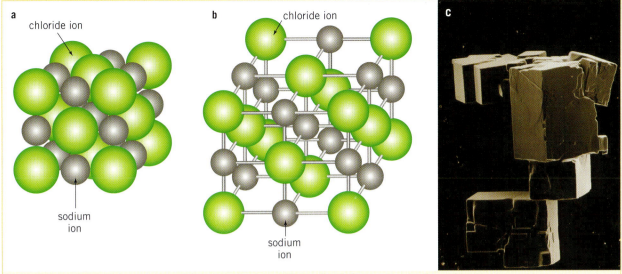

Figure 2.1.4 (a) A close-packed model of part of the sodium chloride crystal lattice. (b) A ball-and-stick model of the sodium chloride crystal lattice. (c) Scanning electron micrograph of sodium chloride crystals.

The most stable arrangement of ions for any particular ionic compound will be the one in which the positively charged ions are packed as closely as possible to the negatively charged ions, and the ions with the same charge are as far apart as possible. This arrangement serves to maximize the electrostatic attraction between the positive and negative ions and minimize the repulsion between like charged ions, thus lowering the overall chemical potential energy of the lattice. There are a number of different ion arrangements that can be generated to meet these criteria, depending on the relative sizes of the ions present and their ratio in the compound. Each arrangement will result in the particular lattice structure found for that compound.

Sodium chloride provides a good example of an ionic lattice. Its structure is shown in figure 2.1.4. Each positive sodium ion is surrounded by six chloride ions, and each chloride ion is surrounded by six sodium ions. A sodium chloride crystal is cubic in shape.

PRAC 2.1
Modelling sodium chloride

DEMO 2.1
Conductivity of ionic compounds

### Section 2.1 Exercises

1. Identify the bonding type in each of the following cases as ionic, covalent or metallic.
   a  Silver
   b  Hydrogen and oxygen
   c  Magnesium and chlorine
   d  Copper and sulfur
   e  Carbon and oxygen
   f  Tin and copper
   g  Aluminium and sulfur

2. Draw electron shell diagrams for each of the following pairs of elements to represent how they interact to form ions. Clearly state the formula of the resulting compound.
   a  Sodium and fluorine to produce sodium fluoride
   b  Calcium and sulfur to produce calcium sulfide
   c  Calcium and chlorine to produce calcium chloride
   d  Aluminium and oxygen to produce aluminium oxide

3. State balanced chemical formulas for each of the following compounds.
   a. Potassium nitrate
   b. Calcium chloride
   c. Sodium hydroxide
   d. Copper(II) sulfate
   e. Ammonium sulfide
   f. Aluminium nitrate

4. State the names of the following compounds.
   a. KCl
   b. $BaSO_4$
   c. $HNO_3$
   d. $Al_2O_3$
   e. $SnI_2$
   f. $Cu_3(PO_4)_2$

5. Determine the charge on the positive ion (cation) for each of the following compounds.
   a. $Mn(SO_4)_2$
   b. $Co(OH)_3$
   c. $NiCO_3$
   d. $Pt(NO_3)_2$
   e. $AuBr_3$
   f. $Ga_2(OH)_2$

6. Determine the charge on the negative ion (anion) for each of the following compounds.
   a. $KBrO_3$
   b. $Na_2SiO_3$
   c. $NH_4AsO_2$
   d. $Ag_2SeO_3$
   e. $LiTeO_4$
   f. $AlIrCl_6$

7. Describe the structure of sodium chloride, with the help of a labelled diagram.

## 2.2 METALLIC BONDING

Observations of the physical properties of metals have led chemists to develop theories to explain these observations. For example, **electrical conductivity** requires the presence of charged particles that are free to move. Hardness and high melting points imply strong bonding. **Malleability** (the ability to be bent without breaking) and **ductility** (the ability to be drawn into a wire) suggest that there is regularity in the structure. From this information, chemists have devised a model to explain and represent the structure and bonding in metal elements in a simple way.

### The nature of the metallic bond

**AS 4.4.1**
Describe the metallic bond as the electrostatic attraction between a lattice of positive ions and delocalized electrons.
© IBO 2007

In this model, metal ions, formed when atoms lose their valence electrons, are arranged in a three-dimensional lattice. This array of ions is surrounded by freely moving electrons that form a 'sea' of mobile electrons. These electrons are said to be **delocalized**, as they are not confined to a particular location but can move throughout the structure. Electrons are attracted to positively charged ions. This electrostatic attraction holds the lattice together, and prevents the ions pushing each other apart due to the electrostatic repulsion of like charges. This type of bonding is called **metallic bonding**.

Why do the metal atoms release their valence electrons to form the sea of electrons? Metal atoms achieve greater stability by releasing their valence electrons. Without their valence electrons, the metal atoms achieve a noble gas configuration—an outer-shell octet of electrons. When non-metals are present, these valence electrons are transferred to the non-metal atoms, giving rise to the ionic

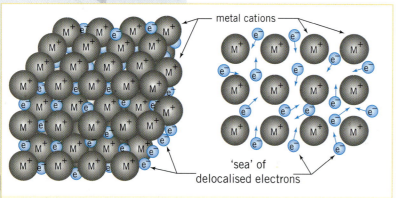

Figure 2.2.1 The structure of a metal may be described as a lattice of cations surrounded by a 'sea' of electrons.

bonding discussed in section 2.1. When only metal atoms are present, the 'lost' valence electrons simply become delocalized within the metallic lattice.

## Important properties of metals

The uses of metals by humans in the past as well as the present centre on two important properties of metals: their ability to conduct electricity and their malleability. You would be reading this book by candlelight if metals had not proved able to conduct electricity, and our bridges and buildings would be far less impressive than they are today.

The electrical conductivity of metals can be explained by the presence of the sea of delocalized electrons that surrounds the lattice of positive metal ions. In the solid state these electrons can move freely and will respond to the application of a potential difference. When a metal is connected to a power supply, electrons enter one end of the metal (the end connected to the negative terminal). The same number of electrons then exits from the other end of the metal (moving towards the positive terminal). Delocalized electrons move freely through the structure, but metal ions vibrate, causing a barrier to the smooth flow of electrons. Some energy is therefore lost, causing the metal to heat as current passes through it.

**DEMO 2.2**
Modelling the structures of metals

**AS 4.4.2**
**Explain the electrical conductivity and malleability of metals.** © IBO 2007

**PRAC 2.2**
Growing metal crystals

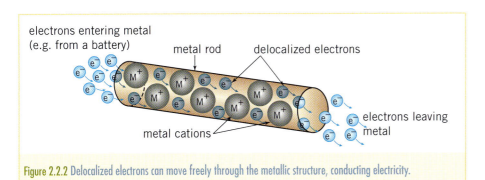

**Figure 2.2.2** Delocalized electrons can move freely through the metallic structure, conducting electricity.

The malleability of a metal is its ability to be beaten or bent into shape without breaking. Once again the sea of delocalized electrons is responsible for this property of metals. When a metal is bent, its lattice of positive ions is displaced and there is a possibility of positive ions coming into contact with other positive ions and repelling each other. The constant movement of the delocalized electrons prevents this from occurring, so the metal bends without breaking.

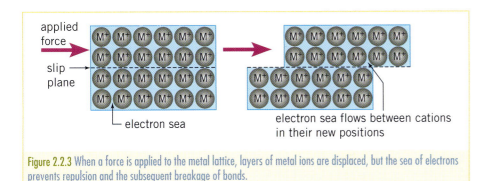

**Figure 2.2.3** When a force is applied to the metal lattice, layers of metal ions are displaced, but the sea of electrons prevents repulsion and the subsequent breakage of bonds.

### CHEM COMPLEMENT

#### Metals with a memory

Metals are malleable and we can readily change their shapes. Some metals can actually 'remember' their original shape and return to it under certain conditions! These 'memory metals' are alloys that have found use in a wide range of applications. One example, Nitinol, an alloy of nickel and titanium, is produced at high temperatures as a fine metal screen. It is then cooled and reshaped into a wire. This wire can be inserted into the bloodstream. As it warms in the body it 'remembers' its original shape and resumes the form of a fine metal screen. This screen can prevent movement of a blood clot through the circulatory system. Another type of memory metal is the brass used in safety devices and automatic switches. The use of memory metals in orthodontic work make attachment of wires to the teeth easy. The alloy used is flexible at room temperature. As it warms in the mouth, the wire becomes less flexible and resumes its original shape, pulling the teeth into position.

### THEORY OF KNOWLEDGE

Models have an important role to play in science because they allow abstract knowledge to be understood using everyday examples. The model of positive ions surrounded by a 'sea' of electrons is often used to describe the structure of a metal. The sea reminding us that the electrons are constantly moving or flowing.

Can you think of any other models used in Chemistry? What do they represent?

### Section 2.2 Exercises

1. Describe what each of the following properties suggests about the nature of the bonding in metals.
   a. Metals conduct electricity.
   b. Most metals have high melting points.
   c. Metals are malleable.

2. Describe how the structure of a metal changes when a force is applied to the metal.

3. a. When referring to the 'electron sea' model for metallic bonding, describe what is meant by the following terms:
      i. delocalized electrons
      ii. lattice of cations.
   b. Explain why the positive ions in a metal do not repel each other.

4. In order for a substance to conduct electricity it must have charged particles that are free to move. State the name of the particles that are responsible for the conduction of electricity in a metal.

5. Use the 'electron sea' model for metallic bonding to explain each of the following observations.
   a. When you touch a piece of metal on a cold day it feels cold.
   b. An empty aluminium drink can be crushed without breaking the metal.
   c. Tungsten used in light globe filaments has a melting point of over 3000°C.

6. In terms of the 'electron sea' model, explain why the electrical conductivity of metals decreases as the metal is warmed.

## 2.3 COVALENT BONDING

A chemical bond forms when outer-shell electrons come close enough to each other to interact and rearrange themselves into a more stable arrangement—one with a lower chemical energy. This chemical energy is the sum of the chemical potential energy of the particles and their kinetic energy. The electrostatic attraction between the positively charged nuclei and the negatively charged electrons is the most significant source of the chemical potential energy. As the two (or more) atoms approach one another, the positively charged nuclei repel one another, as do the negatively charged electrons (see figure 2.3.1). These repulsion forces increase the potential energy of the system. At the same time, however, the oppositely charged particles are attracting one another, causing the potential energy to decrease. At some point, the distance separating the atoms or ions will be such that the repulsive forces of the particles with the same charge exactly balance the attractive forces between oppositely charged particles. It is in this stable arrangement that a chemical bond is formed (see figure 2.3.2).

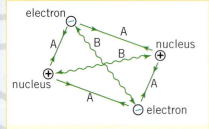

Figure 2.3.1 When the attractive forces (A) are balanced by the repulsive forces (B) between nuclei and electrons, a chemical bond is formed between atoms.

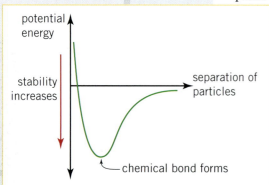

Figure 2.3.2 A new compound is formed at the most stable (lowest potential energy) point.

# The nature of the covalent bond

The simplest of all atoms is hydrogen. It has just a single proton in its nucleus (in the most common isotope) and a single electron occupying the space around it.

Hydrogen exists as **diatomic molecules**—a pair of hydrogen atoms joined together ($H_2$). Why is this a more stable arrangement than single hydrogen atoms, or perhaps a triatomic molecule ($H_3$)? To answer this question, we must return to our consideration of the electrostatic forces occurring within the molecule.

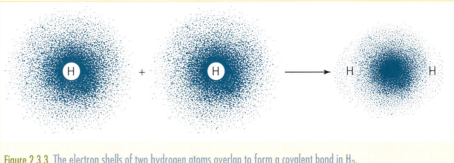

Figure 2.3.3. The electron shells of two hydrogen atoms overlap to form a covalent bond in $H_2$.

A **molecule** may be formally defined as 'a discrete group of non-metal atoms covalently bonded to one another'. Of particular significance in this definition is the fact that molecules contain specific numbers of atoms in a set ratio, an idea first put forward by the English chemist John Dalton in 1803. For example, a molecule of water contains two atoms of hydrogen bonded to one atom of oxygen; the molecule is always the same whether the water is present in the solid, liquid or gaseous phase.

Accurate measurement of the average radii of atoms and molecules is now possible with modern X-ray crystallographic techniques, a luxury certainly not available to Dalton in the late 17th century. The average radius of a hydrogen atom is 1.2 Å ($1.2 \times 10^{-10}$ m), but the distance separating the two nuclei in a hydrogen molecule is just 0.74 Å ($7.4 \times 10^{-11}$ m)! This means that there must be significant overlap of the atomic radii of the individual atoms when the molecule is formed.

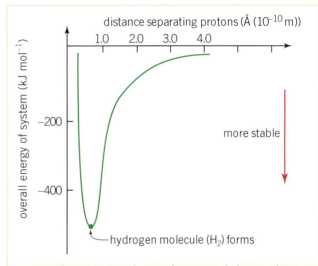

Figure 2.3.4 The variation in total energy of two protons (hydrogen nuclei) as they approach to form a hydrogen molecule.

As the two atoms approach one another, electrostatic attractions and repulsions occur between the positively charged nuclei and the negatively charged electrons. As can be seen in figure 2.3.4, the minimum overall energy of the system occurs at a separation of 0.74 Å, where the greatest amount of energy has been lost to the environment. Effectively, a hydrogen molecule has been formed at this lowest point on the curve, the most stable arrangement of protons and electrons in close proximity to one another. In the formation of the hydrogen molecule, each hydrogen atom has contributed its single electron to occupy the space between the nuclei as a pair.

In 1916 Gilbert Lewis was the first to propose that when a pair of electrons is shared between two atoms, a **covalent bond** forms. The electrons of the bond make up a *bonding pair*. As only one pair of electrons is occupying the space between the two nuclei, a **single covalent bond** has been formed. Any other pairs of valence-shell electrons that do not actually take part in the bond are known as **non-bonding pairs** (**lone pairs**). These non-bonding pairs of electrons can be very important in helping to determine the shape of the molecule, which can have a significant effect on the properties of the substance.

 4.2.1
Describe the covalent bond as the electrostatic attraction between a pair of electrons and positively charge nuclei. © IBO 2007

4.2.2
Describe how the covalent bond is formed as a result of electron sharing. © IBO 2007

**4.2.3**
Deduce the Lewis (electron dot) structures of molecules and ions for up to four electron pairs on each. © IBO 2007

### Lewis structures

Electron shell diagrams, known as **Lewis or electron dot structures**, can be constructed for covalently bonded molecules. All valence electrons are drawn, as these are the electrons that can take part in bonding. Valence electrons that do not take part in bonding (the non-bonding electrons) are also shown. Lewis structures are useful in illustrating how each atom shares electrons to obtain an outer shell of eight electrons.

In a Lewis structure, valence electrons are represented by dots or crosses. As a covalent bond consists of a pair of bonding electrons between the two atoms involved, a dot and a cross between the atoms represent a bond, while pairs of dots or crosses represent non-bonding pairs of electrons. When drawing these diagrams we need to remember that electrons carry a negative charge and so pairs of electrons will tend to repel each other in three-dimensional space. This is important when we begin to look at the shapes of molecules later in this section.

**Diatomic molecules** consist of only two atoms covalently bonded to one another in order to obtain a valence shell of eight electrons. These diatomic molecules include the elemental non-metals (e.g. $O_2$, $N_2$, $Cl_2$). Hydrogen also forms a diatomic molecule, $H_2$, but it needs only two electrons to fill its valence shell. The atoms of the non-metal elements of group 0 (group VIII) of the periodic table already have eight electrons in the valence shell and so have very little tendency to form bonds with other atoms.

Elements such as chlorine and bromine have seven electrons in the valence shell and so will tend to share one further electron in order to obtain eight electrons. This sharing of one pair of electrons results in the formation of a single covalent bond between the atoms. The other six valence electrons do not take part in bond formation—they form three pairs of non-bonding electrons and will repel one another in space as far as possible. Similarly, chlorine with seven valence electrons shares one pair of electrons with hydrogen, which has only one electron, to form a diatomic molecule, HCl, with a single covalent bond.

QuickTime Video
$H_2$ bond formation

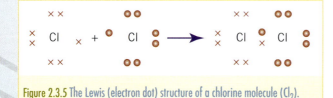

**Figure 2.3.5** The Lewis (electron dot) structure of a chlorine molecule ($Cl_2$).

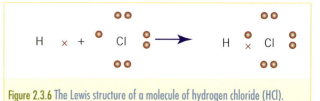

**Figure 2.3.6** The Lewis structure of a molecule of hydrogen chloride (HCl).

The non-metal elements in group 6 of the periodic table, such as oxygen and sulfur, have only six electrons in their outer shells and so must gain a share of two more electrons in order to fulfil the octet rule. In the case of oxygen gas, a diatomic molecule can form between two atoms of oxygen in such a way that four electrons, are shared between them and a **double covalent bond** is formed.

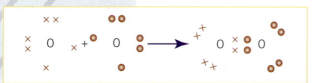

**Figure 2.3.7** The Lewis structure for an oxygen molecule ($O_2$).

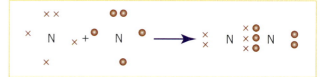

**Figure 2.3.8** The Lewis structure for a molecule of nitrogen gas ($N_2$).

What type of bond might we expect between two nitrogen atoms, considering that nitrogen is in group 5 of the periodic table? In order to fulfil the octet rule, each nitrogen atom must share three electrons. In this case, six bonding electrons form a **triple covalent bond**, leaving only a single pair of non-bonding electrons on each nitrogen atom.

Lewis structures are not restricted to diatomic molecules of elements; they can be used to represent any molecule. We will examine such examples in the section on shapes of molecules.

> **THEORY OF KNOWLEDGE**
> What symbolic function do Lewis structures perform?

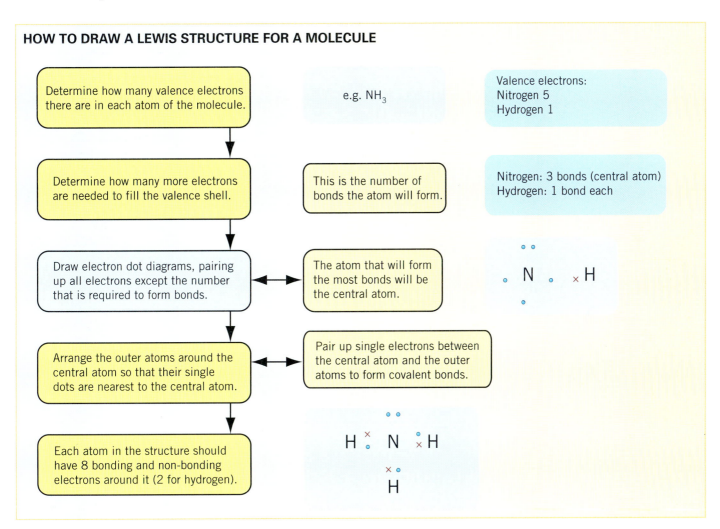

### The relationship between bond length and bond strength

The triple covalent bond between nitrogen atoms is very strong and helps us to understand why nitrogen is such an unreactive gas. It requires a great deal of energy (945 kJ mol$^{-1}$, compared to just 242 kJ mol$^{-1}$ for chlorine gas) to break the triple bond between the atoms and so allow nitrogen to form new bonds with other atoms. The amount of energy required to break a bond is known as the bond dissociation enthalpy, and is generally measured in kilojoules per mole of molecules dissociated. It provides quantitative evidence that the bond strengths vary. Bond enthalpies will be discussed further in chapter 6. Notice also that the more electron pairs that are involved in a covalent bond, the shorter the bond length and the stronger the bond.

**AS 4.2.4**
State and explain the relationship between the number of bonds, bond length and bond strength. © IBO 2007

**TABLE 2.3.1 BOND DISSOCIATION ENTHALPIES AND BOND LENGTHS OF SOME BONDS**

| Bond | Bond dissociation enthalpy (kJ mol$^{-1}$) | Bond length (pm, 1 pm = $10^{-12}$ m) |
|---|---|---|
| H–H | 436 | 74 |
| Cl–Cl | 242 | 199 |
| O=O | 498 | 121 |
| N≡N | 945 | 109 |
| C–C | 364 | 154 |
| C=C | 602 | 134 |
| C≡C | 835 | 120 |
| C–O | 358 | 143 |
| C=O | 799 | 120 |

This relationship between bond length and bond strength can be understood when the definition of bond length is explored. The *bond length* is the distance between two nuclei at the point where a balance is achieved between the attractive force pulling the nuclei together (via the region of electron density between them) and the repulsive force of the two positively charged nuclei pushing each other apart. When there are two pairs of electrons shared between the two nuclei, the attractive force pulling the two nuclei together will be greater, and the balance between repulsion and attraction will be at a shorter bond length. The bond will be stronger due to the larger region of electron density between the two nuclei. In the case of a triple bond, the attractive force pulling the two nuclei together will be greater and the bond length will be even shorter.

This is most obvious when the C–C, C=C and C≡C bonds are compared (see table 2.3.1).

### Shapes of molecules

**4.2.7 Predict the shape and bond angles for species with four, three and two negative charge centres on the central atom using the valence shell electron pair repulsion theory (VSEPR).**
© IBO 2007

Lewis diagrams are of use when we are learning what a covalent bond is, but the most useful representation of molecules is the **structural formula**. In this representation each pair of electrons, bonding and non-bonding pairs alike, is shown as a simple line. Non-bonding pairs can also be shown as two dots (figure 2.3.9). The actual shape of the molecule is also shown. This shape has an important part to play in determining the chemical and physical properties of a molecule.

Recall that each pair of electrons will be repelled from the others as far as possible in three-dimensional space because electrons all carry a negative charge. This is known as the valence shell electron pair repulsion or **VSEPR theory**, and was first put forward by Sidgwick and Powell in 1940.

This theory states that the electron pairs around an atom repel each other. The electrostatic repulsion of pairs of electrons determines the geometry of the atoms in the molecule. The non-bonding pairs and bonding pairs of electrons are arranged around the central atom so as to minimize this electrostatic repulsion between the non-bonding and bonding pairs of electrons. The relative magnitude of the electron pair repulsions is:

non-bonding pair–non-bonding pair > bonding pair–non-bonding pair > bonding pair–bonding pair

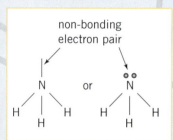
Figure 2.3.9 The two ways of representing non-bonding electron pairs.

The *shape*, according to the VSEPR theory, depends on the number of bonding pairs and non-bonding pairs of electrons on the central atom. The *bond angle* is the angle between the atoms bonded to the central atom. While non-bonding pairs of electrons can be important in determining the overall shape of a molecule, they are not actually considered part of the shape. Shape describes the position of atoms only; however, non-bonding electrons repel other pairs of electrons and so *influence* the final shape of a molecule.

VSEPR—the basic molecular configurations

Negative charge centre or region refers to pairs of electrons on the *central* atom. This includes both the non-bonding pairs and bonding pairs of electrons in single, double or triple bonds. Each double or triple bond is counted as one negative charge centre.

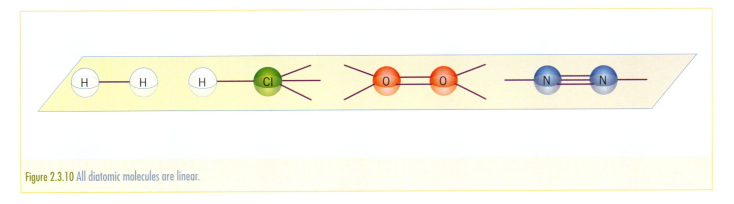

**Figure 2.3.10** All diatomic molecules are linear.

In diatomic molecules there are only two atoms to consider and so the molecule will invariably be linear. All the molecules we have considered to date ($H_2$, $Cl_2$, HCl, $O_2$ and $N_2$) are therefore linear. Polyatomic molecules are those that consist of more than two atoms covalently bonded to one another. This group encompasses the great majority of molecules known, and includes significant ones such as water, carbon dioxide and methane. How do we determine the valence structures and shapes of these molecules?

Consider the unusual case of three pairs of electrons, as is the case in boron trifluoride ($BF_3$). The three electron pairs will repel in such a manner as to form an equilateral triangle with a bond angle of 120° between each pair of bonds. This arrangement of atoms is known as a *trigonal planar* shape.

**DEMO 2.3**
Balloon analogy for electron repulsion

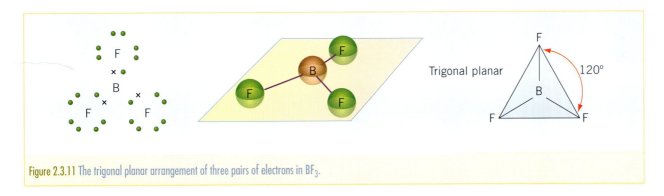

**Figure 2.3.11** The trigonal planar arrangement of three pairs of electrons in $BF_3$.

The most common situation encountered in molecules is the existence of four pairs of electrons, either bonding or non-bonding, surrounding each atom. While it may be tempting to assume that the four pairs of electrons will arrange themselves in a square planar structure with each pair of electrons at 90° to the others, this is not the case because the molecules are not flat. In fact, the most widely spaced arrangement of four pairs of electrons in three-dimensional space is known as the *tetrahedral* arrangement, in which each atom can be imagined to be at the vertex of a regular triangular-based pyramid. The bond angle in this arrangement is 109.5°.

Now let us consider some common substances whose molecules show this tetrahedral arrangement.

VSEPR

CHEMISTRY: FOR USE WITH THE IB DIPLOMA PROGRAMME **STANDARD LEVEL**

**Interactive Animation**
Representations of methane

### Example 1: Methane (CH$_4$)

Carbon is in group 4 of the periodic table and so has four electrons in its outer shell. To gain a share of eight electrons it must form covalent bonds with four hydrogen atoms, as shown in figure 2.3.12. Note that all the valence electrons of both carbon and hydrogen are involved in the formation of bonding pairs of electrons; there are no non-bonding pairs of electrons. The molecule is described as tetrahedral in shape. The H–C–H bond angle is 109.5°.

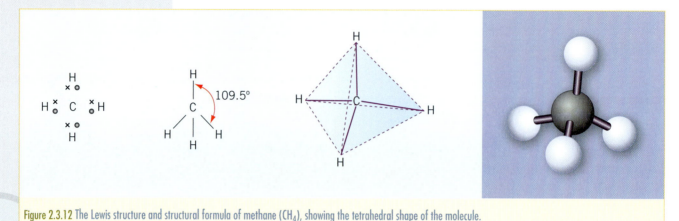

**Figure 2.3.12** The Lewis structure and structural formula of methane (CH$_4$), showing the tetrahedral shape of the molecule.

### Example 2: Ammonia (NH$_3$)

Ammonia, found commonly in household cleaning products, consists of one nitrogen and three hydrogen atoms. The formula is similar to that of boron trifluoride (BF$_3$), and so we might expect ammonia to be trigonal planar in shape. This is not the case, however. Nitrogen is in group 5 of the periodic table and so has five valence electrons. To gain a full valence shell, nitrogen must form covalent bonds with three hydrogen atoms. The remaining two electrons exist as a non-bonding pair. The three hydrogen atoms and the nitrogen atom can be considered to form the vertices of a *trigonal pyramid*. The H–N–H bond angle is 107°. Note that this shape is different from the more common tetrahedron. In the tetrahedron, the central atom occupies the centre of the tetrahedron and the other atoms of the molecule are at the four vertices.

Note also that the bond angle is smaller than that in methane. This is due to the presence of one non-bonding pair of electrons. A non-bonding pair of electrons has a greater power of repulsion than does a bonding pair of electrons, so the hydrogen atoms in ammonia are pushed closer together than they are in methane.

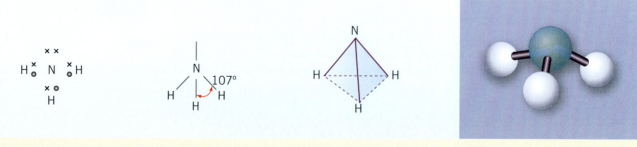

**Figure 2.3.13** The Lewis structure and structural formula of ammonia (NH$_3$), showing the trigonal pyramidal shape of the molecule.

## Example 3: Water ($H_2O$)

Oxygen has six electrons in its valence shell, and so must gain a share of two more to obtain an octet of valence electrons. Thus, oxygen will form two single covalent bonds with two hydrogen atoms. The oxygen atom will still have two pairs of non-bonding electrons and so will be surrounded by four pairs of electrons in all: two bonding and two non-bonding pairs. Even though these four pairs of electrons are arranged tetrahedrally, the water molecule itself is described as being a *V-shaped* or bent molecule. The non-bonding pairs of electrons prevent the molecule from being linear (H–O–H) by repelling the bonding electrons strongly. The bond angle in water is 104.5°, smaller again than ammonia due to the presence of two non-bonding pairs of electrons.

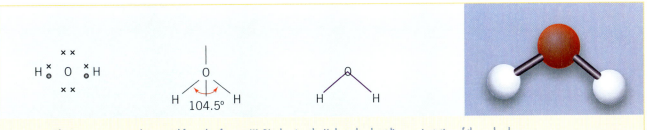

Figure 2.3.14 The Lewis structure and structural formula of water ($H_2O$), showing the V-shaped or bent linear orientation of the molecule.

## Example 4: Carbon dioxide ($CO_2$)

The carbon dioxide molecule is slightly different from the above examples, as in this case multiple bonds are formed. Carbon needs to share four electrons and oxygen two, in order that the atoms of both elements share eight valence electrons. To achieve this, the carbon atom shares two of its valence electrons with each of the oxygen atoms, thus forming a pair of double bonds between carbon and oxygen. As there are no non-bonding electrons, the bonding electrons separate as far as possible and a linear molecule results, with a bond angle of 180°.

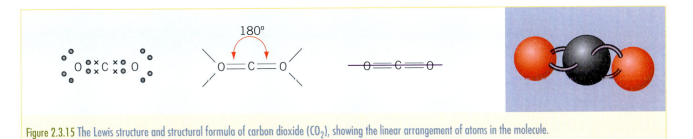

Figure 2.3.15 The Lewis structure and structural formula of carbon dioxide ($CO_2$), showing the linear arrangement of atoms in the molecule.

### TABLE 2.3.2 SUMMARY OF MOLECULAR SHAPES

| Number of negative charge centres on the central atom | Number of bonding pairs on the central atom | Number of non-bonding pairs on the central atom | Example | Bond angle | Shape of molecule | Representation of the shape |
|---|---|---|---|---|---|---|
| 2 | 4 | 0 | $CO_2$ <br> HCN | 180° | Linear | |
| 3 | 3 | 0 | $BF_3$ <br> $SO_3$ | 120° | Trigonal planar | |
| 3 | 4 | 1 | $SO_2$ | 117° | V-shaped | |
| 4 | 4 | 0 | $CH_4$ <br> $CCl_4$ | 109.5° | Tetrahedral | |
| 4 | 3 | 1 | $NH_3$ <br> $NF_3$ | 107° | Trigonal pyramidal | |
| 4 | 2 | 2 | $H_2O$ <br> $H_2S$ | 104.5° | V-shaped or bent linear | |

### Further examples

#### Example 5: The ammonium ion ($NH_4^+$)

The ammonium ion is formed when a hydrogen ion ($H^+$) bonds to the non-bonding pair of an ammonia molecule. Although the bond between the $H^+$ and the nitrogen is a covalent bond, it is formed in a slightly different way from the other covalent bonds in this molecule. The $H^+$ ion has no electrons to share in the bond, so both bonding electrons come from the nitrogen atom (the two non-bonding electrons). This is shown by the arrow in the structural formula.

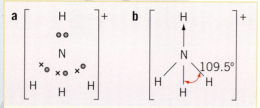

Figure 2.3.16 The (a) Lewis structure and (b) structural formula of the ammonium ion ($NH_4^+$).

A covalent bond which is formed in this way is called a **dative covalent bond**. In $NH_4^+$ the nitrogen is surrounded by four bonding pairs of electrons, rather than three bonding and one non-bonding, so the ammonium molecular ion has the same shape as a methane molecule—it is tetrahedral, with bond angles of 109.5°.

PRAC 2.3
Making molecular models

### Example 6: The hydronium ion ($H_3O^+$)

The hydronium ion is formed when a hydrogen ion ($H^+$) bonds to one of the non-bonding pairs of electrons on a water molecule. Again a dative covalent bond is formed between the $H^+$ and the oxygen atom. The resulting $H_3O^+$ ion is positively charged and has the same shape as an ammonia molecule—it is trigonal pyramidal, with bond angles of 107°.

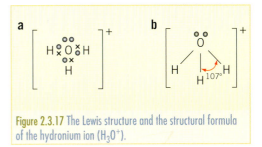

Figure 2.3.17 The Lewis structure and the structural formula of the hydronium ion ($H_3O^+$).

### Example 7: Carbon monoxide (CO)

Carbon atoms need to share four electrons to fill the outer shell, and oxygen atoms must share two electrons to fill the outer shell. How can a molecule be made from just one carbon and one oxygen atom?

In the carbon monoxide molecule the carbon atom shares two pairs of electrons with the oxygen atom and the oxygen atom makes an additional dative covalent bond with the carbon atom. In this way, both atoms gain a full valence shell. Two shared covalent bonds and one dative bond exist between the two atoms. The molecule is, of course, linear.

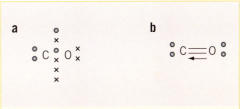

Figure 2.3.18 The Lewis structure and the structural formula of carbon monoxide (CO).

WORKSHEET 2.3
Molecules

### Example 8: Sulfur dioxide ($SO_2$)

The structure of sulfur dioxide is more unusual than those we have met so far. The central atom, sulfur, has six valence electrons. Each oxygen atom has six electrons and needs to make two bonds to fill its valence shell. This gives a total of 10 electrons around the sulfur atom, creating an exception to the octet rule. The sulfur atom forms a double bond with each oxygen atom, leaving two non-bonding electrons in the valence shell of the sulfur atom. These non-bonding electrons repel the bonding pairs, creating a bent linear or V-shaped molecule with a angle of 117°.

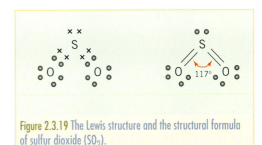

Figure 2.3.19 The Lewis structure and the structural formula of sulfur dioxide ($SO_2$).

## Molecules with more than one central atom

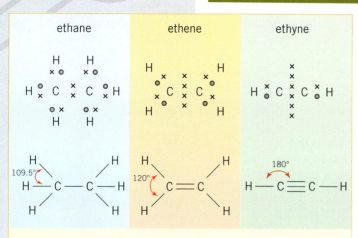

Figure 2.3.20 Structures of ethane ($C_2H_6$), ethene ($C_2H_4$) and ethyne ($C_2H_2$) molecules.

In all of the molecules considered so far, there has been a single central atom to which the other atoms are attached. However, many other molecules we encounter are larger and have several atoms forming the central structure, with other atoms branching from them. How do we determine the valence structures for these more complex molecules? The reassuring answer is that the rules and techniques that we have already looked at work equally well for larger structures. However, we do not generally attempt to describe the shape of such multi-centred molecules, although it will often be the case that the molecules are constructed of a series of tetrahedra joined end to end.

## Electronegativity and bond polarity

So far we have assumed that when atoms share a pair of electrons they share the electrons equally. However, just like two siblings who have been asked to share a bag of sweets, different atoms have different attraction for the electrons to be shared, and so do not share equally.

**Electronegativity** is a measure of the ability of an atom to attract the electrons in a bond.

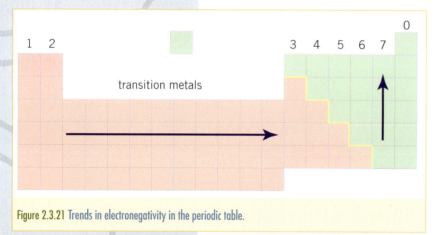

Figure 2.3.21 Trends in electronegativity in the periodic table.

Electronegativity is highest at the top of a group of elements in the periodic table, as the valence shell of electrons is closest to the positively charged nucleus in the smaller atoms. As we move down the group, the valence electrons move further away with each successive shell. Elements in the same period have electrons filling the same valence shell. As we move across the period from left to right, the nuclear charge increases, and so does the attraction between the nucleus and the valence electrons. In summary, electronegativity increases from left to right and bottom to top of the periodic table. Group 0 elements have undefined electronegativities, as they already have a filled outer shell and so have little tendency to attract further electrons. Electronegativity will be discussed in more detail in chapter 3.

**AS 4.1.6**
Predict whether a compound of two elements would be ionic from the position of the elements in the periodic table and from their electronegativity values. ©IBO 2007

Comparisons between electronegativity values can be used to make generalizations about the type of bonding and types of atoms forming a bond. Elements such as fluorine, oxygen and nitrogen have high electronegativities, whereas metals have low electronegativities. For the atoms in a molecule to share the bonding electrons equally, the electronegativities must be identical. This description is true for the diatomic molecules of an element such as $N_2$, $O_2$, $Cl_2$, $H_2$ and so on. However, when the electronegativities are similar (such as for carbon and hydrogen), the sharing of bonding electrons is approximately equal.

## THEORY OF KNOWLEDGE

Linus Pauling (1901–1994), an American chemist, was intrigued by the theories of Lewis and Langmuir. Between 1930 and 1933 in collaboration with colleague Don Yost and armed with an idea about the partial ionic character of covalent bonds, Pauling developed a system for estimating the theoretical strength of bonds. With his new numbers in hand, Pauling could now compare his theoretical numbers to the real behaviour of different pairs of elements as they formed compounds. A greater ionic character meant that one of the atoms had a much stronger ability to attract the electrons in the bond than the other. Pauling then began mapping this electron attracting ability on a scale. Fluorine, with the greatest ability to attract electrons, was at one end of the scale; lithium was towards the other. The bond in the compound they formed, lithium fluoride, was almost 100% ionic. Iodine was somewhere towards the middle of the scale, and the lithium iodide bond therefore had more covalent character. By comparing a number of such pairs, Pauling was able to map a relative property he called electronegativity and assign values to various elements. These values, in turn, could be used to predict the bond type and strength in many molecules, including those for which no experimental data were available.

Pauling's electronegativity scale was imaginative but many scientists criticized his ideas, saying the ideas were not sufficiently grounded in accepted theory. However, the scale went on to become influential, one reason being that it was easily understood by chemists. One daring prediction Pauling made was that fluorine, the most electronegative element, could form compounds with an unreactive noble gas such as xenon. At the time the noble gases were believed not to form bonds with other atoms, and so the creation of a xenon compound would have made history. Experiments were carried out to test his prediction, but they did not yield the results Pauling was hoping for. In 1962 another team of scientists would make international news by producing the xenon compounds $XeF_2$, $XeF_4$ and $XeF_6$ Pauling had said were possible 30 years earlier.

- What issues about the process by which scientific knowledge is produced does Pauling's electronegativity story raise?
- Einstein said 'imagination is more important than knowledge'. Does imagination have a role to play in the development of scientific knowledge?

The greater the difference in the electronegativities of the atoms in a compound, the more uneven will be the sharing of electrons between them. The extreme of unequal sharing is the formation of ions. When ions are formed, one ion loses its valence electrons completely and the other gains valence electrons. Where the difference in electronegativities is great (1.8 and above), the compound is most likely to be ionic. Similarly, the closer the electronegativity values of the two atoms the more likely they are to form a covalent compound by sharing electrons. The bonds formed between atoms with electronegativity differences of between 0.5 and 1.8 are more likely to be *polar covalent*, while those with an electronegativity difference of zero will form *pure covalent bonds*.

Magnesium, for example, has an electronegativity of 1.2 and oxygen has an electronegativity of 3.5. This large difference of 2.3 suggests that magnesium oxide is an ionic compound. This is supported by the location of magnesium in group 2, which is on the left-hand side of the periodic table and therefore makes magnesium a metal. Oxygen is in group 6, which is on the right-hand side of the periodic table among the non-metals.

Magnesium bromide, $MgBr_2$, also has ionic bonding, but the electronegativity difference of 1.6 between magnesium and bromine suggests that the bond could be covalent. In fact, most bonds are somewhere along a bonding type continuum. For this reason electronegativity values are only used as a general guide for identifying bonding type.

**4.2.5
Predict whether a compound of two elements would be covalent from the position of the elements in the periodic table and from their electronegativity values.**
©IBO 2007

### TABLE 2.3.3 ELECTRONEGATIVITIES OF SELECTED ELEMENTS

| Period | Group 1 | | Group 2 | | Group 3 | | Group 4 | | Group 5 | | Group 6 | | Group 7 | | Group 0 | |
|---|---|---|---|---|---|---|---|---|---|---|---|---|---|---|---|---|
| 1 | H | 2.1 | | | | | | | | | | | | | He | – |
| 2 | Li | 1.0 | Be | 1.5 | B | 2.0 | C | 2.5 | N | 3.0 | O | 3.5 | F | 4.0 | Ne | – |
| 3 | Na | 0.9 | Mg | 1.2 | Al | 1.5 | Si | 1.8 | P | 2.1 | S | 2.5 | Cl | 3.0 | Ar | – |
| 4 | K | 0.8 | Ca | 1.0 | Ga | 1.6 | Ge | 1.8 | As | 2.0 | Se | 2.4 | Br | 2.8 | Kr | – |
| 5 | Rb | 0.8 | Sr | 1.0 | In | 1.7 | Sn | 1.8 | Sb | 1.9 | Te | 2.1 | I | 2.5 | Xe | – |
| 6 | Cs | 0.7 | Ba | 0.9 | Tl | 1.8 | Pb | 1.8 | Bi | 1.9 | Po | 2.0 | At | 2.2 | Rn | – |
| 7 | Fr | 0.7 | Ra | 0.9 | | | | | | | | | | | | – |

### Worked example 1

Use electronegativity values and the location of the element in the periodic table to identify the compound made by each pair of elements as either covalent or ionic.

**a** Sodium and sulfur

**b** Sulfur and oxygen

**c** Lead and oxygen

### Solution

| Element pair | Electronegativity | Difference in electronegativity | Position in periodic table | Classification | Bonding |
|---|---|---|---|---|---|
| Sodium and sulfur | 0.9 | 1.6 | Group 1, period 3 | Metal | Ionic |
| | 2.5 | | Group 6, period 3 | Non-metal | |
| Sulfur and oxygen | 2.5 | 1.0 | Group 6, period 3 | Non-metal | Covalent |
| | 3.5 | | Group 6, period 2 | Non-metal | |
| Lead and oxygen | 1.8 | 1.7 | Group 4, period 6 | Metal | Ionic |
| | 3.5 | | Group 6, period 2 | Non-metal | |

### Worked example 2

**AS 4.2.6**
Predict the relative polarity of bonds from electronegativity values. © IBO 2007

Identify which of the following covalently bonded pairs of atoms would have the greatest difference in electronegativities and hence would be the most polar.

**a** Nitrogen and hydrogen

**b** Oxygen and hydrogen

**c** Chlorine and oxygen

**d** Sulfur and chlorine

**Solution**

| Pair of atoms | Electronegativity | Difference in electronegativity |
|---|---|---|
| Nitrogen | 3.0 | 0.9 |
| Hydrogen | 2.1 | |
| Oxygen | 3.5 | 1.4 |
| Hydrogen | 2.1 | |
| Chlorine | 3.0 | 0.5 |
| Oxygen | 3.5 | |
| Sulfur | 2.5 | 0.5 |
| Chlorine | 3.0 | |

Oxygen and hydrogen have the greatest difference in electronegativity (1.4) so, of the pairs given, the oxygen–hydrogen bond is the most polar.

If the electrons are shared unevenly in a covalent bond, the bond is said to be a **polar covalent bond** or a **permanent dipole**. Such a bond can be identified using the symbol δ (delta). In particular, δ– and δ+ are used to indicate a slight negative and a slight positive charge respectively.

Bond polarity

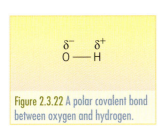

Figure 2.3.22 A polar covalent bond between oxygen and hydrogen.

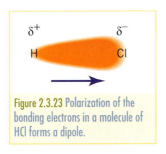

Figure 2.3.23 Polarization of the bonding electrons in a molecule of HCl forms a dipole.

**AS 4.2.8**
**Predict whether or not a molecule is polar from its molecular shape and bond polarities.** © IBO 2007

If a polar covalent bond occurs in a diatomic molecule (that has only one covalent bond), one part of the molecule will be more negative than the other, due to having a larger share of the bonding electrons. This is the case with diatomic molecules such as HCl and HBr. The molecule is then described as a polar molecule.

Molecular polarity

When there is more than one polar covalent bond in a molecule, the shape of the molecule must be considered. It is possible to have molecules that contain polar bonds but overall are non-polar—the permanent dipoles cancel each other out! An example of such a molecule is methane ($CH_4$). This molecule contains four polar C–H bonds, with carbon being slightly more electronegative (2.5) than hydrogen (2.1). In three dimensions, the tetrahedral arrangement of the bonds in this molecule means that the slight positive charges of the hydrogen atoms are perfectly balanced by each other and are cancelled by the partial negative charge on the carbon atom. Indeed, any molecule that is perfectly symmetrical will be non-polar overall.

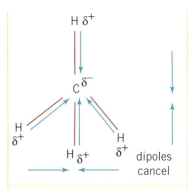

Figure 2.3.24 The symmetry of methane's structure results in an overall non-polar molecule, despite the presence of slightly polar bonds.

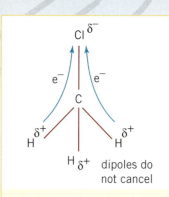

Figure 2.3.25 Chloromethane is not symmetrical and so is a polar molecule.

Chloromethane ($CH_3Cl$) is a similar molecule to methane, but it is polar. Note that this molecule is not symmetrical in three dimensions and so the dipoles cannot cancel each other out. Chlorine (3.0) is more electronegative than either carbon (2.5) or hydrogen (2.1) and so draws the bonding electrons towards it.

Another important example of a non-polar molecule that nevertheless contains polar bonds is carbon dioxide, $CO_2$.

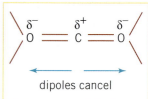

Figure 2.3.26 Like methane, $CO_2$ is a non-polar molecule that contains polar bonds.

### Section 2.3 Exercises

1. State the relationship between the group in the periodic table in which an element is found and its number of valence electrons.

2. All chemical bonds involve electrostatic attraction. State the names of the particles between which the electrostatic attraction occurs in a covalent bond.

3. Of the substances in the following list, identify those that can accurately be described as being composed of molecules.
   - **I** Carbon dioxide
   - **II** Ice
   - **III** Bronze
   - **IV** Potassium bromide
   - **V** Hydrogen chloride

4. State the total number of valence electrons, and the number of non-bonding valence electrons in each of the following molecules.
   - **a** $H_2$
   - **b** $Cl_2$
   - **c** $O_2$
   - **d** HBr
   - **e** $N_2$

5. Draw Lewis structures (electron dot diagrams) for the following linear molecules.
   - **a** HI
   - **b** $F_2$
   - **c** $O_2$
   - **d** $CO_2$
   - **e** $N_2$

6. Consider the information in the table, which shows the bond lengths for the group 7 halogens.

   Explain why the distance between atoms in these diatomic molecules increases as we move down the group.

   | Molecule | Bond length (pm) |
   | --- | --- |
   | F–F | 142 |
   | Cl–Cl | 199 |
   | Br–Br | 229 |
   | I–I | 266 |

7. Consider the bonds C–C, C=C and C≡C.
   - **a** State which of these bonds is the strongest and explain your answer.
   - **b** State which of these bonds is the longest.

8. Draw Lewis structure diagrams for the following substances, and use these to construct structural formulas for each.
   - **a** $H_2O$
   - **b** HBr
   - **c** $CH_3Cl$
   - **d** $CO_2$
   - **e** $NH_3$
   - **f** $H_3O^+$
   - **g** $SO_2$

9. State the shapes of each of the substances in question **8**.

10  Identify a molecule from the list below to match each description. You may use each molecule from the list only once.

   I    Hydrogen sulfide ($H_2S$)        a  Trigonal pyramidal
   II   Chloromethane ($CH_3Cl$)         b  Six non-bonding pairs of electrons
   III  Hydrogen fluoride (HF)           c  Similar to a water molecule in shape
   IV   Chlorine gas ($Cl_2$)            d  Contains double bonds
   V    Ammonia ($NH_3$)                 e  Tetrahedral, polar molecule
   VI   Carbon dioxide ($CO_2$)          f  Linear molecule

11  By first constructing electron dot diagrams as a guide, draw structural formulas for the following substances, and state their shapes and bond angles.

   a  Phosphine ($PH_3$)          b  Hydrogen cyanide (HCN)
   c  Propane ($C_3H_8$)          d  Ammonium ion ($NH_4^+$)
   e  Boron trichloride ($BCl_3$)

12  Determine which of the following pairs of atoms will have the greatest electronegativity difference.

   I    Carbon and hydrogen     II  Hydrogen and oxygen
   III  Sulfur and oxygen       IV  Carbon and chlorine

13  Identify which atom in each of the following bond pairs will carry a slight negative charge (δ−) and which a slight positive charge (δ+).

   a  C–H     b  B–O     c  P–Cl     d  S–H     e  Si–O

14  Copy and complete the table below.

| Molecule name | Structural formula | Are bonds polar? Yes/No | Is the molecule symmetrical? Yes/No | Is the molecule polar overall? Yes/No |
|---|---|---|---|---|
| Oxygen ($O_2$) | | | | |
| Dibromomethane ($CH_2Br_2$) | | | | |
| Carbon disulfide ($CS_2$) | | | | |
| Ammonia ($NH_3$) | | | | |

## 2.4 COVALENT BONDING IN NETWORK LATTICES

Many elements exist in different structural forms. Carbon is one of them. Carbon can be found in its most precious form, diamond, a more useful form, graphite, and as the fullerene molecule ($C_{60}$). The different structural forms of an element are called **allotropes**.

The appearance and physical properties of these three forms of carbon are vastly different. What makes them so different is the way in which their atoms are covalently bonded. Diamond exists as a **covalent network lattice** and graphite as a covalent layer lattice, whereas fullerene, identified in 1985 by Richard Smalley, exists as discrete, soccer-ball-shaped molecules each made up of 60 carbon atoms.

 4.2.9
**Describe and compare the structure and bonding in the three allotropes of carbon (diamond, graphite, C60 fullerene).** © IBO 2007

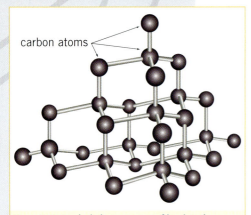

Figure 2.4.1 Tetrahedral arrangement of bonds in the covalent network lattice of diamond.

Figure 2.4.2 Australia's largest diamond, weighing 104.5 carats (around 21 g). It was discovered in the Merlin mine in the Northern Territory.

## Diamond

The structure of diamond consists of a giant covalent network lattice in which each carbon atom is surrounded by four other carbon atoms bonded in a tetrahedral lattice structure. The four valence electrons are all used in the formation of single covalent bonds, so that carbon has no spare electrons left to enable it to conduct electricity. The bonding forces are very strong in all three dimensions, resulting in diamond being an extremely hard substance with a very high sublimation point (approximately 3550°C). At these extremely high temperatures, the bonds between the carbon atoms are overcome, and they have so much energy that the atoms move straight into the gaseous phase.

As the tetrahedral lattice structure is very regular, diamond can be cut in very specific directions known as 'cleavage planes'. If a model of diamond is constructed and held to the eye at particular angles, it is easy to see these cleavage planes. They are the lines along which all of the atoms in the lattice align perfectly. When diamond cutters are practising their art, they carefully identify these cleavage planes so as to optimize the beauty and size of the cut diamond obtained from a raw stone. If the cutter was to strike the uncut stone in the wrong place, it might well shatter.

The phrase 'diamonds are forever', made famous by the James Bond movie of the same name, correctly suggests that diamonds will last forever; they are in fact the hardest known naturally occurring substance. For this reason very small or flawed stones, collectively known as industrial diamonds, are commonly used on the tips of industrial cutting equipment such as drills.

## CHEM COMPLEMENT

### Carat or carob?

The unit by which the mass of a diamond is measured is the carat. The word *carat* comes from the Greek word *kerátion*, meaning 'fruit of the carob' and dates back to when carob seeds were used as weights on scales, because it was assumed that all carob seeds were practically identical, making them perfect as standard weights. This method of measurement worked for many years until this assumption was proved false. A research team at the University of Zurich weighed 550 seeds from 28 carob trees and found they are just as variable in weight as seeds from 63 other species of tree.

The metric carat is exactly 200 mg in mass. It can be divided into 100 points, each of which is 2 mg in mass.

Hundreds of tonnes of rock and ore must be processed to uncover a single one-carat gem-quality diamond. As the carat is a unit of mass and not size, two diamonds of the same carat mass may appear to be different sizes, depending on how the diamond is cut.

It is not true that the larger the carat weight the more valuable the diamond. The four characteristics of diamonds are the four Cs—cut, colour, clarity and carat all contribute to the overall worth of a diamond.

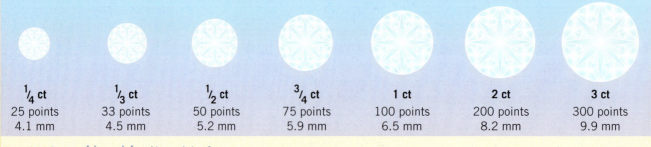

Figure 2.4.3 A range of diamonds from ¼ carat (ct) to 3 ct.

## Graphite

Graphite displays many properties fundamentally different from those of diamond. It is a soft, greasy solid that is a good conductor of electricity and makes an excellent lubricant. Some uses of graphite are shown in figure 2.4.4. Graphite is also used as a dry lubricant to replace oil in certain industrial applications such as the 'dry sump' used in some high-performance racing car engines, as a chemical additive to rubber and certain plastics to make them more flexible, and as a moderator in nuclear reactors, where it slows down neutrons sufficiently so that they can engage in nuclear fission reactions. Graphite is also made into a fibre and used to produce high-strength composite materials that are also light and flexible, and used in such items as tennis racquets and fishing rods.

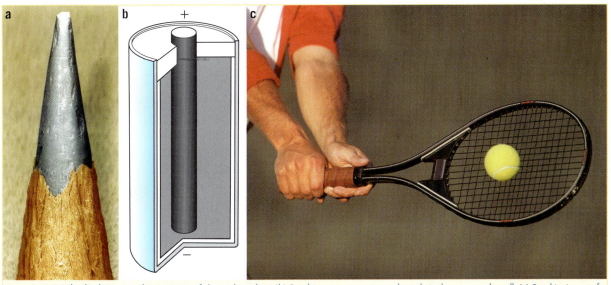

Figure 2.4.4 (a) The 'lead' in a pencil is a mixture of clay and graphite. (b) Graphite serves as an inert electrode in the common dry cell. (c) Graphite is part of the composite material used to make tennis racquets.

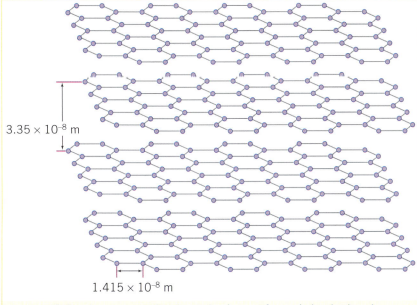

Figure 2.4.5 The layer lattice structure of graphite. Each carbon atom forms only three bonds, so the spare delocalized electron is free to conduct electricity.

In the giant covalent structure of graphite, each carbon atom is bonded to only three others in a series of hexagons that make up a layer (figure 2.4.5). As each carbon atom has four valence electrons, this leaves one unattached electron for each carbon atom. These delocalized electrons are able to move relatively freely between the layers of strongly bonded carbon atoms, and explains why graphite is such a good conductor of electricity. With only weak van der Waals' forces (see p. 58) holding the layers of carbon atoms together, it is easy for one layer to slide past another, explaining the lubricant properties of graphite. Because the bonding within the layers is very strong, graphite has the very high melting point of 3730°C and so can readily be used as a lubricant at the high temperatures generated within engines and other moving mechanical structures.

### CHEM COMPLEMENT

#### Black lead pencils

Surprise, surprise, black lead pencils do not contain any lead! They do contain graphite, however, and their history is a fascinating one. In 1564, a very pure deposit of graphite was found in England. The deposit had to be guarded and mined very sparingly. The graphite was so stable that it could be sawn into thin, square sticks that were used for writing. At first, these sticks were bound in string, but the string was eventually replaced by wood. In the absence of pens, pencils were very important for communication. During the Napoleonic wars, the supply of graphite to France was cut off, so Napoleon commissioned Nicholas-Jacques Conté, an inventor, to find a substitute for his armies. Conté powdered graphite, mixed it with clay and created his pencil 'lead' by firing this mixture in a kiln. This same formula is still used today. Altering the ratio of the soft graphite to the harder clay can vary the hardness of the pencil. Grades such as 6H, HB and 2B are used to indicate the hardness of the point.

Figure 2.4.6 Black lead pencils are graded for hardness.

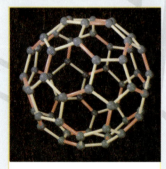

Figure 2.4.7 A model of the buckminsterfullerene $C_{60}$ molecule. Note the alternating pentagons and hexagons, as found in the pattern on a soccer ball.

### Fullerenes (buckyballs)

In the late 1970s, a new form of carbon was manufactured by Australian scientist Dr Bill Burch at the Australian National University in Canberra, although he did not go on to investigate its structure. In 1985, Drs Richard Smalley and Robert Curl at Rice University in Houston, Texas, working with Harry Kroto at the University of Sussex, UK, were the first to determine the structure of this new carbon form, work that was to earn them the Nobel Prize in Chemistry in 1996. The structure of the molecule consisted of a roughly spherical group of carbon atoms arranged in a series of pentagons and hexagons in a pattern similar to that seen in the leather swatches that make up a soccer ball.

In a deviation from systematic chemical nomenclature, Kroto and Smalley named this most unusual molecule buckminsterfullerene, after the designer of geodesic domes, R. Buckminster Fuller. This honour was bestowed upon Fuller because, according to Kroto, 'the geodesic ideas associated with the constructs of Buckminster Fuller had been instrumental in arriving at a plausible structure'. These molecules quickly became known as *buckyballs*. Close inspection of the bonding in fullerenes indicates that each carbon atom is bonded to only three other atoms, resulting in the presence of one free, or delocalized, electron per carbon atom. This characteristic enables fullerenes to conduct electricity, and confers on them a number of other unusual and useful properties. While initially little more than a scientific curiosity, fullerenes have swiftly found roles in a wide range of potential applications involving superconductivity and the production of micro-scale semiconductors and electromechanical devices, and even broad-spectrum lasers.

**TABLE 2.4.1 ALLOTROPES OF CARBON**

| Allotrope | Bonding | Structure | Properties | Uses |
|---|---|---|---|---|
| Diamond | Covalent bonding throughout giant covalent network lattice | Giant covalent network lattice, each carbon surrounded by four other carbon atoms in a tetrahedral arrangement | Very hard, sublimes, non-conductive, brittle | Jewellery, cutting tools, drills |
| Graphite | Covalent bonding in layers and delocalized electrons that allow electrostatic attraction between layers | Giant covalent layer lattice, each carbon bonded to three other carbons; one delocalized electron per carbon atom ($3.35 \times 10^{-8}$ m, $1.415 \times 10^{-8}$ m) | Conductive, lubricant, soft, greasy | Lubricant, pencils, electrodes, reinforcing fibres |
| Fullerenes | Covalent bonding throughout covalent macromolecule | Large, almost spherical shape, hexagonal and pentagonal covalently bonded carbon rings | Not yet fully understood, discovered very recently | Catalyst, medical uses, electromagnetic devices, other possible uses |

### THEORY OF KNOWLEDGE

The discovery of $C_{60}$ fullerenes by Kroto, Curl and Smalley shows how chance or serendipity plays a role in the growth of scientific knowledge. In the process of trying to replicate the formation of carbon molecules near giant red stars, they built a laser to evaporate and analyse graphite. Quite by chance and totally unexpectedly they produced clusters of 60-carbon atoms.

*My advice is to do something which interests you or which you enjoy ... and do it to the absolute best of your ability. If it interests you, however mundane it might seem on the surface, still explore it because something unexpected often turns up just when you least expect it.*

Harold Kroto
Source: http://www.chem.fsu.edu/people/pe_faculty_spotlight_kroto.asp

### CHEM COMPLEMENT

#### Carbon nanotubes

Perhaps of even more interest than carbon buckyballs is the related *carbon nanotube*. In 1991, Japanese scientist Sumio Iijima found a new form of carbon while attempting to produce fullerenes. The carbon he found had a structure resembling a graphite layer that had been rolled into a cylinder. The cap on each end of the cylinder was a half-fullerene molecule. The cylinder has a very small diameter—about a nanometre ($10^{-9}$ metre)—hence the name carbon nanotube. A nanotube can have a width about one ten-thousandth that of a human hair, yet its length can be millions of times greater than its width. Scientists have improved their techniques of manufacturing these tubes and they are now tubes of a single-layer.

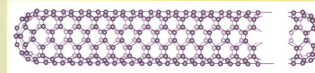

**Figure 2.4.8** The structure of a nanotube: hexagons are joined with pentagons and rolled into a cylinder with a half-fullerene on each end.

## Silicon and silicon dioxide

**AS 4.2.10 Describe the structure and bonding in silicon and silicon dioxide.** © IBO 2007

Like carbon, silicon is a member of group 4 of the periodic table and so is able to form four covalent bonds with other silicon atoms and create a network lattice structure. Silicon crystals are a dark blue-grey colour, and silicon is well known for its ability to behave as a semiconductor when it has had boron, gallium, phosphorus or arsenic added to it. Because silicon is a larger atom than carbon, it is not surprising that the Si–Si bond length (235.2 pm) is considerably longer than that of the C–C bond in diamond (142.6 pm). As the general trend is that the shorter the bond the higher its bond enthalpy will be, this longer bond length means that the Si–Si bond will have a lower bond enthalpy than the C–C bond. Less energy is required to break a Si–Si bond than a C–C bond, making silicon more reactive than diamond.

Silicon dioxide ($SiO_2$), or silica, is a major constituent of sand and is used to make glass. It also has a network lattice structure that is made up of alternating silicon and oxygen atoms. Silica occurs commonly in nature as sandstone, silica sand or quartzite. Silica is one of the most abundant oxide materials in the Earth's crust. In its crystal form silica is known as quartz.

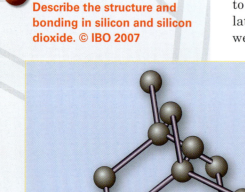

**Figure 2.4.9** The structure of silicon.

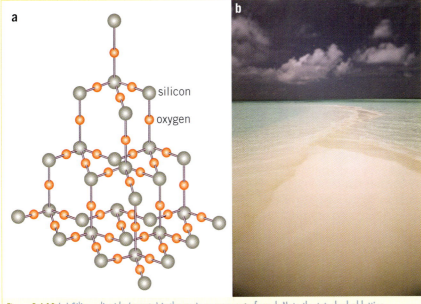

**WORKSHEET 2.4**
Covalent network lattices

**Figure 2.4.10** (a) Silicon dioxide (quartz) is the main component of sand. Note the tetrahedral lattice structure is similar to that of diamond. (b) Beach sands in the Maldives.

### Section 2.4 Exercises

1. **a** Define the term *allotrope*.
   **b** State the names of the different allotropes of carbon and describe one use of each.

2. **a** Describe the structure in diamond.
   **b** Describe the properties of diamond to which this structure gives rise.

3. Describe the structure of fullerene and state its properties.

4. **a** State the type of structure displayed by silicon.
   **b** The Si–Si bond length is 235.2 pm; that of C–C in diamond is 142.6 pm. Explain how this difference in bond length affects its reactivity.

5. **a** Silicon dioxide occurs commonly in nature and in a number of different forms. State the different names for all these forms.
   **b** State the name given to the structure of silicon dioxide.
   **c** List other substances that have this structure.

6. Compare the hardness and conductivity of diamond, graphite and fullerenes.

## 2.5 INTERMOLECULAR FORCES

When water boils, what is happening to the water molecules? They are not breaking into oxygen and hydrogen atoms; rather they are separating from each other to form gaseous water, or steam. Similarly, a candle left outside on a hot day will melt and soon become a puddle of liquid wax.

The addition of heat energy to a molecular substance initially separates the molecules from each other and changes their state. In many cases this requires very little added energy to occur. Ice melts at 0°C and wax can melt at 55°C. The bonds that are being broken are not covalent bonds, they are **intermolecular**

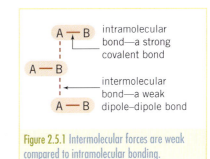

**Figure 2.5.1** Intermolecular forces are weak compared to intramolecular bonding.

forces. The strength of the intermolecular forces in a substance determines the melting and boiling temperatures of the substance. The strength of intermolecular forces is determined by the strength of the electrostatic attraction between the molecules that, in turn, is dependent on whether the molecules are polar or non-polar, big or small.

### van der Waals' forces (London or dispersion forces)

**4.3.1 Describe the types of intermolecular forces (attractions between molecules that have temporary dipoles, permanent dipoles or hydrogen bonding) and explain how they arise from the structural features of molecules. ©IBO 2007**

While many molecules such as water and hydrogen chloride have a permanent dipole, there are many other substances whose molecules are non-polar. This may be because they are elements, they are noble gases, or they have polar covalent bonds that cancel each other out to produce a non-polar molecule. Hydrocarbons are a major part of this group. Any molecule that is made up only of hydrogen and carbon will be non-polar.

These non-polar molecules must be attracted to each other by some electrostatic force because many of them are liquids or solids at room temperature. If there was no force at all between non-polar molecules, they would all be gases to temperatures well below 0°C. However, the melting temperatures of non-polar solids such as waxes and the fact that most non-polar substances are liquids or gases at room temperature suggests that the bonding between these molecules is weak.

van der Waals' forces

To appreciate the source of the weak attractive forces acting between these molecules, we must consider the orientation of all electrons, both bonding and non-bonding, at a particular instant rather than over a period of time. Take the simplest molecule, hydrogen, and consider the orientation of the two bonding electrons. While it is possible that these fast-moving electrons will be symmetrically oriented around the two hydrogen nuclei, it is far more likely that, for an instant of time, they will be found at one side of the molecule. When this happens a temporary or *instantaneous dipole* is formed (see figure 2.5.2).

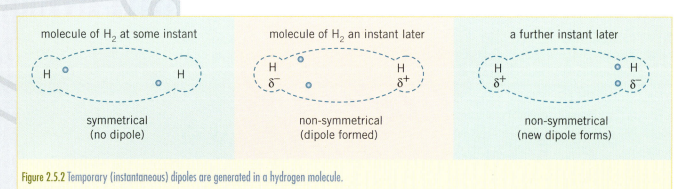

**Figure 2.5.2** Temporary (instantaneous) dipoles are generated in a hydrogen molecule.

Once that temporary dipole has been generated, it will influence the orientation of electrons within the molecules close to it. Electrons in neighbouring molecules will be repelled by the negative end of the temporary dipole, creating a new induced dipole. A moment later, the electrons attain a different orientation, and a new set of interacting dipoles will be generated. Over time, the orientation in three-dimensional space of the electrons of a specific molecule will average out to produce no permanent dipole. However, the weak interactions generated by billions of temporary dipoles will have resulted in a weak overall attractive force. These weak bonding forces are known collectively as **van der Waals' forces** or dispersion forces or London forces. They exist in all molecules, irrespective of any other intermolecular forces that may also be present.

The larger a molecule, and so the greater the number of electrons present, the more pronounced will be these van der Waals' forces. Referring to table 2.5.1, it can be seen that the boiling points of the group 7 elements increase as we move down the group from fluorine to iodine. Melting and boiling points are directly related to the strength of the bonds acting between particles; the stronger the forces of attraction, the higher the boiling point. All four molecules are diatomic and non-polar. Molecules of fluorine have a total of 18 electrons, chlorine 34 electrons, bromine 70 electrons and iodine 106 electrons. As the number of electrons present within the molecule increases, the number and magnitude of instantaneous dipoles increases and hence so do the melting and boiling points of the elements.

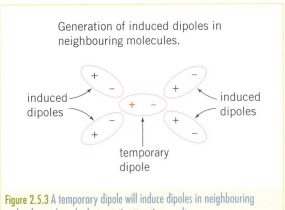

Figure 2.5.3 A temporary dipole will induce dipoles in neighbouring molecules and weak electrostatic attraction results.

| TABLE 2.5.1 MELTING AND BOILING POINTS OF HALOGENS | | | |
|---|---|---|---|
| Molecule | Number of electrons in molecule | Melting point (°C) | Boiling point (°C) |
| Fluorine ($F_2$) | 18 | −220 | −188 |
| Chlorine ($Cl_2$) | 34 | −101 | −35 |
| Bromine ($Br_2$) | 70 | −7 | 58 |
| Iodine ($I_2$) | 106 | 114 | 184 |

As all molecules contain electrons, van der Waals' forces are always present between molecules, though, if the molecules are significantly polarized or the molecule is relatively small, the effects of these van der Waals' forces may be relatively slight.

## Permanent dipole attraction (dipole–dipole attraction)

Polar molecules have a permanent dipole. One 'end' of the molecule is always positive in relation to the other 'end' of the molecule. For example, the hydrogen chloride molecule has a permanent dipole because the chlorine atom is more electronegative than the hydrogen atom, and so the bonding electrons will more often be located nearer to the chlorine than the hydrogen. As electrons have a negative charge, the chlorine end of the molecule will tend to carry more negative charge than the slightly positive hydrogen end.

As these molecules have a positive part and a negative part, they can attract other polar molecules by electrostatic attraction. The strength of this attraction depends on the size of the dipole, and this is determined by the difference in the electronegativities of the atoms in the molecule.

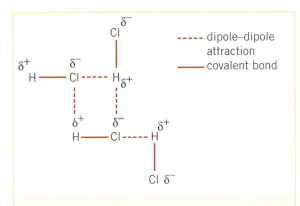

Figure 2.5.4 Permanent dipole attraction between polar HCl molecules.

For example, HCl and HBr are both polar molecules. The difference in electronegativities in HCl is (3.0 − 2.1) = 0.9, and in HBr it is (2.8 − 2.1) = 0.7. HCl is a more polar molecule than HBr.

This example presents an interesting problem. Based on polarities alone, it would be reasonable to expect the boiling point of HCl to be higher than those of HBr. After all, HCl does have a larger dipole than HBr. However, you will recall in our discussion of van der Waals' forces that as all molecules contain

**AS 4.3.2**
**Describe and explain how intermolecular forces affect the boiling points of substances.**
© IBO 2007

electrons, van der Waals' forces are always present between molecules. The difference in number of electrons of HCl and HBr is significant. HCl has 18 electrons and HBr has 36 electrons (twice as many as HCl). This difference in the strength of van der Waals' forces is sufficiently large to overcome the greater strength of the HCl dipole, and HBr has a higher boiling point than HCl.

**TABLE 2.5.2 THE BOILING POINTS OF THE HYDROGEN HALIDES**

| Hydrogen halide | Melting point (°C) | Boiling point (°C) |
| --- | --- | --- |
| Hydrogen fluoride (HF) | −83 | 20 |
| Hydrogen chloride (HCl) | −114 | −85 |
| Hydrogen bromide (HBr) | −87 | −67 |
| Hydrogen iodide (HI) | −51 | −35 |

As discussed above, this table shows that the melting and boiling points of HBr are higher than those of HCl, despite the greater polarity of the HCl molecule. But there is another surprising trend shown in this table; the smaller HF molecule has a higher boiling point than any of the other hydrogen halides. This anomaly will be explained in the next section.

### Hydrogen bonding: a special case of permanent dipole attraction

As we have seen, a permanent dipole is formed when two atoms in a molecule have different electronegativities; the greater the difference in electronegativities, the more polar the resultant bond. The most electronegative elements are all found in the top right-hand corner of the periodic table: fluorine (4.0), oxygen (3.5) and nitrogen (3.0). The combination of any of these three elements with hydrogen (2.1) produces a particularly strong dipole and results in the formation of highly polar molecules such as water, ammonia ($NH_3$) and hydrogen fluoride (HF). The special name of **hydrogen bonding** is given to the intermolecular forces that occur between molecules that contain H–F, H–O and H–N bonds. The memory mnemonic H–FON is useful to help remember which atoms combine in molecules to exhibit hydrogen bonding. In hydrogen bonding, the positive (H) end of a dipole is strongly attracted to the negative (F, O or N) end of the dipole of another molecule.

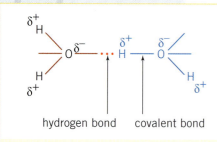

Figure 2.5.5 Hydrogen bonds form between polar water molecules.

Water exhibits hydrogen bonding between its molecules and it is these particularly strong intermolecular bonds that explain some of the unusual properties of water, such as its relatively high melting and boiling points and its low density when frozen. If you have ever fallen awkwardly into the swimming pool and done a 'belly-whacker' you will have been acutely aware of the strong intermolecular forces between water molecules that result in its high surface tension.

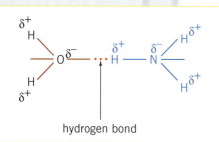

Figure 2.5.6 Hydrogen bonding occurs between the hydrogen of the ammonia molecule and the oxygen of the water molecule.

Another common substance that exhibits hydrogen bonding is ammonia ($NH_3$). The dipole–dipole bond between the hydrogen atom of one molecule and the nitrogen atom on an adjacent molecule is classified as a hydrogen bond. Ammonia has a relatively high boiling point (−33°C), compared with the other group 5 hydrides, although this is much lower than that of water (100°C). The weaker bond in ammonia is due to the smaller dipole, which in turn is due to the smaller electronegativity value of nitrogen (3.0) compared to oxygen (3.5). The somewhat weaker hydrogen bonding is also why ammonia is more volatile than water—it turns into a gas readily and so has quite a strong odour. Ammonia is extremely soluble in water due to the polar nature of both substances.

Hydrogen bonding

The relative strength of hydrogen bonding can be seen clearly when the boiling points of the hydrides (hydrogen compounds) of groups 4, 5, 6 and 7 are graphed (see figure 2.5.7). With the exception of group 4, each group shows the same pattern: an unusually high boiling point for the hydrides of the first element of the group—oxygen, fluorine and nitrogen—and then a sharp drop in boiling point, followed by a steady increase in boiling point as the number of electrons in the molecules increases, due to increasing van der Waals' forces. The pattern for group 4 is different because methane ($CH_4$) is a symmetrical, non-polar molecule.

Other groups of compounds in which the effect of hydrogen bonding on the boiling point can be seen are shown in the following table.

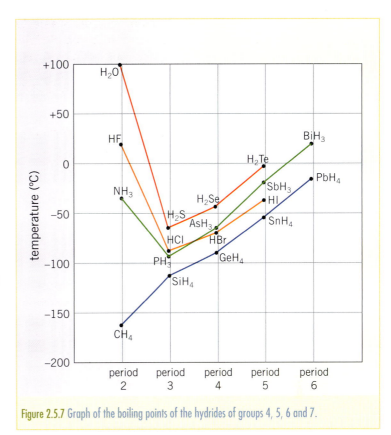

Figure 2.5.7 Graph of the boiling points of the hydrides of groups 4, 5, 6 and 7.

| TABLE 2.5.3 THE EFFECT ON THE BOILING POINT OF HYDROGEN BONDING | | | | |
|---|---|---|---|---|
| **Compound** | $CH_3OCH_3$ | $CH_3CH_2OH$ | $CH_3CH_2CH_3$ | $CH_3CHO$ |
| **Name** | Methoxymethane | Ethanol | Propane | Ethanal |
| **Boiling point (°C)** | −24.8 | 78.4 | 42 | −21 |
| **Molecular mass** | 46 | 46 | 44 | 44 |
| **Type of bonding present** | Permanent dipole attraction | Hydrogen bonding | van der Waals' forces | Permanent dipole attraction |

The molecules are of a similar size and we would therefore expect the van der Waals' forces to be of similar strengths; however, the effect of the presence of hydrogen bonding on the boiling point is dramatic.

WORKSHEET 2.5
Intermolecular forces

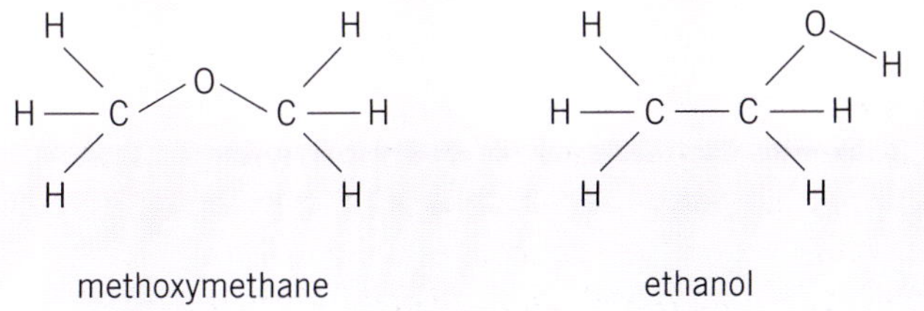

Figure 2.5.8 The structure of methoxymethane in comparison to that of ethanol.

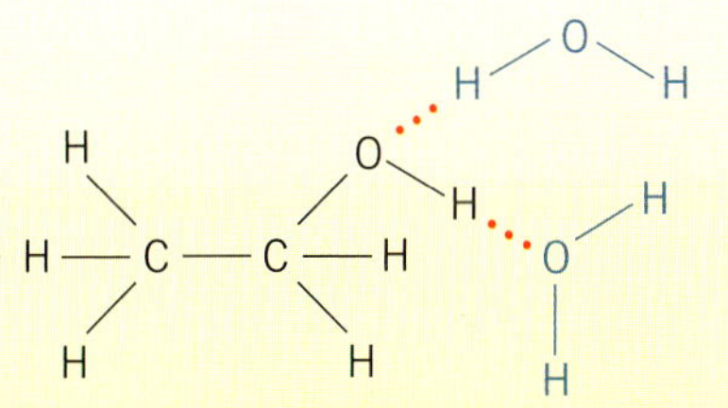

Figure 2.5.9 Hydrogen bonding occurs between the -OH group of ethanol and the water molecules.

### Section 2.5 Exercises

1. For each of the following molecules, identify the strongest type of intermolecular forces (van der Waals' forces, dipole–dipole attractions, hydrogen bonding) they exhibit in the liquid state.
   a. HCl
   b. $O_2$
   c. $NH_3$
   d. $CHCl_3$
   e. $OCl_2$
   f. $SiH_4$
   g. $N_2$
   h. HF
   i. $NBr_3$

2. Hydrocarbons are organic molecules that contain only carbon and hydrogen. The difference in the electronegativities of these two elements is only slight, with the result that hydrocarbons are relatively non-polar molecules. In terms of the intermolecular forces present, explain why methane ($CH_4$) is a gas at room temperature (boiling point of −164°C) while octane ($C_8H_{18}$) is a liquid (boiling point of 126°C).

3. List the following bonds in order from weakest to strongest.
   I   Permanent dipole–dipole bonds between $SO_2$ molecules
   II  Covalent bonds between carbon and hydrogen in $CH_4$
   III van der Waals' forces between atoms of helium
   IV  Permanent hydrogen bonds between HF molecules
   V   Covalent bonds between nitrogen atoms in $N_2$ molecules
   VI  van der Waals' forces between molecules of carbon dioxide

4. On the basis of the intermolecular forces present, determine which molecule in each of the following pairs has the higher boiling point. Explain your answer in each case.
   a. Methanol ($CH_3OH$) or ethanol ($CH_3CH_2OH$)
   b. Chloromethane ($CH_3Cl$) or dichloromethane ($CH_2Cl_2$)
   c. Hydrogen sulfide ($H_2S$) or water ($H_2O$)
   d. Hydrogen fluoride (HF) or hydrogen bromide (HBr)
   e. Ammonia ($NH_3$) or arsenine ($AsH_3$)

## 2.6 PHYSICAL PROPERTIES

Ionic, metallic and covalent bonding are all known as types of strong bonding. Each of these produces a group of compounds or elements with distinctive properties (table 2.6.1).

**AS 4.5.1**
Compare and explain the following properties of substances resulting from different types of bonding: melting and boiling points, volatility, electrical conductivity and solubility in non-polar and polar solvents. © IBO 2007

PRAC 2.4
Investigating the properties of sodium chloride

**TABLE 2.6.1 A COMPARISON OF THE PROPERTIES OF SUBSTANCES RESULTING FROM DIFFERENT TYPES OF BONDING**

| Type of bonding | Ionic | Metallic | Covalent molecular | Covalent network lattice |
|---|---|---|---|---|
| **Melting and boiling points** | Very high | Medium to high | Low | Very high |
| Explanation | Ions are packed closely in the ionic lattice. The ionic bond is strong and it requires large amounts of energy to break the bond. | The size and the charge on the cation determine the melting point. Metallic bonding is stronger than intermolecular forces. | Bonding between the molecules is weak and requires little energy to break. | Covalent bonding throughout the structure requires great amounts of energy to break the bonds. |
| **Volatility (how easily the substance turns into a gas or vapour)** | Low | Low in most cases (mercury is an exception) | High | Low |
| Explanation | Ionic substances are usually solids at room temperature and have very high boiling points, and therefore are difficult to vaporize. | Most metals are solids at room temperature and have high boiling points. Mercury is a liquid at room temperature and is quite volatile due to weak metallic bonding. | In many molecular substances; the intermolecular forces are weak and easily broken so the substance readily becomes a gas. | These substances are usually solids at room temperature and have very high boiling points; therefore they are difficult to vaporize. |
| **Electrical conductivity** | Ionic substances conduct in the liquid state and when dissolved in water. | All metals conduct electricity in the solid state. | Polar molecular substances will conduct electricity when dissolved in water. Pure and non-polar substances do not conduct electricity. | Graphite is the only network lattice element that conducts electricity. |
| Explanation | Ions are present in ionic compounds at all times, but are only free to move and conduct electricity when dissolved in water or when molten. | The sea of delocalized electrons enables electricity to be conducted at all times. | Polar molecules ionize when they dissolve in water, creating ions that are free to move. | The delocalized electrons between the layers of carbon atoms can conduct electricity. In all other network lattices there are no free electrons or other charged particles. |
| **Solubility** | Some dissolve in polar solvents only. Some do not dissolve at all. | Some metals will react with water to form alkaline solutions; however, most metals do not dissolve in polar or non-polar solvents. | Polar molecular substances will dissolve in polar solvents and non-polar molecular substances will dissolve in non-polar solvents. Some small non-polar molecules dissolve slightly in polar solvents. | These substances do not dissolve in any solvents. |
| Explanation | If a substance dissolves it is because the attraction of the ions to the water molecules is greater than the attraction of the ions to each other. | The metals that react with water are excellent reducing agents and undergo a redox reaction with the water. Generally the bonds within a metal are too strong to be broken by water. | Polar molecules are ionized by water as they dissolve. Non-polar molecules bond to non-polar solvent molecules by van der Waals' forces. | The covalent bonds are too strong to be influenced by solvents. |

## CHEM COMPLEMENT

### A mad metal

Mercury is widely used. It can be found in fungicides, pesticides, dental fillings, electrical equipment and mercury vapour lamps, and it is used in extracting metals from their ores. It is, however, highly toxic by absorption through the skin and inhalation of the vapour. Once absorbed, it is only very slowly eliminated from the body. In elevated levels, it can cause kidney and brain damage.

The expression 'mad as a hatter' has its origin in these effects of mercury poisoning. Prior to the 19th century, mercury was used in the preparation of felt for hats. Workers were constantly exposed to mercury vapours and developed nervous system disorders, loss of memory and eventually became insane. They were said to have 'hatter's shakes' or 'hatter's madness', hence the expression 'mad as a hatter'.

Many metals dissolve in liquid mercury at room temperature, forming a family of alloys known as amalgams. The amalgam is initially soft, then hardens. In dentistry, amalgams containing gold, zinc and various other metals are used to fill cavities in teeth. The use of mercury in dental filings has now become controversial, and alternative filings are under investigation.

Mercury and its compounds are, of course, not the only poisons among metals. Barium, cadmium, lead, arsenic and their compounds are all highly toxic and act in a similar way to mercury. There is some evidence that Napoleon and Ivan the Terrible may both have suffered from heavy metal poisoning.

Figure 2.6.1 The Mad Hatter's Tea Party, from Lewis Carroll's *Alice in Wonderland*. Mercury poisoning in hat makers led to the expression 'mad as a hatter'.

**WORKSHEET 2.6**
Physical properties

**WORKSHEET 2.7**
A bonding summary

## Section 2.6 Exercises

1. Identify the charged species that conduct electricity in each of the following substances.
   a. Al(s)
   b. NaCl(aq)
   c. Fe(l)
   d. AgI(l)

2. The melting temperatures of potassium (K) and calcium (Ca) are 63°C and 850°C respectively. The radii of potassium and calcium ions are 133 pm and 99 pm respectively. Explain two reasons why the melting point of potassium is much lower than that of calcium.

3. Use the theoretical model of ionic bonding to explain why calcium chloride conducts electricity very well in the molten state, but is effectively a non-conductor in its solid form.

4. Even though they are very hard, ionic crystals will shatter if significant force is applied across them. In terms of the bonding and arrangement of ions, explain why ionic substances are both hard and brittle.

5. Sodium chloride has a melting point of 801°C, while magnesium oxide melts at 2800°C. In terms of the bonding forces present, explain the reason for the large difference in these values.

6. Consider the data shown in the table below, which lists the group 1 chlorides.

| Name of compound | Melting point (°C) |
|---|---|
| Sodium chloride | 801 |
| Potassium chloride | 770 |
| Rubidium chloride | 722 |
| Caesium chloride | 635 |

   Suggest a possible explanation for the changes in melting point.

7. Of the substances listed below, identify those that are soluble in water.
   I   Silver (Ag)
   II  Calcium chloride ($CaCl_2$)
   III Copper(II) nitrate ($Cu(NO_3)_2$)
   IV  Butane ($C_4H_{10}$)
   V   Methanol ($CH_3OH$)

8. Of the substances listed in question **7**, identify those that would be soluble in a non-polar solvent such as tetrachloromethane, $CCl_4$.

9. a. Describe the structure of graphite.
   b. Explain how the properties of graphite are different from those of diamond.
   c. Identify the structural feature that leads to the 'greasy' nature of graphite.
   d. Explain why graphite is able to conduct electricity.

# Chapter 2 Summary

## Terms and definitions

**Allotropes**  Different structural forms of an element.

**Bond dissociation enthalpy**  The amount of energy (enthalpy) required to break the bonds between 1 mole of bonded atoms.

**Covalent bond**  The electrostatic attraction of one or more pairs of electrons and the nuclei between which they are shared.

**Covalent network lattice**  An arrangement of atoms in a lattice in which there are strong, covalent bonds between the atoms in all three dimensions.

**Dative covalent bond**  A covalent bond in which one atom has contributed two electrons to the bond; the other atom has not contributed any electrons.

**Delocalized electrons**  Electrons that are not confined to a particular location, but are able to move throughout a structure.

**Diatomic molecule**  A molecule that is made up of two atoms.

**Double covalent bond**  A covalent bond made up of two pairs of shared electrons (4 electrons).

**Ductility**  The ability to be drawn into a wire.

**Electrical conductivity**  The ability to allow electricity to pass through a substance.

**Electronegativity**  A measure of the electron attracting power of an atom.

**Electrostatic attraction**  The attraction between positive and negative charges.

**Element**  A substance made up of only one type of atom.

**Empirical formula**  The simplest whole number ratio of elements in a compound.

**Hydrogen bonding**  A type of strong, dipole–dipole bonding occurring between molecules that contain hydrogen bonded to fluorine, oxygen or nitrogen.

**Intermolecular forces**  The electrostatic attraction between molecules.

**Ionic bond**  The electrostatic attraction between a positively charged ion and a negatively charged ion.

**Ionic lattice**  A regular arrangement of ions in which every positive ion is surrounded by negative ions and every negative ion is surrounded by positive ions.

**Lewis structure (electron dot structure)**  A diagram of a molecule or other covalent species in which the outer shell (valence) electrons of the atom are represented by dots or crosses, and the sharing of electrons to form a covalent bond is shown.

**Malleability**  The ability of a metal to be bent or beaten into shape without breaking.

**Metallic bond**  The electrostatic attraction between a positively charged metal ion and the sea of delocalized electrons surrounding it.

**Molecule**  A discrete group of non-metallic atoms, of known formula, covalently bonded together.

**Non-bonding pairs of electrons (lone pairs)**  Pairs of valence electrons not involved in bonding.

**Permanent dipole attraction**  The electrostatic attraction that occurs between polar molecules.

**Polar covalent bond**  A covalent bond in which the electrons are not equally shared between the two nuclei due to a difference in the electronegativities of the elements involved.

**Polyatomic ion**  A group of two or more atoms covalently bonded together with an overall positive or negative charge.

**Strong bonding**  Ionic, metallic or covalent bonding.

**Single covalent bond**  A covalent bond made up of one pair of shared electrons (2 electrons).

**Structural formula**  A formula that represents the three-dimensional arrangement of atoms in a molecule.

**Triple covalent bond**  A covalent bond made up of three pairs of shared electrons (6 electrons).

**Valence electrons**  The electrons in the outermost shell of an atom.

**van der Waals' forces**  The weak attraction between molecules that occurs due to the instantaneous dipoles formed as a result of the random movement of electrons within the molecules (also called dispersion forces or London forces).

**VSEPR theory**  The theory that accounts for the shape of molecules as a result of electrostatic repulsion between each pair of electrons (bonding and non-bonding) to a position as far from the others as possible in three-dimensional space. Also known as the valence shell electron pair repulsion theory.

# Concepts

- All chemical bonding occurs as a result of electrostatic attractions between positive species and negative charges.

- Most atoms will strive to obtain eight electrons in their outer shell when forming chemical bonds. This concept is embodied in the octet rule.

- There are three types of strong bonding that can occur between atoms: ionic, covalent and metallic. Bond type depends on the type of atoms present.

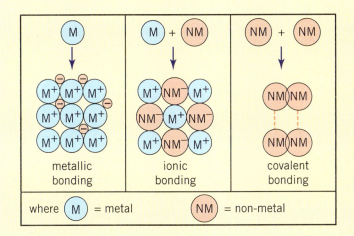

- Ionic bonding occurs between metals and non-metals and involves the transfer of one or more electrons from metal atoms to non-metal atoms, resulting in the formation of positively charged metal cations and negatively charged non-metal anions.

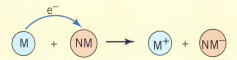

- Vast numbers of anions and cations combine in a fixed ratio to form an ordered, three-dimensional lattice.

- Lattices form in such a way that like charges are as far from each other as possible, while unlike charges are as close together as possible.

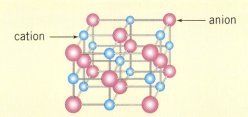

- Knowledge of the names, formulas and charges of ions is necessary in order to write the empirical formulas of ionic compounds.

- A metal consists of a lattice of positively charged ions (cations), surrounded by a 'sea' of mobile delocalized electrons.

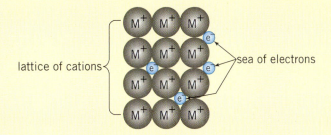

- Metallic bonding is the electrostatic attraction between the metal cations and the delocalized electrons.

- Covalent bonding occurs between non-metal atoms and involves the sharing of one or more electrons between these atoms to (generally) ensure that each atom has eight outer-shell electrons (octet rule), some of which are shared.

- A molecule is defined as a discrete group of covalently bonded non-metal atoms.

- Lewis structures (electron dot diagrams) and structural formulas can be used to represent molecules.

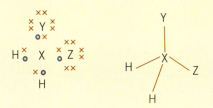

- The VSEPR (valence shell electron pair repulsion) model is used in determining the shapes of molecules and angles between the atoms in a molecule.

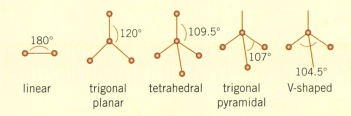

CHEMISTRY: FOR USE WITH THE IB DIPLOMA PROGRAMME **STANDARD LEVEL**

- Electronegativity is the ability of an atom to attract the electrons in a bond. It decreases down groups and increases across periods in the periodic table. Differences in electronegativity produce polarity in bonds.

- Bonding electrons are not always equally shared between atoms. The more electronegative atom gains a larger share and so acquires a slight negative charge, creating a dipole.

$$X^{\delta+} \text{---} Y^{\delta-} \quad \text{electronegativity } Y > X$$

- A molecule may contain polar bonds, but still be non-polar overall, if the dipoles cancel each other out. The symmetry of a molecule will be a major factor in determining the polarity of the molecule.

- Some substances are covalently bonded, but are not molecular, for example, diamond, graphite, silicon and silicon dioxide.

- The polarity of molecules determines the type of attractive forces between molecules (intermolecular forces). These forces are much weaker than the intramolecular bond (the covalent bond within the molecule).

| Molecule | Polarity of molecule | Intermolecular bond type |
|---|---|---|
| H—H Symmetrical | Non-polar | van der Waals' forces only |
| CHCl$_3$ Not symmetrical | Polar | Permanent dipole attraction and van der Waals' forces |
| H$_2$O Not symmetrical | Highly polar | Hydrogen bonding and van der Waals' forces |

- van der Waals' (dispersion or London) forces involve electrostatic attraction between temporary dipoles and are the weakest of the intermolecular bonding forces. Their strength increases with the size of the molecule and the number of electrons present.

- van der Waals' forces occur between all molecules. They are the only intermolecular forces present between non-polar molecules.

- Permanent dipole (dipole–dipole) attractions result from the interactions between polar molecules. These forces are usually stronger than van der Waals' forces.

- Hydrogen bonding is a special case of permanent dipole attraction and occurs between molecules that contain hydrogen bonded to one or more of the electronegative atoms fluorine, oxygen and nitrogen (remembered as 'H–FON').

- The nature of intermolecular bonding forces determines physical properties such as melting and boiling points.

| Property | Explanation |
|---|---|
| Soft<br>Low melting point<br>Many are gases or liquids at room temperature | Weak intermolecular forces |
| Do not conduct electricity when solid or liquid | Molecules are uncharged<br>No delocalized electrons<br>No free-moving charged particles to conduct a current |

- The strength of intermolecular forces depends upon a number of factors such as the relative electronegativity of the elements that are bonded and the size of the molecule.

- The presence of hydrogen bonding between molecules results in unusually high boiling and melting temperatures.

- The difference in electronegativity determines whether a compound of two elements will be more covalent or more ionic in nature. The greater the difference in electronegativity, the more ionic the compound will be.

- The ionic lattice model can be used to explain the characteristic properties of ionic compounds.

| Property | Explanation |
|---|---|
| Hard with high melting point | Strong electrostatic attractions between cations and anions |
| Brittle | Distortion of the lattice brings like charged ions together. These ions repel one another, shattering the lattice. |
| They do not conduct electricity when solid, but are good conductors when molten or in aqueous solution. | Ions are fixed in the lattice and so are unable to move. Melting and dissolving free the ions. These mobile ions are able to conduct a current. |

- The 'electron sea' model can be used to explain the characteristic properties of metals.

| Property | Explanation |
|---|---|
| Hard with high melting point | Strong electrostatic attractions between cations and electrons |
| Malleable and ductile | Non-directional bonding |
| Thermal and electrical conductivity lustre | Presence of mobile delocalized electrons |
| Relatively high densities | Tight packing of ions in an ordered array |

- The properties of diamond, graphite and fullerene ($C_{60}$), differ due to the different structural arrangements of their atoms.

| | Structure | Properties | Explanation |
|---|---|---|---|
| Diamond | | Extremely hard<br>Very high sublimation point<br>Non-conductor of electricity<br>Chemically inert | Very strong covalent bonding in three-dimensional lattice results in hardness and high sublimation point.<br>Non-conductor (and unreactive) as there are no free electrons. |
| Graphite | | Soft and greasy<br>Very high melting point<br>Conductor of electricity<br>Good lubricant | Each carbon atom has one free electron, hence good conductor of electricity.<br>Weak forces between layers, hence 'greasy'.<br>Strong covalent bonds within the layers, hence high melting point. |
| Fullerene, $C_{60}$ (buckyballs) | | Conductor of electricity | Each carbon atom is bonded to only three other atoms, resulting in the presence of one free, or delocalized, electron per carbon atom |

## Chapter 2 Review questions

1. Identify the type of bonding that exists in the following substances as ionic, covalent or metallic.
   a. Water ($H_2O$)
   b. Copper (Cu)
   c. Magnesium fluoride ($MgF_2$)
   d. Sodium iodide (NaI)
   e. Sulfur trioxide ($SO_3$)
   f. Zinc (Zn)
   g. Silicon dioxide ($SiO_2$)
   h. Chromium(III) oxide ($Cr_2O_3$)
   i. Carbon monoxide (CO)
   j. Manganese(IV) sulfide ($MnS_2$)

2. Magnesium can react with nitrogen to form a new compound.
   a. State the electron shell configuration of both magnesium and nitrogen.
   b. Draw an electron shell diagram to represent how magnesium and nitrogen atoms may react to form ions.
   c. State the formula for magnesium nitride.

3. Rubidium chloride is a crystalline solid. It is very hard, and has a high melting point of 722°C. Rubidium chloride dissolves readily in water, and is an excellent conductor of electricity when dissolved. It cannot conduct electricity at all in its solid state, but it is an excellent conductor when melted. Under magnification, it can be seen that the crystal structure of this salt is the same as that for sodium chloride.
   a. Describe the bonding in RbCl.
   b. Explain the properties of RbCl in terms of the ionic bonding model.

4. State whether you would expect magnesium oxide to have a higher or lower melting point than sodium chloride. Explain your answer in terms of the strength of the bond between the constituent particles.

5. State the balanced chemical formulas for each of the following compounds.
   a. Nickel chloride
   b. Calcium carbonate
   c. Copper(II) sulfate
   d. Copper(I) nitrate
   e. Lead(IV) iodide
   f. Potassium hydroxide
   g. Aluminium sulfide
   h. Zinc hydrogensulfate
   i. Iron(III) oxide
   j. Magnesium fluoride

6. State the names of the following compounds.
   a. $Mg(OH)_2$
   b. $KMnO_4$
   c. $Ag_2O$
   d. $PbO_2$
   e. $K_2Cr_2O_7$
   f. $Sn(NO_3)_2$
   g. $H_2S$
   h. $Fe(NO_3)_2$
   i. $Zn_3N_2$
   j. $CaSO_4$

7. Describe the structure and bonding in a metal using the 'electron sea' model. Include a labelled diagram in your answer.

8. Explain the following observations in terms of structure and bonding.
   a. Copper is malleable, but copper(II) oxide is brittle.
   b. Nickel conducts electricity in both solid and liquid states; nickel chloride only conducts when molten.

9. For each of the following substances:
   i. draw the Lewis (electron dot) diagram
   ii. draw the valence structure
   iii. state the shape.
   a. $PH_3$
   b. $CO_2$
   c. $H_2S$
   d. $NH_4^+$
   e. $SO_2$
   f. $CH_4$
   g. $H_3O^+$
   h. $BF_3$
   i. HF
   j. $C_2H_4$
   k. $C_2H_2$

10. The structure of ethanoic acid is shown below.

    There are two carbon–oxygen bonds (blue) in this molecule. State which of these two bonds is:
    a. stronger
    b. longer.

11 Compare a dative covalent bond to a 'shared' (normal) covalent bond and draw three different molecules in which a dative covalent bond can be found.

12 Identify a molecule from the list below to match its description. You may use each molecule from the list only once.
- I   Ammonium ion ($NH_4^+$)
- II  Sulfur dioxide ($SO_2$)
- III Hydrogen fluoride (HF)
- IV  Boron trifluoride ($BF_3$)
- V   Hydrogen cyanide (HCN)
- VI  Carbon dioxide ($CO_2$)

  a  Trigonal planar
  b  Four pairs of non-bonding electrons
  c  Exhibits hydrogen bonding
  d  Contains a triple bond
  e  Tetrahedral
  f  A V-shaped molecule

13 Water ($H_2O$) and hydrogen sulfide ($H_2S$) are both V-shaped molecules with identical valence structures. Water has a boiling point of 100°C, but hydrogen sulfide vaporizes at a temperature of just −62°C. In terms of the intermolecular bonding forces present, explain the significant difference in boiling points between these two substances.

14 The boiling point of phosphine ($PH_3$) is −85°C; that of ammonia ($NH_3$) is −33°C. Explain this difference in boiling points in terms of the intermolecular forces present.

15 Identify each of the following molecules according to the type of intermolecular forces (van der Waals' forces, dipole–dipole attractions, hydrogen bonding) they exhibit in the liquid state.
  a  HCl
  b  $I_2$
  c  $NH_3$
  d  $CHCl_3$
  e  $OCl_2$
  f  $SiH_4$
  g  $N_2$
  h  HF
  i  $NBr_3$

16 Methane, chloromethane and methanol are simple organic molecules with structural formulas as shown.

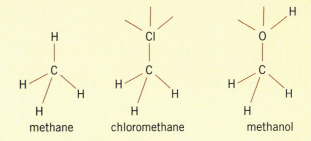

The respective boiling points of these three molecules are −164°C, −24°C and 65°C. Explain this trend in boiling points.

17 List the following compounds in order of their boiling points from lowest to highest. Explain your reasoning in terms of the intermolecular forces present.
- I   Butane ($C_4H_{10}$)
- II  Silicon dioxide
- III Propane ($C_3H_8$)
- IV  Butanol ($C_4H_9OH$)
- V   Sodium chloride
- VI  1-Chlorobutane ($C_4H_9Cl$)

18 Ammonia ($NH_3$) is highly soluble in water, but octane ($C_8H_{18}$), the major component of petrol, is effectively insoluble. Explain the difference in solubility in water of these two substances.

19 By referring to the relevant structural formulas, give concise explanations for each of the following.
  a  Diamond is extremely hard, but graphite is soft and greasy.
  b  Graphite is an excellent lubricant that can be used in high-performance racing-car engines, in which the very high temperatures present would vaporize conventional oil-based lubricants.
  c  Diamond is widely used as an abrasive on the ends of drill bits.
  d  Fullerene is a conductor of electricity; diamond is an excellent insulator.

 www.pearsoned.com.au/schools

Weblinks are available on the Chemistry: For use with the IB Diploma Programme Standard Level Companion Website to support learning and research related to this chapter.

# Chapter 2 Test

## Part A: Multiple-choice questions

1. When the Lewis structure for $HCOOCH_3$ is drawn, how many bonding and how many lone pairs of electrons are present?

   |   | Bond pairs | Lone pairs |
   |---|---|---|
   | A | 8 | 4 |
   | B | 7 | 5 |
   | C | 7 | 4 |
   | D | 5 | 5 |

   © IBO 2002, May P1 Q12

2. Which intermolecular forces exist in dry ice, $CO_2(s)$?
   A  Covalent bonds
   B  Dipole–dipole attractions
   C  van der Waals' forces
   D  Hydrogen bonds

   © IBO 2002, Nov P1 Q11

3. Which of the compounds $H_2O$, $H_2S$, $H_2Se$ and $H_2Te$ has the highest boiling point?
   A  $H_2O$
   B  $H_2S$
   C  $H_2Se$
   D  $H_2Te$

   © IBO 2002, Nov P1 Q13

4. When the following bond types are listed in decreasing order of strength (strongest first), what is the correct order?
   A  Covalent > hydrogen > van der Waals'
   B  Covalent > van der Waals' > hydrogen
   C  Hydrogen > covalent > van der Waals'
   D  Van der Waals' > hydrogen > covalent

   © IBO 2005, Nov P1 Q9

5. Which statement is true for most ionic compounds?
   A  They contain elements of similar electronegativity.
   B  They conduct electricity in the solid state.
   C  They are coloured.
   D  They have high melting and boiling points.

   © IBO 2005, Nov P1 Q10

6. What is the valence shell electron pair repulsion theory (VSEPR) used to predict?
   A  The energy levels in an atom
   B  The shapes of molecules and ions
   C  The electronegativities of elements
   D  The type of bonding in compounds

   © IBO 2005, Nov P1 Q11

7. Which fluoride is the most ionic?
   A  NaF
   B  CsF
   C  $MgF_2$
   D  $BaF_2$

   © IBO 2005, Nov P1 Q12

8. Which molecule has the smallest bond angle?
   A  $CO_2$
   B  $NH_3$
   C  $CH_4$
   D  $C_2H_4$

   © IBO 2006, May P1 Q10

9. Which is a correct description of metallic bonding?
   A  Positively charged metal ions are attracted to negatively charged ions
   B  Negatively charged metal ions are attracted to positively charged metal ions
   C  Positively charged metal ions are attracted to delocalized electrons
   D  Negatively charged metal ions are attracted to delocalized electrons

   © IBO 2006, May P1 Q12

10. What are responsible for the high electrical conductivity of metals?
    A  Delocalized positive ions
    B  Delocalized valence electrons
    C  Delocalized atoms
    D  Delocalized negative ions

    © IBO 2006, Nov P1 Q12

    (10 marks)

 www.pearsoned.com.au/schools

For more multiple-choice test questions, connect to the Chemistry: For use with the IB Diploma Programme Standard Level Companion Website and select Chapter 2 Review Questions.

## Part B: Short-answer questions

1. **a**  **i** Draw Lewis (electron dot) structures for $CO_2$ and $H_2S$ showing all valence electrons.
        (2 marks)

    **ii** State the shape of each molecule and explain your answer in terms of VSEPR theory.
        $CO_2$
        $H_2S$
        (4 marks)

   **b** Identify the strongest type of intermolecular force in each of the following compounds.
       **i** $CH_3Cl$
       **ii** $CH_4$
       **iii** $CH_3OH$
        (3 marks)

   © IBO 2004, May P2 Q3

2. Explain why methanol ($CH_3OH$) has a **much** higher boiling point than ethane ($C_2H_6$).
    (3 marks)

   © IBO 1999, May P2 Q3b

3. **a** Sketch and state the shape of each of the following molecules:
       **i** $SiH_4$
       **ii** $PH_3$
        (2 marks)

   **b** State the bond angle in $SiH_4$ and explain why the bond angle in $PH_3$ is less than in $SiH_4$.
        (2 marks)

   © IBO 2002, May P2 Q6b and c

## Part C: Data-based questions

1. The boiling points of the hydrides of the group 6 elements are shown below.

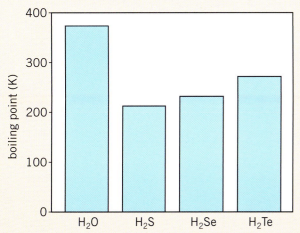

   **a** Explain the trend in boiling points from $H_2S$ to $H_2Te$.
        (2 marks)

   **b** Explain why the boiling point of water is higher than would be expected from the group trend.
        (2 marks)

   © IBO 2003, May P2 Q6a

2. The diagrams below represent the structures of iodine, sodium and sodium iodide.

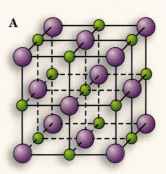

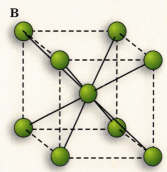

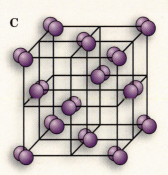

   **a** Identify which of the structures (**A**, **B** and **C**) correspond to iodine, sodium and sodium iodide.
        (1 mark)

   **b** State the type of bonding in each structure.
        (3 marks)

   © IBO 2003, May P2 Q6d

**3** Explain the variation observed in the boiling points of the noble gases, as shown in the table below.

| Element | Boiling point (°C) |
|---------|--------------------|
| Helium  | −269               |
| Neon    | −246               |
| Argon   | −186               |
| Krypton | −152               |
| Xenon   | −108               |
| Radon   | −62                |

(3 marks)

## Part D: Extended-response question

**a** Explain why magnesium has a higher melting point than sodium.

(3 marks)

© IBO 2006, May P2 Q7ai

**b** Draw a Lewis structure of a water molecule, name the shape of the molecule and state and explain why the bond angle is less than the bond angle in a tetrahedral molecule such as methane.

(4 marks)

© IBO 2006, May P2 Q7ci

**c** Explain why water is a suitable solvent for ethanol, but not for ethane.

(2 marks)

© IBO 2006, May P2 Q7cii

**d** Predict and explain the order of the melting points for propanol ($CH_3CH_2CH_2OH$), butane ($C_4H_{10}$) and propanone ($CH_3COCH_3$) with reference to their intermolecular forces.

(4 marks)

© IBO 2006, May P2 Q7d
Total marks = 50

# 3 PERIODICITY

## Chapter overview

This chapter covers the IB Chemistry syllabus Topic 3: Periodicity.

**By the end of this chapter, you should be able to:**

- identify groups and periods within the periodic table.
- explain the relationship between the position of an element in the periodic table and the electron arrangement of the element
- predict the electronic arrangement of any element with $Z \leq 20$ from its position in the periodic table
- predict the properties of an element from its position on the periodic table
- explain trends in properties such as first ionization energy, atomic radius, ionic radius, melting point and electronegativity within both groups and periods of the periodic table
- compare the chemical properties of elements within the alkali metal and the halogen groups of the periodic table
- discuss the ionic or covalent nature and the basic or acidic nature of the oxides of the period 3 elements.

IBO Assessment statements 3.1.1–3.3.2

Scientists in the process of acquiring new knowledge search for recurring patterns in the physical world to help them make generalizations about the nature of matter. The modern periodic table is a diagrammatic representation of the patterns in the chemical and physical properties of the elements. It is an invaluable tool for chemists because it helps them describe and explain the nature of the elements.

From the days of the alchemists to the present day there have always been curious scientists who have sought ways to make new elements. The modern periodic table is the result of many years of work by scientists from all corners of the scientific world.

Our knowledge of the elements went through a period of rapid change in the 1800s. Before 1790 only 27 elements were known to exist. Between 1798 and 1811 a further 17 new elements were discovered and by the end of the 19th century the total number of elements had risen to 60. Along with these discoveries, more and more was being learned about the chemical reactivity and physical properties of the elements. As this knowledge grew so too did the need to classify the elements in some uniform way to reveal general patterns between them.

In 1829, German chemist Johann Döbereiner (1780–1849) proposed patterns in this information and identified several triads. These triads were groups of three similar elements so that when the elements were arranged in order of atomic mass (called atomic weight at the time) the mass of the middle element was close to the average of the mass of the other two elements. Some of Döbereiner's triads are shown in figure 3.0.1.

| lithium | chlorine | calcium |
| sodium | bromine | strontium |
| potassium | iodine | barium |

Figure 3.0.1 Döbereiner's triads.

Many years later, British scientist John Newlands (1837–1898) developed a new method for organizing the elements. Newlands arranged the known elements by atomic mass and noticed recurring patterns in properties. When he organized the elements into groups of seven (starting a new row with each eighth element; figure 3.0.2), he noted that the vertical groupings had similar properties. Newlands referred to this as the *law of octaves*.

| H | Li | Be | B | C | N | O |
| F | Na | Mg | Al | Si | P | S |
| Cl | K | Ca | Cr | Ti | Mn | Fe |
| Co, Ni | Cu | Zn | Y | In | As | Se |
| Br | Rb | Sr | Ce, La | Zr | Di, Mo | Ro, Ru |
| Pd | Ag | Cd | U | Sn | Sb | Te |
| I | Cs | Ba, V | Ta | W | Nb | Au |
| Pt, Ir | Tl | Pb | Th | Hg | Bi | Cs |

Figure 3.0.2 Newlands's octaves.

The work of Döbereiner, Newlands and others in establishing a uniformed sequence of the elements was limited because some of the data available at the time was not reliable and not all the elements had been discovered. The real breakthrough came in 1869 when the Russian chemist Dmitri Mendeleev proposed his version of the periodic table in which the elements were arranged in order of increasing atomic mass and grouped according to their chemical reactivity.

# 3.1 THE PERIODIC TABLE

An insightful feature of Mendeleev's periodic table was that he left gaps for elements that he believed should exist because the elements on either side of the gap matched the expected chemical properties of their groups.

## CHEM COMPLEMENT

### Dmitri Ivanovich Mendeleev

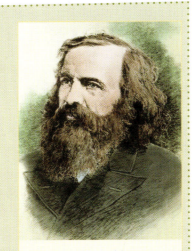

Figure 3.1.1 Dmitri Mendeleev.

The name most commonly associated with the development of the periodic table is that of Dmitri Ivanovich Mendeleev (1834–1907). Mendeleev was born in remote Siberia in 1834 as part of a large family. Education in this part of Russia was very limited, but Mendeleev still managed to develop a passion for science. Poor fortune in business caused Mendeleev's mother to move to Moscow and then to St Petersburg, where she died soon after. At the age of 21 Dimitri was diagnosed with tuberculosis and given a short time to live. Despite all this, Mendeleev managed to regain his good health and establish himself as a talented chemist with an encyclopaedic recall of chemicals and their properties. Mendeleev became a teacher and he accepted a grant to study in Paris. Travel in Europe exposed Mendeleev to the cutting edge of a variety of sciences and, on his return to Russia in 1861, he set about bringing chemistry in Russia up to the standard of these other countries.

Mendeleev wrote several chemistry texts in Russian. It was in writing the second volume of *The Principles of Chemistry* that he faced the problem of deciding in what sequence to discuss the elements. To help organize his thoughts, Mendeleev wrote the names, masses and properties of each element onto a series of cards. He tried many patterns of arranging these cards until he settled upon the arrangement shown in figure 3.1.2. This arrangement was published in 1869 as the 'Periodic system of the elements in groups and series'.

| | | | | | | | |
|---|---|---|---|---|---|---|---|
| | | | K = 39 | Rb = 85 | Cs = 133 | — | — |
| | | | Ca = 40 | Sr = 87 | Ba = 137 | — | — |
| | | | — | ?Yt = 88? | ?Di = 138? | Er = 178? | — |
| | | | Ti = 48? | Zr = 90 | Ce = 140? | ?La = 180? | Tb = 281 |
| | | | V = 51 | Nb = 94 | — | Ta = 182 | — |
| | | | Cr = 52 | Mo = 96 | — | W = 184 | U = 240 |
| | | | Mn = 55 | — | — | — | — |
| | | | Fe = 56 | Ru = 104 | — | Os = 195? | — |
| Typische Elemente | | | Co = 59 | Rh = 104 | — | Ir = 197 | — |
| | | | Ni = 59 | Pd = 106 | — | Pt = 198? | — |
| H = 1 | Li = 7 | Na = 23 | Cu = 63 | Ag = 108 | — | Au = 199? | — |
| | Be = 9,4 | Mg = 24 | Zn = 65 | Cd = 112 | — | Hg = 200 | — |
| | B = 11 | Al = 27,3 | — | In = 113 | — | Tl = 204 | — |
| | C = 12 | Si = 28 | — | Sn = 118 | — | Pb = 207 | — |
| | N = 14 | P = 31 | As = 75 | Sb = 122 | — | Bi = 208 | — |
| | O = 16 | S = 32 | Se = 78 | Te = 125? | — | — | — |
| | F = 19 | Cl = 35,5 | Br = 80 | J = 127 | — | — | — |

Figure 3.1.2 Dmitri Mendeleev's periodic table was published in 1869.

## CHEM COMPLEMENT

### Gallium

Gallium was discovered in 1875 by French chemist Paul Lecoq de Boisbaudran. Mendeleev had not only predicted the existence of gallium, but he had estimated its melting point, mass, valency and density. The new element found by Lecoq matched Mendeleev's prediction in all respects except density. Mendeleev had predicted a density of 4.7, while Lecoq found it to be 5.9. Mendeleev suggested Lecoq recheck his experiment, and sure enough the density was found to be 4.7!

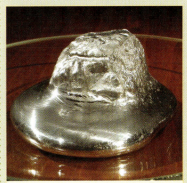

Figure 3.1.3 A partially melted cube of gallium. Gallium melts at 30°C.
© 2007 Theodore Grey periodictable.com

**WORKSHEET 3.1**
*Ordering the elements*

### THEORY OF KNOWLEDGE

Inductive reasoning, forming generalizations from patterns in observations, is an important way of knowing for scientists. Because induction leads to general statements or inferences, it is easy to draw conclusions without sufficient evidence. To overcome this scientists ensure that a number of different examples are looked at in a variety of different circumstances, exceptions are actively sought out, and more evidence is obtained when the results are unexpected.

When Mendeleev found that some of the elements unexpectedly did not fit in order of increasing atomic weight he did not abandon his hypothesis in light of this contradictory evidence. Instead he proposed that the atomic weights of these elements had been incorrectly calculated due to an experimental error. By recalculating these atomic weights he was able to put these elements in the correct place.

- Can you give an example of how you have used inductive reasoning as a way of knowing in Chemistry? What observations did you make and what general conclusion did you draw?
- How certain can we be of the conclusions drawn using inductive reasoning?
- Usually we are not aware of how our mind interprets the senses we experience. For example, the way we sense something can be influenced by how we feel. Can you describe an instance when your emotions interfered with your ability to make accurate observations?
- The tendency to look for patterns and to make generalizations is not unique to scientists. Name other areas of knowledge in which this is important.

## The modern periodic table

There are many different ways of presenting the periodic table and there are several different ways of labelling various sections of the table. Some of the common conventions include the following.

- The elements are arranged in order of increasing atomic number.
- There is a division between metals and non-metals. The *metal* elements are located to the left of the division and the *non-metal* elements to the right. Metals usually have a small number of electrons in their outer shell. Non-metals usually have a large number of electrons in their outer shell.

**AS 3.1.1**
**Describe the arrangement of elements in the periodic table in order of increasing atomic number.** © IBO 2007

- The vertical columns are called **groups**. The groups have been classified in a variety of different ways since 1869. Initially, they were numbered using Roman numerals I to VIII. However, the Arabic system of numbering from 1 to 8 may be used, as can the system recommended by the International Union of Pure and Applied Chemists (IUPAC), which includes the transition elements, and hence the groups are numbered from 1 to 18. IB Chemistry numbers the groups from 1 to 7, with group 8 being referred to as group 0.

**AS 3.1.2**
**Distinguish between the terms *group* and *period*.** © IBO 2007

- Some of the groups have accepted names: group 1 is called the **alkali metals**, group 2 is called the **alkaline earth metals**, group 7 is called the **halogens** and group 0 (8) is called the **noble gases**.
- The horizontal rows are called **periods**. Periods are numbered 1 to 7. An element belongs to the period that matches the number of its outer shell. Elements in the same period have the same number of occupied electron shells.

**DEMO 3.1**
*Examining elements*

- The long metal periods are labelled **transition metals**, **lanthanides** and **actinides**.
- There is difficulty placing hydrogen in the table. It is placed at the top of group 1 because it has some similarities to other group 1 elements, but it is a non-metal.
- Unlike the other noble gases, helium has only 2 electrons in its valence shell, but as this is shell 1, this constitutes a full valence shell.

**Physical properties of the halogens**

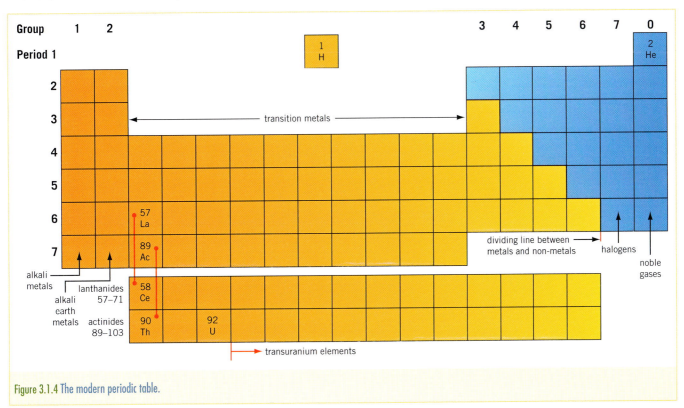

Figure 3.1.4 The modern periodic table.

Groups 3 to 6 contain both metal and non-metal elements. Seven elements—boron (B), silicon (Si), germanium (Ge), arsenic (As), antimony (Sb), tellurium (Te) and polonium (Po)—are called *metalloids* because they have characteristics of both metal and non-metal elements. Most of the elements are solid at room temperature and 1 atm pressure, except for the 11 gaseous elements—hydrogen, nitrogen, oxygen, fluorine, chlorine, and all the group 0 elements—and the two liquid elements, bromine and mercury.

The arrangement of the elements in the periodic table is linked to their electron configurations. Elements with the same outer-shell electron configuration belong to the same group of the periodic table. The alkali metals (group 1), for example, all have one electron in their outer shell. The electron arrangements of some of the alkali metals are:

| | |
|---|---|
| Lithium | 2,**1** |
| Sodium | 2,8,**1** |
| Potassium | 2,8,8,**1** |
| Rubidium | 2,8,18,8,**1** |

**AS 3.1.3**
**Apply the relationship between the electron arrangement of elements and their position in the periodic table up to $Z = 20$.**
© IBO 2007

The similarity in the number of outer-shell electrons of each of these alkali metals leads to similarities in their physical properties and their chemical reactions. They all react readily and need to be stored in oil, they have low melting points and are relatively soft compared to other metals, and all form ions with a charge of 1+.

**AS 3.1.4**
Apply the relationship between the number of electrons in the highest occupied energy level for an element and its position in the periodic table. © IBO 2007

The electron arrangement of an element can, in fact, be predicted from the position of the element on the periodic table. Referring to figure 3.1.4, we can see the following:

- Elements in group 1 have 1 electron in their outer shell. Group 2 elements have 2 electrons in their outer shell. This trend continues across to group 7, whose elements have 7 electrons in the outer shell, and then group 0 with a full outer shell of 8 electrons (2 in the case of helium).
- The number of the outer shell can be read directly from the number of the horizontal period (row) of the periodic table. The elements sodium (Na) to chlorine (Cl) are all in period 3, so their third electron shell is being filled. Argon has 8 electrons in its third electron shell, which fills this shell.
- All electron shells that are lower in energy than the one being filled will be full.

This means that, given a periodic table, the electron arrangement of any element up to $Z = 20$ can be predicted and written.

For example, sulfur is in period 3 and group 6 of the periodic table. It must therefore have 3 shells of electrons, with 6 electrons in the third shell. Its electron arrangement is 2,8,6. Potassium is in the period 4 and group 1, so its electron arrangement is 2,8,8,1.

Potassium and calcium only have 8 electrons in their third electron shell, despite the third shell being able to hold 18 electrons. This is related to the relative energies of the third and fourth electron shells, and will be explained fully in the Higher Level course.

PRAC 3.1
Periodic table overview

WORKSHEET 3.2
Periodic table quiz

### Section 3.1 Exercises

1. State the name of an element:
   a. in the same group as neon
   b. in the same period as sodium
   c. with similar properties to nitrogen
   d. that is a transition metal.

2. Identify each of the following elements.
   a. Element X is in period 4 and group 3.
   b. Element Y is in period 3 and group 2.
   c. Element Z is in period 5 and group 6.

3. The electronic structures of elements A to E are given below.
   A 2,8,3     B 2,8,2     C 2,8,8,2
   D 2,8       E 2,6

   From elements A to E select an element that:
   a. is found in group 3
   b. is found in period 2
   c. is found in group 2, period 4.

4. a. State the group and period to which the element with the electron configuration 2,8,5 belongs.
   b. A cation $X^{2+}$ has the electron configuration 2,8,8. To which group and period of the periodic table does element X belong?

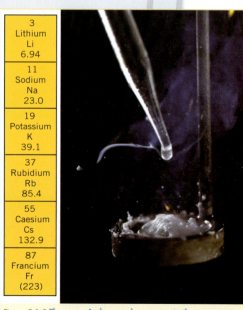
Figure 3.1.5 The group 1 elements have very similar properties. All react vigorously with water.

5. Use the periodic table to determine the electron configuration for:
   a. nitrogen
   b. the nitride ion ($N^{3-}$).

6. State the electron configuration for an element that is in:
   a. group 1 and period 3
   b. group 5 and period 2
   c. group 0 and period 1.

7. An element is in group 2 and period 3. How many electrons are present in its outer shell?

## 3.2 PHYSICAL PROPERTIES OF THE ELEMENTS

Every substance has a set of properties, unique traits or characteristics that are used to identify it in much the same way as a fingerprint is used to identify a person. With today's rapid advances in technology, new substances are continually being synthesized and discovered in nature. An understanding of the unique physical properties of a substance helps chemists recognize it when it appears in a new place, identify its structure and bonding, and group it with other similar substances.

Atomic radii, first ionization energies, electronegativities and melting points are examples of physical properties that experience **periodicity**, a repeating pattern or trend on the periodic table. A **physical property** is a characteristic that can be determined without changing the chemical composition of the substance. For example, the melting point of the element sulfur can be found by determining the temperature at which it turns from a solid to a liquid. The sulfur changes state only; its chemical composition does not alter.

### Trends within the alkali metals and the halogens

The alkali metals (group 1) of the periodic table are very similar in their properties, but they are not identical. There are trends in the way properties vary. This makes it possible to predict what variations will occur between members of the particular group. Table 3.2.1 provides data on the alkali metals (group 1) and the halogens (group 7), and serves to illustrate the trends that exist within some properties of elements of the same group. Similar trends are also evident in other groups of the periodic table. A summary of some trends is shown in figure 3.2.4 on page 85.

**Electronegativity** is a measure of the attraction an atom has for a shared pair of electrons when it is covalently bonded to another atom. As we saw in chapter 2, this ability largely determines how atoms bond together to form compounds. In general, metals have few electrons in their outer shell and so tend to lose these electrons in reactions. They do not tend to attract electrons and have low electronegativities. Non-metals have several outer-shell electrons and so tend to gain electrons in order to complete their outer shell. They have high electronegativities. Electronegativity tends to increase across a period and up a group.

The electronegativities and first ionization energies can be shown on a periodic table. In this way the trends are obvious and the values are easily available for use. See figure 3.2.1 (p. 83).

> **THEORY OF KNOWLEDGE**
> Symbols are an important part of the language of Chemistry. The symbols on the periodic table provide an interesting history of the discovery of the elements and the origins of their names. For example, the element *lead* has the symbol Pb, which comes from the Latin word *plumbum*.
> - Can you give the symbols of any other elements and state their origins?
> - Symbols follow a convention in the way they are written. Outline the conventions used when representing the symbols of elements.

**AS 3.2.1**
Define the terms *first ionization energy* and *electronegativity*.
© IBO 2007

Periodic trends: electronegativity

### TABLE 3.2.1 PROPERTIES OF THE ALKALI METALS AND THE HALOGENS

| Element | Atomic radius ($10^{-12}$ m) | Ionic radius ($10^{-12}$ m) | Electronegativity | First ionization energy (kJ mol$^{-1}$) | Melting point (°C) |
|---|---|---|---|---|---|
| **Alkali metals** | | | | | |
| Lithium | 152 | 68 | 1.0 | 519 | 181 |
| Sodium | 186 | 98 | 0.9 | 494 | 98 |
| Potassium | 231 | 133 | 0.8 | 418 | 64 |
| Rubidium | 244 | 148 | 0.8 | 402 | 39 |
| Caesium | 262 | 167 | 0.7 | 376 | 29 |
| **Halogens** | | | | | |
| Fluorine | 58 | 133 | 4.0 | 1680 | −219 |
| Chlorine | 99 | 181 | 3.0 | 1260 | −101 |
| Bromine | 114 | 196 | 2.8 | 1140 | −7 |
| Iodine | 133 | 219 | 2.5 | 1010 | 114 |

**AS 3.2.4**
Compare the relative electronegativity values of two or more elements based on their positions in the periodic table. © IBO 2007

In figure 3.2.1, the arrows indicate that electronegativity increases from left to right across a period and from bottom to top in a group. If the electronegativities of two elements in the same period are compared, the element on the right will have the greater electronegativity. If the electronegativities of two elements in the same group are compared, the element that is higher in the group will have the greater electronegativity.

**First ionization energy** is the energy required to remove one mole of electrons from one mole of atoms in the gaseous state. The outer-shell electron is the most easily removed, and so ionization energy is a measure of how tightly the outer-shell electrons are held in an atom. In general, electrons in metals are easily removed, so metals have low ionization energies. Non-metals hold electrons strongly, and so have high ionization energies. See figure 3.2.1.

Ionization of an atom can be represented by an equation. The following equation describes the first ionization of sodium:

$$Na(g) \rightarrow Na^+(g) + e^-$$

Periodic trends: ionization energy

How can we explain the differences in electronegativity and first ionization energy within a group? Members of group 1 are all metals and so have low first ionization energies. Why do the first ionization energies vary as we go down the group?

Two main factors determine how tightly an outer-shell electron is held. The force of electrostatic attraction between the positive protons in the nucleus and the outer-shell electron is directly related to the charges and inversely related to the distance between them. As the size of atoms increases, the attractive force on the outer-shell electron decreases. As the nuclear charge increases, the attractive force increases. There is, however, a complication: the outer-shell electrons are 'shielded' from the full nuclear charge by the inner-shell electrons. The concept of **core charge** is used to allow for this shielding. The effective nuclear charge felt by the outer-shell electrons, the core charge, may be found by subtracting the number of inner-shell electrons from the nuclear charge. For example, the core charge of some group 1 elements can be determined by:

> Sodium: 11 protons and 10 inner-shell electrons = core charge of +1
> Potassium: 19 protons and 18 inner-shell electrons = core charge of +1
> Rubidium: 37 protons and 36 inner-shell electrons = core charge of +1

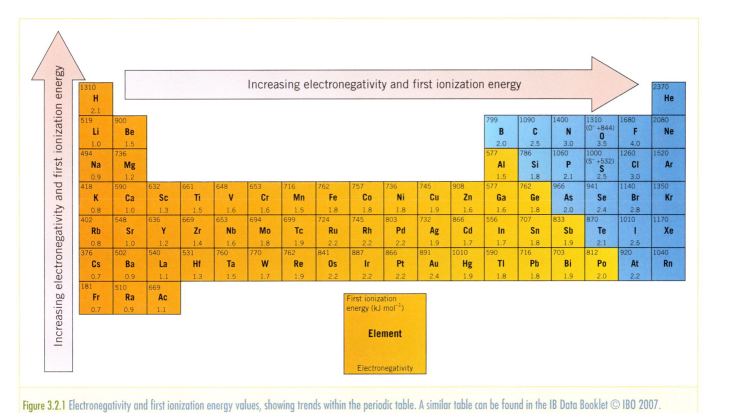

Figure 3.2.1 Electronegativity and first ionization energy values, showing trends within the periodic table. A similar table can be found in the IB Data Booklet © IBO 2007.

Figure 3.2.2 Atomic and ionic radii values showing trends within the periodic table. A similar table can be found in the IB Data Booklet © IBO 2007.

**AS 3.2.2**
Describe and explain the trends in atomic radii, ionic radii, first ionization energy, electronegativities and melting points for the alkali metals (Li → Cs) and the halogens (F → I). © IBO 2007

Periodic trends: atomic radii

The core charge remains constant within a group. This means that within a group the only factor affecting the electrostatic attraction between outer-shell electrons and the nucleus is the distance of the outer shell from the nucleus. As the atomic number increases within a group (going *down* the group), the attractive force between the nucleus and the outer-shell electrons *decreases*.

Atomic radii and ionic radii increase going down groups 1 and 7 because there in an increase in the number of electron shells surrounding the nucleus as you go down these groups (indeed any group). The electron shells account for most of the volume of the atom, so an extra electron shell increases the atomic or ionic radius.

Both first ionization energy and electronegativity decrease down groups 1 and 7 because of the decreasing electrostatic attraction between the outer-shell electrons and the nucleus. This is due to the increasing distance of the outer shell from the nucleus. The smaller the attraction of the outer shell to the nucleus, the easier it is to remove an electron from the outer shell (first ionization energy) and the harder it is to attract an electron to the outer shell (electronegativity).

The melting points of group 1 and group 7 elements differ in the trends they exhibit. This can be attributed to the type of bonding exhibited by each group. You will recall from chapter 2 that the strength of the bonding within a substance governs its melting and boiling points.

**The stronger the bonding within a substance, the higher the melting point.**

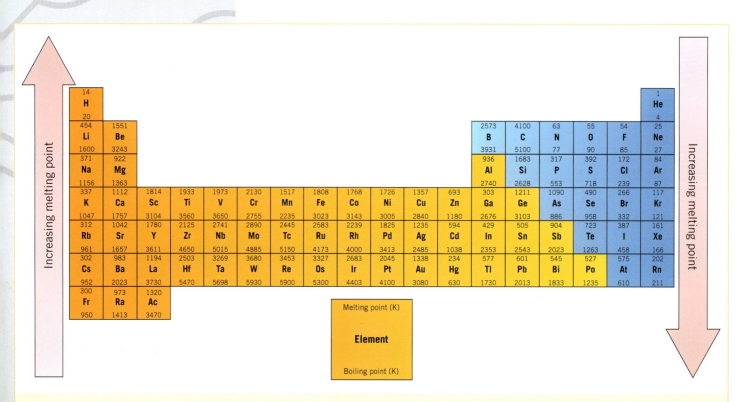

Figure 3.2.3 Melting and boiling points of elements, showing trends within the periodic table. A similar table can be found in the IB Data Booklet © IBO 2007.

The members of group 1 are all metals. As the atomic number increases down the group, the size of the positive metallic ion increases; however, the number of delocalized electrons does not change, nor does the charge on the ion. This results in a weaker electrostatic attraction between the ions and the delocalized electrons and a weaker metallic bond. Consequently, the melting point decreases down the group.

The group 7 elements, the halogens, are non-metals. They exist as diatomic, non-polar molecules between which the only intermolecular bonds are van der Waals' forces. As the atomic number of the halogen molecules increases, the strength of the van der Waals' forces increases significantly and the melting point also increases. See table 2.5.1 page 59.

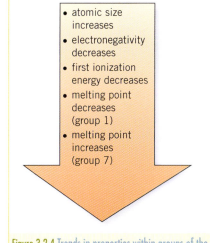

Figure 3.2.4 Trends in properties within groups of the periodic table.

### Trends across period 3

Properties of elements within periods are more variable than properties within groups. Consider the trends shown in figures 3.2.5 to 3.2.7. Marked variations occur because the electron configurations and core charges differ for each element in the period.

Recall that the core charge of an atom may be found by subtracting the number of inner-shell electrons from the nuclear charge. For example, across period 3 the core charge changes as shown:

Sodium: 11 protons and 10 inner-shell electrons = core charge of +1
Aluminium: 13 protons and 10 inner-shell electrons = core charge of +3
Chlorine: 17 protons and 10 inner-shell electrons = core charge of +7

 **3.2.3**
**Describe and explain the trends in atomic radii, ionic radii, first ionization engergy and electronegativities for elements across period 3.**
© IBO 2007

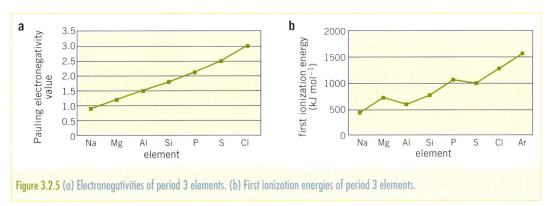

Figure 3.2.5 (a) Electronegativities of period 3 elements. (b) First ionization energies of period 3 elements.

Interactive periodic table

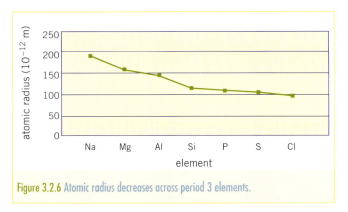

Figure 3.2.6 Atomic radius decreases across period 3 elements.

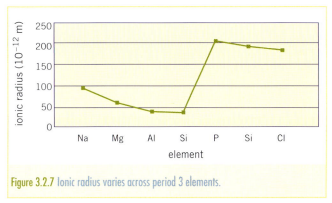

Figure 3.2.7 Ionic radius varies across period 3 elements.

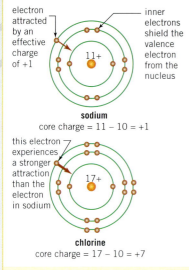

**Figure 3.2.8** Core charge can be used to explain trends within the periodic table.

- atomic radius decreases
- electronegativity increases
- first ionization energy increases
- ionic radius decreases

**Figure 3.2.9** Trends in properties across a period.

The core charge increases across the period. This means that the outer-shell electrons of chlorine therefore experience a greater attraction to the nucleus than does the outer-shell electron of sodium.

**Atomic radius** decreases from left to right across period 3 due to the increasing attraction experienced by the outer-shell electrons. These outer-shell electrons are all in the third electron shell of the atoms; however, as the core charge increases, the electrostatic attraction between the outer-shell electrons and the nucleus increases. This has the effect of pulling the electrons in closer to the nucleus and making the atom smaller.

The trend in ionic radii is not as clear as for atomic radii (see figure 3.2.6). For the metals (sodium to aluminium) in period 3, the ionic radius decreases across the period. Silicon can be represented as a positive ($Si^{4+}$) or negative ($Si^{4-}$) ion. For the non-metals, the ionic radius decreases from the phosphorus ($P^{3-}$) to the chloride ion ($Cl^-$) (figure 3.2.7). A positive ion has a smaller ionic radius than the original atom, due to the loss of the valence electrons, and a negative ion has a larger ionic radius than the original atom, since the addition of extra negative charges introduces more electron–electron repulsion. Negative ions have a larger radius than positive ions, as they have one more shell of electrons than the positive ions.

The increase in first ionization energy and electronegativity from left to right across period 3 can also be explained by the increasing core charge. As the core charge increases, it becomes increasingly difficult to remove an electron from the outer shell of the atom (first ionization energy). Similarly, the increasing electrostatic attraction of outer-shell electrons for the nucleus results in a greater power of attraction for electrons in the outer shell (electronegativity).

## Section 3.2 Exercises

1. State and explain how:
   a. the first ionization energy of strontium compares with that of magnesium
   b. the electronegativity of selenium compares with that of oxygen.

2. a. Explain what is meant by the term *core charge*.
   b. How is core charge used to explain the trend in atomic radius of the period 3 elements?

3. Explain why the electronegativity of fluorine is higher than that of magnesium.

4. a. Compare the atomic radii of magnesium and chlorine.
   b. Explain the difference you have described.

5. a. Compare the first ionization energy of phosphorus and chlorine.
   b. Explain the difference you have described.

6. a. State the trend in the ionic radii from $Na^+$ to $Al^{3+}$.
   b. State the trend in the ionic radii from $Si^{4-}$ to $Cl^-$.
   c. Explain the trend in part **b**.

Summary of periodic trends

**WORKSHEET 3.3**
Periodic table trends

**7** For each of the following pairs, determine whether the atomic/ionic radius of the first particle listed is larger than (>), the same size (=), or smaller than (<) the radius of the second particle.

|   | First particle | >, =, < | Second particle |
|---|---|---|---|
| a | Chlorine atom (Cl) | | Chloride ion (Cl$^-$) |
| b | Aluminium ion (Al$^{3+}$) | | Aluminium atom (Al) |
| c | Calcium atom (Ca) | | Sulfur atom (S) |
| d | Sodium ion (Na$^+$) | | Fluoride ion (F$^-$) |
| e | Magnesium ion (Mg$^{2+}$) | | Calcium ion (Ca$^{2+}$) |
| f | Sulfide ion (S$^{2-}$) | | Potassium ion (K$^+$) |

## 3.3 CHEMICAL PROPERTIES OF ELEMENTS AND THEIR OXIDES

Just as the physical properties exhibit a periodicity within groups and periods, so do the **chemical properties** of the elements and some of their compounds. The chemical properties relate to the electron arrangement of the elements.

### Trends in chemical properties within a group

The alkali metals are well known for their reactive nature. Their tendency to react sometimes violently with water means that they must be stored under oil. Potassium, in particular, is seldom found in secondary school laboratories because of its difficulty in storage and its violent reaction with water.

The reactivity of the alkali metals with water increases down group 1. While lithium metal floats on the surface of the water and reacts slowly, producing some hydrogen gas, sodium reacts more violently, whizzing around on the surface of the water on a layer of hydrogen gas in what is sometimes described as 'hovercraft' motion.

$2Li(s) + H_2O(l) \rightarrow Li_2O(aq) + H_2(g)$

$2Na(s) + H_2O(l) \rightarrow Na_2O(aq) + H_2(g)$

The sodium can be forced to ignite by slowing its progress on the surface of the water by placing it on paper towel (figure 3.3.1). The white smoke in figure 3.3.1 is sodium oxide being carried off as a smoke. Potassium burns spontaneously in water, producing a violet flame.

$2K(s) + H_2O(l) \rightarrow K_2O(aq) + H_2(g)$

All three alkali metals produce an alkaline solution when they react with water.

$Li_2O(s) + H_2O(l) \rightarrow 2LiOH(aq)$

$Na_2O(s) + H_2O(l) \rightarrow 2NaOH(aq)$

$K_2O(s) + H_2O(l) \rightarrow 2KOH(aq)$

Evidence for this alkaline solution can be seen in figure 3.3.1. Phenolphthalein indicator has been added to the water before the reaction with sodium and has turned pink due to the presence of sodium hydroxide.

The increase in reactivity can be explained by the decrease in electrostatic attraction between the outer-shell electron and the positive nucleus of the alkali metal. The further the outer shell is from the nucleus, the more easily it is lost in the reaction with water, and the more spectacular the result.

The increasing reactivity of the alkali metals can also be seen in their reaction with halogens. As lithium, sodium and potassium all lose electrons easily (are good **reducing agents**) and the halogens gain electrons easily (are good **oxidizing agents**, see chapter 10), these reactions are predictably violent.

**3.3.1**
Discuss the similarities and differences in the chemical properties of elements in the same group. © IBO 2007

Figure 3.3.1 Sodium will burn with a yellow flame if its motion on the surface of water is stilled.

DEMO 3.2
Reactions of group 1 and group 2 elements with water

Figure 3.3.2 Potassium burns spontaneously in water with a violet flame.

Reactions with oxygen
Sodium and potassium in water

The product in each case is an ionic compound. The reactions all follow the same pattern, since the alkali metals all form ions with a 1+ charge and the halogens all form ions with a 1− charge.

$$2Na(s) + Cl_2(g) \rightarrow 2NaCl(s)$$
$$2K(s) + I_2(g) \rightarrow 2KI(s)$$
$$2Li(s) + Br_2(g) \rightarrow 2LiBr(s)$$

The smaller the halogen atom, the greater is its ability to gain electrons. This can be explained by the closeness of the outer shell to the nucleus. When halogens are mixed with halide salts such as potassium iodide, KI, the ability to react depends on the relative electron attracting strength of the halogen and how easily the **halide ion** will lose its electron. The larger the halide ion, the less strongly the outer-shell electrons are attracted to the nucleus (due to distance from the nucleus) and so the easier it is to remove an electron from the ion.

A list of the halogens in order of their electron attracting ability and the halide ions in order of their ability to lose an electron can be seen in figure 3.3.3.

Fluorine is the most reactive halogen, but its reactions are too violent to perform in a school laboratory. The next most reactive halogen chlorine, $Cl_2$, will reduce bromide and iodide ions to bromine and iodine; bromine, $Br_2$ will only reduce iodide ions and iodine cannot reduce any of the halide ions.

$$Cl_2(g) + 2I^-(aq) \rightarrow 2Cl^-(aq) + I_2(s)$$
$$Cl_2(g) + 2Br^-(aq) \rightarrow 2Cl^-(aq) + Br_2(s)$$
$$Br_2(g) + 2I^-(aq) \rightarrow 2Br^-(aq) + I_2(s)$$

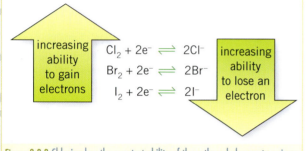

Figure 3.3.3 Chlorine has the greatest ability of these three halogens to gain an electron and the iodide ion has the greatest ability to lose an electron.

Figure 3.3.4 The reaction between chlorine and potassium iodide displaces red-brown iodine from the solution.

### THEORY OF KNOWLEDGE

In a Chemistry textbook, chlorine is described as a yellow-green gas at room temperature with a pungent, irritating odour. It is approximately two and a half times denser than air. When chlorine gas is inhaled, depending on the level of exposure, it can cause irritation to the eyes, skin and throat, a cough, chest tightness, wheezing and severe chemical burns.

During World War I, chlorine gas was deployed as a chemical weapon. *Dulce et decorum est*, written by Wilfred Owen in 1918 is one of the best known poems written in English from World War I.

*Gas! Gas! Quick, boys! — An ecstasy of fumbling,*
*Fitting the clumsy helmets just in time;*
*But someone still was yelling out and stumbling,*
*And flound'ring like a man in fire or lime …*
*Dim, through the misty panes and thick green light,*
*As under a green sea, I saw him drowning.*
*In all my dreams, before my helpless sight,*
*He plunges at me, guttering, choking, drowning.*

- Read the stanza of the poem and comment on the claim that art conveys no knowledge or literal truths that can be verified.
- Consider the language used by the chemist and the poet to describe the effects of chlorine gas. Which is more precise, specific and direct? Which is more suggestive and leaves itself open to the readers' interpretation? In what other ways might the use of language in English differ from that in science?

# The oxides of period 3 elements

 **AS 3.3.2**
**Discuss the changes in nature from ionic to covalent and from basic to acidic of the oxides across period 3.** © IBO 2007

Each of the period 3 elements reacts with oxygen. This reaction was one of the periodic properties that gave Mendeleev confidence in the periodic law. The formulas of some of the **highest oxides** of the period 3 elements and the reaction of the oxide with water are shown in table 3.3.1.

As we saw in chapter 2, the large difference in electronegativity between the metals and oxygen produces an ionic compound, whereas the non-metals form a covalent compound with oxygen. The ionic nature of the oxides decreases from left to right across period 3.

Periodic trends: acid–base properties of oxides

### TABLE 3.3.1 THE OXIDES OF SOME PERIOD 3 ELEMENTS AND THEIR REACTION WITH WATER

| Element | Highest oxide | Reaction | Nature of aqueous solution |
| --- | --- | --- | --- |
| Sodium | $Na_2O$ | $Na_2O(s) + H_2O(l) \rightarrow 2NaOH(aq)$ | Alkaline |
| Magnesium | $MgO$ | $MgO(s) + H_2O(l) \rightarrow Mg(OH)_2(aq)$ | Alkaline |
| Phosphorus | $P_4O_{10}$ | $P_4O_{10}(s) + 6H_2O(l) \rightarrow 4H_3PO_4(aq)$ | Acidic |
| Sulfur | $SO_3$ | $SO_3(g) + H_2O(l) \rightarrow H_2SO_4(l)$ | Acidic |

**DEMO 3.3**
Acidic and basic properties of oxides

The periodicity of properties is very clear in these reactions. Sodium burns in air to produce sodium oxide, $Na_2O$, which reacts easily with water to produce a strongly alkaline solution of sodium hydroxide, NaOH. To the right of sodium in period 3 is magnesium. Magnesium burns with a bright white flame in air to produce magnesium oxide, MgO (see figure 3.3.5), which then reacts with water to make a solution of quite alkaline magnesium hydroxide, $Mg(OH)_2$.

The decrease in metallic nature from left to right across period 3 can be seen from the very different behaviour of aluminium oxide, $Al_2O_3$, compared with the group 1 and group 2 metal oxides. Aluminium oxide can behave as an acid or base—it is **amphoteric** (see chapter 9). Aluminium oxide does not dissolve in water, but will react with acids and bases.

Figure 3.3.5 Magnesium burns in air with a bright white flame to form magnesium oxide.

Figure 3.3.6 Sulfur burns in air with a blue flame to form sulfur dioxide.

Acting as a base: $Al_2O_3(s) + 6HCl(aq) \rightarrow 2AlCl_3(aq) + 3H_2O(l)$

Acting as an acid: $Al_2O_3(s) + 2NaOH(aq) + 3H_2O(l) \rightarrow 2NaAl(OH)_4(aq)$
sodium aluminate

**WORKSHEET 3.4**
Chemical trends

Further to the right in period 3, phosphorus burns violently in air to make phosphorus(V) oxide, $P_4O_{10}$, which then reacts violently with water to make a weakly acidic solution of phosphoric acid, $4H_3PO_4$. Sulfur dioxide is made when sulfur burns in air (see figure 3.3.6) and sulfur trioxide, $SO_3$, is made when sulfur dioxide, $SO_2$, reacts further with oxygen. The violent reaction of sulfur trioxide with water is to be avoided as it produces a 'fog' of the strongly acidic sulfuric acid, $H_2SO_4$.

We can generalize this trend by stating that:
- non-metals form acidic oxides
- metals form basic oxides
- aluminium forms an amphoteric oxide.

Carbon dioxide behaves as an acid in water

**THEORY OF KNOWLEDGE**
In the natural sciences verbs, or command terms, are used to communicate scientific information. How can *describing* and *explaining* the patterns in the periodic table help us understand the physical and chemical properties of matter?

### Section 3.3 Exercises

1. State and explain the trend in reactivity of the alkali metals down the group.

2. State and explain the trend in reactivity of the halogens down the group.

3. a State the balanced chemical equation for the reaction of lithium with water, including states.
   b State the balanced chemical equation for the reaction between the product produced in part **a** and water.
   c State the sort of solution that is produced in the reaction in part **b**.
   d Explain how you could show that this is the type of solution produced.

4. a List the possible halide ions with which chlorine is able to react.
   b List the possible halide ions with which iodine is able to react.

5. Write balanced chemical equations for the reactions of chlorine with the halide ions, $Br^-$ and $I^-$.

6. a Write the balanced chemical equation for the reaction of phosphorus(V) oxide, $P_4O_{10}$, with water.
   b State the sort of solution that is produced.

7. Compare the reactions of:
   a lithium and sodium with water
   b chlorine and bromine gas with sodium
   c fluorine and chlorine gas with bromide ions
   d sodium and potassium with bromine gas.

# Chapter 3 Summary

## Terms and definitions

**Alkali metals**   The name given to group 1 of the periodic table.

**Alkaline earth metals**   The name given to group 2 of the periodic table.

**Amphoteric**   Able to act as an acid or a base.

**Atomic radius**   The distance from the centre of the nucleus to the outermost electron shell.

**Chemical property**   A characteristic that is exhibited as one substance is chemically transformed into another.

**Core charge**   The effective nuclear charge experienced by the outer-shell electrons. Core charge is the difference between the nuclear charge and the number of inner-shell electrons.

**Electronegativity**   A measure of the attraction that an atom has for a shared pair of electrons when it is covalently bonded to another atom.

**First ionization energy**   The amount of energy required to remove one mole of electrons from one mole of atoms in the gaseous state.

**Group**   A vertical column within the periodic table.

**Halide ion**   A negative ion formed when a halogen atom gains one electron.

**Halogens**   The name given to group 7 of the periodic table.

**Highest oxide**   The compound formed with oxygen in which the element is in its highest possible oxidation state.

**Noble gases**   The gaseous elements of group 0 of the periodic table, all of which have a full valence shell.

**Oxidizing agent**   A substance which causes another substance to be oxidized by accepting electrons from it.

**Period**   Horizontal row within the periodic table.

**Periodicity**   The repetition of properties at regular intervals within the periodic table.

**Physical property**   A characteristic that can be determined without changing the chemical composition of the substance.

**Reducing agent**   A substance that causes another substance to be reduced by donating electrons to it.

## Concepts

- The modern periodic table arranges elements horizontally in order of atomic number and vertically in groups with the same outer-shell electron configuration and similar chemical reactivity.

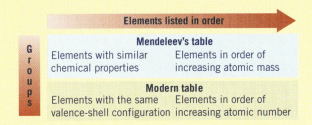

- A group is a vertical column and a period is a horizontal row in the periodic table.

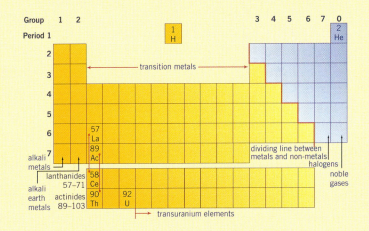

- There are trends in the properties of the elements both in groups and across periods.

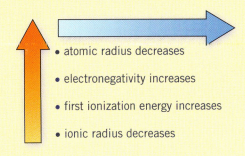

- atomic radius decreases
- electronegativity increases
- first ionization energy increases
- ionic radius decreases

- The patterns in the periodic table are a consequence of the valence-shell electron configurations of the elements. Elements in the same group behave similarly because they have the same valence-shell configuration. Elements in the same period have electrons filling the same electron shell (energy level).

- Patterns in periods are explained by the increasing core charge. Patterns in groups are explained by the increasing size of atoms (atomic radius).
- Electronegativity is a measure of the attraction that an atom has for a shared pair of electrons when it is covalently bonded to another atom. It increases from left to right across a period and up a group within the periodic table.
- First ionization energy is the energy required to remove an electron from a gaseous atom. It is generally low for metals and high for non-metals.
- Melting point is the temperature at which a solid turns into a liquid. The melting point depends on the strength of the bonding within a substance. Melting points decrease down group 1 and increase down group 7.
- For the alkali metals (group 1), the chemical reactivity (ability to lose electrons) increases down the group. This is explained by the decrease in electrostatic attraction, due to the greater distance between the valence electrons and the nucleus.
- For the halogens (group 7), the chemical reactivity (ability to gain electrons) decreases down the group. This is explained by the decrease in electrostatic attraction, due to greater distance between the valence electrons and the nucleus.
- The oxides of the period 3 elements show a distinct periodicity of properties. From left to right across period 3, ionic nature decreases and the basicity of the oxides also decreases.
- Generally speaking, metallic oxides are basic, non-metallic oxides are acidic and aluminium oxide is amphoteric.

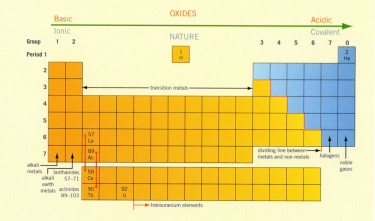

## Chapter 3 Review questions

1. Middle secondary school students are taught that non-metals 'want' to gain electrons to complete their outer shells, while metals 'want' to lose electrons. How does the concept of core charge explain more precisely why non-metal atoms attract electrons more strongly than metal atoms?

2. Give the name, symbol and ground state electron configuration of the element:
   a. in group 2, period 3
   b. with an atomic number 1 greater than that of oxygen
   c. that is the noble gas in period 2
   d. that is the smallest metal in group 3.

3. The electron configuration of a chemical species is 2,8,8.
   a. State the element with this configuration.
   b. State an element with similar properties to the element in part **a**.
   c. Under what circumstances could this electron configuration belong to an element in group 6?
   d. Under what circumstances could this electron configuration belong to an element in group 2?
   e. Is it possible for this electron configuration to belong to an element in period 2?

4. Define the term *electronegativity*.

5. a. State and explain the trend in electronegativities across a period.
   b. State and explain the trend in electronegativities down a group.

6. Consider these elements of period 3:
   Al   Cl   Mg   Na   P   S   Si
   a. Identify each of these elements as a metal or a non-metal.
   b. List these elements in order of increasing electronegativity
   c. List these elements in order of increasing atomic radius.

7  Compare the radii of each of the following pairs of particles and explain the difference between:
   a  a sodium ion and a sodium atom
   b  a chloride ion and a chlorine atom
   c  a chloride ion and a sodium atom.

8  a  List the following in order of increasing size (smallest to largest):
      $Cl^-$, $Cl^+$, $Cl$
   b  Explain why you have chosen this order.

9  Elements exhibit a wide range of chemical and physical properties. Many of these properties can be predicted from the position of the element in the periodic table.

   Four elements are given as examples in the table below. Copy this table and complete it by adding your answers to parts **a** and **b**.

   |  | Electron configuration | Electronegativity |
   |---|---|---|
   | Sodium |  |  |
   | Aluminium |  |  |
   | Phosphorus |  |  |
   | Chlorine |  |  |

   a  State the electron configuration of each element.
   b  Rank the elements in order of electronegativity (1 = highest).

10 a  Explain why the chemical properties of barium and magnesium are similar.
   b  Compare the physical properties of barium and magnesium and explain the differences.

11 For each of the following pairs of elements, compare the electronegativities and state which element would be expected to have the highest electronegativity:
   a  O or S
   b  P or Al
   c  B or Ne
   d  N or Si

12 Explain each of the following statements concisely:
   a  There are eight elements in period 2.
   b  Iodine has a lower electronegativity than fluorine.
   c  Although sulfur has a great atomic mass than sodium, its atomic radius is smaller than that of sodium.

13 Element X is more electronegative and smaller than nitrogen. X has valence electrons in the second shell and forms ions with a charge of 1. Deduce the identity of X and state its name.

14 a  Define the term *first ionization energy*.
   b  Using potassium as an example, state the equation for the reaction that occurs during this ionization process.

15 a  State and explain the trend in first ionization energies across a period.
   b  State and explain the trend in first ionization energies down a group.

16 Below is an outline of the periodic table with a number of elements identified by letters that are *not* their normal symbols. Use these letters when answering the following questions.

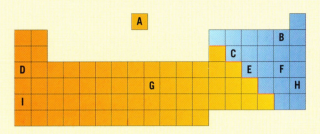

   a  Select a halogen and an alkali metal in the same period.
   b  Which element has the highest electronegativity?
   c  Which element has the lowest first ionization energy?
   d  Which element has an electron configuration ending in 4?
   e  Which element does not readily form bonds?
   f  Select a member of the transition metal series.

17 Identify each of the following elements:
   a  the element with one more proton than titanium
   b  the noble gas that is lighter than neon
   c  the element closest in mass to iodine
   d  the more electronegative element of oxygen and sulfur
   e  the more electronegative element of magnesium and sulfur
   f  the larger atom of aluminium and fluorine

18  The alkali metals all react with water.
    a  Describe what happens as each of lithium, sodium and potassium reacts with water.
    b  State the difference in the reactivity of these alkali metals with water.
    c  State the equation for each of the reactions in part **a**.
    d  Describe what you could do experimentally to show what the product(s) are.

19  Compare the melting points of each of the following pairs of elements and explain the difference:
    a  aluminium and sodium
    b  magnesium and chlorine
    c  bromine and fluorine

20  a  State the trend in reactivity of the halogens down group 7.
    b  Identify the halide ion(s) with which chlorine will react.
    c  State balanced chemical equations for the reaction(s) that occur.

 www.pearsoned.com.au/schools

Weblinks are available on the Chemistry: For use with the IB Diploma Programme Standard Level Companion Website to support learning and research related to this chapter.

## Chapter 3 Test

### Part A: Multiple-choice questions

1  For which element are the group number and the period number the same?
   A  Li
   B  Be
   C  B
   D  Mg
   © IBO 2004, May P1 Q7

2  Which property increases with increasing atomic number for both the alkali metals and the halogens?
   A  atomic radius
   B  electronegativity
   C  ionization energy
   D  melting point
   © IBO 2002, May P1 Q9

3  An element is in group 3 and period 2. How many electrons are present in its outer shell?
   A  2
   B  3
   C  5
   D  6
   © IBO 2002, May P1 Q8

4  Which of the following displacement reactions is possible?
   A  $Br_2(aq) + 2Cl^-(aq) \rightarrow 2Br^-(aq) + Cl_2(aq)$
   B  $I_2(aq) + 2Cl^-(aq) \rightarrow 2I^-(aq) + Cl_2(aq)$
   C  $Cl_2(aq) + 2I^-(aq) \rightarrow 2Cl^-(aq) + I_2(aq)$
   D  $I_2(aq) + 2Br^-(aq) \rightarrow 2I^-(aq) + Br_2(aq)$
   © IBO 2002, Nov P1 Q9

5  On descending a group in the periodic table:
   I    All the atoms have the same number of valence electrons.
   II   Ionization energy increases.
   III  Electronegativity decreases.

   Which of the above statements are correct?
   A  I and II only
   B  I and III only
   C  II and III only
   D  I, II and III
   © IBO 2002, Nov P1 Q8

6  A potassium atom has a larger atomic radius than a sodium atom. Which statement about potassium correctly explains this difference?
   A  It has a larger nuclear charge.
   B  It has a lower electronegativity.
   C  It has more energy levels occupied by electrons.
   D  It has a lower ionization energy.
   © IBO 2005, Nov P1 Q8

**7** Which properties are typical of most non-metals in period 3 (Na to Ar)?

  **I** They form ions by gaining one or more electrons.
  **II** They are poor conductors of heat and electricity.
  **III** They have high melting points.

  **A** I and II only
  **B** I and III only
  **C** II and III only
  **D** I, II and III

  © IBO 2005, Nov P1 Q7

**8** Which reaction results in the formation of a coloured substance?

  **A** $2Li(s) + 2H_2O(l) \rightarrow 2LiOH(aq) + H_2(g)$
  **B** $2Na(s) + Cl_2(g) \rightarrow 2NaCl(s)$
  **C** $Cl_2(g) + 2NaI(aq) \rightarrow 2NaCl(aq) + I_2(s)$
  **D** $Ag^+(aq) + Cl^-(aq) \rightarrow AgCl(s)$

  © IBO 2006, May P1 Q8

**9** Which statement is correct for a periodic trend?

  **A** Ionization energy increases from Li to Cs.
  **B** Melting point increases from Li to Cs.
  **C** Ionization energy increases from F to I.
  **D** Melting point increases from F to I.

  © IBO 2006, May P1 Q7

**10** Which equation represents the first ionization energy of fluorine?

  **A** $F(g) + e^- \rightarrow F^-(g)$
  **B** $F^-(g) \rightarrow F(g) + e^-$
  **C** $F^+(g) \rightarrow F(g) + e^-$
  **D** $F(g) \rightarrow F^+(g) + e^-$

  © IBO 2006, Nov P1 Q7
  (10 marks)

www.pearsoned.com.au/schools

For more multiple-choice test questions, connect to the Chemistry: For use with the IB Diploma Programme Standard Level Companion Website, and select Chapter 3 Review Questions.

## Part B: Short-answer questions

**1 a i** Explain why the ionic radius of chlorine is less than that of sulfur.

  (2 marks)

  **ii** Explain what is meant by the term *electronegativity* and explain why the electronegativity of chlorine is greater than that of bromine.

  (3 marks)

**b** For each of the following reactions in aqueous solution, state **one** observation that would be made, and deduce the equation.

  **i** The reaction between chlorine and sodium iodide.
  (2 marks)

  **ii** The reaction between sodium metal and water.
  (2 marks)

  adapted from IBO 2006, Nov P2 Q4a,b

**2 a i** Define the term *first ionization energy*.
  (2 marks)

  **ii** Write an equation, including state symbols, for the process occurring when measuring the first ionization energy of aluminium.
  (1 mark)

**b** Explain why the first ionization energy of magnesium is greater than that of sodium.
  (2 marks)

**c** Lithium reacts with water. Write an equation for the reaction and state **two** observations that could be made during the reaction.
  (3 marks)

  adapted from IBO 2005, Nov P2 Q4

**3 a** Boron and aluminium are in the same group in the periodic table. Based on their electron configurations, explain why the first ionization energy of boron is greater than that of aluminium.
  (2 marks)

**b** Aluminium and silicon are in the same period. Explain why the first ionization energy of silicon is greater than that of aluminium.
  (2 marks)

  © IBO 2001, Nov P2 Q3

**4 a** Define, in words or with an equation, the first ionization energy of Na.
  (2 marks)

**b** State how the first ionization energy varies down group 1.
  (1 mark)

**c** Li, Na and K react with water. Which of the three reactions will be the most vigorous? Explain this at an atomic level.
  (2 marks)

  adapted from IBO 2000, May P2 Q2

## Part C: Data-based questions

1. Atomic radii are given for the halogens (F → At) and the period 2 elements (Li → F) below.

| Element | Atomic radius ($10^{-12}$ m) |
|---|---|
| F | 58 |
| Cl | 99 |
| Br | 114 |
| I | 133 |
| At | 140 |

| Element | Atomic radius ($10^{-12}$ m) |
|---|---|
| Li | 152 |
| Be | 112 |
| B | 88 |
| C | 77 |
| N | 70 |
| O | 66 |
| F | 58 |

  a  Explain the trend in atomic radii of the halogens.
  (2 marks)

  b  Explain the trend in atomic radii of the period 2 elements.
  (2 marks)

  adapted from IBO 2006, SL Nov P2 Q4a, b

2. Describe and explain the trends in the graph below, which shows the melting points of the elements of period 3.

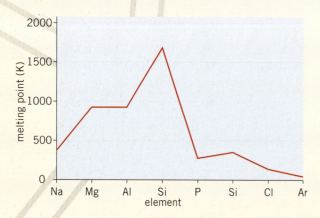

adapted from IBO 2000, Nov P2 Q5
(3 marks)

## Part D: Extended-response question
adapted from IBO SL 1998, May P2 Q5

The physical and chemical properties of elements and their compounds vary across and down the periodic table.

  a  State and explain the variation in atomic radius that occurs:
    i  across a period, such as Na → Cl
    (3 marks)
    ii  down a group, such as Li → Cs
    (2 marks)

  b  Write the formulas of the oxides of sodium, phosphorus and sulfur and describe their acid/base nature.
  (6 marks)

  c  Elements on the left of the periodic table, e.g. Na, are good electrical conductors; those on the right of the periodic table, e.g. S, conduct electricity poorly or not at all.

  Account for these results in terms of the bonding present in these two elements.
  (4 marks)

Total marks 50

# 4 QUANTITATIVE CHEMISTRY

- 1.0 mole of $O_2(g)$ at 0°C and 1.0 atm occupies a volume of 22.4 $dm^3$
- 58.5 g of NaCl(s) is 1.0 mole of NaCl
- 1.0 mole of He(g) occupies a volume of 22.4 $dm^3$ at 0°C and 101.3 kPa
- 1.0 mole of Ag(s) contains $6.02 \times 10^{23}$ Ag atoms
- 1 mole of paper sheets ($6.02 \times 10^{23}$) would be so tall that it would reach from here to the Sun more than a million times!
- A 1.0 mol $dm^{-3}$ NaOH solution contains 40 g of NaOH dissolved in 1.0 $dm^3$ of water

## Chapter overview

This chapter covers the IB Chemistry syllabus Topic 1: Quantitative Chemistry.

### By the end of this chapter, you should be able to:

- recall that the mole is a measure of an amount of substance and that one mole always constitutes the same number of particles
- recall that Avogadro's constant ($6.02 \times 10^{23}$) is equal to the number of particles in one mole of any substance
- use the relationship $n = \dfrac{N}{L}$ to calculate any of the three variables, given the other two variables
- accurately apply such terms as *relative atomic mass*, *relative molecular mass* and *molar mass*
- use the relationship $n = \dfrac{m}{M}$ to calculate any of the three variables, given the other two variables
- use mass or percentage composition data to determine the empirical formulas of compounds
- recall the relationship between empirical and molecular formulas and, given the relevant data, interconvert between them
- deduce chemical equations, given the reactants and products
- relate the balanced equation for a reaction to the amounts of substances reacting
- calculate percentage yield from the theoretical and experimental yields
- recall and use the appropriate steps in carrying out a stoichiometric calculation
- recognize when a problem involves an excess reagent and be able to determine which reactant is limiting
- carry out calculations using various gas laws and the ideal gas equation
- use the relationship $c = \dfrac{n}{V}$ to calculate any of the three variables, given the other two variables
- use the correct units when carrying out calculations involving masses, molar masses, amounts of substance and volumes of solutions.

**IBO Assessment statements 1.1.1 to 1.5.2**

In previous chapters, we have concentrated on developing knowledge of some fundamental, qualitative aspects of chemistry such as atomic structure, bonding and the patterns found in the periodic table. Chemistry is, first and foremost, an experimental and practical science; it can be applied to a wide range of diverse situations. For example, if a chemical engineer is required to produce a batch of several tonnes of aspirin, precisely how much of each reactant is needed to produce the maximum yield of product with the minimum wastage? The basis of such quantitative aspects of chemistry will be considered in this chapter.

## 4.1 THE MOLE CONCEPT AND AVOGADRO'S CONSTANT

The invention of the mass spectrometer revolutionized our understanding of the atom because it provided an accurate method to compare the masses of atoms, while taking into account the various proportions of different isotopes. It gave chemists the basis on which to invent a relative scale of atomic masses that allows them to perform useful and accurate calculations involving masses of reactants and products.

A relative scale is one in which all measurements are compared to one standard or reference measure. For example, if a student is 16 years of age and has a father who is 48 and a younger sister who is 8 years old, he could invent a relative scale of ages of these members of his family. According to this relative scale, he would be the reference and so would be given an arbitrary value of 1 unit, his father would have a value of 3 units (he is three times older than the reference person) and the younger sister 0.5 units (she is half the age of the reference). What would be the relative scale age of the student's mother if she was 40 years of age? Note that relative scales have no units as such because they are merely comparisons of one quantity with another.

The advantage of relative scales is that very large or very small numbers can be compared more easily. To generate the relative scale of atomic masses, chemists chose the most abundant isotope of the element carbon, the carbon-12 isotope ($^{12}$C), and assigned it a relative mass of exactly 12 units. The element carbon was chosen as the reference for a number of important reasons:

- Carbon is very cheap and is widely available.
- It is relatively easy to isolate and purify this isotope.
- Carbon is not toxic in any way.

It was decided to assign a mass of 12 units to carbon, rather than 1 as might have been expected, as this number mirrors the mass number of the isotope. As protons and neutrons are the basic building blocks of atoms (in addition to the very light electrons), the relative atomic mass will closely parallel the number of these fundamental particles in the nucleus of the element. Thus, the lightest of all elements, hydrogen, will have a relative atomic mass of close to 1 (as most hydrogen atoms have only one proton in their nucleus). Hydrogen is approximately one-twelfth as massive as the standard carbon-12 atom. The most common isotope of magnesium, $^{24}$Mg, has 12 protons and 12 neutrons in its nucleus. This isotope will have a relative mass of approximately 24 and so will be about twice as massive as an atom of the carbon-12 isotope.

The fundamental difficulty with this system of relative masses is that it does not have a unit; we still require a connection between these tiny building blocks of matter and useable units of mass such as the gram and kilogram. This relationship is achieved by a concept called the mole.

A **mole** is defined as the amount of substance that contains as many elementary particles as there are atoms in precisely 12 grams of the carbon-12 isotope.

The mole is defined in terms of the number of atoms present in 12 grams of carbon-12 isotope. If you imagine it was somehow possible to count these carbon-12 atoms, one at a time, and place them onto the weighing dish of electronic scales, eventually we will have counted sufficient atoms for their total mass to reach exactly 12 grams. We would now have an amount of substance, in this case of carbon-12 atoms, that would be equal to one mole. The actual number of atoms required is huge and has been determined as approximately $6.02 \times 10^{23}$.

This number is known as **Avogadro's constant**, (symbol, $L$) in honour of the Italian chemist and physicist Amedeo Avogadro. It is important to recognize that a mole is simply a number, just as a dozen is equal to 12 or a gross is equal to 144. We can have a dozen eggs and know that this means 12 eggs, or a dozen shoes and know that there are 12 shoes. The fact that we are counting different substances is immaterial to the fact that we have counted the same number of items. In the unlikely event that we wanted to purchase a mole of eggs from the local supermarket, we would need to purchase $6.02 \times 10^{23}$ eggs.

The number of elementary particles in one mole is known as Avogadro's constant ($L$) where:

$$L = 6.02 \times 10^{23} \text{ mol}^{-1}$$

Avogadro's constant is an almost unimaginably large number. Atoms are so small that it is impossible to see an individual atom, even with the most powerful of modern microscopic techniques. Indeed, if an atom was the size of a grain of sand, a mole of sand grains ($6.02 \times 10^{23}$ grains) would cover the entire land mass of Australia to a depth of approximately 2 metres!

Note that while we refer to one mole of particles, the units are written as mol, just as we might say that 20 grams of material has a mass of 20 g.

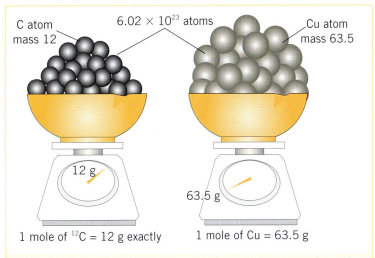

Figure 4.1.1 While one mole of carbon-12 has a mass of 12 g, the same amount of copper has a mass of 63.5 g. The number of atoms of each element is identical.

Figure 4.1.2 One mole of several common substances (table sugar, salt, carbon, copper and helium in the balloon).

**AS 1.1.1 Apply the mole concept to substances.** © IBO 2007

Figure 4.1.3 If an atom was the size of a grain of sand, a mole of atoms would cover the entire land mass of Australia to a depth of 2 metres.

# CHEM COMPLEMENT

## Whose number?

Avogadro's constant, or number, is known to all chemistry students—but was there really someone called Avogadro? His full name was Lorenzo Romano Amedeo Carlo Avogadro, his title Count of Quaregna and Cerrato. Born in Italy in 1776, he worked for three years as a lawyer before leaving the legal profession to study physics and mathematics. This path eventually led him to formulate his hypothesis, which we now know as Avogadro's law: under identical conditions of temperature and pressure, equal volumes of gases contain an equal number of molecules. Avogadro did not live to see his work recognized by the scientific community. He died in 1856 and it was not until five years after his death that the Italian chemist Stanislao Cannizzaro presented Avogadro's work and Avogadro finally received credit for it. It was much later in the 19th century before the actual number of molecules in one mole of a gas, $6.02 \times 10^{23}$, was determined. In German-speaking countries, this number is usually called the Loschmidt number, after Johann Josef Loschmidt (1821–1895), who made an initial estimate of the number in 1865. In most places, however, it is known as Avogadro's number, in spite of the fact that Avogadro never knew how big the number was!

Figure 4.1.4 Amedeo Avogadro (1776–1856).

PRAC 4.1
Determination of Avogadro's constant

### THEORY OF KNOWLEDGE

Consider the following quote by Sir William Osler (1849–1919), a Canadian physician:

> In Science the credit goes to the man who convinces the world, not to the man to whom the idea first occurred.

Avogadro did not live to see his work recognized by the scientific community. Why do you think that many important contributions to science are often not recognized as such at the time? Is this unique to scientific discoveries?

## Calculating numbers of particles

How many atoms in a mole of molecules?

In the well-loved children's game of 'Beetle' a 'beetle' is made up of 1 body, 6 legs, 1 head, 2 eyes and 2 antennae. Together, these 12 body parts make just 1 'beetle'. If we had 10 'beetles', we would have 120 body parts: 10 bodies, 60 legs, 10 heads, 20 eyes and 20 antennae.

Similarly, one molecule of carbon dioxide, $CO_2$, is made up of 3 atoms: 2 oxygen atoms and 1 carbon atom. If we have 10 carbon dioxide molecules, we have 30 atoms (10 carbon and 20 oxygen atoms). Taking the numbers further, if we have 1 dozen carbon dioxide molecules we would have 3 dozen atoms, and if we have 1 mole of carbon dioxide molecules we would have 3 mole of atoms: 1 mole of carbon atoms and 2 mole of oxygen atoms.

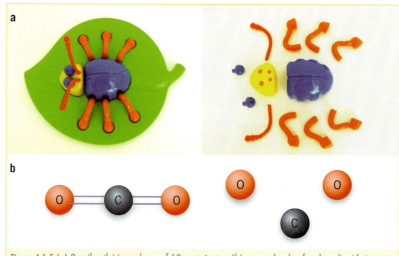

Figure 4.1.5 (a) One 'beetle' is made up of 12 parts just as (b) one molecule of carbon dioxide is made up of three atoms.

## Worked example 1

Calculate the number of mole of the stated particles in each of the following.

**a** Atoms of sulfur in 0.250 mole of sulfur trioxide, $SO_3$

**b** Atoms of oxygen in 0.250 mole of sulfur trioxide, $SO_3$

**c** Atoms (of both types) in 0.250 mole of sulfur trioxide, $SO_3$

### Solution

**a** There is 1 S atom in 1 $SO_3$ molecule
∴ in 0.250 mole of $SO_3$ there is 0.250 mole of S atoms

**b** There are 3 O atoms in 1 $SO_3$ molecule
∴ in 0.250 mole of $SO_3$ there are $3 \times 0.250$
$= 0.750$ mole of S atoms

**c** There are 4 atoms in 1 $SO_3$ molecule
∴ in 0.250 mole of $SO_3$ there are $4 \times 0.250$
$= 1.00$ mole of atoms

Knowing that the number of particles present in one mole of any substance is $6.02 \times 10^{23}$ (Avogadro's constant), it is possible to calculate the number of particles present in a given amount of substance.

**AS 1.1.2 Determine the number of particles and the amount of substance (in moles).** © IBO 2007

Notice the difference between a *number* of particles and an *amount* of substance. If you are asked to calculate a *number* of particles, then you are calculating the quantity represented by $N$ and you will obtain a number with no units. If you are asked to calculate an *amount* of substance, then you are trying to find a *number of mole* of the substance. This is the quantity represented by $n$ and will have the units 'mol'.

$$n = \frac{N}{L}$$

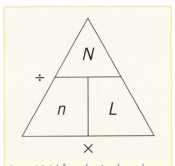

**Figure 4.1.6** A formula triangle can be used to make it easier to rearrange this formula.

where  $n$ = number of mole (amount)
$N$ = number of particles
$L$ = Avogadro's constant

## Worked example 2

Calculate the number of particles present in each of the following.

**a** Atoms of helium in 3.00 mol of He

**b** Molecules of sulfur trioxide in 0.125 mol of $SO_3$

**c** Oxygen atoms in 0.0855 mol of magnesium sulfate ($MgSO_4$)

**WORKSHEET 4.1** Mole calcualtions

### Solution

**a** $N = n \times L$, where $n(He) = 3.00$ mol, $L = 6.02 \times 10^{23}$
∴ $N(He) = 3.00 \times 6.02 \times 10^{23} = 1.81 \times 10^{24}$

**b** $N = n \times L$, where $n(SO_3) = 0.125$ mol, $L = 6.02 \times 10^{23}$
∴ $N(SO_3) = 0.125 \times 6.02 \times 10^{23} = 7.53 \times 10^{22}$

**c** $N = n \times L$, where $n(MgSO_4) = 0.0855$ mol, $L = 6.02 \times 10^{23}$
∴ $N(MgSO_4) = 0.0855 \times 6.02 \times 10^{23} = 5.15 \times 10^{22}$
∴ $N(O) = 4 \times 5.15 \times 10^{22} = 2.06 \times 10^{23}$

### Worked example 3

Calculate the amount of substance (in mol) present in each of the following.

**a** Molecules in $3.01 \times 10^{23}$ molecules of hydrogen gas ($H_2$)

**b** Atoms in $5.0 \times 10^{21}$ atoms of zinc

**c** Ions in $2.55 \times 10^{24}$ formula units of NaCl

**Solution**

**a** $N = n \times L$, where $N = 3.01 \times 10^{23}$, $n(H_2) = ?$, $L = 6.02 \times 10^{23}$

$n(H_2) = \dfrac{N}{L} = \dfrac{3.01 \times 10^{23}}{6.02 \times 10^{23}} = 0.50$ mol of molecules

**b** $N = n \times L$, where $N = 5.0 \times 10^{21}$, $n(Zn) = ?$, $L = 6.02 \times 10^{23}$

$n(Zn) = \dfrac{N}{L} = \dfrac{5.0 \times 10^{21}}{6.02 \times 10^{23}} = 8.3 \times 10^{-3}$ mol of atoms

**c** $N = n \times L$, where $N = 2.55 \times 10^{24}$, $n(ions) = 2n(NaCl) = ?$, $L = 6.02 \times 10^{23}$

$n(NaCl) = \dfrac{N}{L} = \dfrac{2.55 \times 10^{24}}{6.02 \times 10^{23}} = 4.24$

$n(ions) = 2n(NaCl) = 2 \times 4.24 = 8.48$ mol of ions

### Section 4.1 Exercises

**1** Calculate the number of mole of the stated particles in 1.00 mole of sulfuric acid ($H_2SO_4$).

    **a** Atoms of sulfur     **b** Atoms of hydrogen

    **c** Atoms of oxygen     **d** Total number of atoms

**2** Calculate the number of mole of the stated particles in each of the following.

    **a** Magnesium ions in 1.25 mole of magnesium nitrate ($Mg(NO_3)_2$)

    **b** Sulfate ions in 2.50 mole of copper(II) sulfate ($CuSO_4$)

    **c** Sodium ions in 3.25 mole of sodium sulfate ($Na_2SO_4$)

    **d** Aluminium and sulfate ions in 1.25 mole of aluminium sulfate ($Al_2(SO_4)_3$)

**3** Calculate the number of mole of the stated particles in each of the following.

    **a** Nitrogen atoms in 1.50 mole of ammonia ($NH_3$)

    **b** Carbon atoms in 2.00 mole of propane ($C_3H_8$)

    **c** Propane molecules in 0.25 mole of propane ($C_3H_8$)

    **d** Carbon and hydrogen atoms in 1.25 mole of pentane ($C_5H_{12}$)

**4** Determine the number of particles, identified in brackets, present in each of the following.

    **a** 3.50 mol of ammonia molecules ($NH_3$ molecules)

    **b** 0.941 mol of xenon atoms (Xe atoms)

    **c** $3.22 \times 10^{-3}$ mol of manganese nitrate, $Mn(NO_3)_2$ ($Mn^{2+}$ ions)

    **d** $5.661 \times 10^{-5}$ mol of lactic acid ($C_3H_6O_3$ molecules)

**5** Calculate the amount of substance (in mol) present in each of the following.

    **a** $6.02 \times 10^{21}$ molecules of ethyne ($C_2H_2$)

    **b** $1.44 \times 10^{24}$ atoms of neon (Ne)

    **c** $8.93 \times 10^{17}$ molecules of carbon dioxide ($CO_2$)

    **d** 1000 atoms of mercury (Hg)

## 4.2 CALCULATIONS OF MASS AND NUMBER OF MOLE

The **relative atomic mass**, $A_r$, of an element is defined as the weighted mean of the masses of the naturally occurring isotopes, on the scale in which the mass of an atom of the carbon-12 isotope ($^{12}$C) is 12 units exactly. The calculation of relative atomic mass is discussed in chapter 1, Atomic Structure, section 1.2.

**1.2.1**
Define the terms *relative atomic mass* ($A_r$) and *relative molecular mass* ($M_r$). © IBO 2007

Just as relative atomic mass refers to atoms, relative molecular mass refers to molecules. To determine the relative molecular mass, add together the relative atomic masses of the atoms present in the molecule.

The **relative molecular mass**, $M_r$, of a molecule is the sum of the relative atomic masses of the constituent elements, as given in the molecular formula.

The technique used to calculate **relative formula mass** is identical to that used for relative molecular mass. The only difference is that relative formula mass is used for ionic compounds that cannot be described as molecules.

Formula mass calculation

### Worked example 1
Determine the relative molecular mass of the following molecules.

**a** Water (H$_2$O)  **b** Methane (CH$_4$)  **c** Sulfuric acid (H$_2$SO$_4$)

### Solution
**a** $M_r(H_2O) = 2A_r(H) + A_r(O)$
$= 2 \times 1.01 + 16.00$
$= 18.02$

**b** $M_r(CH_4) = A_r(C) + 4A_r(H)$
$= 12.01 + 4 \times 1.01$
$= 16.05$

**c** $M_r(H_2SO_4) = 2A_r(H) + A_r(S) + 4A_r(O)$
$= 2 \times 1.01 + 32.06 + 4 \times 16.00$
$= 98.08$

### Worked example 2
Determine the relative formula mass of the following ionic compounds.

**a** Sodium chloride (NaCl)
**b** Magnesium hydroxide (Mg(OH)$_2$)
**c** Aluminium carbonate (Al$_2$(CO$_3$)$_3$)

### Solution
**a** Relative formula mass of NaCl $= A_r(Na) + A_r(Cl)$
$= 22.99 + 35.45$
$= 58.44$

**b** Relative formula mass of Mg(OH)$_2$ $= A_r(Mg) + 2(A_r(O) + A_r(H))$
$= 24.31 + 2 \times (16.00 + 1.01)$
$= 58.33$

**c** Relative formula mass of Al$_2$(CO$_3$)$_3$ $= 2A_r(Al) + 3(A_r(C) + 3A_r(O))$
$= 2 \times 26.98 + 3 \times (12.01 + 3 \times 16.00)$
$= 233.99$

**AS 1.2.2**
**Calculate the mass of one mole of a species from its formula.**
© IBO 2007

## Molar mass

By defining relative atomic mass and the mole in terms of the same reference, an atom of the carbon-12 isotope, an important link is made so that the mass of one mole of any substance is equivalent to its relative atomic (or molecular or formula) mass, measured in grams.

The mass of one mole of any substance is known as **molar mass** (symbol **M**), where the molar mass is equal to the relative atomic (or molecular or formula) mass measured in units of grams per mole (g mol$^{-1}$).

### Worked example 3

Determine the mass of one mole of each of the following substances.

**a** Iron atoms (Fe)

**b** Carbon dioxide ($CO_2$)

**c** Nitric acid ($HNO_3$)

**d** Hydrated copper(II) sulfate ($CuSO_4.5H_2O$)

### Solution

**a** $A_r(Fe) = 55.85$ ∴ one mole of Fe atoms = 55.85 g

**b** $M_r(CO_2) = A_r(C) + 2A_r(O)$
$= 12.01 + 2 \times 16.00$
$= 44.01$

∴ one mole of $CO_2$ = 44.01 g

**c** $M_r(HNO_3) = A_r(H) + A_r(N) + 3A_r(O)$
$= 1.01 + 14.01 + 3 \times 16.00$
$= 63.02$

∴ one mole of $HNO_3$ = 63.02 g

**d** $M_r(CuSO_4.5H_2O) = A_r(Cu) + A_r(S) + 4A_r(O) + 5 \times (2A_r(H) + A_r(O))$
$= 63.55 + 32.06 + 4 \times 16.00 + 5 \times (2 \times 1.01 + 16.0)$
$= 249.71$

∴ one mole of $CuSO_4.5H_2O$ = 249.71 g

**AS 1.2.3**
**Solve problems involving the relationship between the amount of substance in moles, mass and molar mass.**
© IBO 2007

## Calculations involving masses and the mole

Using the idea of molar mass, we can determine the amount (in mol) of a particular substance in a given mass. Notice that the term *amount* is used in this context to refer specifically to the number of mole of a substance. It would not be used for mass or any other quantity.

$$n = \frac{m}{M}$$

where  $n$ = number of mole (mol)
 $m$ = mass (g)
 $M$ = molar mass (g mol$^{-1}$)

This formula can be transposed to calculate any of the three variables, given the other two values. When performing calculations involving quantitative data, you should always write the relevant formula and record the information provided in terms of the appropriate symbols.

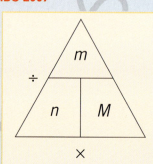

Figure 4.2.1 A formula triangle can be used to make it easier to rearrange this formula.

## Worked example 4

Determine the number of mole present in:

**a** 127.1 g of copper atoms

**b** 11.0 g of carbon dioxide molecules ($CO_2$).

### Solution

**a** $n = \dfrac{m}{M}$, where $n = ?$, $m = 127.1$ g, $M(Cu) = 63.55$ g mol$^{-1}$

$\therefore n(Cu) = \dfrac{127.1}{63.55} = 2.000$ mol

**b** $n = \dfrac{m}{M}$, where $n = ?$, $m = 11.0$ g, $M(CO_2) = 44.01$ g mol$^{-1}$

$\therefore n(CO_2) = \dfrac{11.0}{44.01} = 0.250$ mol

## Worked example 5

Determine the mass of substance present in:

**a** 2.00 mole of iron

**b** 0.500 mole of nitric acid ($HNO_3$).

### Solution

**a** $n = \dfrac{m}{M}$, where $n = 2.00$ mol, $m = ?$, $M(Fe) = 55.85$ g mol$^{-1}$

First we must transpose the formula to make $m$ the subject:

$n = \dfrac{m}{M} \therefore m = n \times M$

$\therefore m(Fe) = 2.00 \times 55.85 = 112$ g

**b** $n = \dfrac{m}{M}$, where $n = 0.500$ mol, $m = ?$, $M(HNO_3) = 63.02$ g mol$^{-1}$

$\therefore m(HNO_3) = n \times M = 0.500 \times 63.02 = 31.5$ g

## Worked example 6

It is determined that 2.117 g of a shiny metallic element is 0.04071 mole of substance. Calculate the molar mass of the element and so determine its identity.

### Solution

$n = \dfrac{m}{M}$, where $n = 0.04071$ mol, $m = 2.117$ g, $M(\text{element}) = ?$

First, we must transpose the formula to make $M$ the subject:

$n = \dfrac{m}{M} \therefore M = \dfrac{m}{n}$

$\therefore M(\text{element}) = \dfrac{2.117}{0.04071} = 52.00$ g mol$^{-1}$

The molar mass of the element is 52.00 g mol$^{-1}$. From the periodic table, it can be determined that the element is chromium.

### Finding the mass of a small number of atoms or molecules

There are two formulas that we have encountered so far which are linked to the amount of material in mol:

$$n = \frac{m}{M} \text{ and } n = \frac{N}{L}$$

If we need to find the mass of a small number of particles—perhaps 100 atoms of copper, or even one molecule of water—we can combine these formulas to perform the calculation in one step. (Two steps can be used if you wish to find the value of $n$.)

$$\frac{m}{M} = \frac{N}{L}$$

$$\therefore m = \frac{N \times M}{L}$$

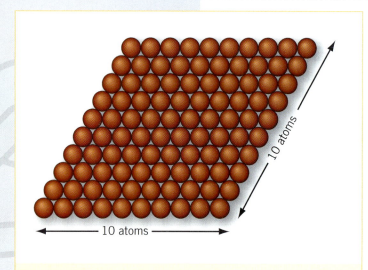

Figure 4.2.2 The mass of 100 atoms of copper is $1.06 \times 10^{-20}$ g.

### Worked example 7

Determine the mass of 100 atoms of copper.

**Solution**

$$m = \frac{N \times M}{L}$$

where $N = 100$, $M(\text{Cu}) = 63.55$ g mol$^{-1}$, $L = 6.02 \times 10^{23}$

$$m(\text{Cu}) = \frac{100 \times 63.55}{6.02 \times 10^{23}}$$
$$= 1.06 \times 10^{-20} \text{ g}$$

## Section 4.2 Exercises

1. Determine the relative molecular mass of the following molecules.
   - **a** $HNO_3$
   - **b** $P_4$
   - **c** $H_2S_2O_3$
   - **d** $C_{12}H_{22}O_{11}$
   - **e** $N_2O_4$

2. Determine the relative formula mass of the following ionic compounds.
   - **a** $Fe_2O_3$
   - **b** $NiSO_4$
   - **c** $KMnO_4$
   - **d** $Al_2(SO_3)_3$
   - **e** $BaCl_2.2H_2O$

3. Compare the terms relative formula mass and relative molecular mass.

4. Determine the mass of one mole of each of the following substances.
   - **a** Copper atoms (Cu)
   - **b** Nitrogen monoxide molecules (NO)
   - **c** Magnesium carbonate ($MgCO_3$)
   - **d** Ozone molecules ($O_3$)
   - **e** Zinc oxide (ZnO)
   - **f** Ethanol molecules ($C_2H_5OH$)

5. Determine the molar mass of each of the substances in question 4.

6 Calculate the amount of substance (in mol) present in each of the following.
   a  48.6 g of magnesium
   b  3.65 g of hydrogen chloride (HCl)
   c  36 g of glucose ($C_6H_{12}O_6$)
   d  7.92 g of sodium phosphate ($Na_3PO_4$)
   e  5.30 kg of gold
   f  150 kg of iron(III) oxide ($Fe_2O_3$)

7 Thiamine, better known as vitamin B1, is essential to the effective metabolism of carbohydrates in the human body. The chemical formula of thiamine is $C_{12}H_{17}ON_4SCl$. Calculate the number of mole of thiamine present in the recommended daily allowance of 0.112 g.

8 Calculate the mass of each of the following.
   a  0.128 mol of cadmium atoms
   b  0.0382 mol of tin(II) sulfate ($SnSO_4$)
   c  4.726 mol of trichloromethane ($CHCl_3$)
   d  500 mol of oleum ($H_2S_2O_7$)
   e  $9.29 \times 10^{-4}$ mol of ascorbic acid ($C_6H_8O_6$)
   f  $7.712 \times 10^{-3}$ mol of arsenic

9 0.08850 mol of a gaseous oxide of carbon has a mass of 2.479 g. Calculate the molar mass of the substance and so determine the formula of the gas.

10 A certain elemental chloride ($XCl_3$) of 0.435 mole has a mass of 71.895 g. Determine the molar mass of the element and thus determine its name.

11 0.03845 mol of a hydrocarbon compound has a mass of 2.775 g. Determine the molar mass of this hydrocarbon.

12 Determine the number of particles present in:
   a  50 g of graphite (pure carbon)
   b  2.0 dm$^3$ (2000 g) of water
   c  10 mg of niacin ($C_6H_4NO_2$) (1 mg = $10^{-3}$g)
   d  50 tonnes of iron (1 tonne = $10^6$ g)

13 Determine the mass of:
   a  1000 atoms of silver
   b  10 atoms of potassium
   c  2 molecules of methane, $CH_4$
   d  1 molecule of water, $H_2O$
   e  1 molecule of glucose, $C_6H_{12}O_6$

14 Vitamin A is chemically known as retinol ($C_{20}H_{30}O$). Carrots do not contain retinol, but they do have substantial amounts of β-carotene, which the body readily converts into retinol. If an average carrot contains the equivalent of 2.4 mg of retinol, determine the:
   a  amount (in mol) of retinol produced from a carrot
   b  number of retinol molecules produced
   c  number of carbon atoms present in this amount of retinol.

## 4.3 EMPIRICAL AND MOLECULAR FORMULAS

One of the many useful applications of the mole concept is to assist in the determination of the formula of newly synthesized compounds. While this is a long and involved process, one of the first steps is to determine the elements present and their **percentage composition** by mass.

### Determining the percentage composition by mass of a compound

To determine the percentage composition by mass of a compound, given its formula, divide the relative atomic mass of each element present by the relative molecular (or formula) mass of the whole compound and express the result as a percentage.

Alternatively, if no formula is given, divide the mass of each element present by the mass of the whole sample and express the result as a percentage.

### Worked example 1

Determine the percentage composition by mass of each element in the following compounds:

**a** sodium chloride, NaCl

**b** potassium carbonate, $K_2CO_3$

### Solution

**a** $M_r(NaCl) = 22.99 + 35.45 = 58.44$

$\%\ Na\ in\ NaCl = \dfrac{A_r(Na)}{M_r(NaCl)} \times \dfrac{100}{1} = \dfrac{22.99}{58.44} \times \dfrac{100}{1} = 39.34\%$

$\%\ Cl\ in\ NaCl = \dfrac{A_r(Cl)}{M_r(NaCl)} \times \dfrac{100}{1} = \dfrac{35.45}{58.44} \times \dfrac{100}{1} = 60.66\%$

(or simply $100 - 39.36 = 60.66\%$)

**b** $M_r(K_2CO_3) = 2 \times 39.10 + 12.01 + (16.00 \times 3) = 138.21$

$\%\ K\ in\ K_2CO_3 = \dfrac{2A_r(K)}{M_r(K_2CO_3)} \times \dfrac{100}{1} = \dfrac{78.20}{138.21} \times \dfrac{100}{1} = 56.58\%$

$\%\ C\ in\ K_2CO_3 = \dfrac{A_r(C)}{M_r(K_2CO_3)} \times \dfrac{100}{1} = \dfrac{12.01}{138.21} \times \dfrac{100}{1} = 8.69\%$

$\%\ O\ in\ K_2CO_3 = \dfrac{3 \times A_r(O)}{M_r(K_2CO_3)} \times \dfrac{100}{1} = \dfrac{48.00}{138.21} \times \dfrac{100}{1} = 34.73\%$

(or simply $100 - 56.58 - 8.69 = 34.73\%$)

### Worked example 2

A compound of 2.225 g, known to be a carbohydrate, was analysed and found to contain 0.890 g of carbon, 0.148 g of hydrogen and the rest was oxygen. Determine the percentage composition by mass of the compound.

### Solution

Mass of oxygen present in compound = 2.225 g − (0.890 + 0.148) g = 1.187 g

$\%\ C = \dfrac{0.890}{2.225} \times \dfrac{100}{1}$ $\qquad \%\ H = \dfrac{0.148}{2.225} \times \dfrac{100}{1}$ $\qquad \%\ O = \dfrac{1.187}{2.225} \times \dfrac{100}{1}$

$\quad\ \ = 40.0\%$ $\qquad\qquad\quad\ \ = 6.70\%$ $\qquad\qquad\quad\ \ = 53.3\%$

If the percentage composition by mass of a compound is known, it is possible to determine the empirical formula of the compound. In effect, the empirical formula is the mole ratio of elements in the compound. It provides information about the ratio in which atoms are bonded, without indicating the actual number of atoms present. To determine these actual numbers we must establish the molecular formula.

The **empirical formula** of a compound is defined as the simplest whole number ratio of atoms of different elements in the compound. The **molecular formula** of a compound is the actual number of atoms of different elements covalently bonded in a molecule.

**AS 1.2.4**
**Distinguish between the terms *empirical formula* and *molecular formula*.** © IBO 2007

## Determining empirical formulas

To determine the empirical formula of a compound, take the following steps.

1. Write the elements present in the compound as a ratio.
2. Underneath each element write its percentage composition (or mass proportion) to give a mass ratio.
3. Divide each percentage (or mass) by the relative atomic mass of the relevant element and calculate this value. This will give a molar ratio of elements in the compound.
4. Divide each ratio obtained by the smallest quotient to obtain a whole number ratio.
5. Express these ratios as an empirical formula.

**AS 1.2.5**
**Determine the empirical formula from the percentage composition or from other experimental data.** © IBO 2007

## Worked example 3

A compound of carbon and hydrogen is analysed and found to consist of 75.0% by mass carbon and 25.0% by mass hydrogen. Determine the empirical formula of the hydrocarbon.

### Solution

| | |
|---|---|
| C : H | Elements in compound |
| 75.0% : 25.0% | Percentage composition (If we assume there is 100 g of sample, this becomes a mass ratio.) |
| $\dfrac{75.0}{12.01} : \dfrac{25.0}{1.01}$ | Divide by $A_r$. |
| 6.245 : 24.75 | Determine the ratio. |
| $\dfrac{6.245}{6.245} : \dfrac{24.75}{6.245}$ | Simplify the ratio. |
| 1 : 3.96 | |
| 1 : 4 | Round up to obtain whole-number ratio. |

The empirical formula of the hydrocarbon is $CH_4$.

Empirical formula determination

PRAC 4.2
Determination of percentage composition and empirical formula of magnesium oxide

## Worked example 4

A student burns a weighed sample of magnesium in excess air to produce a white powder, magnesium oxide. If the original sample of magnesium weighs 1.204 g and the magnesium oxide produced has a mass of 1.996 g, determine the empirical formula of magnesium oxide.

### Solution

Mass of magnesium = 1.204 g
Mass of magnesium oxide = 1.996 g
∴ Mass of oxygen = 0.792 g

| Mg : O | Elements in compound |
|---|---|
| 1.204 : 0.792 | Mass ratio |
| $\dfrac{1.204}{24.31} : \dfrac{0.792}{16.00}$ | Divide by $A_r$ (molar ratio) |
| 0.0495 : 0.0495 | Determine ratio |
| $\dfrac{0.0495}{0.0495} : \dfrac{0.0495}{0.0495}$ | Simplify ratio |
| 1 : 1 | |

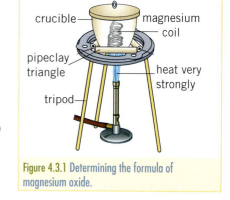

Figure 4.3.1 Determining the formula of magnesium oxide.

The empirical formula of magnesium oxide is MgO.

## Worked example 5

A bright orange crystalline compound of potassium, chromium and oxygen is analysed to determine its percentage composition by mass. A 1.308 g sample of the compound is found to consist of 0.348 g of potassium and 0.498 g of oxygen. Determine the empirical formula of the compound.

### Solution

Mass of chromium in compound = 1.308 − (0.348 + 0.498) = 0.462 g

| K : Cr : O | Elements in compound |
|---|---|
| 0.348 : 0.462 : 0.498 | Mass ratio |
| $\dfrac{0.348}{39.10} : \dfrac{0.462}{52.00} : \dfrac{0.498}{16.00}$ | Divide by $A_r$ (molar ratio) |
| 0.00890 : 0.00888 : 0.0311 | Determine ratio |
| $\dfrac{0.00890}{0.00888} : \dfrac{0.00888}{0.00888} : \dfrac{0.0311}{0.00888}$ | Simplify ratio |
| 1 : 1 : 3.5 | Convert to whole-number ratio |
| 2 : 2 : 7 | |

The empirical formula of the compound is $K_2Cr_2O_7$.

PRAC 4.3
Empirical formula of a hydrate

### Determining molecular formulas

The empirical formula of a compound is the simplest whole-number ratio of atoms in the substance. The molecular formula is defined as the actual number of atoms found in each molecule. It therefore follows that the molecular formula must be a multiple of the empirical formula of the same substance. Table 4.3.1 shows several examples. It is possible for the empirical and molecular formulas to be the same, as is the case with water and carbon dioxide.

To determine the molecular formula of a compound, you must know the empirical formula and the molar mass (or relative molecular mass) of the compound. The molar mass of the empirical formula will be in direct ratio to that of the actual molecular formula.

**AS 1.2.6
Determine the molecular formula when given both the empirical formula and experimental data. © IBO 2007**

**TABLE 4.3.1 A COMPARISON OF EMPIRICAL AND MOLECULAR FORMULAS**

| Compound | Molecular formula | Empirical formula |
|---|---|---|
| Ethene | $C_2H_4$ | $CH_2$ |
| Propene | $C_3H_6$ | $CH_2$ |
| Butene | $C_4H_8$ | $CH_2$ |
| Pentene | $C_5H_{10}$ | $CH_2$ |
| Water | $H_2O$ | $H_2O$ |
| Carbon dioxide | $CO_2$ | $CO_2$ |
| Hydrogen peroxide | $H_2O_2$ | $HO$ |
| Glucose | $C_6H_{12}O_6$ | $CH_2O$ |
| Ethanoic acid | $C_2H_4O_2$ | $CH_2O$ |

## Worked example 6

Given that the empirical formula of a hydrocarbon is CH and its molar mass is 26.04 g mol$^{-1}$, determine its molecular formula.

### Solution

$M(CH) = 13.02$ g mol$^{-1}$, $M(\text{compound}) = 26.04$ g mol$^{-1}$

$$\frac{M(\text{compound})}{M(\text{empirical formula})} = \frac{26.04}{13.02} = 2$$

The molecular formula is 2 times the empirical formula.

∴ Molecular formula of the hydrocarbon is $C_2H_2$.

## Worked example 7

11.0 g of a gaseous hydrocarbon is burnt completely in oxygen, producing 0.750 mole of carbon dioxide and 1.00 mol of water. It was found that 0.0908 mole of the hydrocarbon had a mass of 4.00 g. Calculate the molecular formula of the hydrocarbon.

### Solution

$n = \dfrac{m}{M}$

$M(\text{hydrocarbon}) = \dfrac{m}{n}$

$= \dfrac{4.00}{0.0908}$

$= 44.0$ g mol$^{-1}$

$n(\text{hydrocarbon}) = \dfrac{11.0}{44.0}$

$= 0.25$ mol

*(solution continued on next page)*

$n(\text{C}) = n(\text{CO}_2)$
$= 0.75$ mol

$n(\text{H}) = 2 \times n(\text{H}_2\text{O}) = 2 \times 1.00$
$= 2.00$ mol

C : H
0.75 : 2.00
3 : 8

The empirical formula of the hydrocarbon is $\text{C}_3\text{H}_8$.

$M(\text{C}_3\text{H}_8) = 3 \times 12.01 + 8 \times 1.01$
$= 44.11$

As the molar mass of the hydrocarbon is also 44.0 g mol$^{-1}$, the molecular formula must be $\text{C}_3\text{H}_8$.

### Worked example 8

Hydrazine is a highly reactive substance used principally as a rocket fuel. When it is analysed, its percentage composition is determined to be 87.4% nitrogen and 12.6% hydrogen and its molar mass as 32.06 g mol$^{-1}$. Determine the molecular formula of hydrazine.

### Solution

First calculate the empirical formula of hydrazine.

N : H
87.4% : 12.6%

$\dfrac{87.4}{14.01} : \dfrac{12.6}{1.01}$

6.238 : 12.475

$\dfrac{6.238}{6.238} : \dfrac{12.475}{6.238}$

1 : 2

The empirical formula of hydrazine is $\text{NH}_2$.

$M(\text{NH}_2) = 16.03$ g mol$^{-1}$, $M(\text{compound}) = 32.06$ g mol$^{-1}$

$\dfrac{M(\text{compound})}{M(\text{empirical formula})} = \dfrac{32.06}{16.03} = 2$

The molecular formula is 2 times the empirical formula.

∴ Molecular formula of hydrazine is $\text{N}_2\text{H}_4$.

### Worked example 9

An oxide of phosphorus is made when phosphorus is burnt in air. The oxide is found to be 43.6% phosphorus and 56.4% oxygen by mass. 0.006 mole of the oxide has a mass of 1.703 g. Find the empirical formula and hence the molecular formula of this compound.

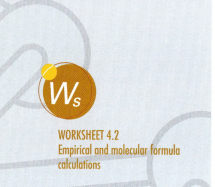

**WORKSHEET 4.2**
Empirical and molecular formula calculations

### Solution

First calculate the empirical formula of the oxide of phosphorus.

P : O

43.6% : 56.4%

$\dfrac{43.6}{30.97} : \dfrac{56.4}{16.00}$

1.408 : 3.525

$\dfrac{1.408}{1.408} : \dfrac{3.525}{1.408}$

1 : 2.5

2 : 5

The empirical formula of this oxide is $P_2O_5$.

$M(P_2O_5) = (2 \times 30.97) + (5 \times 16) = 141.94$ g mol$^{-1}$

and $M(\text{compound}) = \dfrac{m(\text{compound})}{n(\text{compound})} = \dfrac{1.703}{0.006} = 283.83$ g mol$^{-1}$

$\dfrac{M(\text{compound})}{M(\text{empirical formula})} = \dfrac{283.83}{141.94} = 2$

The molecular formula is 2 times the empirical formula.

∴ Molecular formula of phosphorus oxide is $P_4O_{10}$.

## Section 4.3 Exercises

1. Calculate the percentage composition of the first-named element in:
   a. sulfur dioxide ($SO_2$)
   b. calcium carbonate ($CaCO_3$)
   c. zinc sulfate ($ZnSO_4$)
   d. sodium carbonate ($Na_2CO_3$)
   e. aluminium sulfate ($Al_2(SO_4)_3$)

2. Determine the empirical formula for each of the following compounds.
   a. Hydrogen peroxide ($H_2O_2$)
   b. Octane ($C_8H_{18}$)
   c. Potassium thiosulfate ($K_2S_2O_3$)
   d. Fructose ($C_6H_{12}O_6$)
   e. Iron(III) oxide ($Fe_2O_3$)
   f. Butanoic acid ($C_3H_7COOH$)

3. Determine the empirical formula of the compound with the following percentage composition: tin 52.8%, iron 12.4%, carbon 16.0%, nitrogen 18.8%.

4. An oxide of titanium can be used as a brilliant white pigment in high-quality paints. Determine the empirical formula of this compound, given that its percentage composition is 59.9% titanium and 40.1% oxygen.

5. A 2.250 g sample of a hydrocarbon is found to be made up of 1.396 g carbon, 0.235 g hydrogen and the rest is oxygen. Find the empirical formula of this hydrocarbon.

6  A compound has been found to be made up of 5.200 g chromium, 4.815 g sulfur and 9.600 g oxygen. Find the empirical formula of this compound.

7  Carbohydrates are an important class of organic molecules that act as energy storage compounds in living systems. A 0.8361 g sample of a particular carbohydrate is analysed and found to contain 0.4565 g of carbon and 0.0761 g of hydrogen, with the rest of the mass being made up of oxygen. The molar mass of the compound is 88 g mol$^{-1}$. Determine both the empirical and molecular formulas of the compound.

8  Nicotine is a highly toxic chemical that is strongly addictive when ingested, generally through the inhalation of burning tobacco. The relative molecular mass of nicotine is approximately 160. Quantitative analysis of this compound yields the following percentages: carbon 74.1%, hydrogen 8.7%, nitrogen 17.2%. Calculate the empirical and molecular formulas of nicotine.

## 4.4 CHEMICAL EQUATIONS

When chemistry was a young science, there was no system for naming compounds. Names such as oil of vitriol (sulfuric acid), quicklime (calcium oxide) and milk of magnesia (magnesium hydroxide) were used by the early chemists. Although these common names are still used in places, a systematic approach is now more common. Chemical formulas are used as a shorthand representation of compounds. You can already write the formulas of a range of ionic compounds from your work in chapter 2. Your knowledge of chemical formulas will develop as you progress through the IB Chemistry course.

symbols of elements

$BaSO_4$  $H_2O$

number of oxygen atoms per atom of sulfur and barium

number of atoms of hydrogen in a water molecule

**Figure 4.4.1** A chemical formula indicates the types and ratio of atoms present in a compound.

**1.3.1 Deduce chemical equations when all reactants and products are given.** © IBO 2007

Physical and chemical changes
Counting atoms

Chemical reactions are represented by equations using the chemical formulas and symbols of the substances involved in the reaction. The combustion of methane can be represented by the following word equation:

methane + oxygen → carbon dioxide + water

Although word equations can be useful, a full, balanced equation using chemical formulas provides much more information about the reaction. To start, the names of the reactants and products are replaced with their chemical formulas:

$$CH_4 + O_2 \rightarrow CO_2 + H_2O$$

Already it is clear that this equation is more informative than the word equation above, but it is still incomplete.

The **law of conservation of mass** states that 'matter can be neither created nor destroyed, it can only be changed from one form to another'. For chemical equations, this means that there must be the same numbers and types of atoms on both sides of the equation. All that happens in a chemical reaction is that the bonds in the reactants break, the atoms rearrange, and the bonds in the products are formed. In order to reflect the law of conservation of mass, equations must be balanced. This is achieved by adding integer coefficients to the chemical formulas in the equation. If the coefficient is '1', no number is added.

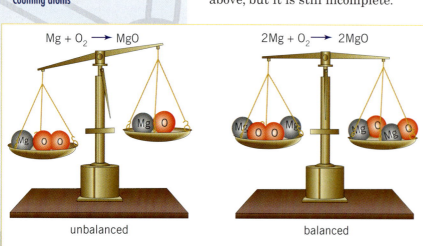
**Figure 4.4.2** In an unbalanced equation, the mass of the reactants does not equal the mass of the products. A balanced equation reflects the fact that the numbers and types of atoms involved in a reaction does not change. They are simply rearranged.

## THEORY OF KNOWLEDGE
### Phlogiston

Up until the 1600s new knowledge about the nature of matter had been acquired from empirical observations, using qualitative evidence. Because none of the gases in air had been discovered, Johann Becher (1635–82) and Georg Stahl (1660–1734) proposed the theory of phlogiston to explain some of the unanswered questions about combustion reactions. Phlogiston was a substance that had no colour, smell, taste or mass and was released during burning. In phlogiston theory:

- A substance loses mass when burned because it loses phlogiston.
- When placed in an airtight container, a candle flame goes out and a mouse will die because the air has become saturated with phlogiston.
- Charcoal leaves hardly any ash when it burns because it is made of pure phlogiston.
- When a metal oxide and charcoal are heated in a sealed container, the air pressure in the container decreases because the metal oxide absorbs phlogiston from the air.

The certainty of phlogiston theory was cast in doubt in the late 1700s, when Frenchman Antoine Lavoisier invented a balance that could accurately measure changes in mass. Using this new technology, Lavoisier was able to collect quantitative data to show that:

- When phosphorus and sulfur were combusted, the final mass was greater than the initial mass.
- When metal oxides were heated with charcoal in a sealed container, the mass of the container was the same before and after combustion.

While the phlogiston theory was qualitatively adequate, it failed when subjected to the more quantitative data that Lavoisier's balance could provide. This new data contradicted the phlogiston theory and laid the foundation for our modern theory of combustion. Lavoisier proposed a relationship to describe how mass behaved in a chemical reaction. Called the law of conservation of mass, it stated that the total mass of the reactants and products in a chemical reaction remains the same.

Many scientists of the day continued to support phlogiston theory and would not accept Lavoisier's evidence or radical way of thinking. They tried to account for his observations by claiming that elements such as sulfur and phosphorus had negative mass and when combusted lost negative mass, causing their mass to increase. Other substances such as wood lost positive mass when burned, causing their mass to decrease. However, with time, more and more scientists became dissatisfied with the inability of the phlogiston theory to explain combustion, and they became more confident in Lavoisier's ideas until eventually these ideas gained sufficient credibility to be accepted.

The story of phlogiston shows how both evolutionary and revolutionary changes occurred in the development of our knowledge of combustion. With new evidence and interpretation, old theories were replaced by newer ones. Although no single universal step by step scientific method captures the complexity of doing science, something universal to all sciences is the demand for explanations to be supported by empirical evidence.

- Why do you think some scientists were reluctant to accept Lavoisier's ideas and tried hard to keep the phlogiston theory alive?
- Why do we no longer believe in the phlogiston theory?
- Attempts to prove a theory by testing it are central to the work of a scientist. Scientists are in a sense sceptics, they like to test the validity of a theory by confirming or disproving it. How did Lavoiser disprove phlogiston theory?
- Why do you think Lavoisier's law of conservation of mass caused such a scientific revolution or paradigm shift?
- Lavoisier has been called the 'father of modern chemistry'. Comment on this claim.

Figure 4.4.3 Antoine Lavoisier, the 'father of modern chemistry'.

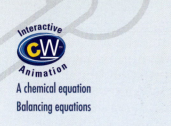

A chemical equation
Balancing equations

**AS** 1.3.3
Apply the state symbols (s), (l), (g) and (aq). © IBO 2007

> **THEORY OF KNOWLEDGE**
> All areas of knowledge have conventions or rules, sets of common understandings to make communication easier.
> 
> - Describe the conventions used when writing a balanced chemical equation.
> - When are 'state' symbols necessary in aiding understanding of a chemical reaction and when are they not important?

**AS** 1.3.2
Identify the mole ratio of any two species in a chemical equation. © IBO 2007

Multiple proportions

Equations can be balanced by looking at each type of atom in turn. For example, the *unbalanced* equation for the combustion of methane is:

$$CH_4 + O_2 \rightarrow CO_2 + H_2O$$

Starting with the carbon, there is one carbon atom on each side, so the equation is balanced with respect to carbon.

There are 4 hydrogen atoms on the left, but only 2 on the right. Placing a '2' in front of $H_2O$ will give 4 hydrogen atoms on the right, and so balance the equation with respect to hydrogen.

$$CH_4 + O_2 \rightarrow CO_2 + 2H_2O$$

There are 2 oxygen atoms on the left, but 4 oxygen atoms on the right. Placing a '2' in front of $O_2$ will give 4 oxygen atoms on each side. The equation is now *fully balanced*.

$$CH_4 + 2O_2 \rightarrow CO_2 + 2H_2O$$

Reactant side (left):   1 carbon, 4 hydrogens, 4 oxygens
Product side (right):  1 carbon, 4 hydrogens, 4 oxygens

Adding symbols to show the physical states of the reactants and products is the final step in equation writing. The **state symbol** (**g**) represents a gas, (**s**) a solid, (**l**) a liquid and (**aq**) means dissolved in aqueous solution (in water). These symbols represent the state of each substance at room temperature, unless otherwise specified. Adding the symbols to the equation above gives:

$$CH_4(g) + 2O_2(g) \rightarrow CO_2(g) + 2H_2O(g)$$

This equation shows that one mole of gaseous methane molecules combines with two mole of gaseous oxygen molecules, producing one mole of gaseous carbon dioxide molecules and two mole of gaseous water molecules. The equation shows the ratio of reactants and products to each other. The coefficients in the equation can be shown as a ratio.

$CH_4 : O_2 : CO_2 : H_2O$
  1  :  2  :  1  :  2

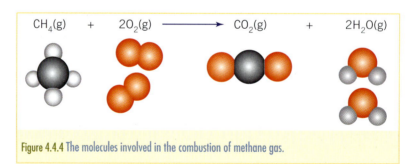

**Figure 4.4.4** The molecules involved in the combustion of methane gas.

Another example is the reaction of solid lithium oxide with aqueous hydrochloric acid, producing aqueous lithium chloride and water. Following the steps below will lead to the full, balanced equation.

Step 1: Write the word equation for this reaction.

   lithium oxide + hydrochloric acid → lithium chloride + water

Step 2: Write the unbalanced formula equation.

   $Li_2O + HCl \rightarrow LiCl + H_2O$

Step 3: Balance the numbers and types of atoms on each side of the equation.

   $Li_2O + 2HCl \rightarrow 2LiCl + H_2O$

Step 4: Add symbols showing the physical states, to give the final equation.

$Li_2O(s) + 2HCl(aq) \rightarrow 2LiCl(aq) + H_2O(l)$

In this equation we can see that one mole of lithium oxide reacts with two mole of hydrochloric acid to produce two mole of lithium chloride and one mole of water.

Balancing chemical equations is a skill related to an area of chemistry known as stoichiometry (pronounced stoy-key-om-uh-tree). Stoichiometry deals with quantitative aspects of chemistry. The word *stoichiometry* comes from the Greek *stoicheion*, meaning element, and *metron*, meaning measure. The types of problems encountered in stoichiometry include calculating the concentration of a solution, working out the mass of product expected from a particular reaction, and finding the volume of oxygen consumed in a combustion reaction. Essentially, any problem that involves finding out 'how much' is part of stoichiometry.

Views of a chemical reaction

**WORKSHEET 4.3**
Balancing chemical equations

### Section 4.4 Exercises

1   Balance the following equations.
    a   $Na(s) + Cl_2(g) \rightarrow NaCl(s)$
    b   $P(s) + O_2(g) \rightarrow P_2O_5(s)$
    c   $SO_2(g) + O_2(g) \rightarrow SO_3(g)$
    d   $P(s) + O_2(g) \rightarrow P_2O_3(s)$
    e   $Fe(s) + Cl_2(g) \rightarrow FeCl_3(s)$
    f   $C_2H_4(l) + O_2(g) \rightarrow CO(g) + H_2O(g)$
    g   $CO(g) + O_2(g) \rightarrow CO_2(g)$
    h   $N_2(g) + H_2(g) \rightarrow NH_3(g)$

2   Balance the following equations.
    a   $C_6H_{12}O_6(aq) \rightarrow C_2H_5OH(aq) + CO_2(g)$
    b   $NH_3(g) + O_2(g) \rightarrow NO(g) + H_2O(l)$
    c   $AlI_3(s) + Cl_2(g) \rightarrow AlCl_3(s) + I_2(s)$
    d   $NaCl(aq) + H_2O(l) \rightarrow NaOH(aq) + H_2(g) + Cl_2(g)$

3   Write balanced equations, including symbols for physical states, for the following reactions.
    a   Iron and oxygen gas ($O_2$) react to produce solid iron(III) oxide.
    b   Silver nitrate and zinc iodide solutions are mixed and form a solution of zinc nitrate and a precipitate of silver iodide.
    c   Heating iron(III) oxide in a stream of hydrogen gas ($H_2$) produces metallic iron and water vapour.
    d   Magnesium nitrate and sodium carbonate solutions are mixed and form a solution of sodium nitrate and a precipitate of magnesium carbonate.
    e   Solid iron(III) hydroxide is heated to give solid iron(III) oxide and water vapour.

4   Write balanced equations, including symbols for physical states, for the following reactions.
    a   In a car engine, liquid octane ($C_8H_{18}$) is burned in the presence of gaseous oxygen ($O_2$), producing gaseous carbon dioxide ($CO_2$), gaseous water ($H_2O$) and heat.
    b   Solid potassium (K) reacts explosively with liquid water, producing a solution of potassium hydroxide (KOH) and hydrogen gas ($H_2$).
    c   Aqueous hydrofluoric acid (HF) cannot be stored in glass because it reacts with sodium silicate ($Na_2SiO_3$), producing aqueous $H_2SiF_6$, aqueous sodium fluoride (NaF) and liquid water.

**5** An early step in the production of sulfuric acid involves the oxidation of sulfur dioxide gas to sulfur trioxide, according to the equation:

$2SO_2(g) + O_2(g) \rightarrow 2SO_3(g)$

How many moles of $SO_3$ will be produced by the oxidation of 10 moles of $SO_2$ by an excess amount of oxygen gas?

**6** Copper(II) oxide reacts with ammonia to produce copper metal, water vapour and nitrogen gas according to the equation:

$CuO(s) + 2NH_3(aq) \rightarrow Cu(s) + 3H_2O(g) + N_2(g)$

What amount of copper metal (in mol) will result from the reduction of 0.335 mol of copper(II) oxide by excess ammonia?

**7** Ammonia is an important industrial chemical and is produced in large quantities by the Haber–Bosch process. The equation for the reaction to produce ammonia is:

$N_2(g) + 3H_2(g) \rightarrow 2NH_3(g)$

Determine the number of moles of nitrogen and hydrogen gas that would be required to react to produce 500 mole of ammonia gas.

## 4.5 MASS RELATIONSHIPS IN CHEMICAL REACTIONS

The mole concept provides a vital connection between numbers of atoms involved in a chemical reaction (via the relevant equation) and the masses of reactants and products involved. Armed with this information, we can begin to calculate and relate amounts of reactants and products in a chemical reaction; that is, we can carry out stoichiometric calculations.

There is a wide range of stoichiometric problems that you are likely to encounter as you progress through IB Chemistry. While this will perhaps be a little confusing at first, the primary difference between most problems revolves around the nature of the reactants (solids, aqueous solutions, gases and even liquids) and the particular products of interest. Close reading of the question will generally lead you to recognize which particular species you have sufficient information for in order to calculate the number of moles present. You may find it useful to highlight or underline the species on which you have information (the 'known') and the one you need to find out some information about (the 'unknown').

### Calculating the mass of a product given the mass of a reactant

Chemistry is an experimental science with theories that have been based on experimental observations. These observations often require measurements of masses of solids, or volumes of solutions or of gases. The calculations that need to be performed on experimental data often form the basis for problems that you may encounter in a theoretical context.

In a chemistry problem you may be given the mass of a particular chemical of known formula, enabling you to use the formula $n = \dfrac{m}{M}$. You will then be generally asked to calculate the theoretical mass that may be obtained of another species, often a product of the reaction. To achieve this, a connection must be made between the two species: the 'known' and the 'unknown'. A balanced chemical equation makes this connection, as it displays the actual number of atoms or molecules or ions involved in the reaction and the ratio in which they are reacting. As we saw in section 4.1, the mole is also, in effect, a number, and so the ratio of atoms is equivalent to the ratio of moles.

## CHEM COMPLEMENT

### National Mole Day

What would you consider to be the most important new concept in chemistry that you have mastered so far? The mole concept would surely be a strong contender. The importance of the mole concept is recognized around the world by the annual celebration of National Mole Day on 23 October from 6.02 am to 6.02 pm. The day was created as a way to foster interest in chemistry.

The National Mole Day Foundation website includes a dictionary of mole terms and jokes; for example, 'Q: How much does Avogadro exaggerate? A: He makes mountains out of molehills'. Also included are the 'Ask Monty Mole' section, an advice column for befuddled molers and examples of Mole Day projects, such as creating a special game board for 'mole-opoly'!

Find out more about National Mole Day by connecting to Pearson Places and selecting the Chapter 4 Web Destinations link.

Because these calculations are based on experimental data, it is important to consider significant figures and try to quote your answers to the correct number of significant figures. For a full explanation of significant figures and how to use them in calculations see chapter 5, section 5.1.

Consider the following problem.

Magnesium metal reacts rapidly and evolves a great deal of light when it is burnt in the presence of oxygen. For this reason, magnesium is used in emergency distress flares on ships and in fireworks displays. A student weighs a 2.431 g strip of magnesium ribbon, places it in a crucible and ignites it. At the end of the reaction a white powder of magnesium oxide remains. Calculate the mass of magnesium oxide formed, given that the equation for this reaction is:

$$2Mg(s) + O_2(g) \rightarrow 2MgO(s)$$

Figure 4.5.1 Magnesium burns readily to produce magnesium oxide—the white 'smoke'.

Note that in this example we are provided with two pieces of information about the magnesium to enable us to calculate the number of mole of this reactant present: its mass (2.431 g) and the fact that it is magnesium (molar mass of 24.31 g mol$^{-1}$). Hence:

$$n(Mg) = \frac{m}{M} = \frac{2.431}{24.31} = 0.1000 \text{ mol}$$

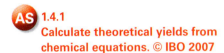

**1.4.1 Calculate theoretical yields from chemical equations. © IBO 2007**

The balanced chemical equation supplies the vital link between the known species, magnesium, and the one we are trying to calculate, the unknown magnesium oxide. The equation shows us that 'two mole of magnesium react with one mole of oxygen gas to produce two mole of magnesium oxide'. The magnesium and the magnesium oxide are in the mole ratio 2:2, which simplifies to 1:1. Hence, from the equation:

$$n(MgO) = \frac{1}{1} \times n(Mg) = \frac{1}{1} \times 0.1000 = 0.1000 \text{ mol}$$

Now we know the number of mole of the product of this reaction, 0.1000 mol of magnesium oxide, and we can calculate the molar mass of this compound. With these pieces of data we can calculate the mass of product:

$n(Mg) = 0.1000$ mol, $m(Mg) = ?$, $M(MgO) = 40.31$ g mol$^{-1}$

$$n(MgO) = \frac{m}{M}$$
$$\therefore m(MgO) = n \times M = 0.1000 \times 40.31 = 4.031 \text{ g}$$

Stoichiometry calculation

The mass of magnesium oxide formed is 4.031 g.

Note that in this calculation we have effectively ignored oxygen as a reactant. The oxygen for this reaction comes from the air, and so there is an effectively inexhaustible supply of this reactant for the reaction. We say that the oxygen is in excess and so will have no bearing on the calculation. It is possible to calculate the mass of oxygen from the air that would be consumed in this reaction, but we were not explicitly asked to calculate this mass in this question.

### A general technique

The principles applied above can be used to solve the great majority of stoichiometric problems.

The steps that will take you through most problems can be summarized as listed on the next page.

(Note that you have not yet learnt to use all of the formulas listed in this box, but you will be able to use them by the end of this chapter.)

> Step 1: Write a balanced chemical equation for the reaction under consideration.
>
> Step 2: List all data given, including relevant units. These data may be masses or volumes and concentrations of aqueous solutions, or volumes and pressure and temperature of gases. Remember to also write down the symbol of the unknown quantity!
>
> Step 3: Convert the data for the *'known'* quantity to moles, using the relevant formula which will be one of:
>
> $$n = \frac{m}{M},\ n = c \times V,\ n = \frac{N}{L},\ n = \frac{V}{V_m},\ n = \frac{PV}{RT}$$
>
> Step 4: Use the chemical equation to determine the mole ratio of the *'unknown'* quantity to the known quantity (that is, $\frac{unknown}{known}$).
> This ratio enables calculation of the number of moles of the unknown quantity. **Remember to divide by the coefficient of the known and multiply by the coefficient of the unknown**.
>
> Step 5: Finally, convert this number of mole back into the relevant units of the unknown. The problem is solved!

When working through a stoichiometry problem it is important that you follow each step sequentially, as set out above, and not be tempted to take short-cuts.

## Examples of calculating the theoretical yield of a reaction

### Worked example 1

When calcium carbonate (limestone) is heated strongly, it decomposes to calcium oxide (quicklime) and carbon dioxide gas. The relevant equation for this reaction is:

$$CaCO_3(s) \rightarrow CaO(s) + CO_2(g)$$

Calculate the mass of quicklime formed when 500 g of limestone is reduced to quicklime.

### Solution

Step 1: Write a balanced equation (already provided).

$$CaCO_3(s) \rightarrow CaO(s) + CO_2(g)$$

Step 2: List all data given, including relevant units.

| Known | Unknown |
|---|---|
| $m(CaCO_3) = 500.0$ g | $m(CaO) = ?$ |
| $M(CaCO_3) = 100.09$ g mol$^{-1}$ | $M(CaO) = 56.08$ g mol$^{-1}$ |

Note that we do not write anything for carbon dioxide, as the question has not requested any information about this particular product.

Step 3: Convert the data given for the 'known' quantity to mole, using the relevant formula.

$$n(CaCO_3) = \frac{m}{M} = \frac{500.0}{100.09} = 4.996 \text{ moles}$$

Figure 4.5.2 Limestone plaster has been produced in ancient kilns such as these for a thousand years.

Step 4: Use the chemical equation to determine the mole ratio of the unknown quantity to the known quantity:

$$CaCO_3(s) \rightarrow CaO(s) + CO_2(g)$$

From the equation, $n(CaO) = \frac{1}{1} \times n(CaCO_3) = 4.996$ mol

Step 5: Convert this number of mole back into the relevant units of the unknown.

$$n(CaO) = \frac{m}{M}$$

$\therefore m(CaO) = n \times M = 4.996 \times 56.08 = 280.2 = 280$ (to 3 significant figures)

The mass of quicklime formed is 280 g.

DEMO 4.1
Products of a decomposition reaction

PRAC 4.4
Decomposition of potassium perchlorate

## Worked example 2

Aluminium is a most important metal for today's society as it is strong yet relatively light. Twenty million tonnes of aluminium are produced worldwide every year in specially designed reaction vessels using the Hall–Héroult process. The relevant equation for the electrolysis reaction between alumina and the 600 kg carbon anode blocks is:

$$2Al_2O_3(l) + 3C(s) \rightarrow 4Al(l) + 3CO_2(g)$$

**a** Calculate the mass of aluminium produced when 1.000 tonne ($1.000 \times 10^6$ g) of alumina undergoes reduction.

**b** What mass of carbon is consumed in the process?

Figure 4.5.3 Production of molten aluminium involves a redox reaction between alumina ($Al_2O_3$) and carbon.

## Solution

Note that in this instance we have been asked two separate, though related, questions and so will need to perform two calculations.

**a** Step 1: Write a balanced equation (already provided).

$$2Al_2O_3(l) + 3C(s) \rightarrow 4Al(l) + 3CO_2(g)$$

Step 2: List all data given, including relevant units.

| Known | Unknown a | Unknown b |
|---|---|---|
| $m(Al_2O_3) = 1.000 \times 10^6$ g | $m(Al) = ?$ | $m(C) = ?$ |
| $M(Al_2O_3) = 101.96$ g mol$^{-1}$ | $M(Al) = 26.98$ g mol$^{-1}$ | $M(C) = 12.01$ g mol$^{-1}$ |

Step 3: Convert the data given for the 'known' quantity to mole, using the relevant formula.

$$n(Al_2O_3) = \frac{m}{M} = \frac{1.000 \times 10^6}{101.96} = 9.808 \times 10^3 \text{ mol}$$

Step 4: Use the chemical equation to determine the mole ratio of the unknown quantity to the known quantity:

$$2Al_2O_3(l) + 3C(s) \rightarrow 4Al(l) + 3CO_2(g)$$

From the equation, $n(Al) = \frac{4}{2} \times n(Al_2O_3)$

$$= \frac{4}{2} \times (9.808 \times 10^3)$$

$$= 1.962 \times 10^4 \text{ mol}$$

WORKSHEET 4.4
Mass–mass stoichiometry

Step 5: Convert this number of mole back into the relevant units of the unknown.

$$n(Al) = \frac{m}{M}$$

$$\therefore m(Al) = n \times M = 1.962 \times 10^4 \times 26.98 = 5.293 \times 10^5 \text{ g}$$

The mass of aluminium produced is 529.3 kg.

**b** To calculate the mass of carbon consumed, we can move straight to Step 4, as we already know the number of mole of alumina consumed.

**2Al$_2$O$_3$(l) + 3C(s) → 4Al(l) + 3CO$_2$(g)**

From the equation, $n(C) = \frac{3}{2} \times n(Al_2O_3) = \frac{3}{2} \times (9.808 \times 10^3) = 1.471 \times 10^4$ mol

$$n(C) = \frac{m}{M}$$

$$\therefore m(C) = n \times M = 1.471 \times 10^4 \times 12.01 = 1.767 \times 10^5 \text{ g}$$

The mass of carbon consumed is 176.7 kg.

Limiting reagents

Limiting reagent

**AS 1.4.2
Determine the limiting reactant and the reactant in excess when quantities of reacting substances are given.** © IBO 2007

## Problems involving limiting or excess reactants

In all of the problems considered to date, sufficient information has been provided to determine the amount (in mol) of only one chemical species (generally a reactant). Sometimes it has been explicitly stated that another chemical is in excess, and so it has not been included in our calculations. But how can we solve a problem where information about more than one reactant is provided? How can we determine which reactant will be completely consumed in the reaction? In these cases, we must determine the amount (in mol) of both reactant species present and use the the molar ratios from the equation to determine which will be completely consumed in the reaction (the **limiting reactant**) and which is present in excess. When solving the problem, the limiting reactant must be used in the calculation.

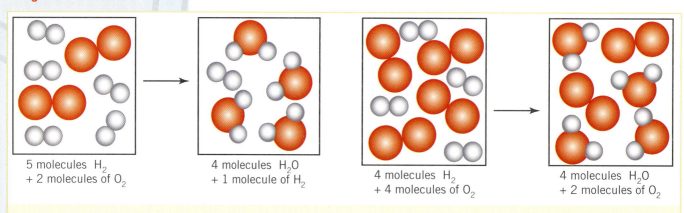

**Figure 4.5.4** Two mixtures of gases for the reaction between hydrogen and oxygen. Can you determine the limiting reagent in each mixture?

## Worked example 3

In an electric furnace, carbon, C, and silicon dioxide, $SiO_2$, react together according to the equation:

$$2C(s) + SiO_2(s) \rightarrow SiC(s) + CO_2(g)$$

If 40.0 g of carbon and 40.0 g of silicon dioxide are placed in an electric furnace, what is the maximum mass of silicon carbide, SiC, that could be produced?

### Solution

Step 1: The equation for the reaction is:

$$2C(s) + SiO_2(s) \rightarrow SiC(s) + CO_2(g)$$

Step 2: List all data given, including relevant units.

| Known 1 | Known 2 | Unknown |
|---|---|---|
| $m(C) = 40.0$ g | $m(SiO_2) = 40.0$ g | $m(SiC) = ?$ |
| $M(C) = 12.01$ g mol$^{-1}$ | $M(SiO_2) = 28.09 + 2 \times 16.00$ | $M(SiC) = 28.09 + 12.01$ |
| | $= 60.09$ g mol$^{-1}$ | $= 40.10$ g mol$^{-1}$ |

Step 3: Convert the data given for the 'known' quantity to mole, using the relevant formula.

$$n = \frac{m}{M}$$

$$\therefore n(C) = \frac{40.0}{12.01} = 3.331 \text{ mol} \quad \text{AND} \quad n(SiO_2) = \frac{40.0}{60.09} = 0.6657 \text{ mol}$$

Step 4a: Determine the reactant in excess.

The equation for this reaction indicates that two mole of C reacts with one mole of $SiO_2$ to produce one mole of SiC.

This can be written as a ratio.

C : $SiO_2$
2 : 1

If all of the C reacts, then

$n(SiO_2)$ required $= \frac{1}{2} \times 3.331 = 1.666$

BUT there is only 0.6657 mol of $SiO_2$ available to react. All of the C cannot react as there is not enough $SiO_2$ $\therefore$ the C is in excess and $n(SiO_2)$ is the limiting factor.

**Remember:** To find $n(SiO_2)$ divide by the coefficient of the known reagent (C) and multiply by the coefficient of the unknown reagent ($SiO_2$).

Step 4b: Use the chemical equation to determine the mole ratio of the unknown quantity to the known quantity. Note that we must use the limiting reactant ($SiO_2$) as the *known* to calculate the mass of product formed.

$$2C(s) + \mathbf{SiO_2(s)} \rightarrow \mathbf{SiC(s)} + CO_2(g)$$

From the equation, $n(SiC) = n(SiO_2) = 0.6657$ mol

**Remember:** To find the factor by which to multiply, divide by the coefficient of the known ($SiO_2$) and multiply by the coefficient of the unknown (SiC).

Step 5: Convert this number of mole back into the relevant units of the unknown.

$$n(SiC) = \frac{m}{M}$$

$$\therefore m(SiC) = 0.6657 \times 40.10 = 26.69 \text{ g}$$

The mass of silicon carbide produced is 26.69 g.

### Worked example 4

Aluminium and iodine react according to the equation

$$2Al(s) + 3I_2(g) \rightarrow 2AlI_3(s)$$

What mass of $AlI_3$ would be formed when 1.350 g of aluminium is heated with 2.534 g of iodine, $I_2$?

### Solution

Step 1: The equation for the reaction is:

$$2Al(s) + 3I_2(g) \rightarrow 2AlI_3(s)$$

Step 2: List all data given, including relevant units.

**Known 1**
$m(Al) = 1.350$ g
$M(Al) = 26.98$ g mol$^{-1}$

**Known 2**
$m(I_2) = 2.534$ g
$M(I_2) = 253.4$ g mol$^{-1}$

**Unknown**
$m(AlI_3) = ?$
$M(AlI_3) = 407.68$ g mol$^{-1}$

Step 3: Convert the data given for the 'known' quantity to mole, using the relevant formula.

$$n = \frac{m}{M}$$

$$\therefore n(Al) = \frac{1.350}{26.98} = 0.05000 \text{ mol} \quad \text{AND} \quad n(I_2) = \frac{2.534}{253.4} = 0.01000 \text{ mol}$$

Step 4a: Determine the reactant in excess.

The equation for this reaction indicates that two mole of Al reacts with three mole of $I_2$ to produce two mole of $AlI_3$.

This can be written as a ratio.

Al : $I_2$
2 : 3

If all of the aluminium reacts, then

$$n(I_2) \text{ required} = \frac{3}{2} \times 0.05000$$
$$= 0.07500$$

There is only 0.01000 mol of $I_2$ available to react. All of the Al cannot react as there is not enough $I_2$, therefore the Al is in excess and $n(I_2)$ is the limiting reactant.

**Remember:** To find $n(I)$, divide by the coefficient of the known reagent (Al) and multiply by the coefficient of the unknown reagent ($I_2$).

Step 4b: Use the chemical equation to determine the mole ratio of the unknown quantity to the known quantity. Note that we must use the limiting reactant to calculate the mass of product formed.

$$2Al(s) + 3I_2(g) \rightarrow 2AlI_3(s)$$

From the equation, $n(AlI_3) = \frac{2}{3} \times n(I_2) = \frac{2}{3} \times 0.01000 = 6.667 \times 10^{-3}$ mol

Step 5: Convert this number of mole back into the relevant units of the unknown.

$$n(AlI_3) = \frac{m}{M}$$

$$\therefore m(AlI_3) = 6.667 \times 10^{-3} \times 407.68 = 2.718 \text{ g}$$

The mass of aluminium iodide produced is 2.718 g.

**Remember:** To find the factor by which to multiply, divide by the coefficient of the known ($I_2$) and multiply by the coefficient of the unknown ($AlI_3$).

## Calculating percentage yield

While the questions that you have seen so far involve a perfect situation in which the full amount of product is made, this is far from the case in many experiments. The amount of product made during an experiment is called the **experimental yield** and it is often significantly less than the **theoretical yield**, the amount you expect to make from your mathematical calculations. One way in which chemists express how successful their preparation of a product has been is to calculate the **percentage yield**.

**AS 1.4.3**
Solve problems involving theoretical, experimental and percentage yield. © IBO 2007

$$\text{Percentage yield} = \frac{\text{experimental yield}}{\text{theoretical yield}} \times \frac{100}{1}$$

### Worked example 5

In the Haber process, ammonia is made when nitrogen gas reacts with hydrogen gas according to the equation:

$$N_2(g) + 3H_2(g) \rightarrow 2NH_3(g)$$

If 45.00 g of ammonia, $NH_3$, is produced when 42.03 g of nitrogen, $N_2$, reacts with excess hydrogen gas, $H_2$, what is the percentage yield of the experiment?

#### Solution

**Calculate the theoretical yield of the experiment.**

Step 1: Write a balanced equation (already provided).

$$N_2(g) + 3H_2(g) \rightarrow 2NH_3(g)$$

Step 2: List all data given, including relevant units.

| Known | Unknown |
|---|---|
| $m(N_2) = 42.03$ g | $m(NH_3) = ?$ |
| $M(N_2) = 28.02$ g mol$^{-1}$ | $M(NH_3) = 17.04$ g mol$^{-1}$ |

Note that because this is the calculation of the theoretical yield, we are ignoring the experimental yield of ammonia, for now.

Step 3: Convert the data for the 'known' quantity given to moles, using the relevant formula.

$$n(N_2) = \frac{m}{M} = \frac{42.03}{28.02} = 1.500 \text{ moles}$$

Step 4: Use the chemical equation to determine the mole ratio of the unknown quantity to the known quantity.

$$\mathbf{N_2}(g) + 3H_2(g) \rightarrow \mathbf{2NH_3}(g)$$

$N_2 : NH_3$
$1 : 2$

From the equation, $n(NH_3) = \frac{2}{1} \times n(N_2) = 3.000$ mol

Step 5: Convert this number of mole back into the relevant units of the unknown.

$$n(NH_3) = \frac{m}{M}$$

$$\therefore m(NH_3) = n \times M = 3.000 \times 17.03 = 51.12 \text{ g}$$

> **THEORY OF KNOWLEDGE**
> Describe the role mathematics plays in our knowledge of the relationships between reactants and products.

The theoretical yield of ammonia is 51.12 g.

**Calculate the percentage yield.**

Experimental yield($NH_3$) = 45.00 g

Theoretical yield($NH_3$) = 51.12 g

$$\text{Percentage yield} = \frac{\text{experimental yield}}{\text{theoretical yield}} \times \frac{100}{1}$$
$$= \frac{45.00}{51.12} \times \frac{100}{1}$$
$$= 88.03\%$$

The percentage yield of ammonia in this experiment was 88.0%.

## Section 4.5 Exercises

1. Elemental iodine can be produced by the following reaction:
   $10HI(aq) + 2KMnO_4(aq) + 3H_2SO_4(aq)$
   $\quad \rightarrow 5I_2(aq) + 2MnSO_4(aq) + K_2SO_4(aq) + 8H_2O(l)$
   Calculate the exact number of mole of each reactant that would be required to produce 0.614 mol of iodine.

2. Propene undergoes incomplete combustion in the presence of a limited amount of oxygen at a temperature of 150°C, to produce carbon monoxide and water vapour according to the equation:
   $2C_3H_6(g) + 9O_2(g) \rightarrow 6CO(g) + 12H_2O(g)$
   Calculate the total number of mole of gaseous products that would result from the incomplete combustion of 1.37 mole of propene.

3. Calculate the mass of hydrogen gas evolved when 6.54 g of zinc is dissolved in an excess of dilute hydrochloric acid according to the equation:
   $Zn(s) + 2HCl(aq) \rightarrow ZnCl_2(aq) + H_2(g)$

4. Potassium metal reacts vigorously when placed into a beaker of water. The reaction liberates so much heat that the hydrogen gas evolved often ignites, burning with a characteristic lilac flame. The equation for the reaction is:
   $2K(s) + 2H_2O(l) \rightarrow 2KOH(aq) + H_2(g)$
   Calculate the mass of hydrogen gas that would be given off by the reaction of 3.91 g of potassium metal in a beaker of water.

5. In some countries, large quantities of coal are burnt to generate electricity, in the process generating significant amounts of the greenhouse gas carbon dioxide.
   The equation for this combustion reaction is:
   $C(s) + O_2(g) \rightarrow CO_2(g)$
   Determine the mass of carbon dioxide produced by the combustion of one tonne ($1.00 \times 10^6$ g) of coal, assuming that the coal is pure carbon.

6. Copper(II) nitrate decomposes on heating according to the equation:
   $2Cu(NO_3)_2(s) \rightarrow 2CuO(s) + 4NO_2(g) + O_2(g)$
   After decomposition of a sample of copper(II) nitrate, 7.04 g of black copper(II) oxide remains. Calculate the original mass of copper(II) nitrate decomposed.

7 Ethene undergoes an addition reaction with water at temperatures of around 300°C in the presence of a phosphoric acid catalyst to produce ethanol, according to the equation:
$$C_2H_4(g) + H_2O(g) \xrightarrow{H_3PO_4} C_2H_5OH(g)$$
What mass of ethene must be reacted to produce 500 kg of ethanol?

8 Aluminium sulfide reacts with hydrochloric acid to produce the foul-smelling hydrogen sulfide, better known as 'rotten egg' gas. The equation for this reaction is:
$$Al_2S_3(s) + 6HCl(aq) \rightarrow 2AlCl_3(aq) + 3H_2S(g)$$
Calculate the mass of hydrogen sulfide that would be generated by the reaction of 9.67 g of aluminium sulfide with an excess of hydrochloric acid.

9 Sulfur burns readily in oxygen gas to produce sulfur dioxide according to the equation:
$$S(s) + O_2(g) \rightarrow SO_2(g)$$
Calculate the mass of sulfur dioxide produced when 6.40 g of sulfur reacts completely with excess oxygen gas.

10 Hydrazine ($N_2H_4$) reacts with oxygen in a rocket engine to form nitrogen and water vapour, in addition to substantial amounts of heat. The equation for this reaction is:
$$N_2H_4(l) + O_2(g) \rightarrow N_2(g) + 2H_2O(g)$$
Calculate the mass of water vapour produced by the complete reaction of 500 kg of hydrazine with excess oxygen.

11 2.5 mol of hydrogen sulfide gas is reacted with 1.8 mol of sulfur dioxide according to the equation:
$$2H_2S(g) + SO_2(g) \rightarrow 3S(s) + 2H_2O(l)$$
Calculate which reactant is in excess and by what amount it is in excess.

12 2.55 g of zinc powder is brought into contact with 5.92 g of chlorine gas. They react to produce zinc chloride according to the equation:
$$Zn(s) + Cl_2(g) \rightarrow ZnCl_2(s)$$
a State which reagent is in excess, and by what mass.
b Calculate the mass of zinc chloride produced.

13 At the temperature of an electric furnace, carbon and silicon dioxide react according to the equation:
$$3C(s) + SiO_2(s) \rightarrow SiC(s) + 2CO(g)$$
48 g of carbon is mixed with 48 g of silicon dioxide.
a State which reagent is in excess.
b Calculate the mass of silicon carbide (SiC) that could be formed.

14 A student (student 1) decided to carry out a precipitation reaction in which a solution containing 1.00 g of silver nitrate was mixed with another solution containing 1.00 g of sodium chloride. The student expected a precipitate of silver chloride.
a Write a balanced chemical equation for the reaction, including states.
b Calculate the mass of precipitate that will be theoretically obtained.

15. Another student (student 2) in the class was also carrying out an investigation on precipitation reactions in which two solutions, each of which contained only 0.50 g of solid, were mixed. A solution of lead(II) nitrate was mixed with a solution of sodium chloride and a precipitate of lead(II) chloride was expected.

   a Write a balanced chemical equation for the reaction, including states.

   b Calculate the mass of precipitate that will theoretically be obtained.

16. a Student 1 (question 14) filtered the silver chloride precipitate and allowed it to dry. The mass of the precipitate that had been collected was 0.732 g. Calculate the percentage yield obtained.

   b Student 2 (question 15) followed the same procedure to collect the precipitate and collected 0.413 g of lead(II) chloride. Calculate the percentage yield obtained.

   c Which student had been more careful in their experimental technique to minimize loss of precipitate?

17. A student weighed a 3.20 g strip of magnesium ribbon, placed it in a crucible and ignited it. At the end of the reaction a white powder of magnesium oxide remained. The equation for the reaction that occurred is:

   $2Mg(s) + O_2(g) \rightarrow 2MgO(s)$

   a Calculate the theoretical mass of magnesium oxide to be formed.

   b The student dropped the crucible when removing it from the clay triangle that was being used during the heating. The crucible did not break, but the magnesium oxide fell out of the crucible. It was swept back into the crucible and found to have a mass of 4.43 g.

   Calculate the percentage yield obtained.

   c Explain the percentage yield obtained in part **b**.

## 4.6 FACTORS AFFECTING AMOUNTS OF GASES

When we attempt to consider the quantitative behaviour of gases, we are concerned with measurement of various properties of a gas. Four measures concern us—pressure, temperature, volume and amount of gas (usually measured in mole).

### Pressure

As gas particles collide with the walls of their container, they exert a force on the walls. This force per unit area exerted on the walls is the pressure exerted by the gas. The fundamental units of pressure will then depend on the units of force and area. The SI unit of pressure is the pascal (Pa), where one pascal is a force of one newton (N) acting over an area of one square metre.

$1\ Pa = 1\ N\ m^{-2}$

The IUPAC has, in the past, used as a standard a pressure of one atmosphere, equivalent to 101 325 Pa. Recently they adopted another unit, one bar, equivalent to $10^5$ Pa as their standard.

$1\ atm = 101\,325\ Pa$
$1\ bar = 10^5\ Pa$

Pressure can be measured using Torricelli's barometer. This uses the height of a column of mercury supported by a gas as a measure of its pressure. The unit, mmHg, is named a torr, after Torricelli. Atmospheric pressure, the pressure due to the particles in the atmosphere, has a mean value at sea level of 760 mmHg.

1 atm = 760 mmHg = 760 torr

The units used for pressure depend on the situation. Although millimetres of mercury (mmHg) and atmospheres (atm) are older units, they are still frequently used. Meteorologists use the unit hectopascals (hPa), where one hectopascal is equivalent to 100 Pa.

1 hPa = 100 Pa

Chemists use pascals, or more frequently kilopascals (kPa), in their work. Conversions between these many units are shown in table 4.6.1 below.

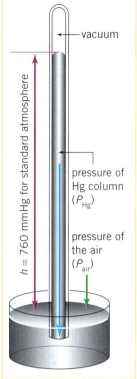

Figure 4.6.1 A torricellian barometer. Mercury flows out of the tube until $P_{Hg} = P_{air}$.

### TABLE 4.6.1 CONVERTING UNITS OF PRESSURE

| Unit given | Unit required | Conversion | Example |
| --- | --- | --- | --- |
| Atmospheres (atm) | Millimetres of mercury (torr) (mmHg) | atm × 760 | 1.20 atm = 1.2 × 760 <br> = 912 mmHg |
| Atmospheres (atm) | Pascals (Pa) | atm × 101 325 | 1.2 atm = 1.2 × 101 325 <br> = 121 590 Pa <br> = $1.21 \times 10^5$ Pa |
| Millimetres of mercury (torr) (mmHg) | Atmospheres (atm) | $\dfrac{mmHg}{760}$ | 820 mmHg = $\dfrac{820}{760}$ <br> = 1.08 atm |
| Millimetres of mercury (torr) (mmHg) | Pascals (Pa) | $\dfrac{mmHg}{760} \times 101\,325$ | 820 mmHg = $\dfrac{820}{760} \times 101\,325$ <br> = $1.09 \times 10^5$ Pa |
| Pascals (Pa) | Millimetres of mercury (torr) (mmHg) | $\dfrac{Pa}{101\,325} = 760$ | 102 800 Pa = $\dfrac{102\,800}{101\,325} \times 760$ <br> = 771 mmHg |

## Temperature

For everyday practical use the Celsius scale is used for temperature. This scale is named after the Swedish astronomer Anders Celsius, who in 1742 proposed the 100 degree scale. This scale has the melting point of ice is at zero degrees Celsius (0°C) and the boiling point of water at 100°C.

In scientific work, an artificial scale for temperature, known as the absolute temperature scale is used. Here, temperatures are quoted in kelvin (K). (Note there is no degree sign before the K.) This scale was proposed by Scottish physicist William Thomson, who later became Baron Kelvin of Langs in 1892. It is an extension of the degree Celsius scale down to −273°C. At this point, known as absolute zero (0 K), it is proposed that all particles will have zero kinetic energy; that is, all motion will have stopped. This then is the lower limit of temperature and should therefore be taken as zero on the temperature scale.

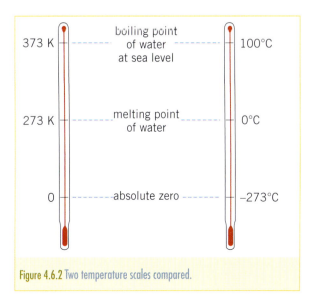

Figure 4.6.2 Two temperature scales compared.

Temperature scales

A change of one kelvin is the same as a change of one degree Celsius, but the two scales have different starting points.

To convert kelvin to degrees Celsius: °C = K − 273
To convert degrees Celsius to kelvin: K = °C + 273

## Volume

Volume, the space occupied by a gas, may be measured using a number of units. The relationship between these units is shown in figure 4.6.3.

### CHEM COMPLEMENT

**Another temperature scale**

In countries that do not use the metric system, such as the USA, everyday measurement of temperature uses the Fahrenheit scale. This scale, named after the German physicist Daniel Fahrenheit was one of the first natural scales and has 32°F as the freezing point of water and 212°F as the boiling point of water. Conversion between the Fahrenheit and Celsius scales is based on the relationship:

$$°C = \frac{(°F - 32)}{1.8} \quad \text{and} \quad °F = (°C \times 1.8) + 32$$

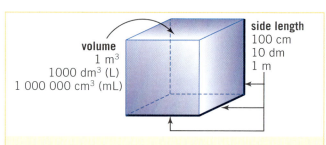

Figure 4.6.3 Volume may be measured using several different units.

1 dm = 10 cm
10 cm
10 cm
10 cm
1 dm³ = 1000 cm³

1 m = 10 dm
10 dm
10 dm
10 dm
1 m³ = 1000 dm³

Figure 4.6.4 Conversion of cubic units.

### Worked example 1

Convert the quantity stated into the unit shown in the brackets.

**a** 1.40 atm (kPa)     **b** 860 mmHg (kPa)
**c** 6.3 m³ (dm³)     **d** 52°C (K)

### Solution

**a** 1.40 atm = 1.40 × 101 325 Pa
           = 141 855 pa
           = 142 kPa (1 kPa = 1000 Pa)

**b** 860 mmHg = $\frac{860}{760}$ × 101 325 Pa
               = 114 657 Pa
               = 115 kPa

**c** 6.3 m³ = 6.3 × 10³ dm³

**d** 52°C = 52 + 273 = 325 K

## Section 4.6 Exercises

**1** Convert the following pressures into the unit shown in the brackets.
     **a** 200 600 Pa (kPa)      **b** 800 mmHg (kPa)
     **c** 120 kPa (Pa)      **d** 36 atm (kPa)

2. Convert the following temperatures into the unit shown in the brackets.
   a. 120°C (K)    b. 37°C (K)    c. 500 K (°C)
   d. 236 K (°C)   e. 25°C (K)

3. Convert the following volumes into the unit shown in the brackets.
   a. 4.1 m³ (dm³)              b. 200 cm³ (dm³)
   c. 50 cm³ (dm³)              d. 520 dm³ (m³)

4. Convert the quantity stated into the unit shown in the brackets.
   a. 790 mmHg (kPa)            b. 589 K (°C)
   c. 5.6 dm³ (cm³)             d. 629 cm³ (m³)
   e. 1.25 atm (kPa)            f. 203 kPa (atm)
   g. 39°C (K)                  h. 22.4 dm³ (cm³)

## 4.7 GASEOUS VOLUME RELATIONSHIPS IN CHEMICAL REACTIONS

The gas quantities, pressure, volume, amount of gas and temperature are related by a series of mathematical expressions known as the gas laws.

### Boyle's law

The Irish chemist Robert Boyle (1627–1691) performed the first quantitative experiments on gases. Boyle studied the relationship between the pressure of a gas and its volume. Using a J-shaped tube closed at one end, he measured the volume and pressure of the trapped gas. He found that, within the limits of his measurements, the product of the pressure and volume was constant.

Boyle's proposal is now known as Boyle's law:

> **The pressure exerted by a given mass of gas at constant temperature is inversely proportional to the volume occupied by the gas.**

This relationship may be represented in a number of ways. Graphically, two plots, shown in figure 4.7.1, may be used. A plot of $V$ versus $P$ produces a hyperbola, indicating an inverse relationship. Plotting $V$ versus $\frac{1}{P}$ produces a straight line with an intercept of zero.

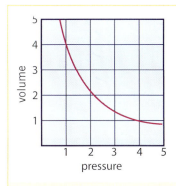
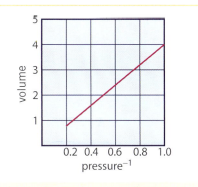

Figure 4.7.1 Two graphical representations of Boyle's law.

---

**THEORY OF KNOWLEDGE**
Theories are constructed to explain observations and help predict the results of future experiments. Kinetic molecular theory explains the properties of ideal gases, hypothetical gases that obey the gas laws. Real gases, those that are found in the atmosphere, and laboratory gases behave as ideal gases at low pressure and high temperature and under these conditions can be described using kinetic molecular theory. Kinetic molecular theory makes the following assumptions about atoms or molecules in ideal gases:

- They are in constant random motion.
- They are widely spaced apart.
- There are no intermolecular forces.
- Collisions are perfectly elastic; when the molecules collide their kinetic energy does not change.
- They have no size, they occupy zero volume.

Kinetic molecular theory demonstrates the important role of mathematics in the development of scientific ideas. Not only does kinetic theory fit the observational data, but mathematical logic gives scientists another way to reliably justify the explanations it provides, and a way to test it for consistency and freedom from contradiction.

**1.4.8**
**Analyse graphs relating to the ideal gas equation.** © IBO 2007

Figure 4.7.2 Robert Boyle stated that for a gas under constant temperature, the volume is inversely proportional to pressure.

Boyle's law

Pressure–volume relationships

Mathematically, the relationship is represented by the equation:

$$PV = k$$

where $k$ is a constant for a given sample of gas at a specified temperature.

Another useful representation of Boyle's law is the equation:

$$P_1V_1 = P_2V_2$$

where  $P_1$ = the initial gas pressure
$P_2$ = the final gas pressure
$V_1$ = the initial gas volume
$V_2$ = the final gas volume.

This relationship applies, provided that the temperature and amount of gas remain constant. Boyle's law is consistent with the kinetic molecular theory. If the volume of the gas container is increased, the particles travel greater distances between collisions with each other and the walls of the container. Fewer collisions with the walls mean decreased force per unit area and hence a decreased pressure.

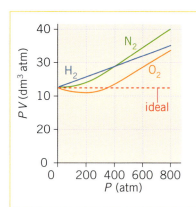

Figure 4.7.3 Graph of $PV$ versus $P$ for several gases. An ideal gas should have a constant value for $PV$.

The kinetic molecular theory applies to an ideal gas. In an ideal gas the gas particles are completely independent. Real gases behave less ideally when the pressure is high. Thus, real gases show some deviation from Boyle's law at high pressures. As shown by the graph of $P$ versus $PV$ in figure 4.7.3, $PV$ is not quite a constant as the pressure increases to values much higher than atmospheric pressure.

### Worked example 1

A perfectly elastic balloon has a volume of 1.2 dm³ at a pressure of 750 mmHg. Assuming the temperature remains constant, what volume will the balloon occupy if the pressure is reduced to 620 mmHg?

### Solution

$P_1$ = 750 mmHg  $\qquad P_2$ = 620 mmHg
$V_1$ = 1.2 dm³  $\qquad V_2$ = ? dm³

(Ensure that units for pressure are the same, and units for volume are the same.)

Boyle's law: $P_1V_1 = P_2V_2$

$$\therefore V_2 = \frac{P_1V_1}{P_2} = \frac{750 \times 1.2}{620} = 1.5 \text{ dm}^3$$

### Charles' law

More than a century after Boyle's findings, the French physicist Jacques Charles (1746–1823) discovered the relationship between the volume and temperature of a gas. Charles designed and flew the first hydrogen gas-filled balloon in 1783. In 1787, he proposed what is now known as Charles' law:

> **At constant pressure, the volume of a given mass of gas is directly proportional to its absolute temperature.**

Figure 4.7.5 shows this relationship graphically. A plot of volume of gas versus temperature (°C) is a straight line. When this graph is extrapolated, regardless of the gas used, the zero occurs at the same temperature: –273°C. Recall that this is the zero point of the absolute temperature scale, so a plot of volume versus temperature (K) is a straight line with a zero intercept.

Notice that for temperatures below 0 K, the gas volume would be negative. Gases cannot have a negative volume! This graph supports the idea of an absolute zero of temperature. If a gas could reach this point (gases liquefy before this point), the particles would lose all of their kinetic energy.

Mathematically, the relationship is represented by the equation:

$$\frac{V}{T} = k$$

Figure 4.7.4 Jacques Charles studied the relationship between gas volume and temperature.

where $k$ is a constant for a given sample of gas at a constant pressure, and $T$ is measured on the Kelvin scale.

Another useful representation of Charles' law is the equation:

$$\frac{V_1}{T_1} = \frac{V_2}{T_2}$$

where  $V_1$ = the initial gas volume
$V_2$ = the final gas volume
$T_1$ = the initial temperature (in K)
$T_2$ = the final temperature (in K)

This relationship applies provided that the pressure and amount of gas remain constant. Charles' law is consistent with the kinetic molecular theory. If the temperature of a gas is increased, the kinetic energy of the particles increases. The particles move faster, colliding with the container walls more often and with greater force. If the initial pressure is to be maintained, the volume of the container must increase.

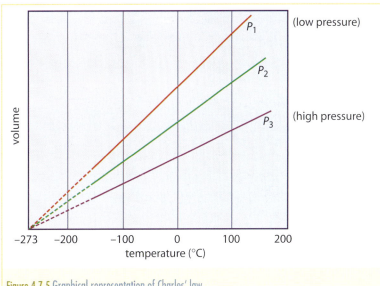

Figure 4.7.5 Graphical representation of Charles' law.

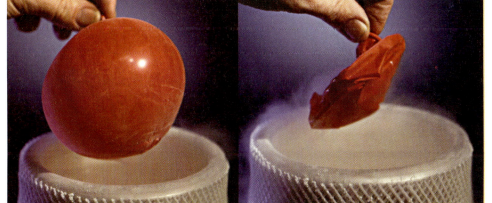

Charles' law

Figure 4.7.6 Charles' law in action. A balloon filled with air is dipped into liquid nitrogen at 77 K. Can you explain why the balloon collapses?

CHEMISTRY: FOR USE WITH THE IB DIPLOMA PROGRAMME **STANDARD LEVEL**

Do real gases obey Charles' law under all conditions? Figure 4.7.7 shows a plot of volume versus temperature for a number of gases. At what temperatures do real gases deviate from Charles' law? Which gases behave most ideally?

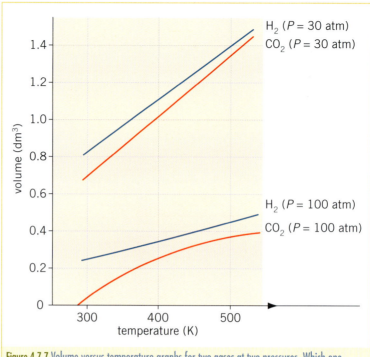

Figure 4.7.7 Volume versus temperature graphs for two gases at two pressures. Which one behaves ideally?

DEMO 4.2
The gas laws in action

### Worked example 2

A partially filled party balloon contains 2.6 dm³ of helium gas at atmospheric pressure and a temperature of 12°C. What volume will the balloon occupy if it warms to a temperature of 20°C at atmospheric pressure?

### Solution

$V_1 = 2.6$ dm³         $V_2 = ?$ dm³
$T_1 = 12 + 273 = 285$ K    $T_2 = 20 + 273 = 293$ K

Ensure that the temperatures are in kelvin.

Charles' law: $\dfrac{V_1}{T_1} = \dfrac{V_2}{T_2}$

$$\therefore V_2 = \dfrac{V_1}{T_1} \times T_2$$

$$= \dfrac{2.6}{285} \times 293$$

$$= 2.7 \text{ dm}^3$$

### Avogadro's law

In 1811, the Italian chemist Amedeo Avogadro (1776–1856) postulated that equal volumes of gases at the same temperature and pressure contain equal numbers of particles. This is now known as **Avogadro's law**, although when first stated it was not accepted by most chemists and was, in fact, neglected for nearly half a century.

Avogadro's law may be expressed mathematically as:

$V = kn$

where $k$ is a constant for a given temperature and pressure.

**AS 1.4.4**
**Apply Avogadro's law to calculate reacting volumes of gases. © IBO 2007**

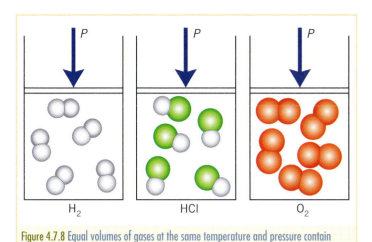

**Figure 4.7.8** Equal volumes of gases at the same temperature and pressure contain equal numbers of molecules.

## Worked example 3

A gas syringe at 25°C is initially filled with 0.00200 mole of carbon dioxide gas which makes the volume of the syringe 50 cm$^3$. What will be the volume of gas in the syringe if the amount of gas in it is raised to 0.00375 mole and temperature and pressure remain constant?

### Solution

$V_1 = 50.0$ dm$^3$                $V_2 = ?$ dm$^3$
$n_1 = 0.00200$ mol            $n_2 = 0.00375$ mol

Avogadro's law: $\dfrac{V_1}{n_1} = \dfrac{V_2}{n_2}$

$$\therefore V_2 = \dfrac{V_1}{n_1} \times n_2$$

$$= \dfrac{50.0}{0.00200} \times 0.00375$$

$$= 93.8 \text{ cm}^3$$

Avogadro's law can also be used to calculate volumes of gases that are made during a reaction. As the volumes of gases under a fixed set of conditions (temperature and pressure) are proportional to the number of mole, a molar ratio taken from an equation involving only gases, can be used to find a ratio of volumes.

Let us consider the combustion of methane under conditions of constant temperature and pressure:

$CH_4(g) + 2O_2(g) \rightarrow CO_2(g) + 2H_2O(g)$

According to the equation, methane and oxygen react in a 1:2 ratio.

Using Avogadro's law:   $n \propto V$

We can conclude that 10 dm$^3$ of methane will react with 20 dm$^3$ of oxygen.

### Worked example 4

Calculate the volume of carbon dioxide that will be released when 1000 dm³ of

**a** ethanol, $C_2H_5OH$

**b** octane, $C_8H_{18}$

undergo complete combustion in a car engine, and hence determine which of the two fuels has greater potential as an environmentally friendly fuel. Assume that the conditions of temperature and pressure are kept constant.

#### Solution

Step 1: Write a balanced chemical equation for the reactions under consideration.

    **a** $C_2H_5OH(g) + 3O_2(g) \rightarrow 2CO_2(g) + 3H_2O(g)$

    **b** $2C_8H_{18} + 25O_2(g) \rightarrow 16CO_2(g) + 18H_2O(g)$

Step 2: List all data given, including relevant units.

    **a** $V(C_2H_5OH) = 1000$ dm³

    **b** $V(C_8H_{18}) = 1000$ dm³

Step 3: Use Avogadro's law and the molar ratios from the chemical equations to find the volume of $CO_2$ in each case.

$C_2H_5OH : CO_2$          $C_8H_{18} : CO_2$
1 : 2                              2 : 16
1000 : 2000            $1000 : \frac{1000}{2} \times 16 = 8000$

Ethanol will produce 2000 dm³ of $CO_2$ and octane will produce 8000 dm³ of $CO_2$ when 1000 dm³ of the fuel is burnt. Because ethanol produces less $CO_2$ per dm³ it is a more environmentally friendly fuel (although energy output of the two fuels should also be considered).

## The ideal gas equation

**1.4.7 Solve problems using the ideal gas equation, $PV = nRT$.**
© IBO 2007

So far we have considered three laws that describe the behaviour of gases.

    Boyle's law:     $V \propto \frac{1}{P}$ at constant $n$ and $T$

    Charles' law:    $V \propto T$ at constant $n$ and $P$

    Avogadro's law: $V \propto n$ at constant $P$ and $T$

Therefore:   $V \propto \frac{nT}{P}$

or          $V = \frac{nRT}{P}$ where $R$ is a constant

Rearranging this gives: $PV = nRT$

This equation is known as the ideal gas equation, or the general gas equation. The constant $R$ is called the universal gas constant. The value of $R$ depends on the units chosen for $V$ and $P$ (remember that $T$ must be measured on the Kelvin scale). The value most commonly used today is based on these units:

    $P$ in kPa     $V$ in dm³     $T$ in K     $n$ in mol

Using these units gives a value of $R$ as 8.31 J K⁻¹ mol⁻¹.

Interactive Animation
Pressure–temperature relationship

## Worked example 5

What volume will 52.0 g of carbon dioxide gas occupy at a temperature of 24°C and 206 kPa?

Applying the ideal gas equation

### Solution

$P = 206$ kPa  
$T = 24 + 273 = 297$ K  
$R = 8.31$ J K$^{-1}$ mol$^{-1}$

$V = ?$  
$m = 52.0$ g

Ideal gas equation: $PV = nRT$

$n = \dfrac{m}{M} \therefore PV = \dfrac{m}{M}RT$

$\therefore V = \dfrac{mRT}{PM}$

$= \dfrac{52.0 \times 8.31 \times 297}{206 \times 44.0}$

$= 14.2$ dm$^3$

---

When using the ideal gas equation, ensure that the correct units are used; that is, $P$ in kPa, $V$ in dm$^3$, $T$ in K.

The ideal gas equation defines the behaviour of an ideal gas. Most gases approach this behaviour at low pressures. We can therefore use this equation to determine one gas quantity (say, $P$) if the other three quantities ($V$, $T$ and $n$) are known.

$$\dfrac{P_1 V_1}{n_1 T_1} = \dfrac{P_2 V_2}{n_2 T_2}$$

where $P_1$ and $P_2$ = the initial and final gas pressures  
$V_1$ and $V_2$ = the volumes  
$T_1$ and $T_2$ = the temperatures  
$n_1$ and $n_2$ = the amounts of gas.

We frequently deal with situations in which the amount of gas is fixed. For this fixed amount of gas, the ideal gas equation reduces to what is sometimes called the combined gas law or combined gas equation.

Given $\dfrac{P_1 V_1}{n_1 T_1} = \dfrac{P_2 V_2}{n_2 T_2}$

If $n_1 = n_2$  $\dfrac{P_1 V_1}{T_1} = \dfrac{P_2 V_2}{T_2}$

**AS 1.4.6**
Solve problems involving the relationship between temperature, pressure and volume for a fixed mass of an ideal gas. © IBO 2007

This equation can be used to solve for any one of the six quantities if the other five are known. Remember when using this equation that the units for $P_1$ and $P_2$ must be the same, the units for $V_1$ and $V_2$ the same, and temperature must be measured on the Kelvin scale.

### Worked example 6

An expandable balloon contains 95.0 dm³ of helium at 760 mmHg and 24°C. What volume will the balloon occupy when the pressure drops to 180 mmHg and the temperature is 11°C?

### Solution

$P_1$ = 760 mmHg          $P_2$ = 180 mmHg
$V_1$ = 95.0 dm³          $V_2$ = ?
$T_1$ = 24 + 273 = 297 K  $T_2$ = 11 + 273 = 284 K

Ensure units for pressure are the same, units for volume are the same and temperatures are in kelvin. The amount of gas is constant; that is, $n_1 = n_2$, so the general gas equation is used.

$$\frac{P_1 V_1}{T_1} = \frac{P_2 V_2}{T_2}$$

$$\therefore V_2 = \frac{P_1 V_1 T_2}{T_1 P_2} = \frac{760 \times 95.0 \times 284}{297 \times 180}$$

$$= 384 \text{ dm}^3$$

WORKSHEET 4.5
Gas calculations

## Molar volume and standard conditions

**AS 1.4.5**
Apply the concept of molar volume at standard temperature and pressure in calculations.
© IBO 2007

Suppose we have 1.00 mole of an ideal gas at 0°C (273 K) and 1.0 atm (101.3 kPa). We can calculate the volume of the gas at these conditions using the ideal gas equation:

$$V = \frac{nRT}{P} = \frac{1.00 \times 8.31 \times 273}{101.3} = 22.4 \text{ dm}^3$$

This volume is known as the **molar volume** ($V_m$) of the gas under the specified conditions.

A gas will expand to fill any container, so it is pointless to specify a gas volume without specifying its temperature and pressure. The conditions chosen here, 0°C and 1 atm (101.3 kPa), are called **standard temperature and pressure (or STP)**, and are often used when comparing gases.

The molar volume at STP calculated as 22.4 dm³ applies to an ideal gas. How do real gases compare? Some values are listed in table 4.7.1. The values are very close to the ideal for some, but others deviate significantly. Can you suggest why?

PRAC 4.5
Determining the molar volume of a gas

| TABLE 4.7.1 MOLAR VOLUMES OF GASES AT STP ||
| Gas | Molar volume (dm³) |
| --- | --- |
| Oxygen | 22.397 |
| Nitrogen | 22.402 |
| Hydrogen | 22.433 |
| Helium | 22.434 |
| Argon | 22.397 |
| Carbon dioxide | 22.260 |
| Ammonia | 22.079 |

Many chemical reactions involve gases. We can use these molar volumes (assuming ideal gas behaviour) to carry out stoichiometric calculations.

## Worked example 7

What volume would 1.20 dm³ of neon at –20°C and 62.0 kPa occupy at STP?

### Solution

Convert to STP (273 K, 101.3 kPa).

$P_1 = 62.0$ kPa  
$V_1 = 1.20$ dm³  
$T_1 = -20 + 273 = 253$ K

$P_2 = 101.3$ kPa  
$V_2 = ?$  
$T_2 = 273$ K

$$\frac{P_1 V_1}{T_1} = \frac{P_2 V_2}{T_2}$$

$$\therefore V_2 = \frac{P_1 V_1 T_2}{T_1 P_2}$$

$$= \frac{62.0 \times 1.20 \times 253}{273 \times 101.3}$$

$$= 0.681 \text{ dm}^3$$

It occupies 0.681 dm³.

## Worked example 8

Phosphorus burns in chlorine according to the equation:

$$P_4(s) + 6Cl_2(g) \rightarrow 4PCl_3(l)$$

What mass of $PCl_3$ is produced when excess phosphorus is burnt in 355 cm³ of chlorine at STP?

### Solution

Recall from section 4.5 the basic steps in a stoichiometry problem.

Step 1: Write a balanced equation for the reaction.

$$P_4(s) + 6Cl_2(g) \rightarrow 4PCl_3(l)$$

Step 2: List all data given, including relevant units.

**Known**  
$V(Cl_2) = 355$ cm³  
$V_m$ at STP = 22.4 dm³

**Unknown**  
$m(PCl_3) = ?$  
$M(PCl_3) = 137.32$ g mol⁻¹

Step 3: Convert the data given about the 'known' to mol, using the relevant formulas.

$$n(Cl_2) = \frac{V}{V_m} = \frac{355}{22400} = 1.585 \times 10^{-2} \text{ mol}$$

Step 4: Use the chemical equation to determine the mole ratio of the unknown quantity to the known quantity.

$$P_4(s) + \mathbf{6Cl_2}(g) \rightarrow \mathbf{4PCl_3}(l)$$

$$n(PCl_3) = \frac{4}{6} \times n(Cl_2) = \frac{4}{6} \times 1.585 \times 10^{-2} = 1.057 \times 10^{-2} \text{ mol}$$

Step 5: Convert this number of mole back into the relevant units of the unknown.

$$m(PCl_3) = n \times M = 1.057 \times 10^{-2} \times 137.32 = 1.451 \text{ g}$$

1.451g of $PCl_3$ is produced.

**WORKSHEET 4.6**
Gas stoichiometry

### Worked example 9

Magnesium burns brightly in air to form magnesium oxide. It is determined that 0.590 g of magnesium burns in oxygen at 19°C and 102.5 kPa pressure. What volume of oxygen is required?

### Solution

Step 1: Write a balanced equation for the reaction.

$$2Mg(s) + O_2(g) \rightarrow 2MgO(s)$$

Step 2: List all data given, including relevant units.

$m = 0.590$ g  $\qquad P = 102.5$ kPa
$T = 19 + 273 = 292$ K  $\qquad V = ?$

Step 3: Convert the data given about the 'known' to mole, using the relevant formulas.

$$n(Mg) = \frac{m}{M} = \frac{0.590}{24.31} = 0.024\,27 \text{ mol}$$

Step 4: Use the chemical equation to determine the mole ratio of the unknown quantity to the known quantity.

$$2Mg(s) + O_2(g) \rightarrow 2MgO(s)$$
$$n(O_2) = \frac{1}{2} \times n(Mg) = \frac{1}{2} \times 0.02\,427 = 0.012\,14 \text{ mol}$$

Step 5: Convert this number of mole back into the relevant units of the unknown.

$$V(O_2) = \frac{nRT}{P} = \frac{0.012\,14 \times 8.31 \times 292}{102.5} = 0.287 \text{ dm}^3$$

### TABLE 4.7.2 GAS RELATIONSHIPS

| Relationship | Formula | Units |
|---|---|---|
| Ideal gas equation | $PV = nRT$ | $P$ in kPa<br>$V$ in dm$^3$<br>$T$ in K |
| For a fixed amount of gas at constant temperature (Boyle's law) | $PV = k$ or $P_1V_1 = P_2V_2$ | $P_1$ and $P_2$ in the same units<br>$V_1$ and $V_2$ in the same units<br>$k$ is a constant |
| For a fixed amount of gas at constant pressure (Charles' law) | $\frac{V}{T} = k$ or $\frac{V_1}{T_1} = \frac{V_2}{T_2}$ | $T$ in K<br>$V_1$ and $V_2$ in the same units<br>$k$ is a constant |
| For a fixed amount of gas (General gas equation) | $\frac{PV}{T} = k$ or $\frac{P_1V_1}{T_1} = \frac{P_2V_2}{T_2}$ | $P_1$ and $P_2$ in the same units<br>$V_1$ and $V_2$ in the same units<br>$T$ in K<br>$k$ is a constant |
| For a gas at constant pressure and temperature (Avogadro's law) | $\frac{V}{n} = k$ or $\frac{V_1}{n_1} = \frac{V_2}{n_2}$ | $n$ in mol<br>$V_1$ and $V_2$ in the same units<br>$k$ is a constant |
| For an amount of gas under conditions of standard temperature and pressure (STP) | $n = \frac{V}{V_m}$ | $V$ in dm$^3$<br>$V_m$ in dm$^3$<br>$n$ in mol |

## Section 4.7 Exercises

1. A 320 cm³ sample of nitrogen gas at 1.5 atm and 10°C was compressed until the pressure reached 2.4 atm. Calculate what the expected volume of the sample would be when the temperature returned to 10°C.

2. If the pressure on a 3.0 dm³ sample of gas is halved, calculate the new volume of the gas (assuming constant temperature).

3. During an experiment to study Boyle's law, the following results were obtained.

| Reading | P (atm) | V (cm³) |
|---|---|---|
| 1 | 0.395 | 588 |
| 2 | 1.180 | 72 |
| 3 | 0.526 | 444 |
| 4 | 0.450 | 516 |
| 5 | 0.921 | 252 |

   a State which two gas measures were kept constant during the experiment.
   b State which reading of volume (1–5) was incorrect.

4. An experiment to investigate Boyle's law was conducted using a fixed mass of ammonia at 25°C. With a pressure of 20.0 kPa the volume was found to be 122 dm³.
   a Predict the expected volume of ammonia at a pressure of 800 kPa.
   b The measured volume at 800 kPa was 2.94 cm³. Account for the difference between this value and the one predicted in part **a**.

5. A balloon can hold 900 cm³ of air before bursting. The balloon contains 750 cm³ of air at 10°C. Will the balloon burst if it is moved to an area with a temperature of 25°C (at the same pressure)?

6. A sample of oxygen gas at 20°C is heated until its temperature reaches 40°C. If the final volume of the oxygen gas is 150 cm³, what was the original volume (assuming constant pressure)?

7. Copy and complete the sketch graphs shown.

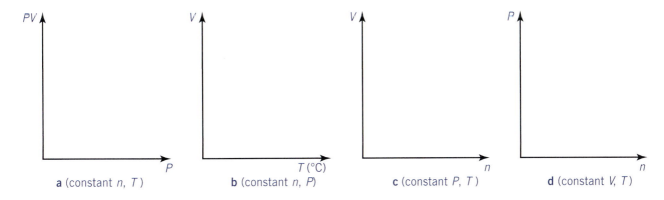

a (constant $n$, $T$)    b (constant $n$, $P$)    c (constant $P$, $T$)    d (constant $V$, $T$)

8  During an experiment to investigate Charles' law, the data shown was collected for a gas sample at constant pressure.

| Reading | V (dm³) | T (°C) |
|---|---|---|
| 1 | 5.42 | 50 |
| 2 | 5.84 | 75 |
| 3 | 6.49 | 100 |
| 4 | 6.68 | 125 |

State which reading (1–4) was incorrect.

9  What is the volume occupied by a 3.00 g sample of hydrogen gas at 28°C and 98.6 kPa?

10  A 0.675 g sample of gas occupies 250.0 cm³ at 25°C and 152 kPa. Determine the molar mass of the gas.

11  A helium-filled weather balloon with a volume of 25.0 dm³ is released at a temperature of 20.0°C and an atmospheric pressure of 101.4 kPa. It reaches a height where the temperature is 10.5°C and atmospheric pressure is 84.2 kPa. Calculate what the volume of the balloon will be at this height.

12  The gas in a scuba diver's tank at 20°C exerts a pressure of $9.0 \times 10^2$ kPa. If the tank is allowed to warm to a temperature of 35°C. Calculate the pressure that the gas in the tank will reach.

13  A sample of neon gas has a volume of 10.0 dm³ at 101 kPa and 0°C. Calculate the temperature (in °C) that is required to reduce the volume of the gas to 1.0 dm³ at 127 kPa.

14  Compare gas samples I and II and state which one contains the greatest number of molecules. (Show your working.)

Sample I:   300 cm³ of $O_2$ at 20°C and 740 mmHg

Sample II:  200 cm³ of $N_2$ at 30°C and 0.75 atm

15  Ammonia is made when hydrogen and nitrogen are reacted together:

$N_2(g) + 3H_2(g) \rightarrow 2NH_3(g)$

Calculate the volume of ammonia that will be produced if 15 dm³ of nitrogen reacts with 45 dm³ of hydrogen with the pressure and temperature kept constant.

## 4.8 SOLUTIONS

**AS 1.5.1**
Distinguish between the terms *solute, solvent, solution* and *concentration* (g dm⁻³ and mol dm⁻³). © IBO 2007

Water is an extremely effective solvent, able to dissolve many different substances due to its high polarity. All water found in nature will contain dissolved substances; that is, water is found as a solution. A **solution** is made up of one or more **solutes** (dissolved substances) dissolved in a **solvent** (bulk substance). When water is the solvent, we call the solution aqueous. Examples of aqueous solutions are everywhere. Many commercial products such as wine, vinegar, domestic cleaners and soft drinks are aqueous solutions. Solutions are clear, which means that you can see through them, but not necessarily colourless. If a solid dissolves in a solvent, it is said to be soluble in that solvent. If it does not dissolve, it is insoluble. There is a spectrum of solubilities between these two extremes.

The **concentration** of a solution is a quantitative expression of how much solute is dissolved in the solution. Concentration can be expressed in $g\,dm^{-3}$ or $mol\,dm^{-3}$. Other units of concentration are used to describe very low concentrations of solutions; these include parts per million, ppm ($mg\,dm^{-3}$) and parts per billion, ppb ($\mu g\,dm^{-3}$).

**1.5.2 Solve problems involving concentration, amount of solute and volume of solution.**
© IBO 2007

## Calculating concentration of a solution

The concentration of a solution can be calculated by dividing the amount of solute (in mol) by the volume of the solution (in $dm^3$).

$$c = \frac{n}{V} \text{ or } n = cV$$

where $c$ = concentration of solution ($mol\,dm^{-3}$)
$n$ = number of mole of solute (mol)
$V$ = volume of solution ($dm^3$)

This formula can be transposed to calculate any of the three variables, given the other two values.

## Worked example 1

Calculate the concentration (in $mol\,dm^{-3}$) of each of the following solutions.

**a** 2.0 mol of NaCl is dissolved in 8.0 $dm^3$ of water.
**b** 0.25 mol of $MgCl_2$ is dissolved in 500 $cm^3$ of water.
**c** 30.0 g of sucrose ($C_{12}H_{22}O_{11}$) is dissolved in 200 $cm^3$ of water.

### Solution

**a** $c = \frac{n}{V}$, where $c = ?$, $n = 2.0$ mol, $V = 8.0\,dm^3$

$\therefore c(NaCl) = \frac{2.0}{8.0} = 0.25\,mol\,dm^{-3}$

**b** With this problem, we must first convert the unit for volume from $cm^3$ to $dm^3$, where $1000\,cm^3 = 1\,dm^3$.

$c = \frac{n}{V}$, where $c = ?$, $n = 0.25$ mol, $V = 500\,cm^3 = 0.500\,dm^3$

$\therefore c(MgCl_2) = \frac{0.25}{0.500} = 0.50\,mol\,dm^{-3}$

**c** To solve this problem, we must first calculate the number of moles of sucrose present.

$n(C_{12}H_{22}O_{11}) = \frac{m}{M}$, where $n = ?$, $m = 30.0$ g, $M(C_{12}H_{22}O_{11}) = 342.34\,g\,mol^{-1}$

$\therefore n(C_{12}H_{22}O_{11}) = \frac{30.0}{342.34} = 0.0876$ mol

It is now possible to calculate the molar concentration of glucose in the solution:

$c = \frac{n}{V}$, where $c = ?$, $n = 0.0876$ mol, $V = 200\,cm^3 = 0.200\,dm^3$

$\therefore c(C_{12}H_{22}O_{11}) = \frac{0.0876}{0.200} = 0.438\,mol\,dm^{-3}$

### CHEM COMPLEMENT

#### A drop in the ocean

To help understand roughly what various concentrations mean, we can equate the concentration of a liquid dye in water as follows: 1% w/w equates to one drop in a teaspoon of water, 1 ppm to a drop in a full bathtub, 1 ppb to one drop in a full-sized swimming pool and 1 ppt (part per trillion) is a drop in 1000 swimming pools. It has been estimated that the world's oceans may contain as much as US$10 trillion worth of gold. Unfortunately, this gold is found at a concentration of just 13 parts per trillion, or just 25 mg in an Olympic-sized swimming pool full of seawater. Two questions to consider: What is the current value of 25 mg of gold, and is it worth extracting the gold from seawater?

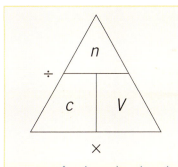

Figure 4.8.1 A formula triangle can be used to make it easier to rearrange this formula.

### Worked example 2

Calculate the amount (in mol) of the substance stated that will be required to make the following solutions.

**a** 250 cm³ of 0.250 mol dm⁻³ H₂SO₄

**b** 10 cm³ of 0.825 mol dm³ NaOH

### Solution

**a** $c = \dfrac{n}{V}$, so $n = cV$ where $c = 0.250$ mol dm⁻³, $V = 0.250$ dm³, $n = ?$ mol

∴ $n(H_2SO_4) = 0.250 \times 0.250 = 0.0625$ mol

**b** $c = \dfrac{n}{V}$, so $n = cV$ where $c = 0.825$ mol dm⁻³, $V = 0.010$ dm³, $n = ?$ mol

∴ $n(NaOH) = 0.825 \times 0.010 = 0.00825$ mol

## Calculating concentrations of components of solutions

It is important to recognize that the molar concentration refers to the amount of solute that has been dissolved in a given volume of the solvent water. For example, a 1.0 mol dm⁻³ solution of hydrochloric acid (HCl) would have 1 mole of HCl molecules per dm³ of water. As HCl ionizes fully in water, the HCl will exist as a 1.0 mol dm⁻³ solution of $H^+$ ions and a 1.0 mol dm⁻³ solution of $Cl^-$ ions. Similarly, a 0.500 mol dm⁻³ solution of copper(II) sulfate would exist in solution as a 0.500 mol dm⁻³ solution of $Cu^{2+}$ and a 0.500 mol dm⁻³ solution of $SO_4^{2-}$ ions.

How do we determine the concentration of individual ions in situations where the cations and anions are not in a 1:1 ratio? As we would expect, the concentration of individual ions in solution is in direct proportion to the chemical formula of the compound. Thus, in a 0.25 mol dm⁻³ solution of $NiCl_2$, the concentration of nickel ions is 0.25 mol dm⁻³, while the chloride ion concentration is twice that, at 0.50 mol dm⁻³. A 2.00 mol dm⁻³ solution of ammonium sulfate (($NH_4)_2SO_4$) exists in solution as ammonium ions at a concentration of 4.00 mol dm⁻³ and sulfate ions as 2.00 mol dm⁻³.

If two or more solutions are mixed and there is no chemical reaction between individual solutes, we can calculate the concentration of each individual ion by determining the total amount of that ion present and dividing by the final volume of solution.

### Worked example 3

Aluminium sulfate is used in large quantities as a flocculating agent in the treatment of wastewater. Calculate the molar concentration of both aluminium and sulfate ions when 40.00 kg of aluminium sulfate is dissolved in 250.0 dm³ of water.

### Solution

$n(Al_2(SO_4)_3) = \dfrac{m}{M}$, where $n = ?$, $m = 40\,000$ g, $M(Al_2(SO_4)_3) = 342.14$ g mol$^{-1}$

$\therefore n(Al_2(SO_4)_3) = \dfrac{40\,000}{342.14} = 116.9$ mol

It is now possible to calculate the molar concentration of aluminium sulfate as a whole in the solution:

$c = \dfrac{n}{V}$, where $c = ?$, $n = 116.9$ mol, $V = 250.0$ dm$^3$

$\therefore c(Al_2(SO_4)_3) = \dfrac{116.9}{250.0} = 0.4676$ mol dm$^{-3}$

$\therefore c(Al^{3+}) = 2 \times 0.4676 = 0.9352$ M and $c(SO_4^{2-}) = 3 \times 0.4676 = 1.403$ mol dm$^{-3}$

## Dilution of solutions

It is reasonably common for chemists to dilute a concentrated solution to produce a solution of a lower specified concentration. As the original solution has merely been diluted with the addition of water, the actual amount (number of moles) of the original solute has not changed. The number of moles of solute in a solution can be determined by applying the formula $n = cV$. This leads us to a formula linking the concentration and volume of an original solution to its diluted form:

$c_1 V_1 = c_2 V_2$

where  $c_1$ = initial concentration of solution
 $V_1$ = initial volume of solution
 $c_2$ = concentration of diluted solution
 $V_2$ = volume of diluted solution

Note that the formula is effectively a ratio and so units of volume must correspond; it is not necessary to convert all volumes to dm$^3$ as long as they are the same.

### Worked example 4

A student adds 200 cm$^3$ of deionized water to 50 cm$^3$ of 0.20 mol dm$^{-3}$ HCl solution. Determine the concentration of the diluted hydrochloric acid solution.

### Solution

**Known**  
$c_1 = 0.20$ mol dm$^{-3}$  
$V_1 = 50$ cm$^3$

**Unknown**  
$c_2 = ?$  
$V_2 = (200 + 50) = 250$ cm$^3$

$c_1 V_1 = c_2 V_2$

$\therefore c_2 = \dfrac{c_1 V_1}{V_2} = \dfrac{0.20 \times 50}{250} = 0.040$ mol dm$^{-3}$

The concentration of the diluted solution is 0.040 mol dm$^{-3}$.

# Using concentrations and volumes of solution to find amounts of products

## Worked example 5

A solution of 20.0 cm³ of 0.250 mol dm⁻³ lead(II) nitrate is added to a large volume of 0.500 mol dm⁻³ potassium iodide solution to produce a brilliant yellow precipitate of lead(II) iodide. Write an equation for the reaction and so determine the mass of precipitate formed.

### Solution

Note that in this problem we are informed that a 'large volume' of potassium iodide solution is used. This indicates that the potassium iodide must be in excess. The term 'in excess' implies that there is a sufficient quantity of this reactant to ensure that all of the lead(II) nitrate solution is consumed in generating the precipitate.

Step 1: Write a balanced equation for the reaction.

$$Pb(NO_3)_2(aq) + 2KI(aq) \rightarrow PbI_2(s) + 2KNO_3(aq)$$

Step 2: List all data given, including relevant units.

**Known**
$c(Pb(NO_3)_2) = 0.250$ mol dm⁻³
$V(Pb(NO_3)_2) = 0.0200$ dm³

**Unknown**
$m(PbI_2) = ?$
$M(PbI_2) = 460.99$ g mol⁻¹

Step 3: Convert the data given for the 'known' to mol, using the relevant formula.

$$c(Pb(NO_3)_2) = \frac{n}{V}$$

$$\therefore n(Pb(NO_3)_2) = cV = 0.250 \times 0.0200 = 5.00 \times 10^{-3} \text{ mol}$$

Step 4: Use the chemical equation to determine the mole ratio of the unknown quantity to the known quantity.

$$\mathbf{Pb(NO_3)_2}(aq) + 2KI(aq) \rightarrow \mathbf{PbI_2}(s) + 2KNO_3(aq)$$

From the equation, $n(PbI_2) = \frac{1}{1} \times n(Pb(NO_3)_2) = 5.00 \times 10^{-3}$ mol

Step 5: Convert this number of mole back into the relevant units of the unknown.

$$n(PbI_2) = \frac{m}{M}$$

$$\therefore m(PbI_2) = 5.00 \times 10^{-3} \text{ mol} \times 460.99 = 2.31 \text{ g}$$

The mass of lead(II) iodide precipitated is 2.31 g.

## Worked example 6

Magnesium reacts readily with dilute hydrochloric acid to produce hydrogen gas, the presence of which can be ascertained by the 'pop' test.

**a** Calculate the volume of 1.50 mol dm⁻³ hydrochloric acid that would be required to completely consume 20.00 g of magnesium.

**b** Calculate the volume of hydrogen gas that would be produced in this reaction at STP ($V_m = 22.4$ dm³ mol⁻¹).

**Solution**

**a** Step 1: Write a balanced equation for the reaction.

$Mg(s) + 2HCl(aq) \rightarrow MgCl_2(aq) + H_2(g)$

Step 2: List all data given, including relevant units.

**Known**
$m(Mg) = 20.00$ g
$M(Mg) = 24.31$ g mol$^{-1}$

**Unknown**
$c(HCl) = 1.50$ mol dm$^{-3}$
$V(HCl) = ?$

Step 3: Convert the data given for the known quantity to mol, using the relevant formulas.

$n(Mg) = \dfrac{m}{M} = \dfrac{20.00}{24.31} = 0.8227$ mol

Step 4: Use the chemical equation to determine the mole ratio of the unknown quantity to the known quantity.

**Mg**(s) + **2HCl**(aq) → MgCl$_2$(aq) + H$_2$(g)

From the equation, $n(HCl) = 2 \times n(Mg) = 1.645$ mol

Step 5: Convert this number of mole back into the relevant units of the unknown.

$c(HCl) = \dfrac{n}{V}$

$\therefore V(HCl) = \dfrac{n}{c} = \dfrac{1.645}{1.50} = 1.10$

The volume of hydrochloric acid required is 1.10 dm$^3$.

Figure 4.8.2 The reaction between magnesium and hydrochloric acid produces bubbles of hydrogen gas.

**b** Step 1: Write a balanced equation for the reaction.

$Mg(s) + 2HCl(aq) \rightarrow MgCl_2(aq) + H_2(g)$

Step 2: List all data given, including relevant units.

**Known**
$m(Mg) = 20.00$ g
$M(Mg) = 24.31$ g mol$^{-1}$

**Unknown**
$V_m(H_2) = 1.50$ mol dm$^{-3}$
$V(H_2) = ?$

Step 3: Convert the data given for the 'known' to mol, using the relevant formula.

$n(Mg) = \dfrac{m}{M} = \dfrac{20.00}{24.31} = 0.8227$ mol

Step 4: Use the chemical equation to determine the mole ratio of the unknown quantity to the known quantity.

**Mg**(s) + 2HCl(aq) → MgCl$_2$(aq) + **H$_2$**(g)

From the equation, $n(H_2) = n(Mg) = 0.8227$ mol

Step 5: Convert this number of mole back into the relevant units of the unknown.

$V(H_2) = n \times V_m$
$\therefore V(H_2) = 0.8227 \times 22.4 = 18.4$

The volume of hydrogen gas released is 18.4 dm$^3$.

DEMO 4.3
Stoichiometry of a precipitation reaction

PRAC 4.6
Precipitation reactions

PRAC 4.7
Chemical detectives

**WORKSHEET 4.7**
Solution stoichiometry

| Examples in which both reactants and products are solutions |

## Worked example 7

Sulfuric acid is a strong acid that reacts completely with the alkali sodium hydroxide to produce water and the salt sodium sulfate. Calculate the volume of $0.486$ mol dm$^{-3}$ sodium hydroxide solution required to completely neutralize $35.5$ cm$^3$ of $0.606$ mol dm$^{-3}$ sulfuric acid.

### Solution

Step 1: Write a balanced equation for the reaction.

$$H_2SO_4(aq) + 2NaOH(aq) \rightarrow Na_2SO_4(aq) + H_2O(l)$$

Step 2: List all data given, including relevant units.

**Known**
$c(H_2SO_4) = 0.606$ mol dm$^{-3}$
$V(H_2SO_4) = 0.0355$ dm$^3$

**Unknown**
$c(NaOH) = 0.486$ mol dm$^{-3}$
$V(NaOH) = ?$

Step 3: Convert the data given for the 'known' to mol, using the relevant formula.

$$c(H_2SO_4) = \frac{n}{V}$$

$\therefore n(H_2SO_4) = 0.606 \times 0.0355 = 0.0215$ mol

Step 4: Use the chemical equation to determine the mole ratio of the unknown quantity to the known quantity.

**$H_2SO_4$(aq) + $2NaOH$(aq)** $\rightarrow Na_2SO_4(aq) + H_2O(l)$

From the equation, $n(NaOH) = \frac{2}{1} \times n(H_2SO_4) = 0.0430$ mol

Step 5: Convert this number of mole back into the relevant units of the unknown.

$$c(NaOH) = \frac{n}{V}$$

$\therefore V(NaOH) = \frac{n}{c} = \frac{0.0430}{0.486} = 0.0885$ dm$^3$

The volume of sodium hydroxide required is $88.5$ cm$^3$.

## Worked example 8

An industrial chemist working for the Environmental Protection Authority wishes to analyse a sample of water for the presence of iron(II) ions. He measures a $400$ cm$^3$ portion of the water and reacts it with $0.0225$ mol dm$^{-3}$ potassium permanganate solution until all of the $Fe^{2+}$(aq) ions present are oxidized to $Fe^{3+}$(aq) ions. The ionic equation for this redox reaction is:

$$MnO_4^-(aq) + 8H^+(aq) + 5Fe^{2+}(aq) \rightarrow Mn^{2+}(aq) + 4H_2O(l) + 5Fe^{3+}(aq)$$

If $4.22$ cm$^3$ of permanganate solution was required to effect the oxidation, determine the concentration of iron(II) ions in the sample of water.

## Solution

Step 1: The equation is supplied.

$$MnO_4^-(aq) + 8H^+(aq) + 5Fe^{2+}(aq) \rightarrow Mn^{2+}(aq) + 4H_2O(l) + 5Fe^{3+}(aq)$$

Step 2: List all data given, including relevant units.

**Known**
$c(MnO_4^-) = 0.0225$ mol dm$^{-3}$
$V(MnO_4^-) = 4.22 \times 10^{-3}$ dm$^3$

**Unknown**
$c(Fe^{2+}) = ?$
$V(Fe^{2+}) = 0.400$ dm$^3$

Step 3: Convert the data given for the 'known' to mol, using the relevant formula.

$$c(MnO_4^-) = \frac{n}{V}$$

$$\therefore n(MnO_4^-) = 0.0225 \times 4.22 \times 10^{-3} = 9.50 \times 10^{-5} \text{ mol}$$

Step 4: Use the chemical equation to determine the mole ratio of the unknown quantity to the known quantity.

$$\mathbf{MnO_4^-}(aq) + 8H^+(aq) + \mathbf{5Fe^{2+}}(aq) \rightarrow Mn^{2+}(aq) + 4H_2O(l) + 5Fe^{3+}(aq)$$

From the equation, $n(Fe^{2+}) = \frac{5}{1} \times n(MnO_4^-) = 4.75 \times 10^{-4}$ mol

Step 5: Convert this number of mole back into the relevant units of the unknown.

$$c(Fe^{2+}) = \frac{n}{V} = \frac{4.75 \times 10^{-4}}{0.400} = 1.19 \times 10^{-3} \text{ mol dm}^{-3}$$

The concentration of iron(II) ions in the water sample was $1.19 \times 10^{-3}$ mol dm$^{-3}$.

## Limiting reagent questions relating to solutions

### Worked example 9

A solution of 18.0 cm$^3$ of 0.496 mol dm$^{-3}$ sodium chloride is added to 24.0 cm$^3$ of 0.288 mol dm$^{-3}$ silver nitrate solution to generate a precipitate of silver chloride. Determine which reactant is in excess and by what amount, and then calculate the mass of precipitate formed.

**WORKSHEET 4.8**
Excess reagent stoichiometry

### Solution

Step 1: The equation for the reaction is:

$$NaCl(aq) + AgNO_3(aq) \rightarrow AgCl(s) + NaNO_3(aq)$$

Step 2: List all data given, including relevant units.

**Known 1**
$c(NaCl) = 0.496$ mol dm$^{-3}$
$V(NaCl) = 0.018$ dm$^3$

**Known 2**
$c(AgNO_3) = 0.288$ mol dm$^{-3}$
$V(AgNO_3) = 0.024$ dm$^3$

**Unknown**
$m(AgCl) = ?$
$M(AgCl) = 143.3$ g mol$^{-1}$

Step 3: Convert the data given for the 'known' quantities to mol, using the relevant formulas.

$$c = \frac{n}{V}$$

$\therefore n(\text{NaCl}) = cV$  AND  $n(\text{AgNO}_3) = cV$
$\quad\quad\quad\quad = 0.496 \times 0.018$ $\quad\quad\quad\quad\quad\quad = 0.288 \times 0.024$
$\quad\quad\quad\quad = 8.93 \times 10^{-3}$ mol $\quad\quad\quad\quad = 6.91 \times 10^{-3}$ mol

Step 4a: Determine the reactant in excess.

The equation for this reaction indicates that NaCl and AgNO$_3$ react in a 1:1 ratio.

NaCl : AgNO$_3$
1 : 1

If all of the NaCl ($8.93 \times 10^{-3}$ mol) reacts, then $n(\text{AgNO}_3)$ that must be available to react $= 8.93 \times 10^{-3}$ mol

BUT there is only $6.91 \times 10^{-3}$ mol of AgNO$_3$ available to react, so the NaCl is in excess and $n(\text{AgNO}_3)$ is the limiting factor.

Step 4b: Use the chemical equation to determine the mole ratio of the unknown quantity to the known quantity. Note that we must use the limiting reactant to calculate the mass of precipitate formed.

NaCl(aq) + **AgNO$_3$**(aq) → **AgCl**(s) + NaNO$_3$(aq)

From the equation, $n(\text{AgCl}) = \frac{1}{1} \times n(\text{AgNO}_3) = 6.91 \times 10^{-3}$ mol

Step 5: Convert this number of mole back into the relevant units of the unknown.

$$n(\text{AgCl}) = \frac{m}{M}$$

$\therefore m(\text{AgCl}) = 6.91 \times 10^{-3} \times 143.32 = 0.990$ g

The mass of silver chloride precipitated is 0.990 g.

## Worked example 10

Arsenic undergoes oxidation by a hot, concentrated solution of sodium hydroxide to produce sodium arsenate and hydrogen gas according to the equation:

2As(s) + 6NaOH(aq) → 2Na$_3$AsO$_3$(s) + 3H$_2$(g)

6.57 g of arsenic is reacted with 250 cm$^3$ of 0.779 mol dm$^{-3}$ sodium hydroxide solution. Determine which reactant is in excess and by what amount.

### Solution

Step 1: The equation for the reaction is as provided.

2As(s) + 6NaOH(aq) → 2Na$_3$AsO$_3$(s) + 3H$_2$(g)

Step 2: List all data given, including relevant units.

**Known 1**
$m(\text{As}) = 6.57$ g
$M(\text{As}) = 74.92$ g mol$^{-1}$

**Known 2**
$c(\text{NaOH}) = 0.779$ mol dm$^{-3}$
$V(\text{NaOH}) = 0.250$ dm$^3$

**Unknown**
$m(\text{H}_2) = ?$
$M(\text{H}_2) = 2.02$ g mol$^{-1}$

Step 3: Convert the data given for the 'known' quantities to mol, using the relevant formulas.

$$n(\text{As}) = \frac{m}{M}$$
$$= \frac{6.57}{74.92}$$
$$= 0.0877 \text{ mol}$$

AND

$$c(\text{NaOH}) = \frac{n}{V}$$
$$\therefore n(\text{NaOH}) = c \times V$$
$$= 0.779 \times 0.250$$
$$= 0.195 \text{ mol}$$

Step 4a: Determine the reactant in excess.

Molar ratio of reactants:

As : NaOH
1 : 3
0.0877 : 3 × 0.0877
= 0.263

If all of the arsenic were to react, 0.263 mol of NaOH would be required. We have only 0.195 mole of NaOH, so NaOH is the limiting reagent.

Step 4b: To determine how much of the As is in excess use the limiting reagent to calculate how much arsenic will react:

**2As**(s) + **6NaOH**(aq) → $2\text{Na}_3\text{AsO}_3$(s) + $3\text{H}_2$(g)

NaOH : As
3 : 1
$0.195 : \dfrac{0.195}{3}$
= 0.065

The amount of arsenic that does not react = 0.0877 − 0.065
= 0.0227 mol

The arsenic is in excess by 0.0227 mol.

## Section 4.8 Exercises

1. Calculate the concentration (in mol dm$^{-3}$) of each of the following solutions.
   a. 0.40 mol of $NH_4NO_3$ is dissolved in 5.00 dm$^3$ of water.
   b. $5.928 \times 10^{-4}$ mol of saccharin ($C_7H_5NO_3S$) is dissolved in 375 cm$^3$ of water.
   c. 2.761 g of $AgNO_3$ is dissolved in 35.5 cm$^3$ of water.
   d. 45 µg of chromium is dissolved in 200 dm$^3$ of wastewater (1 µg = $10^{-6}$ g).

2. Calculate the mass of solute that would be required to make up the following solutions.
   a. 1.50 dm$^3$ of 0.250 mol dm$^{-3}$ $H_2SO_4$
   b. 250 cm$^3$ of 0.125 mol dm$^{-3}$ $CuSO_4$
   c. 425 cm$^3$ of 0.086 mol dm$^{-3}$ $Na_2S_2O_3$
   d. 5.00 cm$^3$ of 1.25 mol dm$^{-3}$ HCl

**3** Determine the concentration (in mol dm$^{-3}$) of chloride ion in each of the following solutions.

    **a** 24.5 g of barium chloride dissolved in 150 cm$^3$ of solution

    **b** 36.8 g of aluminium chloride in a total volume of 250.0 cm$^3$

**4** A solution of 50.0 cm$^3$ of 0.250 mol dm$^{-3}$ copper(II) nitrate is mixed with 80.0 cm$^3$ of 0.275 mol dm$^{-3}$ zinc nitrate. Calculate the concentration of nitrate ions in the resultant solution.

**5** If a mixture of 15.8 g of sodium hydroxide and 22.4 g of potassium hydroxide is dissolved in 800 cm$^3$ of water, determine the concentration of hydroxide ions in the resultant solution.

**6** A laboratory technician is requested to produce 2.0 dm$^3$ of 0.50 mol dm$^{-3}$ nitric acid solution. What volume of concentrated 14 mol dm$^{-3}$ acid will be needed to produce the dilute solution?

**7** A student adds 25.0 cm$^3$ of distilled water to 125 cm$^3$ of 4.55 mol dm$^{-3}$ MgBr$_2$ solution. What will be the resultant concentration of the solution?

**8** What volume of water must be added to 50.0 cm$^3$ of 8.00 mol dm$^{-3}$ ammonia to reduce its concentration to 1.50 mol dm$^{-3}$?

**9** A student places a beaker containing 1500 cm$^3$ of 0.835 mol dm$^{-3}$ sodium chloride solution onto a hotplate and reduces the volume to 600 cm$^3$ by evaporation. Determine the new concentration of the salt solution.

**10** When an iron nail is left in a beaker of copper(II) sulfate solution, the nail will decolourize the solution over a period of time. Copper metal is deposited onto the nail in a redox reaction, the equation for which is:

    CuSO$_4$(aq) + Fe(s) → FeSO$_4$(aq) + Cu(s)

Determine the mass of copper metal that will be deposited from the complete reaction of 200 cm$^3$ of a 0.150 mol dm$^{-3}$ copper(II) sulfate solution.

**11** A student places 3.89 g of copper metal turnings into a 100 cm$^3$ beaker and pours a large volume of 7.20 mol dm$^{-3}$ nitric acid into the beaker. The copper metal is oxidized to copper(II) ions, and nitrogen monoxide gas is given off according to the equation:

    3Cu(s) + 2HNO$_3$(aq) + 6H$^+$(aq) → 3Cu$^{2+}$(aq) + 2NO(g) + 4H$_2$O(l)

Calculate the volume of nitric acid that would be required to completely consume the copper metal in this reaction.

**12** What volume of SO$_2$ would be expected to form at STP when 4.80 g of sulfur was burnt in excess oxygen?

**13** Hydrogen gas may be prepared by the reaction of zinc granules with dilute hydrochloric acid. What mass of zinc is required to produce 2.60 dm$^3$ of hydrogen gas at 102 kPa and 30°C?

**14** What volume of 0.588 mol dm$^{-3}$ sodium hydroxide solution would be required to exactly neutralize 25.00 cm$^3$ of 0.500 mol dm$^{-3}$ nitric acid? The equation for this acid–base reaction is:

$HNO_3(aq) + NaOH(aq) \rightarrow NaNO_3(aq) + H_2O(l)$

**15** A precipitate of nickel carbonate results from the addition of 75.00 cm$^3$ of 0.814 mol dm$^{-3}$ sodium carbonate solution to a beaker containing excess nickel chloride solution. The equation for this reaction is:

$Na_2CO_3(aq) + NiCl_2(aq) \rightarrow NiCO_3(s) + 2NaCl(aq)$

What mass of precipitate will form?

**16** To neutralize 70.4 cm$^3$ of 0.722 mol dm$^{-3}$ nitric acid, a student gradually added a solution of magnesium hydroxide until the pH of the mixture was 7.0. If 58.0 cm$^3$ of the base was required to effect neutralization, what was the concentration of the Mg(OH)$_2$ solution?

The equation for the reaction is:

$Mg(OH)_2(aq) + 2HNO_3(aq) \rightarrow Mg(NO_3)_2(aq) + 2H_2O(l)$

**17** When a solution of 0.307 mol dm$^{-3}$ potassium permanganate is added to 87.1 cm$^3$ of 0.772 mol dm$^{-3}$ tin(II) chloride, the tin(II) ions are oxidized to tin(IV) ions according to the ionic equation:

$2MnO_4^-(aq) + 16H^+(aq) + 5Sn^{2+}(aq) \rightarrow 2Mn^{2+}(aq) + 8H_2O(l) + 5Sn^{4+}(aq)$

What volume of potassium permanganate was required to fully oxidize the tin(II) ions present?

**18** Nitric oxide is produced by the complete oxidation of ammonia, according to the equation:

$4NH_3(g) + 5O_2(g) \rightarrow 4NO(g) + 6H_2O(g)$

If 3.5 mol of ammonia is reacted with 3.0 mol of oxygen, determine which reactant is in excess and the amount of excess reactant that is present.

**19** 5.95 g of iron filings are sprinkled into 450.0 cm$^3$ of 0.932 mol dm$^{-3}$ nickel nitrate solution. The green colour of the nickel solution fades slightly over time as nickel metal is displaced from solution according to the ionic equation:

$Ni^{2+}(aq) + Fe(s) \rightarrow Ni(s) + Fe^{2+}(aq)$

Determine the concentration of nickel ions remaining in solution after all of the iron has been oxidized.

**20** When the blue liquid dinitrogen trioxide is added to a solution of sodium hydroxide, it reacts to give sodium nitrite, according to the equation;

$N_2O_3(l) + 2NaOH(aq) \rightarrow 2NaNO_2(aq) + H_2O(l)$

If 14.5 g of dinitrogen trioxide is added to 35.0 cm$^3$ of 2.31 mol dm$^{-3}$ sodium hydroxide, determine the mass of sodium nitrite produced in the reaction.

# Chapter 4 Summary

## Terms and definitions

**Avogadro's constant**  The number of elementary particles in one mole of a substance. Equal to approximately $6.02 \times 10^{23}$. Symbol: $L$

**Avogadro's law**  Equal volumes of gases at the same temperature and pressure contain equal numbers of particles.

**Concentration**  The amount of solute in a given volume of solution, expressed in g dm$^{-3}$, or mol dm$^{-3}$.

**Empirical formula**  The lowest whole-number ratio of elements in a compound.

**Experimental yield**  The amount of product made during an experiment.

**Ideal gas**  A gas in which the gas particles are completely independent.

**Law of conservation of mass**  During a chemical reaction the total mass of the reactants is equal to the total mass of the products.

**Limiting reactant**  The reactant which that is used up completely in a chemical reaction.

**Molar mass**  The mass of one mole of a substance. Symbol: $M$; unit: g mol$^{-1}$

**Molar ratio**  The ratio in which reactants and products in a chemical equation react; indicated by the coefficients written in front of each of the reactants and products in the equation.

**Molar volume**  The volume occupied by one mole of a gas under a given set of conditions of temperature and pressure.

**Mole**  The amount of substance containing the same number of elementary particles as there are atoms in 12 g exactly of carbon-12. Symbol: $n$; unit: mol

**Molecular formula**  A formula that shows the actual number of atoms of each element present in one molecule of a compound.

**Percentage composition**  The amount of each element in a compound expressed as a percentage.

**Percentage yield**  A calculation of the experimental yield as a percentage of the theoretical yield.

**Reactant in excess**  A reactant which may not be used up completely in a chemical reaction.

**Relative atomic mass**  The weighted mean of the masses of the naturally occurring isotopes of an element on a scale in which the mass of an atom of carbon-12 is taken to be 12 exactly. Symbol: $A_r$

**Relative formula mass**  The sum of the relative atomic masses of the elements as given in the formula for any non-molecular compound.

**Relative molecular mass**  The sum of the relative atomic masses of the elements as given in the molecular formula of a compound. Symbol: $M_r$

**Solute**  The dissolved component of a solution.

**Solution**  A homogeneous mixture of a solute in a solvent.

**Solvent**  A substance, usually a liquid, that is able to dissolve another substance, the solute.

**Standard temperature and pressure (STP)**  A set of conditions applied to gaseous calculations where the temperature is 0°C and the pressure is 1 atm (101.3 kPa).

**State symbols (s), (l), (g), (aq)**  Symbols used in chemical equations to indicate the state of a reactant or product.

**Theoretical yield**  The amount of product that is expected to be produced in a reaction, based on 100% reaction of the reactants.

## Concepts

- The mole is a measure of an amount of substance and one mole always constitutes the same number of particles.

- Avogadro's constant, $L$ ($6.02 \times 10^{23}$), is the term used to describe the number of particles in one mole of any substance.

- The numerical value of relative atomic, molecular and formula masses (no units) are the same as the molar mass (units of g mol$^{-1}$) of an element or compound.

- The percentage composition of a compound can be calculated, given its formula or relevant mass composition data.

- The empirical formula of a compound can be determined by using mass or percentage composition data.

- The molecular formula of a compound is a whole-number multiple of the empirical formula. Given the empirical formula and the molar mass, the molecular formula of a compound can be calculated.

- Balanced chemical and ionic equations provide information about the mole ratios of reactants to products that is required to perform stoichiometric calculations.

- Chemical equations are useful in stoichiometry, but they provide no information about rate or extent of reaction.

- The basic steps in solving stoichiometric problems are the following.

    Step 1: Write a balanced chemical equation for the reaction.

    Step 2: List all data given, including relevant units. Remember to also write down the symbol of the unknown quantity.

    Step 3: Convert the data given to moles, using the relevant formulas; that is:

    $$n = \frac{m}{M},\ n = cV,\ n = \frac{N}{L},\ n = \frac{V}{V_m},\ n = \frac{RT}{PV}$$

    where $n$ = amount of substance (mol)
    $m$ = mass (g), $M$ = molar mass (g mol$^{-1}$)
    $c$ = concentration (mol dm$^{-3}$)
    $V$ = volume of a solution or of a gas (dm$^3$)
    $N$ = number of particles
    $L$ = Avogadro's constant ($6.02 \times 10^{23}$)
    $V_m$ = molar volume of a gas (dm$^3$ mol$^{-1}$)
    $R$ = gas constant (8.31 J K$^{-1}$ mol$^{-1}$)
    $T$ = temperature (K)
    $P$ = pressure (kPa)

    Step 4: Use the chemical equation to determine the mole ratio of the unknown quantity to the known quantity (that is, $\frac{\text{unknown}}{\text{known}}$). This ratio enables calculation of the number of moles of the unknown quantity.

    Step 5: Finally, convert this number of mole back into the relevant units of the unknown. The problem is solved!

- When information is provided about more than one reactant it is likely that one will be in excess and the other is, therefore, the limiting reactant. This reactant must be used in the calculation.

- A percentage yield is an expression of the success of an experiment.

    $$\text{Percentage yield} = \frac{\text{experimental yield}}{\text{theoretical yield}} \times \frac{100}{1}$$

- Avogadro's law states that, at constant temperature and pressure, the volume of a gas is directly proportional to the amount of gas (in mol).

    $V = kn$ where $k$ is a constant

- The ideal (or general) gas equation combines several gas laws and may be expressed as:

    $PV = nRT$

    where $R$ = 8.31 J K$^{-1}$ mol$^{-1}$ and is known as the ideal gas constant. $R$ has this value when the units used are $P$ in kPa, $V$ in dm$^3$, $T$ in K, $n$ in mol.

- Another useful form of the ideal gas equation is:

    $$\frac{P_1 V_1}{n_1 T_1} = \frac{P_2 V_2}{n_2 T_2}$$

- When the amount of gas is fixed ($n_1 = n_2$), this reduces to the combined gas law:

    $$\frac{P_1 V_1}{T_1} = \frac{P_2 V_2}{T_2}$$

- The molar volume of a gas is the volume occupied by one mole of the gas at stated conditions of temperature and pressure. A frequently used set of conditions is standard temperature and pressure (STP):

    $T$ = 0°C, $P$ = 101.3 kPa, $V_m$ = 22.4 dm$^3$ mol$^{-1}$

- A solute dissolves in a solvent to make a solution.

- Concentration is a measure of the amount of solute in a given amount of solution.

# Chapter 4 Review questions

**REVIEW TABLE A REMINDER WHEN CALCULATING THE AMOUNT (IN MOL) OF A SUBSTANCE**

| | |
|---|---|
| $n = \dfrac{m}{M}$ | $n$ = amount in mol<br>$m$ = mass in g<br>$M$ = molar mass in g mol$^{-1}$ |
| $n = \dfrac{N}{L}$ | $N$ = number of particles<br>$L$ = Avogadro's constant = $6.02 \times 10^{23}$ mol$^{-1}$ |
| $n = cV$ | $c$ = concentration in mol dm$^{-3}$<br>$V$ = volume in dm$^3$ |
| $n = \dfrac{V}{V_m}$ | $V$ = volume in dm$^3$<br>$V_m$ = molar volume in dm$^3$<br>(at STP: 0°C, 1 atm, $V_m$ = 22.4 dm$^3$ mol$^{-1}$) |
| $n = \dfrac{PV}{RT}$ | $P$ = pressure in kPa<br>$V$ = volume in dm$^3$<br>$R$ = gas constant (8.31 J K$^{-1}$ mol$^{-1}$)<br>$T$ = temperature in K |

1. **a** Define the term *mole*.
   **b** How many particles are there in a mole of helium atoms?
   **c** What is the special name given to the number of particles in a mole?

2. Determine the number of atoms present in:
   **a** 0.625 mol of $H_2O$ molecules
   **b** 3.042 mol of $SO_3$
   **c** 4.00 g of oxygen molecules
   **d** $6.704 \times 10^{-3}$ g of sodium thiosulfate ($Na_2S_2O_3$).

3. Calculate the amount of substance (in mol) of each of the following.
   **a** 18 g of carbon atoms
   **b** 6.41 g of sulfur dioxide ($SO_2$)
   **c** 10.1 g of chlorine atoms
   **d** 100 g of water ($H_2O$)
   **e** 5000 g of magnesium oxide (MgO)
   **f** 0.20 g of sodium hydroxide (NaOH)

4. 0.950 mol of a certain element has a mass of 112.75 g. Determine the molar mass of the element and thus determine its name.

5. A certain compound of copper is used extensively as a desiccating agent (absorbs water from the surroundings). If the formula of this compound is $CuXO_4$ in its anhydrous state and 0.355 mol has a mass of 56.66 g, determine the identity of the element X.

6. Calculate the mass of each of the following.
   **a** 15 molecules of ammonia, $NH_3$
   **b** 250 molecules of ethane, $C_2H_6$
   **c** 1 molecule of methanol, $CH_3OH$

7. Chloral hydrate ($C_2H_3Cl_3O_2$) is a drug that is used as a sedative and hypnotic. In combination with alcohol, it can result in a person being rendered unconscious (or dead, as it is highly toxic). It is often described as a 'Mickey Finn' in films and detective stories. Determine the number of molecules of chloral hydrate present in 500 mg of the substance.

8. Calculate the percentage by mass of each element present in the following compounds.
   **a** NaBr
   **b** $H_2O_2$
   **c** $Li_2S$
   **d** $CuSO_4$

9. A certain gaseous hydrocarbon has a percentage composition by mass of 80% carbon and 20% hydrogen. The molar mass of the compound is 30 g mol$^{-1}$. Determine both the empirical and molecular formulas of the hydrocarbon.

10. Lead exists as two different oxides: PbO and $PbO_2$. An industrial chemist is supplied with a 3.094 g sample of a substance known to be an oxide of lead. Upon analysis, its elemental mass composition is found to be 2.68 g lead and 0.414 g oxygen. Determine the empirical formula of the oxide and so ascertain which oxide is present.

11. Police confiscate a white crystalline powder they suspect is cocaine. The substance is purified and analysed by a forensic chemist, who establishes its percentage composition by mass as 67.327% carbon, 6.931% hydrogen, 4.620% nitrogen and 21.122% oxygen. The molecular formula of cocaine is known to be $C_{17}H_{21}NO_4$. Were the police correct in their suspicions?

12. Balance the following equations.
    **a** $NaCN(s) + H_2SO_4(aq) \rightarrow Na_2SO_4(aq) + HCN(g)$
    **b** $C_6H_{14}(l) + O_2(g) \rightarrow CO_2(g) + H_2O(l)$
    **c** $NaHCO_3(aq) + H_3C_6H_5O_7(aq)$
    $\rightarrow CO_2(g) + H_2O(l) + Na_3C_6H_5O_7(aq)$

13. Write balanced equations for the following reactions.
    **a** In aqueous solution, hydrogen peroxide ($H_2O_2$) is unstable and decomposes to give water and oxygen gas ($O_2$).

b The first step in extracting zinc from its ore is roasting in air. The solid zinc sulfide (ZnS) reacts with oxygen ($O_2$), producing solid zinc oxide (ZnO) and gaseous sulfur dioxide ($SO_2$).

14 Propane is a major component of liquid petroleum gas (LPG), which is used in some vehicles as an alternative fuel to petrol. Propane undergoes combustion in a plentiful supply of air to produce carbon dioxide and water according to the equation:

$$C_3H_8(g) + 5O_2(g) \rightarrow 3CO_2(g) + 4H_2O(l)$$

Calculate the amount (in mol) of carbon dioxide gas that would be produced from the complete combustion of 4.915 moles of propane.

15 At temperatures in excess of 1000°C, as are commonly found in high-temperature incinerators, nitrogen gas reacts with oxygen to produce nitrogen monoxide, according to the equation:

$$N_2(g) + O_2(g) \rightarrow 2NO(g)$$

Determine the mass of nitrogen gas oxidized in the production of 0.500 g of nitrogen monoxide.

16 Marble chips (calcium carbonate) dissolve readily in hydrochloric acid to produce calcium chloride, carbon dioxide and water. The equation for this reaction is:

$$CaCO_3(s) + 2HCl(aq) \rightarrow CaCl_2(aq) + CO_2(g) + H_2O(l)$$

Calculate the mass of carbon dioxide which would be produced by the complete reaction of 2.778 g of marble chips in excess acid.

17 Ammonia is a very important chemical in a wide range of industries, including fertilizer production and cleaning agents. It is produced industrially by the Haber process according to the equation:

$$N_2(g) + 3H_2(g) \rightarrow 2NH_3(g)$$

500 mol of hydrogen gas and 150 mol of nitrogen gas are reacted with each other.

a Which gas is in excess, and by what amount?
b How many moles of ammonia would be produced (assuming complete reaction)?

18 A student was carrying out an investigation in which the method to produce a dry sample of a precipitate was designed, and the percentage yield obtained was determined. The student mixed 10.0 cm$^3$ of 0.100 mol dm$^{-3}$ sodium sulfate with 10.0 cm$^3$ of 1.00 mol dm$^{-3}$ of lead(II) nitrate. A precipitate of lead(II) sulfate was filtered and allowed to dry. The dry precipitate was weighed and found to have a mass of 0.298 g.

a Write a balanced chemical equation for the reaction.
b Determine which reactant was in excess.
c Calculate the theoretical mass of lead(II) sulfate the student should have obtained.
d Calculate the percentage yield obtained.
e State a possible reason as to why only this percentage was obtained.

19 When bread dough is cooked, the carbon dioxide present in the dough expands. A dough sample contains 0.10 dm$^3$ of $CO_2$ at 25°C. What volume will it expand to when the dough is cooked to 100°C (assuming constant pressure)?

20 A gas sample has a volume of $x$ dm$^3$ at 20°C. If the pressure is halved, what temperature is required to maintain the volume at $x$ dm$^3$?

21 1.297 g of one of the halogens (group 7) was found to occupy a volume of 450.0 cm$^3$ at a pressure of 749 mmHg and a temperature of 23°C. Which halogen was it?

22 Hydrogen gas may be prepared by the reaction of zinc granules with dilute hydrochloric acid. What mass of zinc is required to produce 2.60 dm$^3$ of hydrogen gas at 102 kPa and 30°C?

23 The compound cisplatin ($Pt(NH_3)_2Cl_2$) is used as a possible agent in destroying certain types of tumour cells. It is synthesized by the following reaction:

$$K_2PtCl_4(aq) + 2NH_3(aq) \rightarrow Pt(NH_3)_2Cl_2(s) + 2KCl(aq)$$

What volume of 0.0338 mol dm$^{-3}$ $K_2PtCl_4$ would be required to generate 0.662 g of cisplatin if it were reacted with excess ammonia solution?

24 Hydrazine, a highly reactive compound widely used as a propellant in rocket engines, is prepared by reacting ammonia with a solution of sodium hypochlorite according to the ionic equation:

$$2NH_3(aq) + OCl^-(aq) \rightarrow N_2H_4(aq) + Cl^-(aq) + H_2O(l)$$

Determine the mass of hydrazine produced when 120 dm$^3$ of 7.50 mol dm$^{-3}$ ammonia is reacted with 50.0 dm$^3$ of 12.0 mol dm$^{-3}$ sodium hypochlorite solution.

www.pearsoned.com.au/schools

Weblinks are available on the Chemistry: For use with the IB Diploma Programme Standard Level Companion Website to support learning and research related to this chapter.

# Chapter 4 Test

## Part A: Multiple-choice questions

1. The empirical formula of a compound is $C_2H_4O$. Which molecular formulas are possible for this compound?
   I   $CH_3COOH$
   II  $CH_3CH_2CH_2COOH$
   III $CH_3COOCH_2CH_3$
   A  I and II only
   B  I and III only
   C  II and III only
   D  I, II and III
   © IBO 2006, Nov P1 Q1

2. Calcium carbonate decomposes on heating as shown below.
   $$CaCO_3 \rightarrow CaO + CO_2$$
   When 50 g of calcium carbonate are decomposed, 7 g of calcium oxide are formed. What is the percentage yield of calcium oxide?
   A  7%   B  25%   C  50%   D  75%
   © IBO 2006, Nov P1 Q2

3. Sodium reacts with water as shown below.
   $$\_\_Na + \_\_H_2O \rightarrow \_\_NaOH + \_\_H_2$$
   What is the total of all the coefficients when the equation is balanced using the smallest possible whole numbers?
   A  3   B  4   C  6   D  7
   © IBO 2006, Nov P1 Q3

4. Which contains the same number of ions as the value of Avogadro's constant?
   A  0.5 mol NaCl
   B  0.5 mol $MgCl_2$
   C  1.0 mol $Na_2O$
   D  1.0 mol MgO
   © IBO 2006, May P1 Q1

5. The equation for a reaction occurring in the synthesis of methanol is
   $$CO_2 + 3H_2 \rightarrow CH_3OH + H_2O$$
   What is the maximum amount of methanol that can be formed from 2 mol of carbon dioxide and 3 mol of hydrogen?
   A  1 mol   B  2 mol   C  3 mol   D  5 mol
   © IBO 2006, May P1 Q3

6. Which solution contains 0.1 mol of sodium hydroxide?
   A  1 cm³ of 0.1 mol dm⁻³ NaOH
   B  10 cm³ of 0.1 mol dm⁻³ NaOH
   C  100 cm³ of 1.0 mol dm⁻³ NaOH
   D  1000 cm³ of 1.0 mol dm⁻³ NaOH
   © IBO 2006, May P1 Q4

7. The relative molecular mass ($M_r$) of a compound is 60. Which formulas are possible for this compound?
   I   $CH_3CH_2CH_2NH_2$
   II  $CH_3CH_2CH_2OH$
   III $CH_3CH(OH)CH_3$
   A  I and II only
   B  I and III only
   C  II and III only
   D  I, II and III
   © IBO 2005, Nov P1 Q2

8. A cylinder of gas is at a pressure of 40 kPa. The volume and temperature (in K) are both doubled. What is the pressure of the gas after these changes?
   A  10 kPa   B  20 kPa
   C  40 kPa   D  80 kPa
   © IBO 2006, May P1 Q14

9. Which change will have the greatest effect on the pressure of a fixed mass of an ideal gas?

   |   | Volume | Temperature (K) |
   |---|--------|-----------------|
   | A | Doubles | Halves |
   | B | Doubles | Doubles |
   | C | Halves | Halves |
   | D | Halves | Remains constant |

   © IBO 2001, May P1 Q15

10. The temperature (in K) is doubled for a sample of gas in a flexible container while the pressure on it is doubled. The final volume of the gas compared with the initial volume will be:
    A  the same.           B  twice as large.
    C  four times as large. D  half as large.
    © IBO 1999, May P1 Q15
    (10 marks)

 www.pearsoned.com.au/schools

For more multiple-choice test questions, connect to the Chemistry: For use with the IB Diploma Programme Standard Level Companion Website and select Chapter 4 Review Questions.

## Part B: Short-answer questions

1. Methylamine can be manufactured by the following reaction.
   $$CH_3OH(g) + NH_3(g) \rightarrow CH_3NH_2(g) + H_2O(g)$$
   In the manufacturing process 2000 kg of each reactant are mixed together.
   a  Identify the limiting reactant, showing your working.
   (2 marks)

**b** Calculate the maximum mass, in kg, of methylamine that can be obtained from this mixture of reactants.

(2 marks)

© IBO 2006, Nov P2 Q2c

**2** A 1.00 g sample of an organic compound contains 0.400 g carbon, 0.0666 mol hydrogen and $2.02 \times 10^{22}$ atoms of oxygen. Determine the empirical formula of the compound.

(3 marks)

© IBO 2001, Nov P2 Q6a

**3** Butane, $C_4H_{10}$, and but-2-ene, $C_4H_8$, are both colourless gases at 70°C. Calculate the volume that 0.0200 mol of butane would occupy at 70°C and $1.10 \times 10^5$ Pa.

(3 marks)

© IBO 2006, Nov P2 Q5aiii

**4** A student was asked to make some copper(II) sulfate-5-water ($CuSO_4 \cdot 5H_2O$) by reacting copper(II) oxide (CuO) with sulfuric acid.

  **a** Calculate the molar mass of copper(II) sulfate-5-water.

(1 mark)

  **b** Calculate the amount (in mol) of copper(II) sulfate-5-water in a 10.0 g sample.

(1 mark)

  **c** Calculate the mass of copper(II) oxide needed to make this 10.0 g sample.

(1 mark)

© IBO 2002, May P2 Q3

**5** Explain, in terms of molecules, what happens to the **pressure** of a sample of hydrogen gas if its volume is halved and the temperature kept constant.

(3 marks)

© IBO 2000, May P2 Q3b

**6** Zinc reacts with hydrochloric acid according to the following equation:

$Zn(s) + 2HCl(aq) \rightarrow ZnCl_2(aq) + H_2(g)$

What volume of 2.00 mol dm$^{-3}$ hydrochloric acid is needed to react completely with 13.08 g of zinc?

(3 marks)

## Part C: Data-based questions

**1** An organic compound **A** contains 62.0% by mass of carbon, 24.1% by mass of nitrogen, the remainder being hydrogen.

  **a** Determine the percentage by mass of hydrogen and the empirical formula of **A**.

(3 marks)

  **b** Define the term *relative molecular mass*.

(2 marks)

  **c** The relative molecular mass of **A** is 116. Determine the molecular formula of **A**.

(1 mark)

© IBO 2006, Nov P2 Q1a

**2** Aspirin, $C_9H_8O_4$, is made by reacting ethanoic anhydride, $C_4H_6O_3$ ($M_r = 102.1$), with 2-hydroxybenzoic acid, $C_7H_6O_3$ ($M_r = 138.1$), according to the equation:

$2C_7H_6O_3 + C_4H_6O_3 \rightarrow 2C_9H_8O_4 + H_2O$

  **a** If 15.0 g 2-hydroxybenzoic acid is reacted with 15.0 g ethanoic anhydride, determine the limiting reagent in this reaction.

(2 marks)

  **b** Calculate the maximum mass of aspirin that could be obtained in this reaction.

(2 marks)

  **c** If the mass obtained in this experiment was 13.7 g, calculate the percentage yield of aspirin.

(1 mark)

© IBO 1998, Nov P2 Q2

## Part D: Extended-response question

Indigo is a blue dye that contains only carbon, nitrogen, hydrogen and oxygen.

  **a** 2.036 g of indigo was completely oxidized to produce 5.470 g of carbon dioxide and 0.697 g of water. Calculate:

  **i** the percentage by mass of carbon in indigo

(2 marks)

  **ii** the percentage by mass of hydrogen in indigo.

(2 marks)

  **b** If the percentage by mass of nitrogen in the indigo sample is 10.75%, determine the empirical formula of indigo.

(4 marks)

  **c** If the molar mass is approximately 260 g mol$^{-1}$, determine the molecular formula of indigo.

(2 marks)

© IBO 2001, Nov (HL) P2 Q2

Total marks = 50

# 5 MEASUREMENT AND DATA PROCESSING

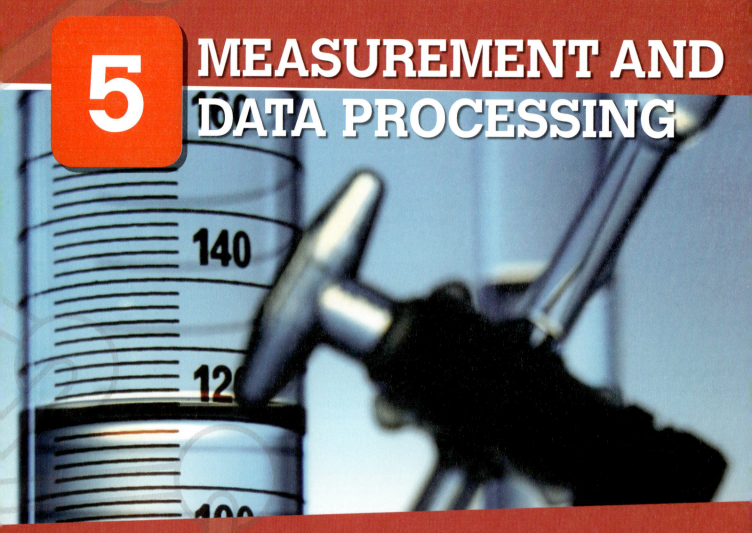

## Chapter overview

This chapter covers the IB syllabus Topic 11: Measurement and Data Processing.

**By the end of this chapter, you should be able to:**
- distinguish between precision and accuracy
- state the difference between random uncertainties and systematic errors
- give examples of systematic errors and random uncertainties and describe how the effects of random uncertainties may be minimized
- recognize the appropriate number of significant figures for the result of a calculation and state that result accordingly
- determine the uncertainties in results and state them as absolute and percentage uncertainties
- sketch graphs to represent dependencies between variables and interpret graphs in terms of the dependencies they show
- use experimental data to construct graphs
- recognize that a best-fit line is appropriate for a set of data and draw such a line or curve through the data points on a graph
- use graphs to determine the values of physical quantities.

IBO Assessment statements 11.1.1 to 11.3.4

For a quantity to have an *exact* value, it must either be defined or be obtained by counting. Examples of such exact quantities include the number of centimetres in one metre (defined as 100), and the number of pages in this book. We can be certain about such quantities. However, all *measured* quantities have an inherent uncertainty. People using various kinds of instruments make measurements. There is an uncertainty in such measurements because all instruments have limitations, and the people operating the instruments have varying technological skills.

Figure 5.0.1 A student performing an acid–base titration.

## 5.1 UNCERTAINTY AND ERROR IN MEASUREMENT

The **accuracy** of a measurement is an expression of how close the measured value is to the 'correct' or 'true' value. Accuracy in measured values is sometimes expressed as a **percentage difference** of the measured value compared to the true (or accepted) value. This type of expression is often found in the data processing (discussion) section of a laboratory report.

$$\% \text{ difference} = \frac{\text{experimental value} - \text{accepted value}}{\text{accepted value}} \times \frac{100}{1}$$

The **precision** of a set of measurements refers to how closely the individual measurements agree with one another. Thus, precision is a measure of the *reproducibility* or consistency of a result. The type of instrument used, in part, determines the precision of a measurement. For example, a balance that measures masses to the nearest one-hundredth of a gram (for example, 12.00 g) will allow greater precision than one that measures only to the nearest tenth of a gram (12.0 g).

**AS 11.1.2**
**Distinguish between precision and accuracy.** © IBO 2007

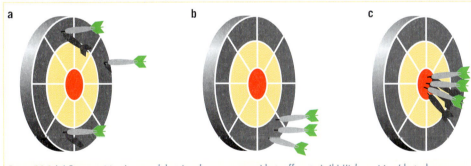

Figure 5.1.1 (a) Poor precision (scattered darts) and poor accuracy (darts off centre). (b) High precision (darts close together) and poor accuracy. (c) High precision and high accuracy (darts centred).

The difference between 12.00 g and 12.0 g is best understood by considering what values might be rounded up or down to give these values. A measurement of 12.0 g could be the result of 11.95 g being rounded up or 12.04 g being rounded down; that is, it is equal to any value between 11.95 g and 12.04 g, to 3 significant figures. In comparison, 12.00 g could be the result of 11.995 g being rounded up or 12.004 g being rounded down; that is, it is equal to any value between 11.995 g and 12.004 g to 4 significant figures. This is a much smaller range of possible values (a range of 0.009 g compared to a range of 0.09 g), so the answer to 4 significant figures is a more accurate value.

Figure 5.1.2 (a) A balance that measures masses to the nearest one-hundredth of a gram (for example, 12.00 g) will allow greater precision than one (b) that measures only to the nearest tenth of a gram (12.0 g).

Consider some results obtained by students who have been asked to find the mass of a beaker. The results of two successive weighings by the students are as follows.

| Student A | Student B | Student C |
|---|---|---|
| 95.800 g | 96.024 g | 96.002 g |
| 95.600 g | 96.028 g | 96.000 g |
| Av. value: 95.700 g | Av. value: 96.026 g | Av. value: 96.001 g |

The true mass of the beaker is 96.002 g. Student B's results are more precise than student A's results because the range is smaller (96.028 – 96.024 = 0.004 compared with 95.800 – 95.600 = 0.200). However, neither set of results is very accurate. Student C's results are more accurate and are also more precise than those of either of the other students. Results that are highly accurate are usually precise as well; however, not all precise measurements are accurate.

## Random uncertainty and systematic errors

**AS 11.1.4
State random uncertainty as an uncertainty range (+/–).
© IBO 2007**

The precision of such measurements is sometimes expressed as an **uncertainty**, using a plus/minus notation to indicate the possible range of the last digit. Using this notation for a particular balance, the measured mass might be reported as 18.34 ± 0.01. This indicates that the mass is between 18.33 g and 18.35 g.

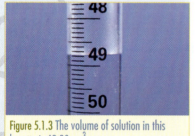

Figure 5.1.3 The volume of solution in this burette is 48.80 cm$^3$.

### Worked example 1

A burette reading may be recorded as 48.80 ± 0.05 cm$^3$. What values may be possible for this reading?

### Solution

48.80 + 0.05 = 48.85   and   48.80 – 0.05 = 48.75

The burette reading is between 48.75 cm$^3$ and 48.85 cm$^3$.

In figure 5.1.3 it is easy to see that the volume of solution is 48.80 cm$^3$; however, in some solutions the meniscus is not as easy to read. It is important to remember that for water or aqueous solutions the volume is read from the bottom of the meniscus, and to avoid parallax error your eyes should be level with the meniscus of the solution (see figure 5.1.5). The meniscus of a column of mercury is inverted when compared to water or aqueous solutions, so the figure level with the *top* of the meniscus should be read. In most thermometers the meniscus is too small to worry about this difference; however, if you have occasion to read a pressure value from a column of mercury this is a point to keep in mind.

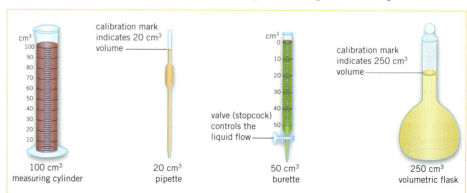

Figure 5.1.4 Some laboratory glassware used to measure liquid volumes.

Sometimes, during an experimental investigation, a student may make a **mistake** such as:

- misreading a scale or a digital reading
- using different balances for a number of related measurements
- wrongly transferring raw data to a table of results
- pressing the wrong buttons on a calculator or making arithmetic errors in mental calculations
- failing to carry out a procedure as described in the method.

These mistakes may be sources of error in the final results, but they are *not* considered as experimental errors. They may be identified when you discuss your results, but cannot be used as an 'excuse' for bad results. Good experimental practice would require that you repeat an experiment when such a mistake has occurred.

It's important to distinguish mistakes from errors: mistakes can be avoided, whereas errors can be minimized but not entirely avoided, because they are part of the process of measurement. Data that is mistaken cannot be used in calculations and should be discarded. It is important to be critical of your results so that mistakes can be detected. There is little value in a set of results that have all been used 'just because they were there'! Care should be used to avoid mistakes wherever possible. Data that contains errors can be useful, if the sizes of the errors can be estimated.

Poor accuracy in measurements is usually associated with an error in the system—a **systematic error**. Using a balance that has been incorrectly zeroed (for example, so that the zero reading is in fact a mass of 0.1 g) will produce measurements that are inaccurate (below their true value). Systematic errors are always biased in the same direction. The incorrectly zeroed balance will always produce measurements that are below their true value. Repeating the measurements will not improve the accuracy of the result. Systematic errors can be the result of:

- poorly calibrated instruments
- instrument parallax error (reading a scale from a position that is not directly in front of the scale)
- badly made instruments
- poorly timed actions (such as the reaction time involved in clicking a stopwatch).

Poor precision in measurements is associated with **random uncertainty**. These are the minor uncertainties inherent in any measurement. Error associated with estimating the last digit of a reading is a random uncertainty. This type of uncertainty is such that the value recorded is sometimes higher than the true value, sometimes lower.

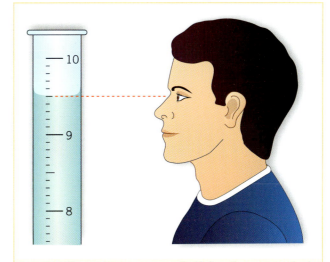

Figure 5.1.5 When reading a meniscus you should always have your eyes level with the meniscus, and for aqueous solutions the volume is read from the bottom of the meniscus.

**AS 11.1.1**
**Describe and give examples of random uncertainties and systematic errors.** © IBO 2007

**AS** 11.1.3
Describe how the effects of random uncertainties may be reduced. © IBO 2007

**Repeating the measurement a number of times and averaging the results reduces the effect of random uncertainty.**

The improvement of your accuracy by repeating measurements as many times as possible can be explained by statistical theory. If you make a large number of independent measurements, such as performing a titration or finding the mass of a sample, the results when plotted on a graph will produce a symmetrical bell-shaped curve. This curve is known as the *normal distribution curve*.

According to statistical theory, and assuming that there are no systematic errors, an infinite number of measurements would produce a bell curve with the exact, true value of what is being measured exactly in the centre of the distribution. So it follows that the more measurements that are made, the closer the mean will be to the true value.

Random uncertainties can be magnified by the inexperience of the student operating the equipment. This is a major reason for repeating a titration experiment until at least three *concordant* results are obtained. Concordant results should all fall within a range that is equal to the random uncertainty in the readings, for example $0.1 + 0.1 = 0.2$ cm$^3$. When a set of results is concordant, it is clear that the student is using the equipment precisely and the results are within the random uncertainty limits of the equipment. When a set of results has a range that is greater than the random uncertainty in the readings, then the error can be estimated to be ± half the spread of the averaged values. This value will take the place of the random uncertainty of the equipment.

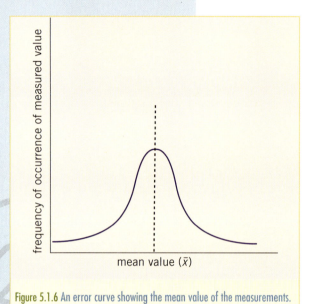

Figure 5.1.6 An error curve showing the mean value of the measurements.

Random uncertainties that are commonly encountered in Chemistry experiments include:

- *mass* uncertainties from a digital balance, usually ±1 in the last decimal place of the reading
- *volume* uncertainties, which depend on the glassware being used and its quality
- *temperature* uncertainties, which are dependent on the range of the thermometer, although generally the uncertainty will be equal to half of the scale gradations
- *time* uncertainties when using a stopwatch. If the stopwatch measures to the nearest 0.01 s, then the uncertainty in the reading is ± 0.005 s. However, human reaction times are much slower than this uncertainty and are about 0.2 s, so this should be used as the uncertainty in a time measured with a stopwatch.

Temperature probes are being used in many school laboratories these days to increase the efficiency of making large numbers of measurements. These temperature probes are very accurate as well as being easy to use. The random uncertainty in temperature when a temperature probe is used may be as small as ± 0.01°C or ± 0.1°C.

Note that in situations in which two readings are needed to find a volume (such as a volume delivered from a burette) or a mass (where mass is measured by difference), the random error doubles.

> **THEORY OF KNOWLEDGE**
> In what way does our increasing dependence on technology influence what raw data in an experiment we regard as valuable or important?

| TABLE 5.1.1 RANDOM UNCERTAINTIES IN VOLUMETRIC GLASSWARE | | |
|---|---|---|
| **Equipment type** | **Size of equipment** | **Random uncertainty*** |
| Pipette | 20.00 cm$^3$ | ±0.05 cm$^3$ |
|  | 10.00 cm$^3$ | ±0.03 cm$^3$ |
| Burette | 50.00 cm$^3$ | ±0.08 cm$^3$ per reading |
| Measuring cylinder | 10 cm$^3$ | ±0.2 cm$^3$ |
|  | 25 cm$^3$ | ±0.5 cm$^3$ |
|  | 50 cm$^3$ | ±1 cm$^3$ |
|  | 100 cm$^3$ | ±1 cm$^3$ |
|  | 250 cm$^3$ | ±2 cm$^3$ |
|  | 500 cm$^3$ | ±5 cm$^3$ |
| Volumetric flask | 100 cm$^3$ | ±0.1 cm$^3$ |
|  | 250 cm$^3$ | ±0.15 cm$^3$ |
|  | 500 cm$^3$ | ±0.25 cm$^3$ |
|  | 1000 cm$^3$ | ±0.4 cm$^3$ |

*These values were taken from the manufacturer's specifications for a particular brand of glassware measured at 20°C. In a practical investigation you should always check the uncertainty quoted on the glassware that you are using and you should never heat volumetric glassware.

## Titration experiments

A titration is an analysis experiment that is commonly carried out in secondary school laboratories. It may be used in acid–base chemistry (see chapter 9) to find the amount of an acid or a base in a sample, or it may be used in redox chemistry (see chapter 10) to find the amount of an oxidizing agent or reducing agent present. A titration uses a range of specialized glassware, the random uncertainties of which are shown in table 5.1.1. If you have not yet performed a titration, then you may find figures 5.1.7, 5.1.8 and 5.1.9 useful in visualizing the equipment being discussed.

A titration is the procedure of adding one solution to another until the reaction between them is just complete. In an acid–base titration, an accurately measured volume (an aliquot) of one solution is drawn into the pipette, using a pipette filler, until it reaches a mark etched above the bulb section. The solution is then allowed to drain into a conical flask.

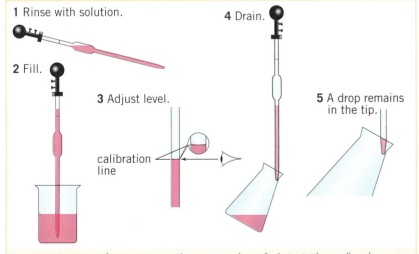

Figure 5.1.7 Correct use of a pipette ensures that an exact volume of solution (with a small random uncertainty) is delivered into the flask during a titration.

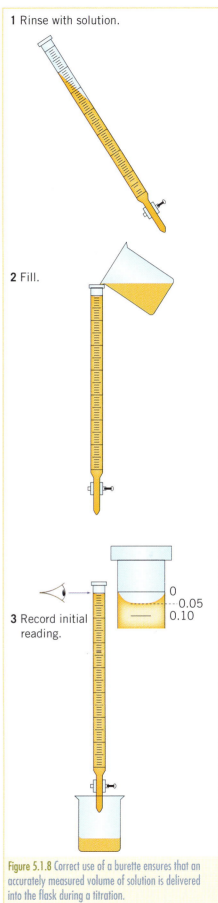

Figure 5.1.8 Correct use of a burette ensures that an accurately measured volume of solution is delivered into the flask during a titration.

A few drops of an acid–base indicator are added to the flask. The second solution is added drop by drop from a burette. Burettes are usually graduated from $0.00\ cm^3$ to $50.00\ cm^3$, with markings for each $0.10\ cm^3$. Initial and final readings are made, allowing the volume delivered by the burette to be determined. This volume is called the *titre*. The completion of the reaction is marked by a colour change in the indicator. Successful analysis therefore relies on the accurate measurement of volumes.

Volumetric (or standard) flasks are flasks that hold an accurately measured known volume of solution. Typical volumes might be $100.0\ cm^3$ or $250.0\ cm^3$. Volumetric flasks are used for the preparation of solutions whose concentration is to be accurately known. Such solutions are known as *standard solutions*.

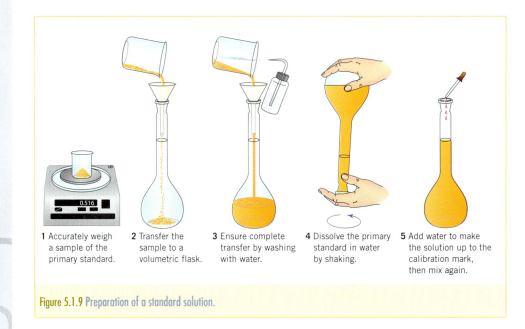

1. Accurately weigh a sample of the primary standard.
2. Transfer the sample to a volumetric flask.
3. Ensure complete transfer by washing with water.
4. Dissolve the primary standard in water by shaking.
5. Add water to make the solution up to the calibration mark, then mix again.

**Figure 5.1.9** Preparation of a standard solution.

## CHEM COMPLEMENT

### Otto Schott

Otto Schott was born in Germany in 1851. His father was a master in window glass making and he followed in the family trade. Schott studied chemical technology at technical college and at the universities of Wijrzbur and Leipzig. His extra knowledge in the field of glass technology led him to develop new lithium-based glasses that not only improved the quality of glass for windows, but also brought some of the optical qualities of glass up to modern standards.

Between 1887 and 1893, Schott developed borosilicate glass and took a vital step forward in the development of laboratory technology. This glass was notable for its high tolerance to heat and substantial resistance to sudden changes in temperature and exposure to chemicals. He used this glass to make a variety of items, including thermometers and laboratory glassware. Although the development of this glass paved the way for all the varieties of glassware that we use in our laboratory today, it was the application of Schott's borosilicate glass to the new incandescent gas lamps that were first sold in 1894 that brought him fame. This lighting technology was so popular, that the increased demand for borosilicate glass caused Schott's company to quickly become an industrial power, the profits of which increasingly came from exporting its goods to other parts of the world.

## Significant figures

The uncertainty of a measurement may be indicated by the use of the plus/minus notation. An alternative method is to indicate the *certainty* of the measurement by the use of **significant figures**. In a measured or calculated value, significant figures are all those digits that are certain plus one estimated (uncertain) digit.

Consider the following examples.

- A temperature reported as 25.06°C contains 4 significant figures.
- A burette reading of 22.57 cm$^3$ also contains 4 significant figures.
- A mass reading of 0.023 g contains 2 significant figures.

Note that significant figures are not the same as decimal places. While the number 0.023 has three decimal places, only two digits (2 and 3) are significant. The two zeros are not significant.

When counting significant figures, the following rules apply:

1. All non-zero digits count as significant figures.

2. Zeros may or may not be significant, depending on whether they are part of the measured value (captive or trailing zeros), or simply used to position a decimal point (leading zeros).

    - **Leading zeros** are those that precede the first non-zero digit. Leading zeros are not counted as significant figures. In the number 0.023 the two zeros simply indicate the position of the decimal point. The number could be rewritten as $2.3 \times 10^{-2}$. This number therefore contains only 2 significant figures.

    - **Captive zeros** are zeros between non-zero integers. Captive zeros always count as significant figures. The number 5.008 contains 4 significant figures.

    - **Trailing zeros** are zeros to the right of a number. Trailing zeros are significant if the number contains a decimal point. The number 0.3000 contains 4 significant figures. If there is no decimal point, the number of significant figures may be unclear.

        Consider, for example, the number 400. It is unclear whether the two zeros here are significant. Was the measurement taken to the nearest unit, ten or hundred? To avoid this confusion the number should be written in *standard form* or scientific notation. A number written in standard form is expressed as a number greater than 1 but less than 10 multiplied by $10^x$, where $x$ is an integer. The number 400, in standard form, would be written as $4 \times 10^2$ (1 significant figure), $4.0 \times 10^2$ (2 significant figures), or $4.00 \times 10^2$ (3 significant figures), depending on the actual measurement made.

3. Significant figures are only applied to measurements and calculations involving measurements. They do not apply to quantities that are inherently integers or fractions (for example, a stoichiometric ratio such as 2 or ½ mole), defined quantities (for example, one metre equals 100 centimetres), or conversion factors (multiplying by 100 to get a percentage or adding 273 to convert °C to K).

**11.1.5 State the results of calculations to the appropriate number of significant figures. © IBO 2007**

Significant figures
Significant figures calculations

A calculated quantity cannot be more precise than the data used in the calculation. By considering the number of significant figures in each measured quantity used in a calculation, we can determine the number of significant figures in a calculated result.

The following rules apply:

- *Multiplication and division*   The result should have the same number of significant figures as the factor with the least number of significant figures.

    For example, consider the calculation, $2.54 \times 2.6 = 6.604$. The limiting factor (2.6) has 2 significant figures, so the result would be expressed to 2 significant figures, that is, 6.6.

- *Addition and subtraction*   The result should have the same number of decimal places as the number used with the fewest decimal places. Note that when adding and subtracting we are interested in decimal places, not significant figures.

    For example, consider the calculation $3.467 + 4.5 + 3.66 = 11.627$. The limiting number (4.5) has one decimal place, so the result would be expressed to one decimal place (11.6).

When completing calculations involving the use of data, such as relative atomic masses, and physical constants, such as Avogadro's constant, these data and constants should be quoted to at least the same number of significant figures used in the measurements. That is, data and physical constants should not determine the number of significant figures in the calculated result. In most cases, using the values given for relative atomic masses on page 403 and Avogadro's constant on page 404 should be sufficient to avoid this problem. Note that these values are exactly the same as those quoted in the IB Data Booklet.

### Worked example 2

Find the mass of 0.735 mol of lead(II) nitrate.

### Solution

Step 1:  Find the molar mass of lead(II) nitrate.

$$M(Pb(NO_3)_2) = 207.19 + 2 \times 14.01 + 6 \times 16.00$$
$$= 331.21 \text{ g mol}^{-1}$$

(2 and 6 are stoichiometric ratios)
(answer is given to the same number of decimal places as the data)

Step 2:  Find the mass of 0.735 mol of $Pb(NO_3)_2$.

$$m(Pb(NO_3)_2) = n \times M$$
$$= 0.735 \text{ mol} \times 331.21 \text{ g mol}^{-1}$$
$$= 243.43935 \text{ g}$$

The smallest number of significant figures is three (0.735), so the answer should be rounded off to 3 significant figures.

So $m(Pb(NO_3)_2) = 243$ g.

In most calculations you will need to round off numbers to obtain the correct number of significant figures.

The following rules should be applied for *rounding off*.

- In a series of calculations, carry the extra digits through to the final result, and then round off.
- If the digit to be dropped is 4 or less, drop it and all the following digits. For example, 78.9432 rounds to 78.94 (4 significant figures) or 78.9 (3 significant figures).
- If the digit to be dropped is 5 or greater, the preceding digit is increased by 1. For example, 57.66 rounds to 57.7 (3 significant figures) or 58 (2 significant figures).
- When rounding, use only the first digit to the right of the last significant figure. Do not round off sequentially. For example, 6.248 rounds to 6.2 (2 significant figures). It is not correct to round to 6.25, and then to 6.3.

**WORKSHEET 5.1**
Precision, accuracy and significant figures

## CHEM COMPLEMENT

### Prefixes used with units

Many units are too large or too small to be useful in a range of situations. It would be confusing for a child to have to learn that their ruler is 0.30 m long even before they learn what a decimal place is. However, a centimetre is a small length that is easily visualized, and the length of a ruler as 30 cm is understandable to even a very young child.

Prefixes are used to convert units to the size that suits the task. We use milli- and kilo- in everyday life when we are referring to one-thousandth of a unit (millimetre) or one thousand times that unit (kilometre). But there are many more prefixes that are reserved for specialist use. These are shown in table 5.1.2.

SI prefixes

**TABLE 5.1.2 PREFIXES**

| Multiple | Prefix | Symbol | Multiple | Prefix | symbol |
|---|---|---|---|---|---|
| $10^{24}$ | yotta | Y | $10^{-3}$ | milli | m |
| $10^{21}$ | zeta | Z | $10^{-6}$ | micro | μ |
| $10^{18}$ | exa | E | $10^{-9}$ | nano | n |
| $10^{15}$ | peta | P | $10^{-12}$ | pico | p |
| $10^{12}$ | tera | T | $10^{-15}$ | femto | f |
| $10^{9}$ | giga | G | $10^{-18}$ | atto | a |
| $10^{6}$ | mega | M | $10^{-21}$ | zepto | z |
| $10^{3}$ | kilo | k | $10^{-24}$ | yocto | y |

Obviously we would have little use for a yottametre in our everyday travels—even the moon is only 0.4 gigametre away!

## Section 5.1 Exercises

1. A student performed an analysis of the sulfur content of a sample of plant food and after several repetitions had the following set of results:
14.35%, 14.40%, 14.37%, 14.34%

   The actual amount of sulfur present in the sample was 14.70%. What would you deduce about the accuracy and precision of these results? Explain your answer.

2. In the plant food analysis described in question **1**, the average of the results was 14.37%. Describe the accuracy of these results by finding the percentage difference between the actual amount of sulfur (14.70%) and the experimental average.

3. 'Systematic errors and random uncertainties cannot be prevented in an experiment, but they can be reduced by repeating the experiment many times.' Discuss this statement.

4. If you were measuring the mass of reactants on a digital balance, what systematic error might be associated with those measurements?

5. A student was performing a titration in which the following set of results was obtained.

| Experiment number | 1 | 2 | 3 | 4 | 5 | 6 | 7 |
|---|---|---|---|---|---|---|---|
| Final volume ($cm^3 \pm 0.08\ cm^3$) | 21.43 | 42.80 | 27.27 | 20.52 | 42.90 | 22.42 | 42.80 |
| Initial volume ($cm^3 \pm 0.08\ cm^3$) | 0.00 | 21.43 | 0.20 | 0.50 | 20.52 | 1.00 | 21.42 |
| Volume of titre ($cm^3 \pm 0.2\ cm^3$) | | | | | | | |

   a Complete the table of results by calculating the volume of titre for each experiment. (Volume of titre = final volume − initial volume)

   b Discuss whether *all* the results should be averaged in order to find the average titre.

   c Compare the range of the results with the random uncertainty of the volume of the titre. State which value should be given as the uncertainty in the average titre.

6. State the number of significant figures in each of the following numbers.

   a 3.875     b 24.08     c 100.23     d 0.045
   e 34.000    f 0.0005    g 100.0      h 0.0250

7. Calculate the following numbers when rounded off to the number of significant figures indicated in brackets.

   a 0.957        (2 significant figures)
   b 34.65        (3 significant figures)
   c 22.2245      (4 significant figures)
   d 0.003 064    (3 significant figures)
   e 0.003 064    (2 significant figures)
   f 1.0007       (3 significant figures)
   g 15.275       (4 significant figures)
   h 0.000 5849   (3 significant figures)
   i 2.4350       (3 significant figures)
   j 0.825        (2 significant figures)

8  Use scientific notation (exponential notation) to express the number 478 000 000 to:

   a  1 significant figure
   b  2 significant figures
   c  3 significant figures
   d  4 significant figures

9  Perform the following mathematical operations and express each result to the correct number of significant figures:

   a  $4.185 \times 103.62 \times (25.27 - 24.15)$

   b  $\dfrac{6.378 - 4.223}{6.378} \times \dfrac{100}{1}$ (Note that this is a percentage error calculation so $\dfrac{100}{1}$ is considered an exact number.)

   c  $\dfrac{8.5 \times 200.46}{6.842 + 3.005}$

   d  $0.1345 + 2.04 - 1.998$

## 5.2 UNCERTAINTY IN CALCULATED RESULTS

The **absolute uncertainty** is the size of an uncertainty, including its units. It may be equal to half the limit of the reading, where the limit of the reading is the markings on the scale. In other pieces of equipment, the absolute uncertainty is larger than half the limit of the reading (see table 5.1.1).

When stating an uncertainty as an absolute uncertainty, the actual value is quoted after the reading. For example, the volume of sodium hydroxide in a 20.00 cm$^3$ bulb pipette would be quoted as 20.00 ± 0.05 cm$^3$. The absolute uncertainty is a feature of the apparatus being used, rather than the amount being measured in it. The quality of the glassware also influences the absolute uncertainty. A class B 20.00 cm$^3$ pipette has a greater uncertainty (±0.05 cm$^3$) than a class A 20.00 cm$^3$ pipette (±0.03 cm$^3$).

In contrast, the **percentage uncertainty** changes with the amount of material that you are measuring. Percentage uncertainty is found by dividing the absolute uncertainty by the measurement that is being made.

Percentage uncertainty = $\dfrac{\text{absolute uncertainty}}{\text{measurement}} \times \dfrac{100}{1}$

If we consider the example of a burette with absolute uncertainty of ±0.16 cm$^3$ (2 × ±0.08 cm$^3$ due to two readings needed for the one volume), if the measurement taken (the volume of the titre) is small, the absolute uncertainty will produce a much greater percentage error than if the volume of the titre is large. This is one of the reasons why, in a titration experiment, the preferred volume for the titre is around 20.00 cm$^3$. (Also with a volume of titre of 20.00 cm$^3$, two titrations can be performed before refilling the burette.)

The size of the absolute uncertainty in comparison to the measurement to be taken is also a guiding factor in choosing equipment for an experiment. For example, it is important to choose the smallest measuring cylinder that will hold (and measure) the required volume of solution in order to minimize the percentage uncertainty. This can be seen in table 5.2.2.

**AS 11.2.1**
**State uncertainties as absolute and percentage uncertainties.**
© IBO 2007

**PRAC 5.1**
Determining the composition of copper oxide by analysis

### TABLE 5.2.1 ABSOLUTE UNCERTAINTY AND PERCENTAGE UNCERTAINTY

| Volume of titre (cm³) | Absolute uncertainty (cm³) | Percentage uncertainty |
|---|---|---|
| 3.35 | ± 0.2* | $\frac{0.2}{3.35} \times \frac{100}{1} = \pm 6\%$ |
| 19.80 | ± 0.2* | $\frac{0.2}{19.80} \times \frac{100}{1} = \pm 1\%$ |
| 35.00 | ± 0.2* | $\frac{0.2}{35.00} \times \frac{100}{1} = \pm 0.6\%$ |

*Note that uncertainties are always quoted to only 1 significant figure, so ±0.16 cm³ rounds up to ±0.2 cm³.

### TABLE 5.2.2 CHOICE OF MEASURING CYLINDER AND PERCENTAGE UNCERTAINTY

| Volume to be measured (cm³) | Volume of measuring cylinder (cm³) | Absolute uncertainty (cm³) | Percentage uncertainty |
|---|---|---|---|
| 10 | 100 | ± 1 | $\frac{1}{10} \times \frac{100}{1} = \pm 10\%$ |
| 10 | 50 | ± 1 | $\frac{1}{10} \times \frac{100}{1} = \pm 10\%$ |
| 10 | 25 | ± 0.5 | $\frac{0.5}{10} \times \frac{100}{1} = \pm 5\%$ |
| 10 | 10 | ± 0.2 | $\frac{0.2}{10} \times \frac{100}{1} = \pm 2\%$ |

From table 5.2.2, it can be seen that large errors are introduced if a small volume such as 10 cm³ is measured in a large measuring cylinder. The best measuring cylinder for such a volume is a 10 cm³ one, although if an accurate volume is required, it may be better to use a pipette, which would only have an absolute uncertainty of ±0.03 cm³ and hence a percentage uncertainty of ±0.3%.

### Calculating uncertainties in results

The majority of calculations involved in processing experimental results involve addition, subtraction, multiplication and division. The following rules should be followed in order to determine the absolute uncertainty, and hence the percentage uncertainty in results.

**11.2.2**
Determine the uncertainties in results. © IBO 2007

- When *adding* measurements, the uncertainty in the sum is equal to the sum of the absolute uncertainties in each measurement taken (i.e. *add* the *absolute uncertainties*).

- When *subtracting* measurements, the uncertainty in the difference is equal to the sum of the absolute uncertainties in each measurement taken (i.e. *add* the *absolute uncertainties*).

- When *multiplying and dividing*, add the percentage uncertainties of the measurements being multiplied or divided. The absolute uncertainty then is the percentage of the final answer (i.e. *add* the *percentage uncertainties* and convert to absolute uncertainty if required).

- When *taking an average* of a set of values, such as a set of titration data, the absolute uncertainty in the average will either be equal to the random uncertainty in the data (±0.2 cm³ for burette volumes) or it may be a larger number if the range of data is greater than the random uncertainty (±0.2 cm³) due to an inexperienced operator (see p. 164).

## Worked example

Calculate the percentage uncertainty in each of the following calculations.

**a** $12.5 \pm 0.16$ cm$^3$ of HCl is added from a burette to $4.0 \pm 0.5$ cm$^3$ of water that had been measured in a measuring cylinder. Calculate the absolute uncertainty and hence the percentage uncertainty in the final solution.

**b** A sample of copper has been produced during a multi-stage experiment. The 250 cm$^3$ beaker was weighed at the start of the experiment and found to have a mass of $180.15 \pm 0.02$ g. Several days later the beaker and copper was found to have a mass of $183.58 \pm 0.02$ g. Calculate the absolute uncertainty and hence the percentage uncertainty in the final mass of copper.

**c** Calculate the percentage uncertainty, and hence the absolute uncertainty, in the concentration of a solution of sodium carbonate, Na$_2$CO$_3$, that has been prepared by the dissolution of $2.346 \pm 0.002$ g of sodium carbonate in a $250.0 \pm 0.15$ cm$^3$ volumetric flask.

**WORKSHEET 5.2**
Treatment of uncertainties

## Solution

**a** Total volume of solution = $12.5 + 4.0 = 16.5$ cm$^3$

Absolute uncertainty = $\pm(0.16 + 0.5) = \pm 0.66$ cm$^3$

Percentage uncertainty = $\dfrac{0.66}{16.5} \times \dfrac{100}{1} = \pm 4\%$

**b** Mass of copper = $183.58 - 180.15 = 3.43$ g

Absolute uncertainty = $\pm(0.02 + 0.02) = \pm 0.04$ g

Percentage uncertainty = $\dfrac{0.04}{3.43} \times \dfrac{100}{1} = \pm 1\%$

**c** $n(\text{Na}_2\text{CO}_3) = \dfrac{m(\text{Na}_2\text{CO}_3)}{M(\text{Na}_2\text{CO}_3)}$

$= \dfrac{2.346}{2 \times 22.99 + 12.01 + 3 \times 16.00} = \dfrac{2.346}{105.99}$

$= 0.022\,13$ mol (to 4 significant figures)

Absolute uncertainty in $n(\text{Na}_2\text{CO}_3) = \dfrac{0.002}{105.99} = 2 \times 10^{-5}$ mol

Percentage uncertainty in $n(\text{Na}_2\text{CO}_3) = \dfrac{2 \times 10^{-5}}{0.022\,13} \times \dfrac{100}{1} = 0.09\%$

$c(\text{Na}_2\text{CO}_3) = \dfrac{0.022\,13}{0.250} = 0.0885$ mol dm$^{-3}$

Absolute uncertainty in $V$(solution) = 0.15 cm$^3$

Percentage uncertainty in $V$(solution) = $\dfrac{0.15}{250} \times \dfrac{100}{1} = 0.06\%$

Percentage uncertainty in $c(\text{Na}_2\text{CO}_3) = 0.09\% + 0.06\% = 0.15\%$

Absolute uncertainty in $c(\text{Na}_2\text{CO}_3) = 0.15\%$ of $0.0885 = 0.0003$ mol dm$^{-3}$

The concentration of the sodium carbonate solution is $0.0885 \pm 0.0003$ mol dm$^{-3}$.

## Section 5.2 Exercises

1. State the range of values indicated by each of the following uncertainties.
   a. $25.84 \pm 0.02$ cm$^3$
   b. $14.4 \pm 0.3$ cm$^3$
   c. $250.0 \pm 0.5$ cm$^3$

2. Calculate the percentage uncertainty in each of the following measurements, given the absolute uncertainty stated.
   a. $11.06 \pm 0.04$ cm$^3$
   b. $21.38 \pm 0.04$ cm$^3$
   c. $85 \pm 5$ cm$^3$
   d. $0.325 \pm 0.005$ g
   e. $24.50 \pm 0.05$°C

3. Calculate the absolute uncertainty in each of the following measurements, given the percentage uncertainty stated.
   a. $25.0$°C $\pm 2\%$
   b. $10.00$ cm$^3 \pm 5\%$
   c. $2.507$ g $\pm 1\%$
   d. $25.50$ cm$^3 \pm 5\%$

4. The following temperature measurements were taken using different thermometers during calorimetry experiments (see chapter 6). In each case calculate the temperature change and the absolute uncertainty in the temperature change.

| Experiment | Initial temperature (°C) | Final temperature (°C) | Change in temperature (°C) | Absolute uncertainty in temperature change (°C) |
|---|---|---|---|---|
| a | $25.0 \pm 0.5$ | $48.0 \pm 0.5$ | | |
| b | $15 \pm 1$ | $20 \pm 1$ | | |
| c | $20.00 \pm 0.05$ | $45.00 \pm 0.05$ | | |

5. A number of solutions were added together in different experiments. Find the absolute uncertainty and hence the percentage uncertainty in each of the final solutions.

| Experiment | Volume of solution 1 (cm$^3$) | Volume of solution 2 (cm$^3$) | Total volume (cm$^3$) | Absolute uncertainty in volume (cm$^3$) | Percentage uncertainty in volume |
|---|---|---|---|---|---|
| a | $25.0 \pm 0.5$ | $15 \pm 1$ | | | |
| b | $2.02 \pm 0.02$ | $12.00 \pm 0.02$ | | | |
| c | $150 \pm 5$ | $25.0 \pm 0.5$ | | | |

| Volume of measuring cylinder (cm$^3$) | Absolute uncertainty (cm$^3$) |
|---|---|
| 250 | 2 |
| 500 | 5 |

6. In an experiment students are required to add 400 cm$^3$ of water to their reaction mixture. Unfortunately there is only one 500 cm$^3$ measuring cylinder in the laboratory. While student A waited patiently and used the 500 cm$^3$ measuring cylinder, student B was keen to get on with the experiment and added the 400 cm$^3$ of water by twice measuring out 200 cm$^3$ in a 250 cm$^3$ measuring cylinder.

   By considering the absolute uncertainties given in the table at left, explain whether student B has introduced more or less uncertainty into the experiment than student A.

7. 0.500 mol of potassium hydroxide with a percentage uncertainty of 0.2% was dissolved to make a solution of volume 250 cm$^3$ with a percentage uncertainty of 1.5%. Calculate the concentration of the solution and the percentage uncertainty in this concentration.

**8** 1.325 g of sodium chloride is weighed out on a balance for which the absolute uncertainty is 0.005 g. It is dissolved to make 100.0 cm$^3$ of solution in a volumetric flask with absolute uncertainty of 0.5 cm$^3$. Calculate the concentration of the sodium chloride solution, the percentage uncertainty and the absolute uncertainty in the concentration.

## 5.3 GRAPHICAL TECHNIQUES

Before you can plot an accurate graph, you need a good table of data. Quantitative and qualitative data should be recorded in different tables due to the very different nature of what you will be recording. **Qualitative data** should consist of observations that would enhance the interpretation of results. The term **quantitative data** refers to numerical measurements of the variables associated with the investigation. Quantitative data should be recorded together with units and uncertainties in the measurements.

A table of quantitative date should be ruled up neatly with columns clearly headed. The heading of a column should include a heading (what has been measured), the appropriate units for the measurement and the uncertainty in the measurements. The figures in the column should be consistent with the uncertainty in terms of significant figures, and the number of decimal places quoted in readings from the same apparatus should be consistent. Figure 5.3.1 shows a table of quantitative results that has been set out appropriately. You may notice that the second column contains processed results, rather than raw data. This is acceptable as long as the processed results are clearly distinguishable.

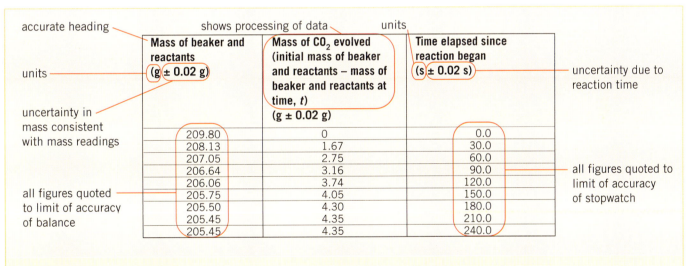

Figure 5.3.1 This table of quantitative results illustrates all the features that should be present in a correct table of quantitative results.

### Sketching graphs

A *fair experiment* involves the control of all variables except for one. The **independent variable** is changed during the course of the experiment and as a result the **dependent variable** usually changes. The independent variable is always allocated to the horizontal axis and the dependent variable to the vertical axis. This relates back to mathematics: $y = mx + c$, where $y$ is a function of $x$.

## THEORY OF KNOWLEDGE

We rely so much on our sense of sight that it is reflected in figures of speech such as 'insight', 'seeing is believing' and 'I'll believe it when I see it'. It is not suprising, then, that graphs are very useful for analysing data and presenting a visual display of the relationship between the independent and dependent variables. They can provide very powerful interpretations of reality. A linear graph, sloping upwards from (0, 0) indicates that the two variables are **directly proportional** to one another. As the independent variable increases, so does the dependent variable.

If the line is horizontal, this is because the dependent variable is not changing as the independent variable changes. For example, the concentration of the independent variable may be being measured and the concentration of the dependent variable is not changing. This may be because the independent variable being measured is a catalyst. It is needed to provide a lower activation energy for the reaction, but if its concentration is above a certain value, a change in its concentration will not affect the amount of product (dependent variable) that is made. This relationship is clearly seen in enzyme catalysis, which is discussed in the Higher Level section of Option B.

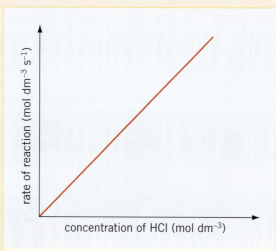

Figure 5.3.2 A graph of concentration of HCl against rate of reaction. As the concentration of HCl increases, the rate of reaction increases.

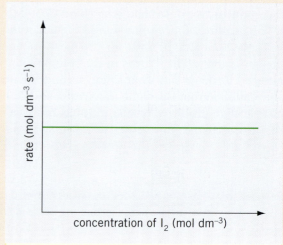

Figure 5.3.3 In the reaction between propanone and iodine, as the concentration of $I_2$ changes the rate of reaction does not change.

**11.3.1**
**Sketch graphs to represent dependencies and interpret graph behaviour.** © IBO 2007

A graph can be used to identify data points that do not obey the relationship and are therefore in error. It can also be used to determine a constant in an equation that relates the two variables. In many cases, this constant is equal to the slope of the graph.

In some cases the graph may be a curve. This indicates that the dependent variable is not directly proportional to the independent variable; instead, it may be **inversely proportional**. An example of this is found in the reaction between sodium thiosulfate and hydrochloric acid (see practical investigation 7.1).

$$Na_2S_2O_3(aq) + 2HCl(aq) \rightarrow S(s) + SO_2(g) + 2NaCl(aq) + H_2O(l)$$

When the concentration of sodium thiosulfate is graphed against the time taken for enough sulfur to be made to make the solution opaque, the graph is a curve (figure 5.3.4a). If the concentration of sodium thiosulfate is graphed against the inverse of the time taken, the graph is a straight line (figure 5.3.4b), indicating that the concentration of sodium thiosulfate is inversely proportional to the time taken for the solution to become opaque.

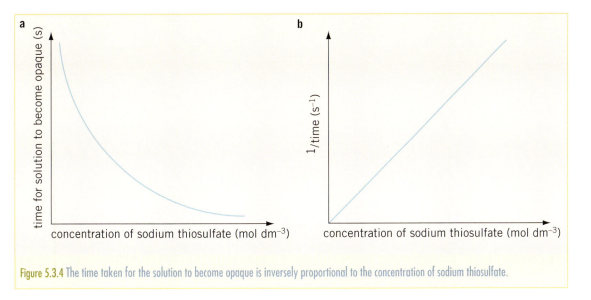

Figure 5.3.4 The time taken for the solution to become opaque is inversely proportional to the concentration of sodium thiosulfate.

In your study of gases (chapter 4), you looked at the relationships between pressure, volume, temperature and amount of gas (mol). These relationships are shown in figure 5.3.5. The volume of a gas is inversely proportional to the pressure of the gas, directly proportional to the temperature, and directly proportional to the amount of gas (mol).

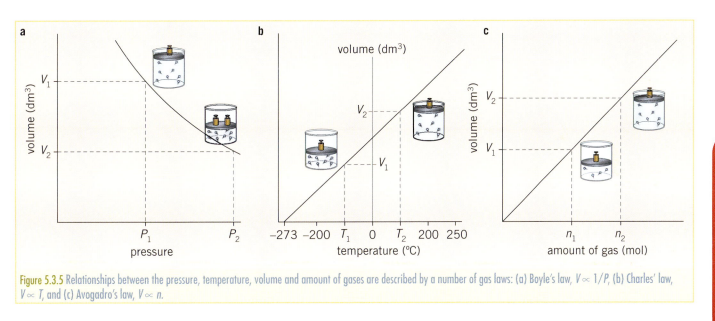

Figure 5.3.5 Relationships between the pressure, temperature, volume and amount of gases are described by a number of gas laws: (a) Boyle's law, $V \propto 1/P$, (b) Charles' law, $V \propto T$, and (c) Avogadro's law, $V \propto n$.

In the study of equilibrium (chapter 8) you will see that a graph can indicate whether a reaction has reached equilibrium (figure 5.3.6). An even more complicated graph can give an insight into the changes that have been made to the system and how the system has reacted in response (figure 5.3.7).

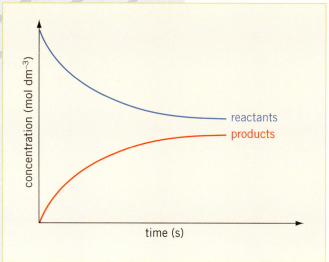

Figure 5.3.6 When the gradient of a concentration against time graph for both reactants and products becomes equal to zero, we have evidence that the system has reached equilibrium.

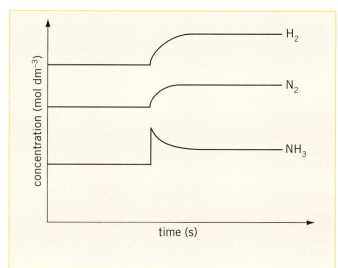

Figure 5.3.7 When there is a sharp rise in the concentration of ammonia, there is a corresponding gradual rise in the concentrations of hydrogen and nitrogen as the system responds to the change.

In the study of acids and bases at higher level you will see that a titration curve shows the equivalence point of a titration, and can be analysed to determine whether the reactants are strong or weak acids or bases. Figure 5.3.8 shows an example of a titration curve (or pH curve) for the reaction between a strong acid and a strong base.

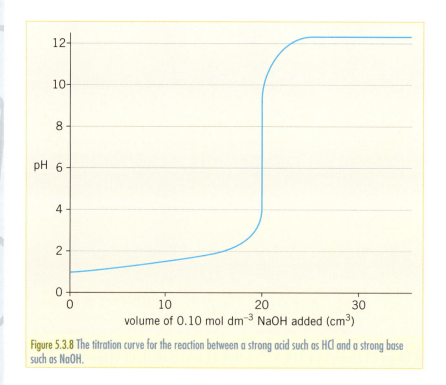

Figure 5.3.8 The titration curve for the reaction between a strong acid such as HCl and a strong base such as NaOH.

AS 11.3.2 Construct graphs from experimental data. © IBO 2007

## Using experimental data to construct a graph

When you plot a **scattergraph** using experimental data, it is important to make the graph as large as possible. To make the graph at least the size of half a page is a good idea. This allows accuracy in plotting the points and will make it easier to read values from your graph if required. The graph should always have a title that relates the independent and dependent variables, and should be drawn in pencil (to allow for corrections to be made easily). It is important to label the axes with the name of the types of measurements that are being plotted and their units. The points should be plotted by drawing a cross or dot with a circle around it to show the point clearly. If more than one set of data is plotted on the same axes, then a key may be required, to show which points correspond to which set of data.

Consider the following set of data obtained from an experiment involving the decomposition of $NO_2$:

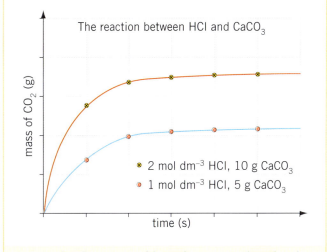

**Figure 5.3.9** When plotting two sets of data on the same axes, a key or legend should be used.

| Time (±1 s) | $[NO_2]$ (mol dm$^{-3}$) | $[NO]$ (mol dm$^{-3}$) | $[O_2]$ (mol dm$^{-3}$) |
|---|---|---|---|
| 0 | 0.100 | 0 | 0 |
| 50 | 0.079 | 0.021 | 0.011 |
| 100 | 0.065 | 0.035 | 0.018 |
| 150 | 0.055 | 0.045 | 0.023 |
| 200 | 0.048 | 0.052 | 0.026 |
| 250 | 0.043 | 0.057 | 0.029 |
| 300 | 0.038 | 0.062 | 0.031 |
| 350 | 0.034 | 0.066 | 0.033 |
| 400 | 0.031 | 0.069 | 0.035 |

The horizontal axis for time needs to be marked off at 50-second intervals, while the vertical axis needs to be able to accommodate much smaller numbers. It would be a good idea to increase all the values for concentration by a factor of 100, so that the largest value is 10 and the smallest is 1.1. The scale can then be drawn with the smallest marking being equal to $0.1 \times 10^{-2}$ mol dm$^{-3}$. The label on the axis will need to indicate that all values are 100 times larger than their actual value. This is done by adding '$\times 10^2$' to the units label on the axis.

$$2NO_2(g) \rightarrow 2NO(g) + O_2(g)$$

A **best-fit line or curve** is a line that fits the majority of the data points plotted. A transparent ruler is most useful for drawing a line of best fit, as it allows you to see the points that fall underneath the ruler. At least five data points are needed on the scattergraph in order to be able to draw a valid trend line as a line of best fit. In figure 5.3.11 the best-fit line passes through the majority of points, but not all. This is quite common and occurs because the points are derived by experiment and will not always be perfect. When drawing a best-fit line or curve, try to balance the points that do not fit evenly on either side of the line. In this way you should be able to obtain a reasonably accurate gradient for the line.

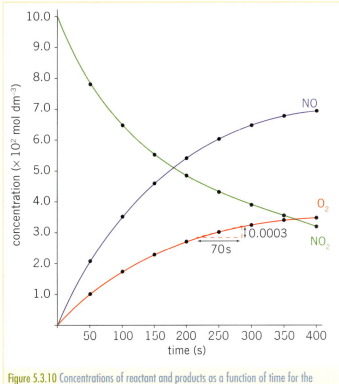

**Figure 5.3.10** Concentrations of reactant and products as a function of time for the reaction.

**AS 11.3.3**
**Draw best-fit lines through data points on a graph.** © IBO 2007

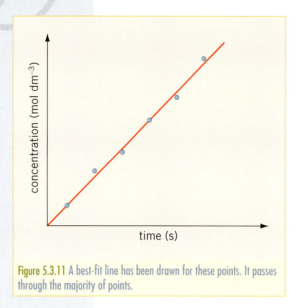

Figure 5.3.11 A best-fit line has been drawn for these points. It passes through the majority of points.

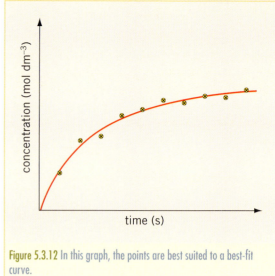

Figure 5.3.12 In this graph, the points are best suited to a best-fit curve.

### Using graphs to determine the values of physical quantities

**AS 11.3.4**
**Determine the values of physical quantities from graphs.**
© IBO 2007

One of the important functions of a graph is to provide further information about the relationship between the quantities being measured. The **gradient** or slope of a straight line often provides valuable information. When you study kinetics (chapter 7) you will find that the gradient of a graph of concentration against time gives you the rate of the reaction.

In figure 5.3.13 the change in concentration (the rise) is measured over a particular time period (the run). The gradient is found by dividing the rise by the run:

Rise = 0.14 − 0.07 = 0.07 mol dm$^{-3}$

Run = 120 − 60 = 60 s

Gradient = $\dfrac{0.07}{60}$ = 0.0012 mol dm$^{-3}$ s$^{-1}$

Figure 5.3.13 The gradient of the line is found by measuring the change.

In some situations a line must be extrapolated to find a required value. This may be an extrapolation back to the *x*- or *y*-axis, or only back to a certain point. In the example shown in figure 5.3.14, a calibration curve has been constructed for a calorimeter (see chapter 6). The calorimeter does not have good thermal insulation, so after the reaction is complete, the contents of the calorimeter cool quite quickly. It must be assumed that heat was also being lost during the reaction, so the graph is extrapolated to find the highest temperature that the calorimeter and contents would have reached had the thermal insulation been adequate.

In figure 5.3.15 the graph has been plotted in order to find the value of the Arrhenius constant, *A*. The quantities *k*, *T* (temperature) and $\dfrac{-E_a}{R}$ are related to each other in an equation called the Arrhenius equation. This is studied in Higher Level Kinetics; however, the equation gives a good example of extrapolation to find a particular value that is required.

**WORKSHEET 5.3**
Graphs

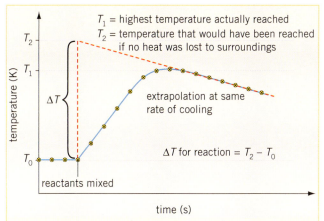

**Figure 5.3.14** The line is extrapolated back to the start of the experiment to show the maximum temperature that would have been achieved if the thermal insulation had been adequate.

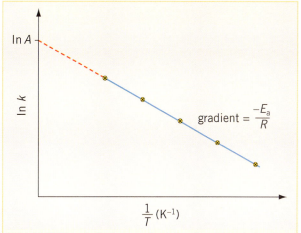

**Figure 5.3.15** When the line is extrapolated, the value of $A$ can be found because the y-intercept has the value of $\ln A$.

### Section 5.3 Exercises

1. Consider the following pair of graphs relating the volume and pressure of a gas.

   State the relationship between volume and pressure.

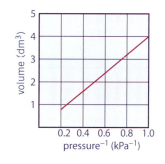

2. Sketch a graph that would show that volume of a gas is directly proportional to number of mole of the gas.

3. Consider the following sketch graphs.

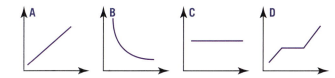

   a. State which graph shows that there is no relationship between the two quantities.
   
   b. State which graph indicates that the two quantities are proportional to each other.
   
   c. State which graph indicates that the two quantities are inversely proportional to each other.

4. An experiment was carried out to investigate the solubility of potassium iodide in 100 g of water. The following results were obtained.

| Temperature (°C) | 0 | 10 | 20 | 30 | 40 | 50 | 60 | 70 |
|---|---|---|---|---|---|---|---|---|
| Solubility (g/100 g water) | 129 | 145 | 163 | 180 | 198 | 215 | 232 | 250 |

   Graph the solubility of potassium iodide against temperature.

5. Plot these points on graph paper and draw a best-fit line for the points.

| Temperature (°C) | 0 | 20 | 40 | 60 | 80 | 100 |
|---|---|---|---|---|---|---|
| Volume (dm³) | 20 | 21 | 23 | 24 | 26 | 27 |

# Chapter 5 Summary

## Terms and definitions

**Absolute uncertainty**   The size of an uncertainty, including its units.

**Accuracy**   An expression of how close the measured value is to the 'correct' or 'true' value.

**Best-fit line or curve**   A line or curve drawn on a graph, which represents the general trend in the measurements but which does not pass through all the data points.

**Captive zeros**   Zeros between non-zero integers.

**Dependent variable**   The variable that changes as a result of the independent variable changing during an experiment.

**Directly proportional**   As the independent variable increases the dependent variable also increases.

**Gradient**   The slope of the line. Calculated by dividing the change in data from the vertical axis (the rise) by the change in data from the horizontal axis (the run).

**Independent variable**   The variable that is deliberately changed during the course of an experiment.

**Inversely proportional**   The dependent variable is proportional to $\frac{1}{\text{independant variable}}$.

**Leading zeros**   Zeros that precede the first non-zero digit.

**Mistake**   A measurement that does not fit the general trend of the results due to some form of carelessness or lack of skill on the part of the operator.

**Percentage difference**   This is found by $\frac{\text{experimental value} - \text{accepted value}}{\text{accepted value}} \times \frac{100}{1}$.

**Percentage uncertainty**   This is calculated by $\frac{\text{absolute uncertainty}}{\text{measurement}} \times \frac{100}{1}$.

**Precision**   An expression of how closely a group of measurements agree with one another.

**Qualitative data**   Observations made during an investigation that would enhance the interpretation of results. These include observations of changes of colour, production of a gas, liquid or solid, and release or absorption of heat.

**Quantitative data**   Numerical measurements of the variables associated with an investigation.

**Random uncertainty**   A minor uncertainty that is inherent in any measurement. An example is the error associated with estimating the last digit of a reading.

**Scattergraph**   A graph on which points are plotted individually (as would experimental data points).

**Significant figures**   All those digits that are certain plus one estimated (uncertain) digit.

**Systematic error**   An error in the system which is usually associated with poor accuracy in measurements.

**Trailing zeros**   Zeros to the right of a number.

**Uncertainty**   An expression of the range of values between which a measurement may fall.

## Concepts

- While *accuracy* is a measure of how close a set of results are to the 'true' or accepted value, *precision* measures how closely a set of results agrees with one another.

- A set of results may be precise without being accurate; however, it is unlikely that they will be accurate without being precise.

- Systematic errors are related to the system and are the result of poor calibration or accuracy in measurements, whereas random uncertainties are inherent in any measurement.

- Examples of systematic errors include parallax error and a scale that is not calibrated correctly (does not return to zero).

- Examples of random uncertainties include uncertainties in measurement of mass or volume or temperature.

- Random uncertainties can be minimized by repeating measurements several times, but systematic errors cannot be minimized in this way.

- A mistake is the result of operator error and such a result should not be used for subsequent calculations.

- The appropriate number of significant figures for a calculation is equal to the number of significant figures in the least accurate piece of data used in the calculation.

- Significant figures are only applied to measurements and calculations involving measurements. They do not apply to quantities that are inherently integers or fractions, defined quantities or conversion factors.

- An absolute uncertainty is expressed as ± a quantity with units, while a percentage uncertainty expresses the absolute uncertainty as a fraction of the measurement.
- A sloping straight-line graph shows that the two variables are proportional to each other. A curve suggests that there may be an inversely proportional relationship between the variables and a graph of $\frac{1}{\text{variable}}$ is needed to confirm this.
- Scattergraphs can be used to give a visual representation of experimental data. They should be drawn carefully and as clearly as possible.
- A best-fit line or curve does not pass through every point in a set of data, but shows the general trend exhibited by the data.
- The value of a physical quantity, such as rate of reaction, may be found from the gradient (slope) of a straight-line graph.

## Chapter 5 Review questions

1. Explain whether it is possible to obtain a set of experimental results that are precise, but not accurate.

2. In an attempt to check the accuracy of a measuring cylinder, a student filled the measuring cylinder to the 10 cm³ mark using water delivered from a burette and then read the volume delivered. The results of five trials are shown below.

| Trial | Volume shown by the measuring cylinder (cm³) | Volume shown by the burette (cm³) |
|---|---|---|
| 1 | 10 | 10.55 |
| 2 | 10 | 10.48 |
| 3 | 10 | 10.50 |
| 4 | 10 | 10.49 |
| 5 | 10 | 10.53 |
| Average | 10 | 10.51 |

Explain whether the measuring cylinder is accurate.

3. Explain the difference between random uncertainties and systematic errors in an experimental technique such as measuring a volume of solution in a measuring cylinder.

4. Describe how the random uncertainty in a burette could be minimized during an experiment.

5. Perform each of the following calculations and state the answer to the correct number of significant figures.

   a  $n(CaCO_3) = \dfrac{34.6}{40.08 + 12.01 + 3 \times 16.00}$

   b  Average titre = $\dfrac{15.06 + 15.20 + 15.15 + 15.09}{4}$

   c  Total mass of products = 124.5 + 23.43 + 1200

6. Convert the following absolute uncertainties to percentage uncertainties.

   a  14.36 ± 0.08 cm³

   b  0.014 ± 0.002 g

   c  21.0 ± 0.5 °C

7. Convert the following percentage uncertainties to absolute uncertainties.

   a  25.00 cm³ ± 0.5%

   b  124.3 g ± 2%

   c  0.257 mol ± 5%

8. Calculate the absolute uncertainty in the result of each of the following calculations. Remember to quote the result to the correct number of significant figures.

   a  $m(\text{copper}) = (54.894 \pm 0.002) - (51.200 \pm 0.002)$ g

   b  $\Delta T(\text{solution}) = ((35.6 \pm 0.5) - (14.6 \pm 0.5))$ °C

   c  $m(\text{beaker + contents}) = (102.23 \pm 0.02) + (2.05 \pm 0.02) + (10.99 \pm 0.02)$ g

9. For each of the calculations in question 8 calculate the percentage uncertainty in the result.

10. Calculate the percentage error in the result for each of the following calculations. Remember to quote the result to the correct number of significant figures.

    a  $n(\text{HCl}) = \dfrac{0.324 \pm 0.002}{36.46}$

    b  $n(\text{NaOH}) = (0.997 \pm 0.005) \times \dfrac{50.0 \pm 3.75}{1000}$

    c  $c(\text{NaCl(aq)}) = \dfrac{(0.00999 \pm 0.00003) \text{ mol}}{(250.00 \pm 0.15) \text{ dm}^3}$

11. For each of the calculations in question 10 calculate the absolute uncertainty in the result.

12. Consider the following graphs of temperature against volume for gases at a series of different pressures.

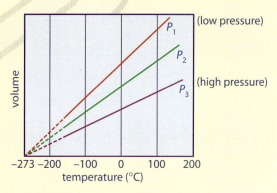

State the relationship between temperature and volume of a gas at constant pressure.

13 For each of the following relationships sketch a graph that would illustrate that relationship.
   a At constant temperature, the number of mole of a gas is directly proportional to the volume of the gas.
   b At constant temperature, the pressure of a gas is inversely proportional to the volume of the gas.
   c To convert temperature in degrees Celsius to kelvin, 273 is added to the temperature in degrees Celsius.
   d As the surface area of a solid increases, the rate of reaction increases.
   e As a reaction progresses with time, the mass of reactant A decreases.

14 The following graph shows the changes in the electrical conductivity of a reaction mixture that occurred as 20.00 cm$^3$ of 0.0500 mol dm$^{-3}$ Ba(OH)$_2$ solution was titrated with a sulfuric acid solution of unknown concentration.

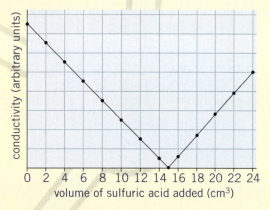

   a Write a balanced equation for the reaction.
   b Describe how the electrical conductivity of the reaction mixture changed as sulfuric acid was added.
   c Suggest a reason why the electrical conductivity of the reaction mixture changed as sulfuric acid was added.

15 A student was investigating the effect of changing the concentration of hydrochloric acid on the time taken for the reaction between CaCO$_3$ and HCl to be completed. The student obtained the following results.

| Concentration of HCl (mol dm$^{-3}$) | Time taken for reaction to be completed (s) |
|---|---|
| 0.05 | 580 |
| 0.10 | 540 |
| 0.15 | 420 |
| 0.20 | 440 |
| 0.25 | 380 |
| 0.30 | 280 |
| 0.35 | 240 |
| 0.40 | 200 |
| 0.45 | 110 |
| 0.50 | 50 |

Plot a scattergraph of the student's results and draw a best-fit line to show the relationship between the concentration of the HCl and the time taken for the reaction to be completed.

16 The graph below shows a tangent drawn to the curve at time $t = 0$ minutes and $t = 1.5$ minutes. The gradient (slope) of the tangent gives a value for the rate of reaction at that time.

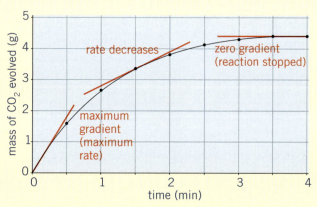

   a Calculate the gradient of the two tangents to the curve and state the units for the gradients.
   b State the point in time at which the gradient has the greater value.

 www.pearsoned.com.au/schools

Weblinks are available on the Chemistry: For use with the IB Diploma Programme Standard Level Companion Website to support learning and research related to this chapter.

# Chapter 5 Test

## Part A: Multiple-choice questions

1. The mass of a length of magnesium strip was measured four times by a student. The masses are:

    2.342 g
    2.344 g
    2.343 g
    2.342 g

    The student averages these masses but does not take into account some white powder that was left on the pan of the balance by another student.

    The average of the masses is:

    **A** precise and accurate.
    **B** precise but not accurate.
    **C** accurate but not precise.
    **D** not accurate and not precise.

2. Which one of the following measurements is stated correctly to 3 significant figures?

    **A** 0.25 $cm^3$
    **B** 0.025 $cm^3$
    **C** 25.0 $cm^3$
    **D** 2.250 $cm^3$

3. The concentration of a 2.50 $dm^3$ solution containing 0.7500 mol of sodium chloride is:

    **A** 0.3 mol $dm^{-3}$
    **B** 0.30 mol $dm^{-3}$
    **C** 0.300 mol $dm^{-3}$
    **D** 0.3000 mol $dm^{-3}$

4. A mass of copper oxide gained in an experiment is quoted as 2.050 ± 0.002 g. This means that the mass of the copper oxide is within the range:

    **A** 2.052–2.024 g
    **B** 2.048–2.052 g
    **C** 2.48–2.50 g
    **D** 2.050–2.052 g

5. In an experiment involving the decomposition of potassium manganate(VI), the oxygen produced was collected in a syringe. The temperature in the laboratory that day was 15° and the atmospheric pressure was 102.02 kPa. The volume of gas was 40.5 $cm^3$ and the gas constant has a value of 8.314 J $K^{-1}$ $mol^{-1}$. The amount of oxygen in the syringe is best quoted as:

    **A** $1.7 \times 10^{-3}$ mol
    **B** $1.67 \times 10^{-3}$ mol
    **C** $1.667 \times 10^{-3}$ mol
    **D** $1.6676 \times 10^{-3}$ mol

6. If 8 $cm^3$ of water measured in a 100 $cm^3$ measuring cylinder has an absolute uncertainty of ±1 $cm^3$, the percentage uncertainty in the volume will be:

    **A** 1%
    **B** 1.25%
    **C** 8%
    **D** 12.5%

7. A 5.0 $cm^3$ sample with a percentage uncertainty of 4% of 8.0 mol $dm^3$ nitric acid is diluted to 500 $cm^3$ when it is added to 450 $cm^3$ of water with a percentage uncertainty of 1%. The final concentration of the nitric acid solution will be

    **A** 0.08 mol $dm^{-3}$ ± 4%
    **B** 0.08 mol $dm^{-3}$ ± 4%
    **C** 0.08 mol $dm^{-3}$ ± 5%
    **D** 0.08 mol $dm^{-3}$ ± 9%

8. If 1.660 ± 0.002 g of potassium iodide is dissolved in a solution of volume 50 ± 1 $cm^3$, the uncertainty in the concentration of the solution will be:

    **A** 0.002%
    **B** 1.002%
    **C** 0.2%
    **D** 2%

9. The graph for rate of reaction against concentration of reactant A has been plotted for a certain reaction. The trend line is a horizontal line (parallel with the horizontal axis). The most likely reason for this line to be horizontal is that:

    **A** the student has made a mistake in recording the data.
    **B** the rate of reaction is inversely proportional to the concentration of reactant A.
    **C** the rate of reaction is directly proportional to the concentration of reactant A.
    **D** the rate of reaction is independent of the concentration of reactant A.

**10** A student is investigating the effect of surface area on the rate of the reaction between calcium carbonate and hydrochloric acid. During the investigation, the student varied the size of the calcium carbonate pieces and kept the concentration of the hydrochloric acid constant. The time taken for the reaction to be completed was measured in each case.

Which of the following graphs has the axes labelled correctly to accurately show the results of this experiment?

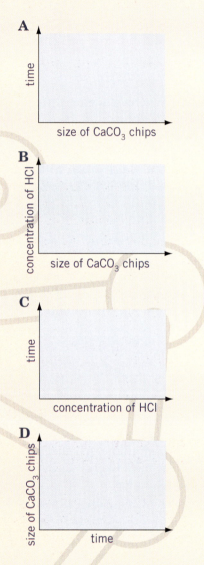

(10 marks)

 www.pearsoned.com.au/schools

For more multiple-choice test questions, connect to the Chemistry: For use with the IB Diploma Programme Standard Level Companion Website and select Chapter 5 Review Questions.

## Part B: Short-answer questions

**1** A student wishes to reduce the effect of systematic errors in an experiment. Describe two ways in which this could be achieved.

(2 marks)

**2** During an acid–base titration experiment a student titrates a hydrochloric acid solution of unknown concentration with a 0.0485 mol dm$^3$ sodium carbonate solution. The table below shows the results of the titrations.

| Experiment | 1 | 2 | 3 | 4 | 5 | 6 |
|---|---|---|---|---|---|---|
| Final volume (cm$^3$ ± 0.08 cm$^3$) | 18.45 | 36.90 | 19.10 | 37.35 | 19.40 | 37.77 |
| Initial volume (cm$^3$ ± 0.08 cm$^3$) | 0.10 | 18.45 | 0.05 | 19.10 | 10.00 | 19.40 |
| Volume of titre (cm$^3$) | | | | | | |

**a** State the random uncertainty in the volume of each titre (figure is missing from the table).

(1 mark)

**b** Calculate the volume of each titre and hence determine whether any results should be omitted from the calculation of the average.

(3 marks)

**c** Calculate the average of the titres and state the uncertainty in this value.

(2 marks)

**3** The reaction between ammonium chloride and sodium nitrite in aqueous solution can be represented by the following equation:

$NH_4Cl(aq) + NaNO_3(aq) \rightarrow N_2(g) + 2H_2O(l) + NaCl(aq)$

The graph below shows the volume of nitrogen gas produced at 30-second intervals from a mixture of ammonium chloride and sodium nitrite in aqueous solution at 20°C.

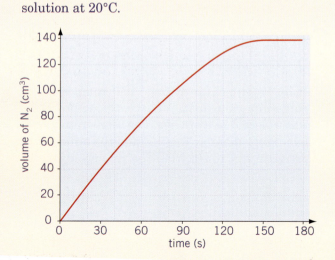

**a** The gradient of the curve represents the rate of formation of nitrogen in this experiment. State how the rate of formation of nitrogen changes with time.

(2 marks)

**b** Explain why the volume of $N_2$ eventually remains constant.

(1 mark)

*adapted from IBO 2004 SL Paper 2 May 04 Q1(a)*

**4** The graph below shows the change in pH when aqueous sodium hydroxide is added to 20 cm$^3$ of aqueous hydrochloric acid.

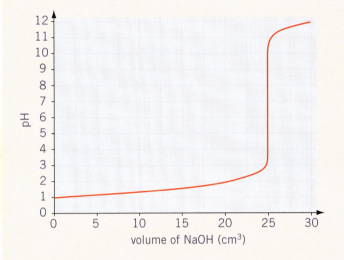

**a** Referring to the graph, describe how the pH changes as the volume of NaOH added increases.

(3 marks)

**b** State the change in pH when the volume of NaOH added is equal to 25 cm$^3$.

(1 mark)

*adapted from IBO 2003 SL Paper 2 May Q7(c)*

## Part C: Data-based questions

**1** An experiment was carried out to determine the effect of concentration on the mass of lead iodide precipitated in the reaction between potassium iodide and lead nitrate. The following quantitative data was collected.

| Concentration potassium iodide (mol dm$^{-3}$) | Mass lead iodide ($\pm$ 0.02 g) |
|---|---|
| 0.5 | 1.00 |
| 0.4 | 0.73 |
| 0.3 | 0.50 |
| 0.2 | 0.30 |
| 0.1 | 0.09 |

**a** Plot a graph of the results and draw a best-fit line through the points.

(4 marks)

**b** Identify the independent and dependent variables.

(2 marks)

**c** Describe the relationship between the independent and dependent variables.

(1 mark)

**2** A graph of ln $k$ vs $1/T$ can be used to find the activation energy for the reduction reaction between peroxodisulfate(IV) and iodide ions.

**a** Plot a graph of the data and draw a line of best fit through the points.

| ln $k$ | $1/T$ (K$^{-1}$) |
|---|---|
| −5.82 | $3.39 \times 10^{-3}$ |
| −5.19 | $3.30 \times 10^{-3}$ |
| −4.95 | $3.25 \times 10^{-3}$ |
| −4.30 | $3.18 \times 10^{-3}$ |
| −4.16 | $3.12 \times 10^{-3}$ |
| −3.61 | $3.07 \times 10^{-3}$ |

(6 marks)

**b** Determine the activation energy for this reaction, given that the slope = $-E_a \div 8.314$ J$^{-1}$ K mol$^{-1}$.

(3 marks)

**c** If the known value for the activation energy is 53.0 kJ mol$^{-1}$, calculate the percentage difference.

(2 marks)

Total marks = 43

# 6 ENERGETICS

## Chapter overview

This chapter covers the IB Chemistry syllabus Topic 5: Energetics.

**By the end of this chapter, you should be able to:**

- recognize the terms *endothermic* and *exothermic* in the context of the enthalpy, or heat content, of a chemical reaction
- construct and interpret the main features of enthalpy level diagrams for endothermic and exothermic reactions: the enthalpy ($H$) of reactants or products, and the determination of $\Delta H$ and activation energy, $E_a$
- be familiar with the terms *specific heat capacity*, *heat of combustion* and *heat of neutralization*
- calculate the change in enthalpy ($\Delta H$) for a reaction, based on experimental data
- use thermochemical equations to calculate the amount of energy involved in the reaction of a given amount of reactant
- recall that experimental thermochemical data is obtained using calorimeters and other simple apparatus
- determine the magnitude and sign of the enthalpy of a series of related thermochemical equations using Hess's law
- define the term *average bond enthalpy*
- use average bond enthalpies to explain why some reactions are exothermic and others are endothermic.

IBO Assessment statements 5.1.1 to 5.4.2

Many of the reactions with which we are familiar produce energy. Burning methane in a Bunsen burner in the laboratory, burning petrol in our cars or burning the fuel used to propel the space shuttle into orbit are all examples of energy-producing, or exothermic, reactions. An injury on the sports field may need ice. This is where an instant chemical icepack can be useful. This is an example of an endothermic reaction that absorbs energy from the surroundings.

Figure 6.0.1 The burning of methane in a Bunsen burner is an example of an exothermic reaction.

## 6.1 EXOTHERMIC AND ENDOTHERMIC REACTIONS

Any moving object possesses *kinetic energy*. As the particles that make up matter are constantly in motion, they too carry a measurable quantity of kinetic energy. When a chemical reaction occurs, bonds in reactants are broken and new bonds are formed in products. Every bond between atoms or ions has an amount of energy stored within the bonds. This energy is known as *chemical potential energy*. Thus, the total energy of a particular compound is the sum of its chemical potential energy and its kinetic energy. This amount of energy is called the **enthalpy** (symbol **H**) or heat content. It is very difficult to directly measure the heat content, or enthalpy, of individual reactants or products and so chemists generally measure the *change in enthalpy* ($\Delta H$) that occurs when a reaction proceeds. Whenever a chemical reaction occurs, there is a change in the energy of the system, as the total energy of the reactants is not equal to the total energy of the products. This difference in energy is usually either absorbed from or released to the surrounding environment as heat. Thus, measuring the change in the temperature of the surroundings caused by a reaction system provides an indirect measure of the energy change of the reaction.

### Standard conditions

The temperature, pressure of gaseous reactants and products, concentrations of solutions and the physical states (s, l, g) of the reactants all influence the value of an enthalpy change. As a result, enthalpy changes are stated under standard conditions with the symbol $\Delta H^\ominus$. At *standard conditions*, gases are at a constant pressure of 1 atm (101.3 kPa), solution concentrations are 1 mol dm$^{-3}$ and the substance should be in its standard state; that is, the state that it would normally be in at the given temperature. If the temperature is other than 298 K (25°C), the temperature must be stated. The standard enthalpy change would then have the symbol $\Delta H^\ominus_T$ where $T$ is the stated temperature. If the symbol $\Delta H^\ominus$ is used, it is taken that the temperature is 298 K (25°C).

Figure 6.1.1 A fireball from a liquid petroleum gas (LPG) explosion.

Components of internal energy

### Exothermic and endothermic reactions

The **standard enthalpy change of reaction**, $\Delta H^\ominus$, can be defined as the difference between the enthalpy of the products and the enthalpy of the reactants under standard conditions.

If the enthalpy of the products is less than that of the reactants, a reaction will lose energy, primarily as heat to its surroundings. We say that the reaction is *exothermic*. An **exothermic reaction** causes the temperature of the surroundings to *increase*. All **combustion reaction**s are exothermic as, by definition, they release heat energy as a product of the conversion of

**AS 5.1.1**
**Define the terms *exothermic reaction, endothermic reaction* and *standard enthalpy change of reaction* ($\Delta H^\ominus$). © IBO 2007**

**AS 5.1.2**
**State that combustion and neutralization are exothermic processes. © IBO 2007**

**AS 5.1.3**
**Apply the relationship between temperature change, enthalpy change and the classification of a reaction as endothermic or exothermic.** © IBO 2007

Airbags—a fast exothermic reaction

DEMO 6.1
An exothermic reaction that doesn't burn
DEMO 6.2
An endothermic reaction

PRAC 6.1
Exothermic and endothermic reactions

reactants to products. Combustion reactions will be discussed further on page 199. **Neutralization reaction**s are also examples of exothermic reactions. The reaction between hydrochloric acid and sodium hydroxide is an example of a neutralization reaction:

$$HCl(aq) + NaOH(aq) \rightarrow NaCl(aq) + H_2O(l)$$

Neutralization reactions will be discussed in chapter 9 'Acids and Bases'.

Figure 6.1.2 Combustion reactions are exothermic.

Figure 6.1.3 Ice packs utilize endothermic reactions to treat injuries.

If the enthalpy of the products is greater than that of the reactants, a reaction will absorb energy as heat from its surroundings. We say that the reaction is *endothermic*. An **endothermic reaction** causes the temperature of the surroundings to *decrease*. The ice packs used by athletes as immediate first-aid treatment for injuries on the field are examples of the use of endothermic reactions.

Mathematically, the change in enthalpy of a reaction is the enthalpy of the products minus the enthalpy of the reactants:

$$\Delta H = H_P - H_R$$

For an *exothermic reaction*, the enthalpy of the reactants is greater than that of the products, therefore the enthalpy change ($\Delta H$) will be negative (less than zero): $H_P < H_R$, so $\Delta H < 0$.

Similarly, for an endothermic reaction the enthalpy of the reactants is less than that of the products, so $\Delta H$ will be positive (greater than zero): $H_P > H_R$, so $\Delta H > 0$.

When a chemical reaction is written with an associated change in enthalpy value, it is termed a **thermochemical equation**. Thermochemical equations for the combustion of methane (natural gas) and for the dissolution of ammonium nitrate in a chemical ice pack are shown below:

$$CH_4(g) + 2O_2(g) \rightarrow CO_2(g) + 2H_2O(l) \qquad \Delta H = -890 \text{ kJ mol}^{-1}$$

$$NH_4NO_3(s) + aq \rightarrow NH_4^+(aq) + NO_3^-(aq) \qquad \Delta H = +25 \text{ kJ mol}^{-1}$$

## Enthalpy level diagrams

An **enthalpy level (energy profile) diagram** is a useful tool for visualizing what happens to the enthalpy of a reaction as it proceeds. The total enthalpy of the reactant species is labelled as $H_R$ and that of the products $H_P$. In converting the reactants to products, the reactant particles must collide with sufficient energy and the correct orientation to break the bonds between them and so allow for the formation of new products. If the particles do not collide with sufficient force, they merely bounce apart without change. When a collision takes place, some of the kinetic energy of the particles is converted into vibrational energy. It is only when the vibrational energy is sufficient to overcome the bonds holding the particles together that a chemical reaction can be initiated. This amount of energy is known as the **activation energy** ($E_a$) for the reaction and is shown as a 'hump' on the diagram.

On the enthalpy level diagrams in figure 6.1.4, the distance between the enthalpy of the reactants ($H_R$) and the top of the activation energy hump shows the amount of energy that must be provided to initiate a chemical reaction. This amount of energy (from the enthalpy of reactants to the top of the 'hump') is called the activation energy, $E_a$. A reaction can only proceed once the bonds within the reactants are overcome, which is why the activation energy must always have a positive value. Once the reactant bonds are broken, the particles can rearrange themselves to generate the products of the reaction. This process of bond making always releases energy and so the energy profile drops from the top of the activation energy hump to the energy level of the products ($H_P$).

The relative **stabilities** of the reactants and products can be deduced from the enthalpy level diagram.

**The higher the enthalpy the less stable the substance.**

In an exothermic reaction, the products have a lower enthalpy than the reactants, and so are more stable (figure 6.1.4b). In an endothermic reaction, the products have a higher enthalpy than the reactants, so are less stable (figure 6.1.4a).

A large activation energy means that the bonds within the reactants must be strong and, as a consequence, a considerable amount of energy will be required to break them. When a reaction has a large activation energy the collision between reactants needs to be very energetic if the reaction is to proceed. For example, the reaction of nitrogen with oxygen gas to produce nitrogen monoxide occurs very slowly indeed at room temperature:

$$N_2(g) + O_2(g) \rightarrow 2NO(g) \qquad \Delta H = +180.6 \text{ kJ mol}^{-1}$$

If the temperature is increased to in excess of 1000°C, as occurs in high temperature incinerators and the combustion chambers of car engines, the reaction proceeds at a measurable rate and the toxic waste gas NO is formed. This increase in rate occurs because, as the temperature increases, the energy of the molecules and of their collisions increases, so more collisions in a given time have energy that is greater than or equal to the activation energy, which will then result in products being formed.

 5.1.4
**Deduce, from an enthalpy level diagram, the relative stabilities of reactants and products and the sign of the enthalpy change for the reaction.** © IBO 2007

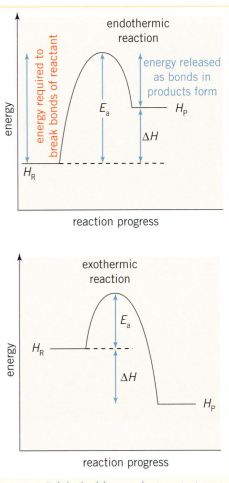

Figure 6.1.4 Enthalpy level diagrams showing activation energies ($E_a$).

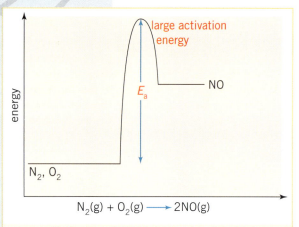

**Figure 6.1.5** The formation of NO(g) from atmospheric $N_2$ and $O_2$ has very large activation energy, resulting in a slow rate of reaction at room temperature.

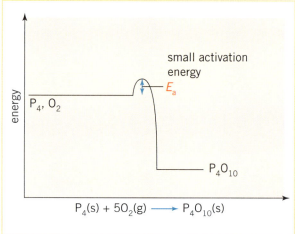

**Figure 6.1.6** Phosphorus(V) oxide forms spontaneously when phosphorus burns in air.

Some chemical reactions occur readily, with only a small amount of energy required to initiate them. For example, the element phosphorus exists in approximately 10 different allotropic forms, broadly classified as white, red and black. White phosphorus, the most reactive of the allotropes, was originally used for the tips of matches, but its tendency to spontaneously ignite while in the user's trouser pocket made it most unpopular! This role was soon taken over by the less reactive form, red phosphorus, which is why the 'safety' matches now in use are still often coloured red. White phosphorus reacts spontaneously with air to produce phosphorus(V) oxide:

$$P_4(s) + 5O_2(g) \rightarrow P_4O_{10}(s) \qquad \Delta H = -3008 \text{ kJ mol}^{-1}$$

The activation energy is very small—little energy is needed to break the bonds of the reactants.

This highly exothermic reaction makes white phosphorus a dangerous compound, and it must be stored under water to ensure that it does not come into contact with air. Numerous fires have been caused when phosphorus under storage has not been closely monitored and the substance has inadvertently come into contact with air.

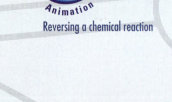

Interactive Animation
Reversing a chemical reaction

**Figure 6.1.7** White phosphorus is highly reactive and was used in incendiary bombs in World War II.

## CHEM COMPLEMENT

### A little too reactive!

In 1846, the Italian chemist Asconio Sobrero reacted glycerol with a mixture of sulfuric and nitric acids to make an explosive liquid called nitroglycerin. It was so unstable, particularly in its impure form, that even a small bump could cause it to explode, with catastrophic consequences. Even though this new explosive was many times more powerful than conventional gunpowder, it was far too dangerous to be practical. Some years later, the Swedish scientist Alfred Nobel learnt how to manage this unstable chemical more safely through his invention of dynamite. But what was it that made nitroglycerin so dangerously unpredictable?

The thermochemical equation for the decomposition of nitroglycerin is:

$$4C_3H_5(NO_3)_3(l) \rightarrow 12CO_2(g) + 10H_2O(g) + 6N_2(g) + O_2(g); \Delta H = -1456 \text{ kJ mol}^{-1}$$

The equation provides the information that the reaction is highly exothermic and produces a large amount of gaseous products. It is also the case that this reaction proceeds very quickly, with the result that the combination of heat and large volumes of expanding gases causes an explosion. But this alone does not explain why this substance is so unstable. The fact that nitroglycerin explodes so readily is an indication that the activation energy for its decomposition reaction is very small.

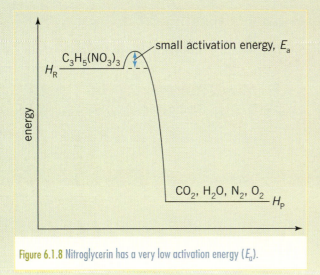

Figure 6.1.8 Nitroglycerin has a very low activation energy ($E_a$).

## THEORY OF KNOWLEDGE

Probably no Swede is as well-known throughout the world as Alfred Nobel (1833–1896), the benefactor of the Nobel Prize. This international prize is awarded every year in Chemistry, Physics, Medicine, Literature and Peace to individuals or groups who, as Nobel stated in his will, 'have made a major contribution to the acquisition of knowledge that is of the 'greatest benefit to mankind'.

When Nobel died on 10 December 1896, his will attracted much attention and criticism. First, it was not common practice at that time to donate large sums of money to science and charitable causes and, second, because he stipulated that the Nobel Prize be an international award. Many critics believed that because he was Swedish only Swedes should be eligible.

- How could you assess the kind of knowledge that is of the greatest benefit to mankind? What sort of 'evidence' or 'support' could you seek?

- Our cultural identity, traditions and experiences affect our beliefs. Outline some of the possible issues that an international award such as the Nobel prize might raise.

## Other exothermic and endothermic reactions

As the definition of an endothermic reaction can be that it is one that absorbs energy and an exothermic reaction is one that releases energy, we can consider chemical reactions other than those which occur in the laboratory as exothermic or endothermic. Consider the ionization of an element. Energy is taken in by the atoms of the element and they lose an electron, forming a positive ion. For example:

$$Na(g) \rightarrow Na^+(g) + e^- \qquad \Delta H \text{ is positive}$$

This reaction may be accurately described as endothermic.

Similarly, a reaction that shows a change of state may be endothermic or exothermic. Melting, evaporation and sublimation are endothermic changes of state. They require heat energy to be added in order for them to occur. Vaporization and condensation are exothermic changes of state. They release heat energy as they occur. Some examples are listed:

$H_2O(s) \rightarrow H_2O(l)$      $\Delta H$ is positive

$H_2O(l) \rightarrow H_2O(g)$      $\Delta H$ is positive

$I_2(s) \rightarrow I_2(g)$      $\Delta H$ is positive

$H_2O(g) \rightarrow H_2O(l)$      $\Delta H$ is negative

$H_2O(l) \rightarrow H_2O(s)$      $\Delta H$ is negative

$I_2(g) \rightarrow I_2(s)$      $\Delta H$ is negative

Bond breaking is an endothermic process and bond making is exothermic. When two atoms come together to make a molecule, energy is released. Similarly when a bond in a reactant breaks, energy is required. For example:

$Cl_2(g) \rightarrow Cl(g) + Cl(g)$      $\Delta H$ is positive

$Cl(g) + Cl(g) \rightarrow Cl_2(g)$      $\Delta H$ is negative

**TABLE 6.1.1 A SUMMARY OF EXOTHERMIC AND ENDOTHERMIC REACTIONS**

| Type of thermo-chemical reaction | Exothermic | Endothermic |
| --- | --- | --- |
| Enthalpy change $\Delta H = H_P - H_R$ | Enthalpy of products < enthalpy of reactants | Enthalpy of products > enthalpy of reactants |
| Sign of $\Delta H$ | Negative | Positive |
| Change in temperature of surroundings | Increases | Decreases |
| Enthalpy level diagram | 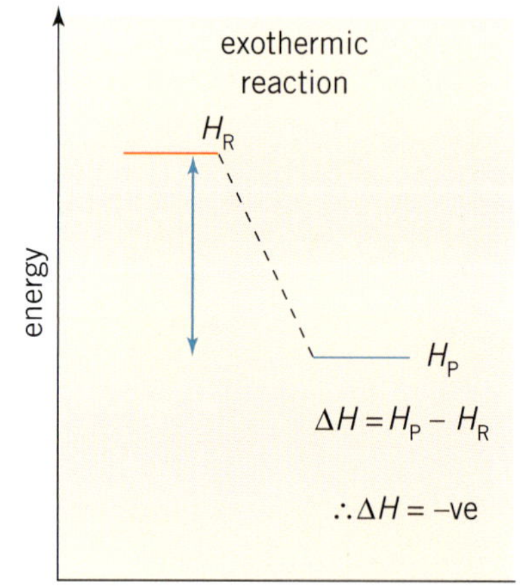 exothermic reaction; $\Delta H = H_P - H_R$; $\therefore \Delta H = -ve$ | 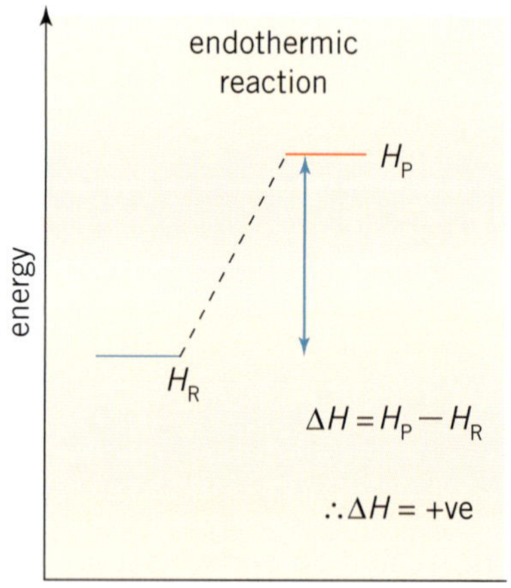 endothermic reaction; $\Delta H = H_P - H_R$; $\therefore \Delta H = +ve$ |

## Section 6.1 Exercises

1 Classify each of the following reactions as either exothermic or endothermic.
   **a** $2H_2O(l) \rightarrow 2H_2(g) + O_2(g)$; $\Delta H = +572$ kJ mol$^{-1}$
   **b** $NH_4NO_3(s) + aq \rightarrow NH_4NO_3(aq)$; $\Delta H = +25$ kJ mol$^{-1}$
   **c** $Mg(s) + Cl_2(g) \rightarrow MgCl_2(s)$; $\Delta H = -644$ kJ mol$^{-1}$

d  NaOH(aq) + HCl(aq) → NaCl(aq) + H$_2$O(l); $\Delta H$ = –57 kJ mol$^{-1}$
e  2C$_2$H$_5$OH(l) + 7O$_2$(g) → 4CO$_2$(g) + 6H$_2$O(l); $\Delta H$ = –2734 kJ mol$^{-1}$
f  NaCl(s) → Na$^+$(g) + Cl$^-$(g); $\Delta H$ = +230 kJ mol$^{-1}$

2  Consider the reactions in question **1**. State which of the six reactions listed has the greatest difference between the enthalpy of the reactants and the enthalpy of the products.

3  When a spark is introduced to a vessel containing a mixture of hydrogen and oxygen gases, they react explosively to form water vapour.
   a  State the role of the spark in this chemical reaction.
   b  State which chemical bonds are broken when this reaction is initiated.
   c  State which bonds are generated when the product is formed.
   d  State whether this reaction is endothermic or exothermic. Explain your reasoning.

4  When 10.0 g of ammonium nitrate is dissolved in 100 cm$^3$ of water, the temperature of the solution decreases from 19.0°C to 10.5°C.
   a  Compare the total enthalpy of the reactants with that of the products and state which is greater.
   b  Classify this reaction as endothermic or exothermic.
   c  State which chemical bonds were broken when the ammonium nitrate dissolved in the water.

5  The enthalpy level diagram for the combustion of methane is shown below. Calculate the magnitude and sign of the:
   a  activation energy ($E_a$)
   b  change in enthalpy ($\Delta H$).

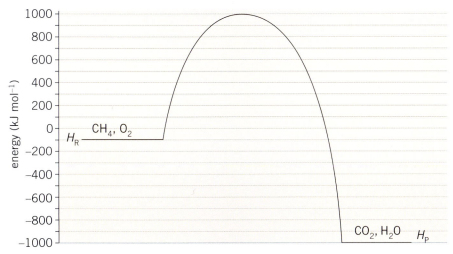

6  Draw an enthalpy level diagram for a reaction in which the total enthalpy of the reactants is 50 kJ mol$^{-1}$, the total enthalpy of products is 120 kJ mol$^{-1}$ and the activation energy for the forward reaction is 120 kJ mol$^{-1}$. Label the diagram clearly to show $\Delta H$ and $E_a$. Is this reaction exothermic or endothermic?

7  The standard enthalpy change of reaction, $\Delta H°$, for 2H$_2$ + O$_2$ → 2H$_2$O is –572 kJ mol$^{-1}$.
   a  State in what state the hydrogen and oxygen would be.
   b  State the value of the pressure of the hydrogen and oxygen at 25°C.

8 Consider the reaction represented by the following equation:

   A + 2B → C

   In this reaction, the total enthalpy of the reactants is 80 kJ mol$^{-1}$, the total enthalpy of products is –90 kJ mol$^{-1}$ and the activation energy for the forward reaction is 120 kJ mol$^{-1}$.
   
   a Draw a diagram of the energy profile for this reaction. Label the diagram clearly to show $\Delta H$ and $E_a$.
   
   b State whether the forward reaction is endothermic or exothermic.
   
   c Calculate the change in enthalpy ($\Delta H$).
   
   d Calculate the amount of energy released as the products of the reaction are formed.
   
   e Calculate the activation energy of the reverse reaction: C → A + 2B.

9 When pellets of solid sodium hydroxide are dissolved in water, the temperature of the solution formed rapidly increases.
   
   a Compare the total enthalpy of the solid NaOH with that of the solution, and state which is greater.
   
   b Classify this reaction as endothermic or exothermic.
   
   c Describe and name the chemical bonds that were broken when the sodium hydroxide dissolved in the water.

10 With reference to the energy profile diagram shown below, state the magnitude and sign of the activation energy ($E_a$) and change in enthalpy ($\Delta H$) for the reaction of nitrogen and hydrogen to produce ammonia.

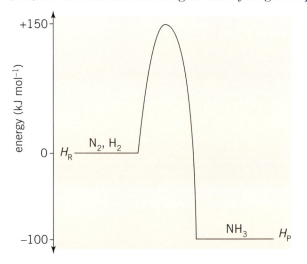

## 6.2 CALCULATION OF ENTHALPY CHANGES

Water absorbs a great deal of energy for a relatively small increase in temperature. The ability of a given mass of water to absorb heat, known as the *specific heat capacity*, is greater than for almost any other common substance, due to the hydrogen bonds between water molecules. Its ability to store large quantities of heat makes water an excellent refrigerant (also a common application of ammonia). Water possesses physical and chemical properties that set it apart from most other substances.

The **specific heat capacity** (symbol $c$) of a substance is defined as the amount of heat energy required to raise the temperature of 1.00 g of the substance by 1.00°C or 1.00 K.

Figure 6.2.1 A large amount of energy is required to boil a pan of water.

Many calculations of **heat energy change** and enthalpy change of reaction involve aqueous solutions or water being heated by a chemical reaction. The specific heat capacity of water (and of aqueous solutions) is 4.18 J °C$^{-1}$ g$^{-1}$.

$$q = m \times c \times \Delta T$$

where  $q$ = heat energy change in J
$m$ = mass of substance in g
$c$ = specific heat capacity of the substance in J °C$^{-1}$ g$^{-1}$ or J K$^{-1}$ g$^{-1}$
$\Delta T$ = change in temperature in °C or K

The quantities represented in this equation must be related to each other. The heat energy change, $q$, relates the energy absorbed or released by a reacting substance to a measured amount of water or another substance whose temperature is changing. The value for specific heat capacity that is used in the equation must be for the water or other substance being heated, and the mass used must also be for the substance that is being heated.

For example, if the heat energy change, $q$, is being measured for a 100 cm$^3$ solution in which a neutralization reaction is occurring, the value of $c$ will be 4.18 J °C$^{-1}$ g$^{-1}$, as an aqueous solution can be considered to have the same specific heat capacity as water. The mass used in the calculation will be the mass of the solution. As the density of water (and most dilute aqueous solutions) is 1 g cm$^{-3}$, the mass of a 100 cm$^3$ solution will be 100 g. The temperature change will be the change in temperature of the solution.

**AS** 5.2.1
Calculate the heat energy change when the temperature of a pure substance is changed.
© IBO 2007

**TABLE 6.2.1 SPECIFIC HEAT CAPACITIES FOR A RANGE OF SUBSTANCES**

| Substance | Specific heat capacity (J °C$^{-1}$ g$^{-1}$) |
| --- | --- |
| Water | 4.18 |
| Ammonia | 2.06 |
| Ethanol | 2.46 |
| Aluminium | 0.889 |
| Iron | 0.448 |
| Copper | 0.386 |
| Gold | 0.128 |

## Worked example 1

A beaker containing 100 cm$^3$ of water is heated using a Bunsen burner. Over a period of 10 minutes, the temperature of the water increased from 15.0°C to 80.0°C. Determine the heat energy change of the water in the beaker.

WORKSHEET 6.1
Specific heat capacity

### Solution

Note that the density of water is 1.0 g cm$^{-3}$, therefore the same numerical value can be used for the mass of water (in g) as is given for the volume of water (in cm$^3$).

$q = m \times c \times \Delta T$
∴ $q = 100 \times 4.18 \times (80.0 - 15.0)$
∴ $q = 25\,916$ J = 25.9 kJ

## Worked example 2

The same Bunsen burner was used to heat a 500 g block of copper for 10 minutes. Assuming that the same amount of energy is transferred from the Bunsen burner to the copper, as was in example 1, determine the highest temperature that the block of copper, initially at 15.0°C, would reach.

### Solution

From table 6.2.1, the specific heat capacity of copper = 0.386 J °C$^{-1}$ g$^{-1}$

$q = m \times c \times \Delta T$
∴ $25\,916 = 500 \times 0.386 \times \Delta T$
∴ $\Delta T = \dfrac{25\,916}{500 \times 0.386} = 134$°C

The highest temperature that the block of copper can reach is (134 + 15) = 149°C.

Notice that the lower specific heat capacity of copper (0.386 J °C$^{-1}$ g$^{-1}$ compared with 4.18 J °C$^{-1}$ g$^{-1}$ for water) enables the copper to be heated to a higher temperature by the same amount of energy.

## Experimental methods for measuring heat energy changes of reactions

### Simple calorimetry

**AS 5.2.2**
**Design suitable experimental procedures for measuring heat energy changes of reactions. © IBO 2007**

The heat energy released by the combustion of a fuel can be measured using a simple set of apparatus. This is shown in figure 6.2.2. The heat released by the combustion reaction transfers somewhat incompletely to the can of water. The initial temperature of the water and the highest temperature reached are measured, as is the volume of water in the can. The change in mass of the fuel can be determined by measuring the initial mass of the fuel and subtracting the final mass. In this way the heat energy change can be related to the amount of fuel that was consumed, and this will produce an enthalpy change for the combustion reaction.

Figure 6.2.2 A simple apparatus for measuring the heat of combustion of a fuel.

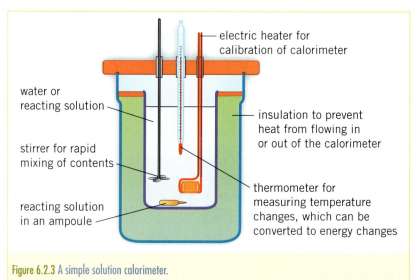

Figure 6.2.3 A simple solution calorimeter.

Enthalpy of solution
Bomb calorimetry
Heats of reaction

**AS 5.2.4**
**Evaluate the results of experiments to determine enthalpy changes. © IBO 2007**

Reactions in solution can be performed in a simple calorimeter, which may be as simple as a polystyrene (styrofoam) cup with a lid. The insulation provided by the polystyrene prevents the loss of heat to, or the addition of energy from, the surroundings. The reaction may be carried out in the calorimeter with the volume of water accurately known. The initial and final temperatures are measured and recorded, as are the amounts of reactants used.

Both of the methods described here have limitations. In the case of the combustion reaction, a lot of heat produced in the reaction is lost to the surroundings due to the simplicity of the apparatus. In addition, this method does not make any allowance for absorption of heat energy by the can. Similarly the method using the simple calorimeter does not make any allowance for the heat absorbed by the polystyrene container, so the heat energy change will be inaccurate.

# Heat of combustion

It is common to measure the *heat of combustion* of a fuel. This is the amount of heat produced by the combustion of one mole of the fuel. The heat of combustion of a fuel is not necessarily the same as its enthalpy of combustion when written as a balanced chemical equation. For example, the heat of combustion of octane is 5464 kJ mol$^{-1}$, but the enthalpy of the reaction, shown below, is twice this value because the balanced equation involves 2 mole of octane.

$$2C_8H_{18}(g) + 25O_2(g) \rightarrow 16CO_2(g) + 18H_2O(l) \quad \Delta H = -10\,928 \text{ kJ mol}^{-1}$$

Many fuels commonly used are complex mixtures of substances: petrol, diesel, coal and wood, for example. Heats of combustion may therefore be provided in units of kJ g$^{-1}$, or MJ dm$^{-3}$ rather than the unit of kJ mol$^{-1}$.

**TABLE 6.2.2 HEATS OF COMBUSTION OF SOME HYDROCARBONS**

| Hydrocarbon | Heat of combustion | |
| --- | --- | --- |
| | kJ mol$^{-1}$ | kJ g$^{-1}$ |
| Methane | 889 | 55.6 |
| Ethane | 1557 | 51.9 |
| Propane | 2217 | 50.4 |
| Butane | 2874 | 49.6 |
| Octane | 5464 | 47.9 |

**TABLE 6.2.3 HEATS OF COMBUSTION OF SOME FUELS**

| Common fuels | Heat of combustion | |
| --- | --- | --- |
| | kJ g$^{-1}$ | MJ dm$^{-3}$ |
| Hydrogen (liquid) | 142 | 286 |
| Ethanol | 29.7 | 19.6 |
| Petrol | 47.3 | 29.0 |
| Diesel | 44.8 | 32.2 |
| LPG | 51 | 22.2 |
| Wood | 15 | N/A |
| Coal (brown, dried) | 25 | N/A |
| Coal (black, dried) | 32 | N/A |

## Worked example 3

A student filled a small spirit burner with methanol and measured its mass. The spirit burner was then positioned under a tin can as shown in figure 6.2.2 and the wick of the burner was lit. After the temperature of the water had risen by approximately 20°C the flame was snuffed out and the final mass of the spirit burner and methanol was measured. The student's results are shown in table 6.2.4. Uncertainties will be considered and calculated in this example.

PRAC 6.2
The heat of combustion of an alcohol

**TABLE 6.2.4 RESULTS OF ETHANOL BURNING EXPERIMENT**

|  | Initial measurement | Final measurement |
|---|---|---|
| Mass of spirit burner and methanol (g ± 0.001 g) | 238.860 | 236.610 |
| Temperature of water in can (°C ± 1°C) | 23 | 47 |
| Volume of water in can (cm³ ± 2 cm³) | 200 | |

Determine the heat energy change during this experiment and hence determine the **heat of combustion** of methanol.

### Solution

$\Delta T = 47 - 23 = 24 \pm 2°C$

Note that this value for the temperature limits the answer to 2 significant figures.

Note: The temperature increased, therefore the reaction is exothermic, so $\Delta H$ is negative.

$q = m \times c \times \Delta T$
$\phantom{q} = 200 \times 4.18 \times 24$
$\phantom{q} = 20\,064 \text{ J} = 20 \text{ kJ}$

Calculation of percentage error in this result:

Mass of water $= \dfrac{2}{200} \times \dfrac{100}{1} = 1\%$

Specific heat capacity is a constant, therefore no error.

Temperature $= \dfrac{2}{24} \times \dfrac{100}{1} = 8\%$

Total percentage error $= 1 + 8 = 9\%$

Absolute error in results $= \dfrac{9}{100} \times 20 \text{ kJ}$
$\phantom{Absolute error in results} = 2 \text{ kJ}$

The heat energy change during this experiment was $20 \pm 2$ kJ.

$m(\text{methanol}) = 238.860 - 236.610 = 2.250 \pm 0.002$ g

Percentage error in mass $= \dfrac{0.002}{2.250} \times \dfrac{100}{1} = 0.9\%$

$n(\text{methanol}) = \dfrac{m}{M} = \dfrac{2.25}{32.05} = 0.0702 \pm 0.9\%$ mol

Heat of combustion $= \dfrac{q}{n} = \dfrac{20\,064}{0.0702} = 285\,812$ J $= 2.9 \times 10^2$ kJ mol$^{-1}$

Percentage error in heat of combustion $= 9 + 0.9 = 10\%$

Absolute error in heat of combustion $= 0.3 \times 10^2$ kJ mol$^{-1}$

The heat of combustion of methanol has been experimentally determined here to be $-2.9 \times 10^2 \pm 0.3 \times 10^2$ kJ mol$^{-1}$.

This value is significantly lower than the literature value of $-715$ kJ mol$^{-1}$. One of the most likely causes of error is the loss of heat from the methanol flame to the surroundings.

Other assumptions made in this calculation include the assumption that the heat energy from the spirit burner is only heating the water in the can, whereas the can itself will be absorbing some heat, as will the thermometer and any equipment such as the clamp that is used to hold the can up.

## CHEM COMPLEMENT

### Diesel versus petrol engines

Internal combustion engines have been built since the middle of the 19th century, but it was not until 1885 that the German engineer Karl Benz built the first practical automobile to be powered by a petrol-driven engine. In these engines, a mixture of air and petrol is drawn into the combustion chamber, where it is compressed by the upward stroke of the piston. When the petrol–air mixture is fully compressed, a spark from the spark plug ignites the vapour. The heat produced by the combustion reaction expands the gaseous products of the reaction dramatically and this increased pressure forces the piston back down the cylinder. The flash point of petrol is low (−43°C), while its ignition temperature at 390°C is too high for it to be ignited without the aid of a spark.

The diesel engine is similar, although a diesel motor has no need of a spark plug or the associated distributor and alternator. The ignition temperature of diesel fuel is sufficiently low at 260°C to ignite spontaneously at the point of maximum compression within the cylinder of the motor. This is the reason why diesel engines run on a compression ratio of from 14 to 25, while most petrol motors function from 8 to a maximum of 12. The invention of this so-called internal combustion engine by Rudolf Diesel in 1893 made him a millionaire within 5 years.

Figure 6.2.4 A diesel engine. How does it differ from a petrol-driven engine?

## THEORY OF KNOWLEDGE

On the methodology of science, Nobel said 'from observation one goes on to experimentation … based on analogies and inductions of empirical laws'.

In chapter 4 a *law* was described as an expression of a relationship that has been established by experiment. In science, laws that are supported by evidence from observations are said to be *empirical*.

Scientists use inductive reasoning to formulate laws. Specific data and observations are combined into general rules or relationships. Deductive reasoning, on the other hand, starts with general statements and leads to a specific conclusion. Generalized statements can be distinguished from specific statements because they use words such as 'all' or 'none', 'always' or 'never'. Each general statement makes the claim about every single instance. For example, 'All alcohols absorb heat energy during vaporization' means that every single alcohol will absorb heat energy when it vaporizes.

- Distinguish between the knowledge claims made using inductive reasoning in science and deductive reasoning in mathematics.
- In the example below identify which knowledge claim is made using inductive and which is made using deductive reasoning.

  Example 1: Alcohols A, B and C release heat energy when combusted. Therefore all alcohols release heat energy when combusted.

  Example 2: All alcohols release heat energy when heated. A is an alcohol. Therefore A releases heat energy when combusted.

- Write your own simple argument on this topic, using inductive and deductive reasoning.

**AS 5.2.3**
Calculate the enthalpy change for a reaction using experimental data on temperature changes, quantities of reactants and mass of water. © IBO 2007

### Calculating enthalpy changes using thermochemical equations

A thermochemical equation is one that gives the relevant $\Delta H$ value for that particular equation exactly as it is written. For example, the thermochemical equation for the combustion of methane (natural gas)

$$CH_4(g) + 2O_2(g) \rightarrow CO_2(g) + 2H_2O(l) \qquad \Delta H = -890 \text{ kJ mol}^{-1}$$

literally means that when 1 mole of methane reacts with 2 mole of oxygen to produce 1 mole of carbon dioxide gas and 2 mole of liquid water, 890 kJ of energy is released. If the equation were doubled, the $\Delta H$ value would also double. If the equation were reversed, the $\Delta H$ value would change sign. These changes are shown in the equations below.

$$2CH_4(g) + 4O_2(g) \rightarrow 2CO_2(g) + 4H_2O(l) \qquad \Delta H = -1780 \text{ kJ mol}^{-1}$$
$$CO_2(g) + 2H_2O(l) \rightarrow CH_4(g) + 2O_2(g) \qquad \Delta H = +890 \text{ kJ mol}^{-1}$$

Thermochemical equations can be used to perform a range of stoichiometric calculations involving heats of reaction. Questions such as what amount of fuel is required to generate a given amount of heat energy, how much ammonium nitrate must be added to an instant ice pack to ensure its effectiveness, and how much energy is provided by particular foods can be answered using the data provided by thermochemical equations.

**THEORY OF KNOWLEDGE**
Outline the conventions used when representing thermochemical equations.

**WORKSHEET 6.2**
Calculation of enthalpy changes

### Worked example 4

Calculate the amount of energy released when 500 cm³ of methane gas at STP reacts with excess air according to the equation:

$$CH_4(g) + 2O_2(g) \rightarrow CO_2(g) + 2H_2O(l) \qquad \Delta H = -890 \text{ kJ mol}^{-1}$$

#### Solution

$$n(CH_4) = \frac{V}{V_m} = \frac{0.500}{22.4} = 0.0223 \text{ mol}$$

From the equation, 1 mol of $CH_4$ releases 890 kJ of energy.

So 0.0223 mol of $CH_4$ releases $x$ kJ.

$$\frac{890}{1} = \frac{x}{0.0223}$$

$$\therefore x = 890 \times 0.0223 = 19.9 \text{ kJ}$$

The energy released by the combustion reaction is 19.9 kJ.

### Worked example 5

What mass of propanol must be burnt in excess air to produce $1.00 \times 10^4$ kJ of energy, according to the following equation?

$$2C_3H_7OH(l) + 9O_2(g) \rightarrow 6CO_2(g) + 8H_2O(l) \qquad \Delta H = -4034 \text{ kJ mol}^{-1}$$

PRAC 6.3
Calibration of a calorimeter

PRAC 6.4
Determining enthalpy changes in chemical reactions

#### Solution

From the equation, 2 mol of propanol yields 4034 kJ of energy, so $x$ mol of propanol yields 10 000 kJ.

$$\frac{2}{4034} = \frac{x}{10\,000}$$

$$x = \frac{2 \times 10\,000}{4034} = 4.96 \text{ mol}$$

$$m(C_3H_7OH) = n \times M(C_3H_7OH) = 4.96 \times 60.11 = 298 \text{ g}$$

## Worked example 6

Ethyne (acetylene) is used as the fuel in oxyacetylene torches used by welders. Given that the complete combustion of 8.55 g of ethyne produces 429 kJ of energy, calculate the $\Delta H$ value for the equation:

$$2C_2H_2(g) + 5O_2(g) \rightarrow 4CO_2(g) + 2H_2O(l)$$

### Solution

$n(C_2H_4) = \dfrac{m}{M} = \dfrac{8.55}{28.06} = 0.305$ mol

The equation shows 2 mol of ethyne.

0.305 mol of $C_2H_4$ releases 429 kJ
$\therefore$ 2 mol of $C_2H_4$ releases $x$ kJ

By ratio, $\dfrac{x}{2} = \dfrac{429}{0.305} \quad \therefore x = \dfrac{2 \times 429}{0.305} = 2813 = 2.81 \times 10^3$ kJ

$\therefore \Delta H = -2.81 \times 10^3$ kJ mol$^{-1}$

(Note that the sign is negative because energy was given out.)

Figure 6.2.5 The combustion of ethyne generates enough heat to melt steel.

## Section 6.2 Exercises

1. 350 g of ethanol is heated using 22.3 kJ of energy. Calculate the increase in temperature of the ethanol, given that its specific heat capacity is 2.46 J °C$^{-1}$ g$^{-1}$.

2. Calculate the amount of heat energy released when 1000 kg of iron cools from 550°C to 24.0°C, given that the specific heat capacity of iron is 0.448 J °C$^{-1}$ g$^{-1}$.

3. Calculate the mass of butane gas that would be needed to heat 724 cm$^3$ of water from an initial temperature of 7.44°C to 50.7°C. The thermochemical equation for the combustion of butane is:

   $2C_4H_{10}(g) + 13O_2(g) \rightarrow 8CO_2(g) + 10H_2O(l) \qquad \Delta H = -5748$ kJ mol$^{-1}$

4. A student dissolves 12.2 g of ammonium nitrate in 85.0 cm$^3$ of water at an initial temperature of 28.3°C. Calculate the temperature the water will reach after the substance has fully dissolved, given that the thermochemical equation for the dissolution reaction is:

   $NH_4NO_3(s) \xrightarrow{\text{water}} NH_4^+(aq) + NO_3^-(aq) \qquad \Delta H = +25$ kJ mol$^{-1}$

5. Calculate the amount of energy released when 100 g of propane gas reacts with excess air according to the equation:

   $C_3H_8(g) + 5O_2(g) \rightarrow 3CO_2(g) + 4H_2O(g) \qquad \Delta H = -2217$ kJ mol$^{-1}$

6. Calculate the amount of energy released when 1.00 dm$^3$ (806 g) of butanol reacts with excess air according to the equation:

   $2C_4H_9OH(g) + 12O_2(g) \rightarrow 8CO_2(g) + 10H_2O(g) \qquad \Delta H = -5354$ kJ mol$^{-1}$

**7** Calculate the mass of pentane that would need to undergo complete combustion to yield 500 kJ of energy, given that the equation for its combustion is:

$$C_5H_{12}(g) + 8O_2(g) \rightarrow 5CO_2(g) + 6H_2O(l) \qquad \Delta H = -3509 \text{ kJ mol}^{-1}$$

**8** Diborane ($B_2H_6$) is a highly reactive boron hydride that has been used as a rocket fuel. Calculate its enthalpy of combustion, given that 1.00 kg of diborane produces 73.7 MJ according to the equation:

$$B_2H_6(g) + 3O_2(g) \rightarrow B_2O_3(s) + 3H_2O(g)$$

**9** Methanol is sometimes used as a fuel additive in heavily modified racing cars as it has a very high octane rating. The thermochemical equation for the combustion of methanol is:

$$2CH_3OH(l) + 3O_2(g) \rightarrow 2CO_2(g) + 4H_2O(l) \qquad \Delta H = -1450 \text{ kJ mol}^{-1}$$

  **a** Calculate the heat of combustion of methanol in:
   **i** kJ per mol
   **ii** kJ per gram
   **iii** MJ per $dm^3$, given that $d(CH_3OH) = 0.787$ g $cm^{-3}$.

  **b** Calculate the volume of $CO_2$ that would be produced by the combustion of 15.0 $dm^3$ of methanol at 35°C and 0.98 atm pressure.

## 6.3 HESS'S LAW

A fundamental principle in science is the law of conservation of energy, which states that energy can neither be created nor destroyed but merely changed in state. In 1840, the German chemist Germain Henri Hess (1804–1850) used this starting point to propose what became known as **Hess's law**. This law can be stated as 'the heat evolved or absorbed in a chemical process is the same, whether the process takes place in one or in several steps'.

Hess's law can be used to determine the enthalpy of a reaction by manipulating known thermochemical equations that could be used as a reaction pathway to the desired reaction.

So, if the enthalpy of reaction for A + B → C + D cannot be measured, but the enthalpies of another two reactions

  A + B → X   and   X → C + D

can be measured, then the enthalpies of these two reactions can be added to find the enthalpy of the original reaction A + B → C + D.

It is important to remember that the combination of reactants and of products in the thermochemical reactions being manipulated must be the same as those in the desired reaction for the enthalpies to be equal.

The reactions can be put into an enthalpy cycle (see figure 6.3.1) that shows the relationship between the three reactions.

In figure 6.3.1 it can be seen that the enthalpy of reaction for A + B → C + D, $\Delta H_1$ is equal to the sum of $\Delta H_2$ and $\Delta H_3$.

**AS 5.3.1**
Determine the enthalpy change of a reaction which is the sum of two or three reactions with known enthalpy changes.
© IBO 2007

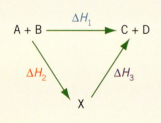

Figure 6.3.1 The enthalpy cycle for determining the enthalpy of the reaction A + B → C + D.

Interactive Animation
Hess's law

## Worked example 1

Calculate the enthalpy of reaction for the formation of $NO_2(g)$ according to the equation:

$N_2(g) + 2O_2(g) \rightarrow 2NO_2(g)$  $\Delta H = ?$

given the following thermochemical equations.

$N_2(g) + O_2(g) \rightarrow 2NO(g)$  $\Delta H = +180$ kJ mol$^{-1}$

$2NO_2(g) \rightarrow 2NO(g) + O_2(g)$  $\Delta H = +112$ kJ mol$^{-1}$

### Solution

The formation of nitrogen dioxide occurs in two steps: nitrogen and oxygen gases react to form nitrogen monoxide, which is then further oxidized to nitrogen dioxide. The required thermochemical equation can be derived by adding the first equation to the reverse of the second, eliminating the intermediate product, nitrogen monoxide.

$N_2(g) + O_2(g) \rightarrow 2NO(g)$  $\Delta H = +180$ kJ mol$^{-1}$

$2NO(g) + O_2(g) \rightarrow 2NO_2(g)$  $\Delta H = -112$ kJ mol$^{-1}$

(Note that the sign of $\Delta H$ is now negative because the equation has been reversed.)

The equations can now be added, along with their respective $\Delta H$ values. The $NO(g)$ is on both sides of the equation and so can be cancelled and the equation simplified to give:

$N_2(g) + O_2(g) + \cancel{2NO(g)} + O_2(g) \rightarrow \cancel{2NO(g)} + 2NO_2(g)$  $\Delta H = (+180 - 112)$ kJ mol$^{-1}$

$N_2(g) + 2O_2(g) \rightarrow 2NO_2(g)$  $\Delta H = +68$ kJ mol$^{-1}$

WORKSHEET 6.3
Hess's law

## Worked example 2

Calculate the enthalpy of reaction for the formation of phosphorus trichloride from white phosphorus according to the equation:

$P_4(s) + 6Cl_2(g) \rightarrow 4PCl_3(g)$  $\Delta H = ?$

given the following thermochemical data.

$PCl_3(g) + Cl_2(g) \rightarrow PCl_5(g)$  $\Delta H = -137$ kJ mol$^{-1}$

$P_4(s) + 10Cl_2(g) \rightarrow 4PCl_5(g)$  $\Delta H = +716$ kJ mol$^{-1}$

### Solution

To eliminate phosphorus pentachloride ($PCl_5$) from the equation system, we must reverse the first equation and multiply it by four:

$4 \times [PCl_5(g) \rightarrow PCl_3(g) + Cl_2(g)]$  $\Delta H = 4 \times (+137$ kJ mol$^{-1})$

$P_4(s) + 10Cl_2(g) \rightarrow 4PCl_5(g)$  $\Delta H = +716$ kJ mol$^{-1}$

The two equations can now be added and simplified.

$P_4(s) + \cancel{10}^{6}Cl_2(g) + \cancel{4PCl_5(g)} \rightarrow \cancel{4PCl_5(g)} + 4PCl_3(g) + \cancel{4Cl_2(g)}$

$\Delta H = +716 + 4 \times (+137)$ kJ mol$^{-1}$

$P_4(s) + 6Cl_2(g) \rightarrow 4PCl_3(g)$  $\Delta H = +1264$ kJ mol$^{-1}$

### Worked example 3

Calculate the standard enthalpy of reaction for the oxidation of nitrogen(II) oxide to form nitrogen(IV) oxide

$NO(g) + O(g) \rightarrow NO_2(g)$

given the following thermochemical data:

(1) $2O_3(g) \rightarrow 3O_2(g)$      $\Delta H^\circ = -427$ kJ mol$^{-1}$

(2) $O_2(g) \rightarrow 2O(g)$      $\Delta H^\circ = +495$ kJ mol$^{-1}$

(3) $NO(g) + O_3(g) \rightarrow NO_2(g) + O_2(g)$      $\Delta H^\circ = -199$ kJ mol$^{-1}$

### Solution

The combination of equations 2 (halved and reversed) and 3 provide the correct reactants (NO and O) for this reaction.

$-\frac{1}{2} \times (2)$      $O(g) \rightarrow \frac{1}{2} O_2(g)$      $\Delta H^\circ = -247.5$ kJ mol$^{-1}$

(3)      $NO(g) + O_3(g) \rightarrow NO_2(g) + O_2(g)$      $\Delta H^\circ = -199$ kJ mol$^{-1}$

When these two equations are added together ozone, $O_3(g)$ is also present.

$O(g) + NO(g) + O_3(g) \rightarrow \frac{3}{2} O_2(g) + NO_2(g)$      $\Delta H^\circ = -446.5$ kJ mol$^{-1}$

Ozone is eliminated by adding equation 1 (halved and reversed)

$-\frac{1}{2} \times (1)$      $\frac{3}{2} O_2(g) \rightarrow O_3$      $\Delta H^\circ = +213.5$ kJ mol$^{-1}$

Add the last two equations together and simplify to give the required oxidation equation of NO.

$O(g) + NO(g) + \cancel{O_3(g)} + \cancel{\frac{3}{2} O_2(g)} \rightarrow \cancel{\frac{3}{2} O_2(g)} + NO_2(g) + \cancel{O_3(g)}$

$\Delta H^\circ = (-446.5 + 213.5)$ kJ mol$^{-1}$

$O(g) + NO(g) \rightarrow NO_2$      $\Delta H^\circ = -233$ kJ mol$^{-1}$

## Section 6.3 Exercises

**1** Calculate $\Delta H^\circ$ for the reaction:

$3FeO(s) + \frac{1}{2} O_2(g) \rightarrow Fe_3O_4(s)$

given the following data.

$Fe(s) + \frac{1}{2} O_2(g) \rightarrow FeO(s)$      $\Delta H^\circ = -272$ kJ mol$^{-1}$

$3Fe(s) + 2O_2(g) \rightarrow Fe_3O_4(s)$      $\Delta H^\circ = -1118$ kJ mol$^{-1}$

**2** Calculate $\Delta H^\circ$ for the reaction:

$S(s) + O_2(g) \rightarrow SO_2(g)$

given the following data.

$S(s) + \frac{3}{2} O_2(g) \rightarrow SO_3(g)$      $\Delta H^\circ = -395.2$ kJ mol$^{-1}$

$2SO_2(g) + O_2(g) \rightarrow 2SO_3(g)$      $\Delta H^\circ = -198.2$ kJ mol$^{-1}$

3. Calculate $\Delta H^\circ$ for the reaction:
$2NO_2(g) \rightarrow N_2O_4(g)$
given the following data.

$N_2(g) + 2O_2(g) \rightarrow 2NO_2(g)$     $\Delta H^\circ = +67.7$ kJ mol$^{-1}$
$N_2(g) + 2O_2(g) \rightarrow N_2O_4(g)$     $\Delta H^\circ = +9.7$ kJ mol$^{-1}$

4. Calculate $\Delta H^\circ$ for the reaction:
$2N_2(g) + 5O_2(g) \rightarrow 2N_2O_5(g)$
given the following data.

$H_2(g) + \frac{1}{2}O_2(g) \rightarrow H_2O(l)$     $\Delta H^\circ = -285.8$ kJ mol$^{-1}$

$N_2O_5(g) + H_2O(l) \rightarrow 2HNO_3(l)$     $\Delta H^\circ = -76.6$ kJ mol$^{-1}$

$\frac{1}{2}N_2(g) + \frac{3}{2}O_2(g) + \frac{1}{2}H_2(g) \rightarrow HNO_3(l)$     $\Delta H^\circ = -174.1$ kJ mol$^{-1}$

5. Calculate $\Delta H^\circ$ for the reaction:
$2C(s) + H_2(g) \rightarrow C_2H_2(g)$
given the following data.

$C_2H_2(g) + \frac{5}{2}O_2(g) \rightarrow 2CO_2(g) + H_2O(l)$     $\Delta H^\circ = -1300$ kJ mol$^{-1}$

$C(s) + O_2(g) \rightarrow CO_2(g)$     $\Delta H^\circ = -394$ kJ mol$^{-1}$

$H_2(g) + \frac{1}{2}O_2(g) \rightarrow H_2O(l)$     $\Delta H^\circ = -286$ kJ mol$^{-1}$

6. Using the equations below,

$C(s) + O_2(g) \rightarrow CO_2(g)$     $\Delta H^\circ = -390$ kJ mol$^{-1}$
$Mn(s) + O_2(g) \rightarrow MnO_2(s)$     $\Delta H^\circ = -520$ kJ mol$^{-1}$

calculate $\Delta H$ (in kJ mol$^{-1}$) for the following reaction:
$MnO_2(s) + C(s) \rightarrow Mn(s) + CO_2(g)$

7. Using the following equations,

$O_2(g) + H_2(g) \rightarrow 2OH(g)$     $\Delta H^\circ = +77.9$ kJ mol$^{-1}$
$O_2(g) \rightarrow 2O(g)$     $\Delta H^\circ = +495$ kJ mol$^{-1}$
$H_2(g) \rightarrow 2H(g)$     $\Delta H^\circ = +435.9$ kJ mol$^{-1}$

calculate $\Delta H^\circ$ for the reaction:
$O(g) + H(g) \rightarrow OH(g)$

## 6.4 BOND ENTHALPIES

The *bond enthalpy* is the amount of energy required to break one mole of bonds. For example when the C–H bonds in methane are broken to make carbon atoms and hydrogen atoms according to

$CH_4(g) \rightarrow C(g) + 4H(g); \Delta H = +1662$ kJ mol$^{-1}$

**AS 5.4.1**
**Define the term *average bond enthalpy*. © IBO 2007**

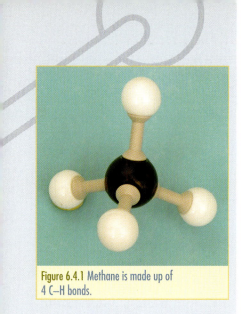

Figure 6.4.1 Methane is made up of 4 C–H bonds.

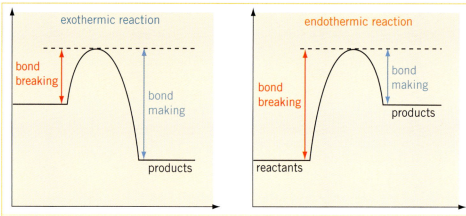

Figure 6.4.2 In an exothermic reaction bond making releases more energy than bond breaking requires and in an endothermic reaction, bond making releases less energy than bond breaking requires.

**AS 5.4.2** Explain, in terms of average bond enthalpies, why some reactions are exothermic and others are endothermic. © IBO 2007

Breaking and forming bonds in a chemical reaction

As there are 4 C–H bonds in methane, this gives a bond enthalpy value of 415.5 kJ mol$^{-1}$.

The enthalpy of a bond is not exactly the same in all compounds containing that bond, so an average value is given. For the C–H bond this value is 412 kJ mol$^{-1}$.

The **average bond enthalpy** is the amount of energy required to break one mole of bonds in the gaseous state averaged across a range of compounds containing that bond.

In the course of a chemical reaction energy is required to break the bonds of reactants and energy is released by the formation of the bonds of the products.

**Bond breaking is endothermic, bond making is exothermic.**

If more energy is released in the formation of bonds in the products than was required to break the bonds of the reactants, then the reaction is exothermic.

If more energy is required to break the bonds of the reactants than is released in the formation of bonds in the products, then the reaction is endothermic.

The enthalpy of reaction can be calculated using bond enthalpies. Since these are average bond enthalpies, the calculation gives only an approximate result.

**Enthalpy of reaction = $\Sigma D$(bonds broken) − $\Sigma D$(bonds formed)**

where $\Sigma$ represents the sum of the terms and $D$ represents the bond enthalpy per mole of bonds.

Note: the value for bonds formed is negative because bond making releases energy.

| TABLE 6.4.1 SOME AVERAGE BOND ENTHALPIES | | | |
|---|---|---|---|
| Bond | Average bond enthalpy (kJ mol$^{-1}$) | Bond | Average bond enthalpy (kJ mol$^{-1}$) |
| H–H | 436 | C–H | 412 |
| C–C | 348 | N–H | 388 |
| C=C | 612 | O–H | 463 |
| C≡C | 837 | S–H | 338 |
| O–O | 146 | Cl–H | 431 |
| O=O | 496 | C–O | 360 |
| Cl–Cl | 242 | C=O | 743 |
| H–Br | 366 | C–N | 305 |
| Br–Br | 193 | C–Br | 276 |

## Worked example 1

Use average bond enthalpies (table 6.4.1) to find the heat of combustion of methane.

$CH_4(g) + 2O_2(g) \rightarrow CO_2(g) + 2H_2O(g)$

### Solution

$\Delta H = [4D_{C-H} + 2D_{O=O}] - [2D_{C=O} + 4D_{H-O}]$
$= [4 \times 412 + 2 \times 496] - [2 \times 743 + 4 \times 463]$
$= 2640 - 3338$
$= -698 \text{ kJ mol}^{-1}$

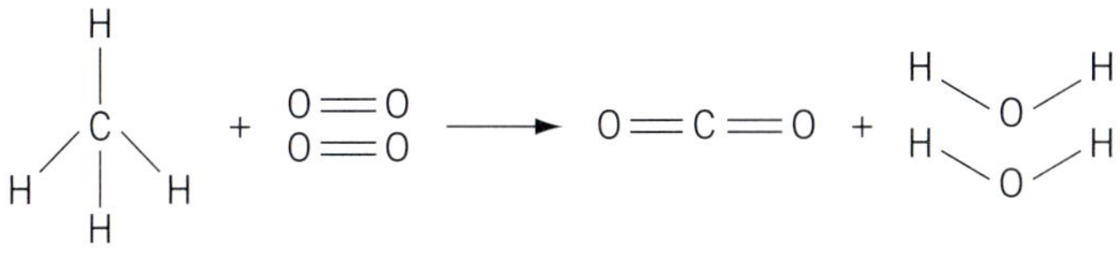

This theoretical value of the heat of combustion of methane is only 78% accurate in comparison to the literature value of $-890 \text{ kJ mol}^{-1}$.

## Section 6.4 Exercises

1. Explain, making reference to bond enthalpies, how a reaction could be endothermic.

2. Why is the average bond enthalpy of C=C greater than C–C?

3. Use average bond enthalpies to calculate the heat of reaction, $\Delta H$, for the following reaction:
   $H_2(g) + Cl_2(g) \rightarrow 2HCl(g)$

4. Use average bond enthalpies to calculate the heat of combustion of methanol.
   $2CH_3OH(l) + 3O_2(g) \rightarrow 2CO_2(g) + 4H_2O(l)$

5. a  Use average bond enthalpies to calculate the heat of combustion of ethane.
   b  Also using average bond enthalpies, calculate the heat of combustion of ethene.
   c  Explain why the heat of combustion of ethene is less exothermic than that of ethane.

6. Without performing the calculations or referring to tables of combustion data, state which compound you would expect to have a greater (more negative) heat of combustion, ethyne, $C_2H_2$, or ethene, $C_2H_4$. Explain your answer.

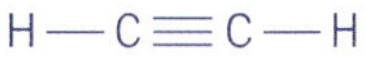

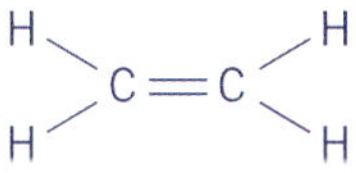

7. a  Use average bond enthalpies to find the heat of combustion of ethanoic acid, $CH_3COOH$.

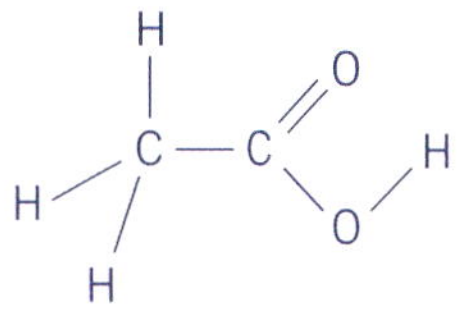

   b  Also using average bond enthalpies, find the heat of combustion of ethanal, $CH_3CHO$.

   c  With reference to the bonding in these two compounds, explain the difference in their heats of combustion.

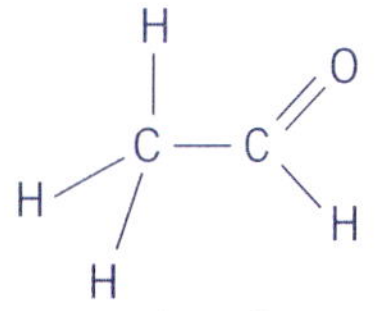

# Chapter 6 Summary

## Terms and definitions

**Activation energy**   The energy required to break the bonds of the reactants and hence allow a reaction to progress. Symbol: $E_a$

**Average bond enthalpy**   The amount of energy required to break one mole of bonds averaged across a range of compounds containing that bond.

**Combustion reaction**   An exothermic reaction in which a fuel is oxidized by oxygen.

**Endothermic reaction**   A reaction that absorbs heat energy from the surroundings, where $H_{products} > H_{reactants}$ (i.e. $\Delta H > 0$).

**Enthalpy level diagram (energy profile diagram)**   A diagram that shows the relative stabilities of reactants and products as well as activation energy, $E_a$, and the change in enthalpy of the reaction.

**Enthalpy**   The energy or heat energy content of a substance. Symbol: $H$; unit: kJ mol$^{-1}$

**Exothermic reaction**   A reaction that releases heat energy to the surroundings, where $H_{products} < H_{reactants}$ (i.e. $\Delta H < 0$).

**Heat energy change**   The energy released or absorbed during a reaction. Symbol: $\Delta H$

**Heat of combustion**   The amount of energy produced by the combustion of one mole of a substance. Symbol: $\Delta H_c$; unit: kJ mol$^{-1}$ or kJ g$^{-1}$

**Hess's law**   A law which states that the heat evolved or absorbed in a chemical process is the same, whether the process takes place in one or in several steps.

**Neutralization reaction**   An exothermic reaction in which an acid reacts with an alkali.

**Specific heat capacity**   The energy required to raise the temperature of 1 g of a substance by 1°C. Symbol: $c$; unit: J °C$^{-1}$ g$^{-1}$ or J K$^{-1}$ g$^{-1}$

**Stability**   Internal energy of a reactant or product, low energy = high stability.

**Standard enthalpy change of reaction**   The difference between the enthalpy of the products and the enthalpy of the reactants under standard conditions. Symbol: $\Delta H^\circ$

**Thermochemical equation**   A chemical equation that includes the enthalpy change $\Delta H$.

## Concepts

- Chemical reactions involve a change in the total energy, or enthalpy, of the reactants as they change into products:

  $\Delta H = H_{products} - H_{reactants}$

- If energy is absorbed from the surroundings as a reaction proceeds, the reaction is described as endothermic and $\Delta H$ is positive. If energy is released into the surroundings, the reaction is exothermic and $\Delta H$ is negative.

- Enthalpy changes are illustrated using enthalpy level (energy profile) diagrams.

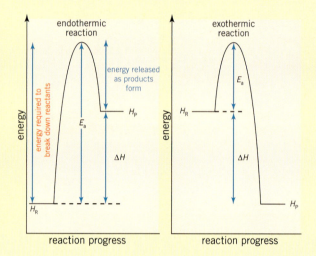

- The activation energy ($E_a$) of a reaction is the amount of energy required to initiate the reaction. This energy is used to break the bonds in the reactant molecules.

- The specific heat capacity of a substance is a measure of the amount of heat energy required to raise the temperature of 1 g of the substance by 1°C:

  $q = m \times c \times \Delta T$

- Thermochemical equations can be used to determine the heat energy produced or absorbed by a chemical reaction.

- Hess's law can be used to find the enthalpy change of a reaction, given two or three other reactions that provide the same reactants and products.

- The average bond enthalpy is the amount of energy required to break one mole of bonds averaged across a range of compounds containing that bond.

- If the energy released by bond making is greater than that used by bond breaking then a reaction is exothermic and if the energy released by bond making is less than that used by bond breaking then a reaction is endothermic

## Chapter 6 Review questions

1. Gas hotplates require a piezo ignition spark generator to light the gas (mainly methane) before the element can be used for cooking.
   a. What is the role of the spark in this chemical reaction?
   b. Which chemical bonds are broken when this reaction is initiated?
   c. Which bonds are generated when the product is formed?
   d. Is this reaction endothermic or exothermic? Explain your reasoning.

2. State which of the following reactions are exothermic and which are endothermic.
   a. $I_2(s) + Cl_2(g) \rightarrow 2ICl(g)$  $\Delta H = +35$ kJ mol$^{-1}$
   b. $N_2(g) + 3H_2(g) \rightarrow 2NH_3(g)$  $\Delta H = -92.3$ kJ mol$^{-1}$
   c. $BCl_3(l) + 6H_2(g) \rightarrow B_2H_6(g) + 6HCl(g)$  $\Delta H = +315$ kJ mol$^{-1}$
   d. $Ag^+(aq) + 2NH_3(aq) \rightarrow Ag(NH_3)_2^+(aq)$  $\Delta H = -111$ kJ mol$^{-1}$
   e. $N_2(g) \rightarrow 2N(g)$  $\Delta H = +946$ kJ mol$^{-1}$

3. Calculate the enthalpy of reaction for the equation:
   $C(s)$ graphite $\rightarrow C(s)$ diamond
   given the following data:
   $C(s)$ graphite $+ O_2(g) \rightarrow CO_2(g)$
   $\Delta H = -393$ kJ mol$^{-1}$
   $C(s)$ diamond $+ O_2(g) \rightarrow CO_2(g)$
   $\Delta H = -395$ kJ mol$^{-1}$

4. Oxygen gas can react with iron metal in a variety of ways, as shown by the following equations:
   $Fe(s) + \frac{1}{2}O_2(g) \rightarrow FeO(s)$  $\Delta H = -272$ kJ mol$^{-1}$
   $2Fe(s) + \frac{3}{2}O_2(g) \rightarrow Fe_2O_3(s)$  $\Delta H = -824$ kJ mol$^{-1}$
   $3Fe(s) + 2O_2(g) \rightarrow Fe_3O_4(s)$  $\Delta H = -1118$ kJ mol$^{-1}$
   Use this data to determine the enthalpy of reaction of the following processes.
   a. $2Fe(s) + O_2(g) \rightarrow 2FeO(s)$
   b. $2Fe_2O_3(s) \rightarrow 4Fe(s) + 3O_2(g)$
   c. $3FeO(s) + \frac{1}{2}O_2(g) \rightarrow Fe_3O_4(s)$

5. Consider the reaction represented by the equation:
   $2P + Q \rightarrow 3R$
   The total enthalpy of the reactants is 100 kJ, the total enthalpy of the products is −150 kJ and the activation energy for the forward reaction is 80 kJ.
   a. Sketch a diagram of the energy profile for this reaction. Label the diagram clearly to show $\Delta H$ and $E_a$.
   b. Is the forward reaction endothermic or exothermic?
   c. Calculate the amount of energy released as the products of the reaction are formed.
   d. Calculate the activation energy of the reverse reaction: $3R \rightarrow 2P + Q$

6. Sketch a diagram of the energy profile for a reaction in which the total enthalpy of the reactants is 25 kJ, the total enthalpy of products is −80 kJ and the activation energy for the forward reaction is 135 kJ. Label the diagram clearly to show $\Delta H$ and $E_a$.

7. Calculate the amount of heat energy absorbed from the surroundings when 25.0 g of sodium sulfate decahydrate crystallizes from water, given the equation:
   $Na_2SO_4 \cdot 10H_2O(s) \xrightarrow{water} Na_2SO_4(aq) + 10H_2O(l)$
   $\Delta H = +79.0$ kJ mol$^{-1}$

8. How much energy is required to convert 1.00 dm$^3$ (1000 g) of boiling water into steam?
   $H_2O(l) \rightarrow H_2O(g)$  $\Delta H = +44.0$ kJ mol$^{-1}$

9. Using the specific heat capacity data from table 6.2.1, calculate the amount of energy:
   a. required to raise the temperature of 20.0 cm$^3$ of water from 11.0°C to 85.0°C
   b. given off as 45.0 cm$^3$ of ethanol [$d = 0.785$ g cm$^{-3}$] cools from 37.0°C to 4.0°C
   c. absorbed by 1000 kg of iron as it is heated from 25.0°C to 1000.0°C.

10. 1200 cm$^3$ of water is heated with 200 kJ of energy. If the initial temperature of the water is 13.3°C, determine its final temperature after heating.

11. Calculate the energy released when 10.0 g of carbon reacts with excess oxygen according to the equation:
    $C(s) + O_2(g) \rightarrow CO_2(g)$  $\Delta H = -393.4$ kJ mol$^{-1}$

12 Calculate the amount of energy absorbed when 250 g of $Fe_3O_4$ reacts with excess carbon monoxide according to the thermochemical equation:

$Fe_3O_4(s) + CO(g) \rightarrow 3FeO(s) + CO_2(g)$
$\Delta H = +38.0$ kJ mol$^{-1}$

13 When an electric current is passed through water, the water will decompose to its constituent elements according to the equation:

$2H_2O(l) \rightarrow 2H_2(g) + O_2(g)$  $\Delta H = +572$ kJ mol$^{-1}$

Calculate the amount of electrical energy required to produce 500 cm$^3$ of oxygen gas at STP.

14 Anhydrous copper sulfate, $CuSO_4$, is a hydroscopic substance—it absorbs water from the surroundings according to the equation:

$CuSO_4(s) + 5H_2O(l) \rightarrow CuSO_4 \cdot 5H_2O(s)$

When 5.73 g $CuSO_4$ reacts with water, it releases 4.81 kJ of thermal energy. Calculate the $\Delta H$ for reaction above.

15 The heat of combustion of hexane is 4158 kJ mol$^{-1}$. Calculate $\Delta H$ for the reaction:

$2C_6H_{14}(l) + 19O_2(g) \rightarrow 12CO_2(g) + 14H_2O(l)$

16 Use average bond enthalpy data to find $\Delta H$ for the reaction:

$2C_6H_{14}(l) + 19O_2(g) \rightarrow 12CO_2(g) + 14H_2O(l)$

Compare your answer to the answer for question 15 and explain why there could be a difference.

17 Calculate the heat of combustion of butanol, given the following:

$2C_4H_9OH(l) + 12O_2(g) \rightarrow 8CO_2(g) + 10H_2O(l)$
$\Delta H = -5354$ kJ mol$^{-1}$

18 What mass of propane gas would be needed to boil 2.00 dm$^3$ of water if the initial temperature of the water was 16.6°C? The thermochemical equation for the combustion of propane is:

$C_3H_8(g) + 5O_2(g) \rightarrow 3CO_2(g) + 4H_2O(l)$
$\Delta H = -2217$ kJ mol$^{-1}$

19 Use average bond enthalpies to find the value of $\Delta H$ (in kJ mol$^{-1}$) for the reaction below.

$C_2H_4(g) + H_2(g) \rightarrow C_2H_6(g)$

20 Two campers had a methanol-fuelled camping stove. They had only 50.0 g of methanol left, but still had one day of camping to go.

  a How many mole of methanol did the campers have left to burn?

  b If the heat of combustion of methanol is $-715$ kJ mol$^{-1}$, how much energy would that amount of methanol be able to produce?

  c If the temperature of the water they are using is initially 15°C, theoretically, what volume of water will the campers be able to boil with their remaining methanol?

  d In reality, will the campers actually be able to boil the amount of water you have stated in your answer to part c of this question? Explain your answer.

 www.pearsoned.com.au/schools

Weblinks are available on the Chemistry: For use with the IB Diploma Programme Standard Level Companion Website to support learning and research related to this chapter.

# Chapter 6 Test

## Part A: Multiple-choice questions

1. Which statement about exothermic reactions is **not** correct?
   A They release energy.
   B The enthalpy change ($\Delta H$) is negative
   C The products have a greater enthalpy than the reactants
   D The products are more stable than the reactants
   © IBO 2001, May P1 Q16

2. In an experiment to measure the heat change when a small amount of sodium hydroxide is dissolved in water, $x$ g of sodium hydroxide was dissolved in $y$ g of water, giving a temperature rise of $z$°C. The specific heat capacity of water is $c$ J g$^{-1}$ K$^{-1}$. Which expression should be used to calculate the heat change (in J)?
   A $cxyz$
   B $cxy$
   C $cyz$
   D $cxz$
   © IBO 2001, May P1 Q17

3. Some average bond enthalpies (in kJ mol$^{-1}$) are as follows:
   H–H = 436, Cl–Cl = 242, H–Cl = 431
   What is the enthalpy change (in kJ) for the decomposition of hydrogen chloride?
   $2HCl \rightarrow H_2 + Cl_2$
   A –184
   B +184
   C +247
   D –247
   © IBO 2001, May P1 Q18

4. When the solids Ba(OH)$_2$·8H$_2$O and NH$_4$SCN are mixed, a solution is formed and the temperature decreases. Which statement about this reaction is correct?
   A The reaction is exothermic and $\Delta H$ is negative.
   B The reaction is exothermic and $\Delta H$ is positive.
   C The reaction is endothermic and $\Delta H$ is negative.
   D The reaction is endothermic and $\Delta H$ is positive.
   © IBO 2002, May P1 Q16

5. Using the information below:
   $H_2(g) + O_2(g) \rightarrow H_2O_2(l)$     $\Delta H = -187.6$ kJ
   $2H_2(g) + O_2(g) \rightarrow 2H_2O(l)$     $\Delta H = -571.6$ kJ
   find the value of $\Delta H$ for the following reaction.
   $2H_2O_2(l) \rightarrow 2H_2O(l) + O_2(g)$
   A –196.4
   B –384.0
   C –759.2
   D –946.8
   © IBO 2002, May P1 Q17

6. What energy changes occur when chemical bonds are formed and broken?
   A Energy is absorbed when bonds are formed and when they are broken.
   B Energy is released when bonds are formed and when they are broken.
   C Energy is absorbed when bonds are formed and released when they are broken
   D Energy is released when bonds are formed and absorbed when they are broken.
   © IBO 2003, May P1 Q15

7. An equation for a reaction in which hydrogen is formed is
   $CH_4 + H_2O \rightarrow 3H_2 + CO$    $\Delta H^\ominus = +210$ kJ
   Which energy change occurs when 1 mol of hydrogen is formed in this reaction?
   A 70 kJ of energy are absorbed from the surroundings.
   B 70 kJ of energy are released to the surroundings.
   C 210 kJ of energy are absorbed from the surroundings.
   D 7210 kJ of energy are released to the surroundings.
   © IBO 2006, May P1 Q16

8. The equations and enthalpy changes for two reactions used in the manufacture of sulfuric acid are:
   $S(s) + O_2(g) \rightarrow SO_2(g)$     $\Delta H^\ominus = -300$ kJ
   $2SO_2(g) + O_2(g) \rightarrow 2SO_3(g)$     $\Delta H^\ominus = -200$ kJ
   What is the enthalpy change, in kJ, for the reaction below?
   $2S(s) + 3O_2(g) \rightarrow 2SO_3(g)$
   A –100
   B –400
   C –500
   D –800
   © IBO 2006, May P1 Q17

**9** Which statement is correct for an endothermic reaction?

**A** The products are more stable than the reactants and $\Delta H$ is positive.

**B** The products are less stable than the reactants and $\Delta H$ is negative.

**C** The reactants are more stable than the products and $\Delta H$ is positive.

**D** The reactants are less stable than the products and $\Delta H$ is negative.

© IBO 2006, Nov P1 Q17

**10** Which equation represents an exothermic process?

**A** $F^-(g) \rightarrow F(g) + e^-$

**B** $F_2(g) \rightarrow 2F(g)$

**C** $Na(g) \rightarrow Na^+(g) + e^-$

**D** $I_2(g) \rightarrow I_2(s)$

© IBO 2006, Nov P1 Q18

(10 marks)

http://www.pearsoned.com.au/schools

For more multiple-choice test questions, connect to the Chemistry: For use with the IB Diploma Programme Standard Level Companion Website and select Chapter 6 Review Questions.

## Part B: Short-answer questions

**1** Methylamine can be manufactured by the following reaction.

$CH_3OH(g) + NH_3(g) \rightarrow CH_3NH_2(g) + H_2O(g)$

**a** Define the term *average bond enthalpy*.

(2 marks)

**b** Use information from table 6.4.1 (p. 208) to calculate the enthalpy change for this reaction.

(4 marks)

© IBO 2006, Nov P2 Q2ab

**2** Calculate the enthalpy change $\Delta H_4$ for the reaction:

$C + 2H_2 + \frac{1}{2}O_2 \rightarrow CH_3OH \quad \Delta H_4$

using Hess's law and the following information.

$CH_3OH + 1\frac{1}{2}O_2 \rightarrow CO_2 + 2H_2O$
$\Delta H_1 = -676$ kJ mol$^{-1}$

$C + O_2 \rightarrow CO_2 \quad \Delta H_2 = -394$ kJ mol$^{-1}$

$H_2 + \frac{1}{2}O_2 \rightarrow H_2O \quad \Delta H_3 = -242$ kJ mol$^{-1}$

(4 marks)

© IBO 2006, May P2 Q1c

**3 a** Using values from table 6.4.1 (p. 208), calculate the enthalpy change for the following reaction:

$CH_4(g) + Br_2(g) \rightarrow CH_3Br(g) + HBr(g)$

(3 marks)

**b** Sketch an enthalpy level diagram for the reaction.

(2 marks)

**c** Without carrying out a calculation, suggest, with a reason, how the enthalpy change for the following reaction compares with that of the reaction in part **a**.

$CH_3Br(g) + Br_2(g) \rightarrow CH_2Br_2(g) + HBr(g)$

(2 marks)

© IBO 2005, Nov P2 Q5biii–v

## Part C: Data-based question

In order to determine the enthalpy change of reaction between zinc and copper(II) sulfate, a student placed 50.0 cm$^3$ of 0.200 mol dm$^{-3}$ copper(II) sulfate in a polystyrene beaker. The temperature was recorded every 30 seconds. After two minutes 1.20 g of powdered zinc was added. The solution was stirred and the temperature recorded every half minute for the next 14 minutes. The results obtained were then plotted to give the following graph:

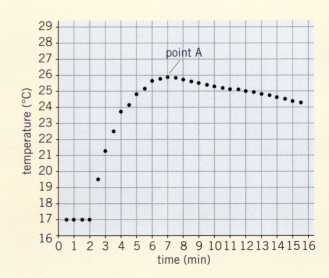

**a** Write the equation for the reaction taking place. (1 mark)

**b** Determine which of the two reagents was present in excess. (2 marks)

**c** The highest temperature is reached at point A. Explain what is happening in the system at this point. (1 mark)

**d** By drawing a suitable line on the graph, estimate what the rise in temperature would have been if the reaction had taken place instantaneously. (2 marks)

**e** Calculate how much heat was evolved during the reaction. Give your answer to **three** significant figures. (2 marks)

**f** What is the enthalpy change of reaction in kJ mol$^{-1}$? (1 mark)

**g** The accepted value for the enthalpy change of reaction is $-218$ kJ mol$^{-1}$. What is the percentage error for the value obtained in this experiment? (1 mark)

**h** Suggest one reason why there is disagreement between the experimental value and the accepted value. (1 mark)

© IBO 1999, Nov P2 Q1

## Part D: Extended-response question

**a** Explain what is meant by the term *standard enthalpy change of reaction*. (1 mark)

**b** Describe an experiment to determine the enthalpy change of the reaction between dilute hydrochloric acid and aqueous sodium hydroxide. Show how the value of $\Delta H$ would be calculated from the data obtained. (9 marks)

**c** Draw an enthalpy level diagram for the neutralization reaction above. Indicate on your diagram the enthalpy change of the reaction and hence compare the relative stabilities of reactants and products. (4 marks)

© IBO 2000, May P2 Q5a
Total marks 52

# 7 KINETICS

## Chapter overview

This chapter covers the IB Chemistry syllabus Topic 6: Kinetics.

**By the end of this chapter, you should be able to:**
- define the term *rate of reaction*
- describe suitable experimental methods for measuring rates of reaction
- analyse rate experiment data
- recall the kinetic particle theory of matter and how it enables an explanation of the three principal physical states of matter: solid, liquid and gas
- recall the fundamental concepts associated with collision theory, including the effect of temperature on the distribution of kinetic energies of particles
- sketch and explain qualitatively Maxwell–Boltzmann distribution curves for a fixed amount of gas at a range of temperatures
- describe the effect of a catalyst on the activation energy of a chemical reaction
- sketch and explain Maxwell–Boltzmann distribution curves for reactions with and without a catalyst
- recall five factors that affect the rate of a reaction: temperature, concentration in aqueous solutions, pressure in the gas phase, available surface area of a solid and the use of catalysts
- explain qualitatively the effect of each of these factors on reaction rate in terms of particle movement and interaction.

IBO Assessment statements 6.1.1 to 6.2.7

Chemical reactions exhibit a wide range of rates. From a dramatic explosion that destroys an old building to a reaction that you can watch in the laboratory, to even slower reactions such as the setting of concrete or the rusting of an old truck in a farm paddock, the rate of a chemical reaction is important to us. How quickly will the milk go off if it's left out on the bench instead of being put in the fridge? Is this reaction likely to proceed explosively, or is it safe to perform in the school laboratory? Reaction kinetics are particularly important in biochemistry where a rate may be as slow as the ageing process that takes us from infancy to old age or as fast as the operation of the enzyme *catalase*, which can convert 40 000 000 molecules of hydrogen peroxide per second into water and oxygen.

## 7.1 RATES OF REACTION

The term used in Chemistry to describe the rate at which a reaction proceeds is *chemical kinetics*. **Rate of reaction** is defined as the 'change of concentration of reactants or products with time'.

**AS 6.1.1**
**Define the term *rate of reaction*.**
© IBO 2007

$$\text{Rate} = -\frac{\text{concentration of R at time } t_2 - \text{concentration of R at time } t_1}{t_2 - t_1}$$

$$= -\frac{\Delta[R]}{\Delta t}$$

where R is the reactant and $\Delta[R]$ means the change in the concentration of R.

The negative sign is needed at the start of the expression to ensure that the rate of reaction has a positive value.

In figure 7.1.1(a) the concentration of reactants over time is graphed. The rate of reaction over the 20-second period between $t = 10$ seconds and $t = 30$ seconds is calculated as follows:

$$\text{Rate} = \frac{-(0.2 - 0.6)}{30 - 10}$$

$$= \frac{0.4}{20}$$

$$= 0.02 \text{ mol dm}^{-3} \text{ s}^{-1}$$

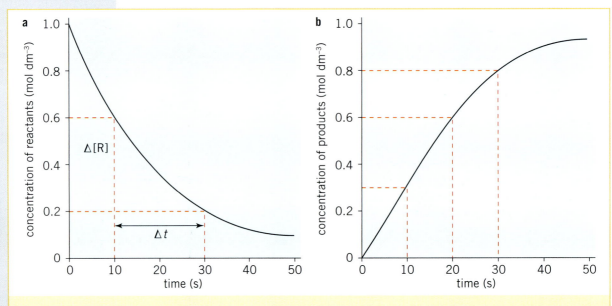

Figure 7.1.1 As a reaction proceeds (a) the concentration of reactants decreases and (b) the concentration of products increases.

Interactive
Animation
Rates of change

If the change of concentration of products with time is being measured, then

$$\text{Rate} = \frac{\text{concentration of P at time } t_2 - \text{concentration of P at time } t_1}{t_2 - t_1}$$

$$= \frac{\Delta[P]}{\Delta t}$$

where P is the product and $\Delta[P]$ means the change in the concentration of P.

In figure 7.1.1(b) the concentration of products over time is graphed. The rate of reaction over the 10-second period between $t = 0$ seconds and $t = 10$ seconds is calculated as follows:

$$\text{Rate} = \frac{(0.3 - 0)}{10 - 0}$$

$$= \frac{0.3}{10}$$

$$= 0.03 \text{ mol dm}^{-3} \text{ s}^{-1}$$

Notice that the rate of reaction decreases as the reaction proceeds. This would be expected for most reactions.

## Experimental methods for measuring rate of reaction

A range of experimental methods may be used to measure the rate of a reaction. The method chosen depends very much on the nature of the reaction itself.

**AS 6.1.2**
**Describe suitable experimental procedures for measuring rates of reactions. © IBO 2007**

### A change in gas volume or a change in mass

If the reaction produces a gas, this gas will escape as the reaction progresses.

For example, in the decomposition of hydrogen peroxide, oxygen gas is produced according to the equation:

$$2H_2O_2(l) \rightarrow 2H_2O(l) + O_2(g)$$

The volume of oxygen being produced can be measured by collecting it in a gas syringe attached to the reaction vessel (figure 7.1.2). The volume of oxygen is read at timed intervals, recorded and graphed. Similarly, the volume of water displaced in a gas delivery tube can be recorded.

Alternatively, the reaction vessel (minus the syringe) can be placed on a top-loading balance and the mass of the apparatus and solution read over time, recorded and graphed. In this case the mass loss will be equal to the mass of gas produced. (See example 1 on the following page.)

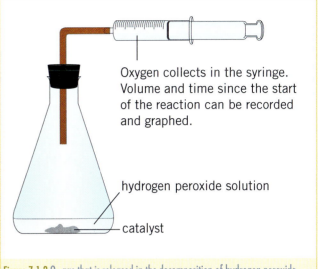

Figure 7.1.2 $O_2$ gas that is released in the decomposition of hydrogen peroxide, $H_2O_2$, is collected in a gas syringe.

### Titrimetric analysis

Small samples of the reaction mixture can be removed at regular intervals and the reaction in the sample stopped or slowed significantly by a procedure called *quenching*. Quenching may be achieved by adding ice-cold water or stopping the reaction by adding another reagent. The concentration of a reactant or product can be determined by titration.

For example, the reaction between hydrogen peroxide and acidified potassium iodide produces iodine according to the equation:

$$H_2O_2(aq) + 2H^+(aq) + 2I^-(aq) \rightarrow I_2(aq) + 2H_2O(l)$$

The concentration of iodine in a quenched sample of the reaction mixture can be found by titrating with standard sodium thiosulfate solution.

$$I_2(aq) + 2S_2O_3^{2-}(aq) \rightarrow 2I^-(aq) + S_4O_6^{2-}(aq)$$

### A change in gas pressure

If the equipment is available, a gaseous reaction can be monitored by measuring the change in gas pressure using a gas pressure sensor. In this case the volume must be kept constant, as should the temperature.

### Colorimetry

This method is particularly useful when one of the reactants or products is coloured. The colour of the reaction mixture will change with time and a colorimeter can be used to monitor the depth of colour in the mixture. A colorimeter passes light through the reaction mixture. This light is of a wavelength that will be absorbed by the reaction mixture. The amount of light that passes through the colorimeter cell (and therefore is not absorbed by the solution) indicates the depth of colour and, therefore, the amount of coloured reactant or product. The colorimeter can be calibrated with solutions of known concentrations, so that the digital readout from the colorimeter can be converted to a concentration of the coloured species.

For example, the reaction between purple permanganate(VII) ($MnO_4^-$) ions and colourless ethanedioate ions ($C_2O_4^{2-}$) can be monitored by colorimetry, due to the decrease in purple colour as the reaction proceeds:

$$2MnO_4^-(aq) + 16H^+(aq) + 5C_2O_4^{2-}(aq) \rightarrow 2Mn^{2+}(aq) + 10CO_2(g) + 8H_2O(l)$$

    purple                        colourless     colourless

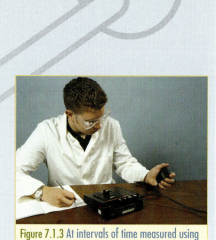

Figure 7.1.3 At intervals of time measured using a stopwatch the digital colorimeter reading is recorded.

The change in the amount of light absorbed when a solid precipitate is formed can also be measured using a colorimeter. The amount of light absorbed is proportional to the concentration of the solid precipitate. Alternatively the time taken for the solid precipitate to obscure a cross marked on the paper under the beaker can be measured (see p. 223)

### A change in electrical conductivity

If ions are produced or used up during a reaction, the electrical conductivity will change. With a suitable data-logging probe this change in conductivity can be measured and graphed to show the change in concentration of the ionic species being produced or used up during the reaction. As in the case of the colorimeter, a conductivity cell, or conductivity data-logging probe can be calibrated by measuring the conductivity of solutions with known concentrations of ions. The electrical conductivity is proportional to the concentration of ions.

In the example of the reaction between permanganate ($MnO_4^-$) ions and ethanedioate ions ($C_2O_4^{2-}$) there are 23 reactant particles that are ions and only 2 product particles that are ions. As the reaction progresses, the concentration of ions in the solution decreases dramatically and so will the electrical conductivity of the solution.

### A change in light absorption

If reactants and products have a different colour, the time taken for the reaction to change colour could be measured.

6.1.3
Analyse data from rate experiments. © IBO 2007

### Example 1

Consider the reaction of marble chips (calcium carbonate) with dilute hydrochloric acid:

$$CaCO_3(s) + 2HCl(aq) \rightarrow CaCl_2(aq) + CO_2(g) + H_2O(l)$$

If 10.0 g of marble is placed into an open beaker containing 200 cm$^3$ of 1.0 mol dm$^{-3}$ hydrochloric acid, the calcium carbonate reacts rapidly to produce carbon dioxide, which bubbles away. The progress of the reaction can be followed by measuring the mass of the container every 30 seconds until the CaCO$_3$ is completely consumed, and then determining the mass of CO$_2$(g) evolved. Typical results obtained are shown in table 7.1.1 and graphed in figure 7.1.4.

| TABLE 7.1.1 REACTION OF MARBLE CHIPS WITH EXCESS HYDROCHLORIC ACID | | |
| --- | --- | --- |
| Time (s) | Mass of beaker and reactants (g) | Mass of $CO_2$ evolved (g) |
| 0 | 209.8 | 0 |
| 30 | 209.1 | 1.7 |
| 60 | 207.0 | 2.8 |
| 90 | 206.6 | 3.2 |
| 120 | 206.0 | 3.8 |
| 150 | 205.7 | 4.1 |
| 180 | 205.5 | 4.3 |
| 210 | 205.4 | 4.4 |
| 240 | 205.4 | 4.4 |

Referring to figure 7.1.4, we can see that much of the carbon dioxide gas is evolved within the first minute of the reaction and that after about 3 minutes very little further gas is produced. If we remember that the rate of reaction is 'the change in concentration of reactants or products with time', we can see that the gradient of the curve indicates the rate of reaction at any particular moment. As the curve is at its steepest at the beginning of the reaction, we can state that the reaction rate is at a maximum at this point, and slowly decreases until, at about 4 minutes, the gradient of the slope is effectively zero. At this point, the calcium carbonate, the limiting reactant, has been completely consumed and no more carbon dioxide will be produced.

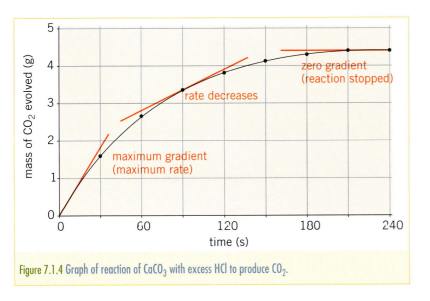

Figure 7.1.4 Graph of reaction of $CaCO_3$ with excess HCl to produce $CO_2$.

### Example 2

Another experimental technique that can be used to measure the rate of reaction is to measure the volume of a gas that is being released. The equipment shown in figure 7.1.2 could be used for this experiment. Consider the decomposition of hydrogen peroxide, using manganese(IV) oxide, $MnO_2$, as a catalyst:

$$H_2O_2(aq) \rightarrow 2H_2O(l) + O_2(g)$$

Table 7.1.2 shows some typical results that could be obtained for this reaction and they are graphed in figure 7.1.5.

| TABLE 7.1.2 THE VOLUME OF $O_2$ GAS COLLECTED DURING THE DECOMPOSITION OF $H_2O_2$ ||
|---|---|
| Time (min) | Volume of oxygen (cm³) |
| 0 | 0 |
| 1 | 35 |
| 2 | 65 |
| 3 | 85 |
| 4 | 100 |
| 5 | 108 |
| 6 | 114 |
| 7 | 118 |
| 8 | 120 |
| 9 | 120 |
| 10 | 120 |

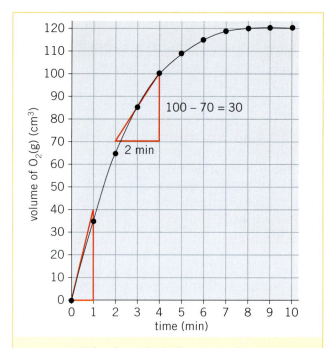

Figure 7.1.5 Graph of production of $O_2$ by decomposition of $H_2O_2$.

This graph shows a very similar shape to that in figure 7.1.4. As the reaction proceeds, the rate of reaction decreases until 9 minutes, at which time the reaction has reached completion and the rate is equal to zero.

The **initial rate of reaction** is the rate of reaction at time $t = 0$. This is measured by drawing a tangent to the curve at the start of the graph and measuring the gradient (slope) of the tangent. In figure 7.1.5 this has been done and the gradient of the tangent to the curve is equal to 40 cm³ min⁻¹.

The **instantaneous rate of reaction** at time $t = 3$ minutes is found in a similar manner. A tangent to the curve is drawn and its gradient is calculated. In figure 7.1.5 the tangent has been drawn at $t = 3$ min and its gradient is found by using the gradient formula: $\frac{100-70}{4-2} = \frac{30}{2} = 15$ cm³ min⁻¹. As expected, this instantaneous rate is less than the initial rate of reaction.

The technique for finding the gradient of a tangent to a graph is described in section 5.3 (p. 180).

In this example, the **average rate of reaction** for a time period is found by dividing the volume of oxygen produced by the time during which the oxygen has been produced. For example, the average rate of reaction during the first 5 minutes of this reaction is equal to $\frac{108}{5} = 21.6$ cm³ min⁻¹. Compare this with the average rate of reaction during the second 5 minutes, which is equal to $\frac{120-108}{5} = \frac{12}{5} = 2.2$ cm³ min⁻¹, and we have ample proof that the rate of reaction is decreasing as the reaction progresses.

The time taken for a solid to appear can also be used to determine the rate of reaction. In practical investigation 7.1 the reaction between sodium thiosulfate and hydrochloric acid is investigated.

$Na_2S_2O_3(aq) + 2HCl(aq) \rightarrow S(s) + SO_2(g) + 2NaCl(aq) + H_2O(l)$
sodium                                         sulfur
thiosulfate

One of the products of this reaction is sulfur. The amount of sulfur produced during the reaction finally makes the solution opaque. The point at which this occurs can be determined by placing the reaction beaker over a page with a black cross on it. When the cross can no longer be seen, the reaction time is recorded. As the concentration of HCl decreases, the reaction time increases, that is the rate of reaction decreases.

Figure 7.1.6 The reaction between sodium thiosulfate and hydrochloric acid. When the amount of sulfur is great enough, the cross underneath the beaker cannot be seen.

PRAC 7.1
Measuring the rate of reaction

Rates of reaction

**THEORY OF KNOWLEDGE**
What would watching your favourite sports game on TV be like without being able to watch 'slow motion' reruns? Chemists regard chemical reactions in a similar way. A desire to explain why some chemical reactions take place but others don't have led to the development of increasingly advanced technology for measuring the rate of a reaction. Ahmed H. Zewail developed one of the world's fastest cameras to study atoms and molecules in 'slow motion' during a chemical reaction to see what happens when chemical bonds break and form. The camera uses laser technology with light flashes on a time scale at which the reactions actually happen, femtoseconds, fs (1 fs = $10^{-15}$ s). The October 1999 press release from the Royal Swedish Academy of Science said that Zewail's contribution 'means that we have reached the end of the road: no chemical reactions take place faster than this'.

Comment on the certainty of the claim that no chemical reaction can be faster.

### Section 7.1 Exercises

1 Dinitrogen pentoxide decomposes over time according to the equation:

   $2N_2O_5(g) \rightarrow 4NO_2(g) + O_2(g)$

   A quantity of $N_2O_5$ was dissolved in tetrachloromethane at 60°C and its concentration was measured over a period of time. The results are shown in the table below.

   | Reaction time (s) | Concentration of $N_2O_5$ (mol dm$^{-3}$) |
   |---|---|
   | 0 | 2.25 |
   | 400 | 1.43 |
   | 800 | 0.88 |
   | 1200 | 0.46 |
   | 1600 | 0.22 |
   | 2000 | 0.13 |
   | 2400 | 0.08 |

   a Draw a graph of these results.
   b Determine the rate of reaction between the following times.
      i $t = 0$ seconds and $t = 400$ seconds
      ii $t = 400$ seconds and $t = 800$ seconds
      iii $t = 1600$ seconds and $t = 2000$ seconds
   c State and explain during which period of 400 seconds was the rate of reaction the smallest value.

2 Nitrogen dioxide, $NO_2$, decomposes to form nitric oxide, NO, and oxygen, $O_2$, when heated. The table below shows the concentrations of $NO_2$, NO and $O_2$ in mol dm$^{-3}$ over a period of 400 seconds.

| Time (s) | [$NO_2$] (mol dm$^{-3}$) | [NO] (mol dm$^{-3}$) | [$O_2$] (mol dm$^{-3}$) |
|---|---|---|---|
| 0 | 0.0100 | 0 | 0 |
| 50 | 0.0079 | 0.0021 | 0.0011 |
| 100 | 0.0065 | 0.0035 | 0.0018 |
| 150 | 0.0055 | 0.0045 | 0.0023 |
| 200 | 0.0048 | 0.0052 | 0.0026 |
| 250 | 0.0043 | 0.0057 | 0.0029 |
| 300 | 0.0038 | 0.0062 | 0.0031 |
| 350 | 0.0034 | 0.0066 | 0.0033 |
| 400 | 0.0031 | 0.0069 | 0.0035 |

a Using a key and different colours for the three curves, draw a graph of each of the three concentrations against time on the same set of axes.

b Explain why the concentration of the $NO_2$ decreases while the concentrations of NO and $O_2$ increase.

c Describe what you notice about the rate of increase of [$O_2$] compared with the rate of increase of [NO]. Explain this in terms of the reaction.

3 The rate of reaction between acidified solution of potassium iodide and hydrogen peroxide solution is investigated.

$H_2O_2(aq) + 2H^+(aq) + 2I^-(aq) \rightarrow I_2(aq) + 2H_2O(l)$

At intervals of 15 minutes, a small sample of the mixture is taken, quenched and titrated to determine the concentration of iodine.

$I_2(aq) + 2S_2O_3^{2-}(aq) \rightarrow 2I^-(aq) + S_4O_6^{2-}(aq)$

The concentration of sodium thiosulfate solution used is 0.20 mol dm$^{-3}$ and the results are shown below.

| Time (min) | Volume of $S_2O_3^{2-}$ (cm$^3$ ± 0.2 cm$^3$) |
|---|---|
| 0 | 0 |
| 15 | 15.0 |
| 30 | 24.0 |
| 45 | 28.0 |
| 60 | 30.0 |
| 75 | 30.0 |

Calculate the rate of the reaction during the following periods.

a The first 15 minutes

b The second 15 minutes

c The final 15 minutes

4  The gas sulfuryl dichloride dissociates according to the following equation:
$SO_2Cl_2(g) \rightarrow SO_2(g) + Cl_2(g)$

The rate of this reaction can be followed by monitoring the pressure of the gases in the reaction vessel and the concentration is then calculated from the pressure.

The following table gives experimental measurements of concentration of $SO_2Cl_2$ with time.

| Time (s) | $[SO_2Cl_2(g)]$ (mol dm$^{-3}$) |
|---|---|
| 0 | 0.50 |
| 500 | 0.43 |
| 1000 | 0.37 |
| 2000 | 0.27 |
| 3000 | 0.20 |
| 4000 | 0.15 |

a  Plot the graph of concentration of $SO_2Cl_2$ against time for this experiment.

b  State and explain how the reaction rate can be calculated at different times using this graph.

As the concentration of $SO_2Cl_2$ decreases, the average rate of reaction changes. The following table records some average rates of reaction for different concentrations of $SO_2Cl_2$.

| $[SO_2Cl_2]$ (mol dm$^{-3}$) | Average rate of reaction (mol dm$^{-3}$ s$^{-1}$) |
|---|---|
| 0.45 | $1.35 \times 10^{-4}$ |
| 0.39 | $1.17 \times 10^{-4}$ |
| 0.34 | $1.02 \times 10^{-4}$ |
| 0.28 | $8.4 \times 10^{-5}$ |
| 0.23 | $6.9 \times 10^{-5}$ |
| 0.18 | $5.4 \times 10^{-5}$ |

c  Plot the graph of average reaction rate against concentration of $SO_2Cl_2$ using the data in this table.

d  Using the graph you plotted in part **c**, describe the relationship between concentration of $SO_2Cl_2$ and reaction rate.

## 7.2 COLLISION THEORY

Collision theory explains rates of reaction on a molecular level. In order to fully appreciate collision theory, we should first consider the kinetic theory of gases.

### Kinetic molecular theory

We can observe that gases have distinctive properties, and that gases show a uniformity of behaviour. The **kinetic molecular theory** is a model, a set of proposals, used to explain these properties and behaviour.

  6.2.1
Describe the kinetic theory in terms of the movement of particles whose average energy is proportional to temperature in kelvin. © IBO 2007

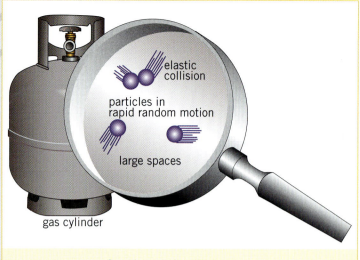

Figure 7.2.1 Gases consist of particles in rapid, random motion.

Kinetic energies of gas molecules

Gases are seen to consist of moving particles. These particles are molecules or atoms (in the case of the noble gases). The kinetic molecular theory states that gas particles:

- are in a continual state of rapid, random motion. They travel in straight lines, colliding with one another and with the walls of the container. They therefore change their speed and direction continually, and the movement may be described as chaotic
- have negligible volume compared to the volume of the gas
- are assumed to exert no forces on each other; that is, they are assumed to experience no forces of attraction or repulsion between them
- collide elastically; that is, energy is transferred during collisions, but no energy is lost.
- have an average kinetic energy that is directly proportional to the absolute temperature (kelvin) of the gas, and depends only on the temperature. (The absolute temperature scale is discussed on page 129.)

Consider the particles (atoms, ions or molecules) making up a substance. These particles have three different modes of movement available to them: vibrational, rotational and translational. Thus a particle is able to vibrate on the spot, spin on its axis or it may actually move from one place to another. Combinations of

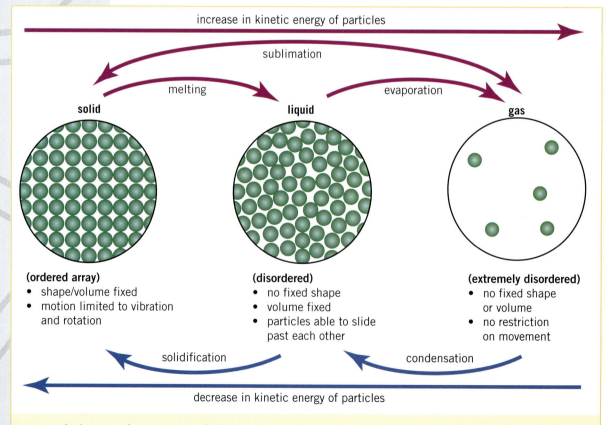

Figure 7.2.2 The three states of matter are interconvertible by heating or cooling. When changes of state occur, the temperature remains constant.

all three movements are possible, and the greater the degree of motion, the higher the kinetic energy. According to the kinetic particle theory, particles in the solid state are fixed into position in close proximity to one another and have only limited ability to move. They are able to vibrate, but have only limited rotational energy and very little, if any, translational energy. As the solid is heated, the particles increase their kinetic energy, resulting in greater degrees of vibrational and rotational energy. Eventually, the particles break free from the bonds holding them in place and are able to slide past one another. The material has changed state and is now a liquid.

**6.2.5**
**Sketch and explain qualitatively Maxwell–Boltzmann energy distribution curves for a fixed amount of gas at different temperatures and its consequences for changes in reaction rate.** ©IBO 2007

The kinetic energy of a particle relates to both its mass and velocity. Within a sample of gas at a particular temperature, the gas particles do not all have the same velocity and therefore do not all have the same kinetic energy. The distribution of particle velocities in a gas is shown in figure 7.2.3. This graph was derived independently by James Maxwell, a Scottish physicist, and Ludwig Boltzmann, an Austrian physicist, who did much of the theoretical work on the kinetic molecular description of gases. This distribution is known, therefore, as the **Maxwell–Boltzmann energy distribution curve**.

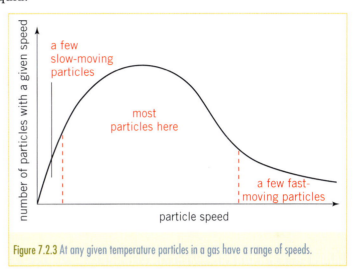

Figure 7.2.3 At any given temperature particles in a gas have a range of speeds.

From the graph in figure 7.2.3, we can see that a small fraction of particles have either very low or very high velocities. The bulk of particles have intermediate velocities. Figure 7.2.4 shows how this distribution changes with temperature. As the temperature is increased, both the average velocity (reflected by the curve's peak) and the spread of velocities increase (the curve flattens out). The area under the curve represents the total numbers of particles present, so this remains constant for a given gas sample, no matter what happens to the temperature.

Distribution of molecular speeds

WORKSHEET 7.1
Maxwell–Boltzmann curves

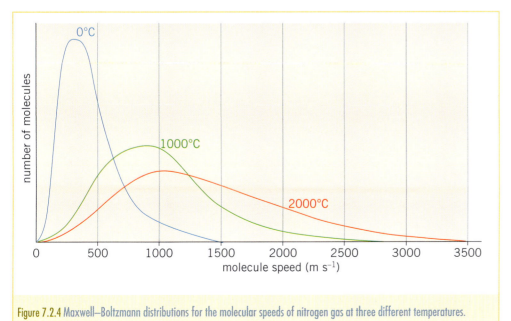

Figure 7.2.4 Maxwell–Boltzmann distributions for the molecular speeds of nitrogen gas at three different temperatures.

**AS** 6.2.3
**Describe the collision theory.**
© IBO 2007

Molecular collisions

## Collision theory

In section 7.1 we saw that the rate of reaction decreased as the reaction progressed. Why? What factors were affecting the rate of reaction? These important questions can be addressed, at least in part, by consideration of the **collision theory** of reactions. According to collision theory, for a reaction to occur, the reactant particles must:

- collide with each other
- collide with sufficient energy to break the bonds between them
- collide with the correct orientation to break the bonds between them and so allow the formation of new products.

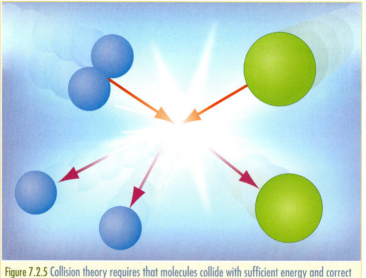

Figure 7.2.5 Collision theory requires that molecules collide with sufficient energy and correct orientation to break their bonds.

The minimum amount of energy a collision must possess for a reaction to occur is called the **activation energy**, $E_a$. As can be seen in figure 7.2.6, only a small proportion of the particles at a given temperature will have sufficient kinetic energy to overcome this activation energy barrier.

**AS** 6.2.2
**Define the term** *activation energy, $E_a$*. © IBO 2007

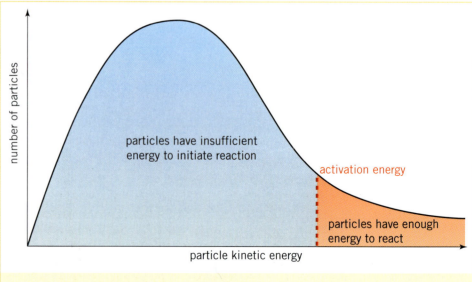

Figure 7.2.6 Only a small number of high-speed particles have sufficient energy to overcome the activation energy barrier.

In addition to colliding with sufficient energy to provide the activation energy, particles must also collide with the correct orientation if a reaction is to occur. For example, if the highly electronegative chlorine molecule is to react with an ethene molecule, one of the chlorine atoms must collide with the double bond of the ethene so that it can capture the electrons of the double bond and so form a new molecule of 1,2-dichloroethane. Collisions between the molecules in other orientations will not result in bond formation, regardless of the energy of the collision, as is shown in figure 7.2.7.

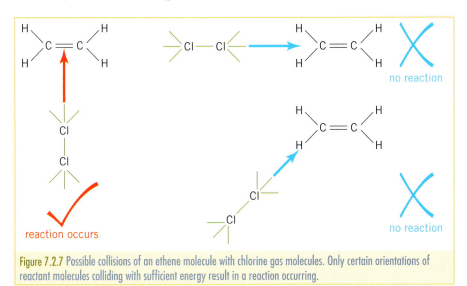

**Figure 7.2.7** Possible collisions of an ethene molecule with chlorine gas molecules. Only certain orientations of reactant molecules colliding with sufficient energy result in a reaction occurring.

## The effect of temperature on reaction rate

We know from experience that the temperature at which a reaction proceeds has a dramatic effect on reaction rate. When food is refrigerated, the rate at which it spoils is slowed considerably. If the food is placed into the freezer, it can keep for months at a time. Conversely, to hard-boil an egg at 100°C will take approximately 10 minutes, while it takes considerably longer to cook the egg if the temperature is reduced to 80°C. Observation of numerous experiments has shown that an increase in temperature almost invariably significantly increases the rate of reaction.

 **6.2.4
Predict and explain, using collision theory, the qualitative effects of particle size, temperature, concentration and pressure on the rate of a reaction. © IBO 2007**

As the temperature increases, the average particle speed increases and, importantly, the number of particles with elevated speeds increases markedly. An important consequence of this is that a greater proportion of particles have sufficient energy at the higher temperature to overcome the activation energy barrier and so possibly initiate reaction. Figure 7.2.4 shows the Maxwell–Boltzmann distribution for the molecular speeds of nitrogen gas at three different temperatures. Recall that the area under all three curves is constant, as the total number of particles in the sample has not changed, and that the higher the temperature, the greater the spread of speeds of particles.

If particles are moving more rapidly, it stands to reason that they will collide more frequently. However, merely increasing the frequency of collisions has only a marginal effect on the overall rate of the reaction. Of far greater importance is that a much larger proportion of particles now have energies in excess of the activation energy for the reaction. Indeed, it can be shown experimentally that an increase in temperature of just 10°C generally results in a doubling of the rate of reaction. In gaseous systems, the increase in rate can be significantly greater than this, with a fourfold increase or more in the rate at which the reaction proceeds.

Rates of reaction

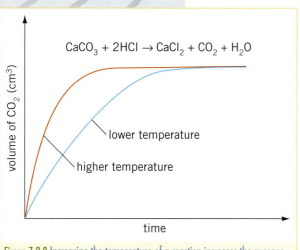

Figure 7.2.8 Increasing the temperature of a reaction increases the average rate of the reaction.

In figure 7.2.8 the volume of carbon dioxide produced is plotted against time for two different temperatures. When the two curves are compared, we can see that at the higher temperature the initial rate of reaction is greater and the graph flattens out in a shorter time. However, the volume of carbon dioxide produced (the point at which the graph flattens out) does not change.

- The initial rate of reaction is greater at the higher temperature because a much larger proportion of particles now have energies in excess of the activation energy for the reaction. The proportion of collisions that are successful has increased.
- The graph flattens out in a shorter time at the higher temperature because the flattening out occurs when the reaction is over. The reaction is proceeding at a greater rate, so it is completed in a shorter amount of time.
- The same volume of carbon dioxide is produced in both reactions because the amount of the limiting reactant has not changed from one reaction to the next.

## CHEM COMPLEMENT

### The Iceman

In September 1991, Erika and Helmut Simon were walking in the Alps near the Austrian border with Italy when they discovered the body of what they thought was a dead mountaineer. Incidents of people dying in the mountains and being enclosed in ice are not uncommon, and bodies can sometimes emerge from glaciers up to 30 years after death. The rate at which their bodies decompose is slowed dramatically by the fact that they are enclosed in ice. Closer examination of the body and the items with it (an unusual ice pick with a head of copper and a handle of yew, a small dagger, and a bow and arrows with flint heads) revealed that the man had died some considerable time ago. It was eventually established that the man had died approximately 5300 years ago in the early Bronze Age. His body, tools and even fragments of clothing, were so well preserved that scientists were able to establish in considerable detail what he ate and his lifestyle. In a strange twist of fate, Helmut Simon, the man who discovered the body, died in the Austrian Alps some years later while hiking in the same area.

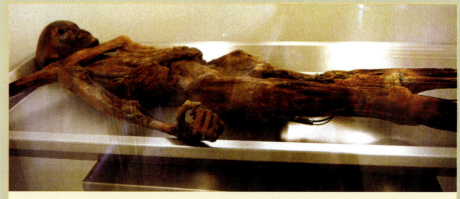

Figure 7.2.9 Ötzi the Iceman, preserved for 5300 years. Ice significantly slowed the rate of his decomposition.

## The effect of concentration (pressure) on reaction rate

If the concentration of reactants is increased, there are more particles present per unit volume and so they will collide more frequently with one another. It is important to recognize that the kinetic energy of the particles has not changed, there are simply more of them and so a greater number of collisions will take place per unit time. An increase in concentration can be achieved in solution by adding more reactant, and in a gaseous system by decreasing the volume of the reaction vessel or increasing the pressure on it. It is not possible to deduce exactly how the rate of a reaction will change with a change in concentration; this can only be determined by experiment. One reason for this is that the concentration of reactants decreases steadily as the reaction proceeds, as more and more reactant is converted into product. Furthermore, the rate at which a reaction proceeds may be directly proportional to the concentration of only one reactant species (a 'first-order' reaction) or two or more reactants.

DEMO 7.1
The blue bottle experiment

WORKSHEET 7.2
Rate of reaction

In figure 7.2.10 the volume of carbon dioxide produced in the reaction between hydrochloric acid and calcium carbonate is plotted against time for two different concentrations of hydrochloric acid. The calcium carbonate is the limiting reactant in this case. When the two curves are compared, we can see that at the higher concentration of HCl, the initial rate of reaction is greater and the graph flattens out in a shorter time. However, the total volume of carbon dioxide produced (the point at which the graph flattens out) does not change.

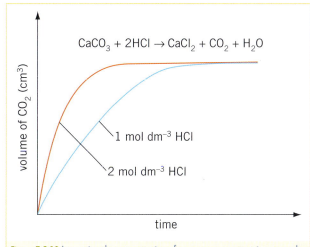

Figure 7.2.10 Increasing the concentration of an aqueous reactant increases the average rate of the reaction.

- The initial rate of reaction is greater when the concentration of HCl is greater (2 mol dm$^{-3}$) because there are more particles present per unit volume and so they will collide more frequently with one another. The number of collisions that occur per unit time has increased and so the initial rate of reaction is greater than for the lower concentration (1 mol dm$^{-3}$).

- The graph flattens out in a shorter time when the concentration of HCl is greater because the reaction is proceeding at a greater rate and so the reaction is completed in less time.

- The same volume of carbon dioxide is produced in both reactions because the amount of the limiting reactant has not changed from one reaction to the next, so the maximum volume of carbon dioxide that can be produced has not changed.

When considering the effect of a change in volume on a reaction in which one or more reactants are aqueous, the limiting reactant must first be identified. In figure 7.2.11 three curves are shown for the reaction between calcium carbonate and hydrochloric acid. In this experiment, however, the amount of HCl is the limiting reactant (rather than calcium carbonate as in the earlier examples). Curve I shows how the reaction proceeds when a fixed volume of 2 mol dm$^{-3}$ HCl is added to excess CaCO$_3$. In curve III, the volume of HCl has been kept constant, but the concentration of HCl has been halved to 1 mol dm$^{-3}$. In curve II, the volume of HCl has been doubled and the concentration of HCl is 1 mol dm$^{-3}$.

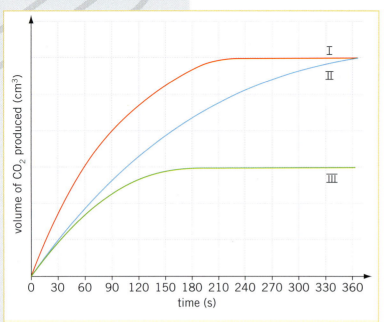

Figure 7.2.11 The effect of decreasing the concentration of HCl (curves I and III) and of increasing the volume of HCl (curve II) on the rate of reaction.

Compared to curve I, curve III has a lower initial rate of reaction, but levels off at about the same time and produces about half the volume of $CO_2$. These observations can be explained by the lower concentration combined with the fixed volume of the HCl used. As HCl is the limiting reactant in this reaction, halving the concentration halves the amount of HCl molecules available to react, so only half the volume of $CO_2$ can be produced. The rate of reaction is lower in curve III than in curve I because there are fewer particles of HCl present per unit volume and so they will collide less frequently with the $CaCO_3$.

Compared to curve I, curve II has a lower initial rate of reaction, takes much longer to level off and it produces the same volume of $CO_2$ when it does finally level off. These observations can be explained by the lower concentration of HCl combined with an increased volume of HCl used. The amount of HCl available to react is the same for curves I and II, so the final volume of $CO_2$ produced will be the same; however, the lower concentration of HCl in the case of curve II means that the rate of reaction is lower than for curve I.

If both reactants are aqueous, then the effect of increasing the volume of one reactant, even without changing the concentration of that reactant, will be to decrease the rate of reaction, since the concentration of the second reactant will be decreased by the addition of more solution. This effect could be investigated in practical 7.2.

PRAC 7.2
Factors affecting the rate of reaction

### THEORY OF KNOWLEDGE

'The concentration of hydrochloric acid affects the rate of its reaction with magnesium' is an example of a knowledge claim.

- Outline the process you would follow to provide reliable evidence to support this claim.
- Part of the process of evidence collection involves using inductive reasoning, collecting specific data and from that data drawing a general conclusion about the effect of concentration on rate. How would you improve the certainty of the conclusions you made? Consider the importance of controlling variables, repeatability, variety and consistency (coherence).
- Outline some of the limitations or problems associated with relying on our sense perception in an investigation such as this.
- There are similarities and differences in the methods of gaining knowledge and the ways of knowing in the natural and human sciences. Construct a table to compare similarities and differences in the methodology used by a psychologist to investigate the effect of the workload pressure on students taking IB Chemistry and a chemist investigating the effect of pressure on the volume of a gas. Comment on the reliability of the knowledge produced by each.

## The effect of surface area on reaction rate

If a solid is involved in a reaction, an increase in its surface area will increase the rate at which the reaction will proceed. Recall that for a reaction to proceed, the reactant molecules must come into contact with one another. This occurs readily in the gaseous or aqueous phases, but if a solid is involved, it is only those particles that are exposed at the surface that are available for reaction. The surface area of a solid can be increased dramatically if the solid is reduced to smaller particles. A cube of side lengths 1 cm will have a volume of 1 cm$^3$ and a surface area of 6 cm$^2$ (see figure 7.2.13). If the cube is then cut in half in all three dimensions, the volume remains constant, but the total surface area of all eight cubes increases to 12 cm$^2$. If the process is repeated once more, the surface area increases to 24 cm$^2$. The surface area to volume ratio is highest for a fine powder. Just one gram of alumina ($Al_2O_3$), for example, has been calculated to have a surface area of 50 m$^2$, sufficient to cover the floor of a large classroom!

Figure 7.2.12 The rapid combustion of coal dust can lead to explosions in mines.

The principle that an increase in surface area of reactants will increase the rate of reaction is readily demonstrated. If a lump of metallic zinc is dropped into dilute hydrochloric acid, it will slowly dissolve and bubbles of hydrogen gas will evolve. If the same quantity of zinc in powder form is added to the acid, the reaction is so vigorous that an explosion may result. The reaction is exothermic, and it is this heat combined with the very rapid evolution of highly combustible hydrogen gas that can result in an explosion. On an atomic level, the powdered zinc has a much greater surface area than the lump, and so there is a much larger area on which the reaction can take place, increasing the frequency of collisions between the zinc and acid.

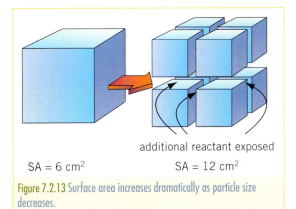

Figure 7.2.13 Surface area increases dramatically as particle size decreases.

PRAC 7.2
Factors affecting the rate of reaction

Figure 7.2.14 An increase in surface area increases the rate of reaction.

**AS 6.2.6**
Describe the effect of a catalyst on a chemical reaction.
© IBO 2007

**DEMO 7.2**
A dramatic decomposition of hydrogen peroxide

**Action at a catalyst surface**

**Catalysis**

**PRAC 7.3**
Catalysing the decomposition of hydrogen peroxide

## Catalysts

A **catalyst** is any substance that causes a reaction to proceed more quickly without actually being consumed itself. Catalysts are widely used in industry, where it is desirable to generate the maximum possible quantities of product as quickly as possible to maximize profits. Indeed, there are many industrial processes that would simply not be viable without the use of catalysts, and a great deal of research is undertaken by industries to find the most effective and cheapest catalyst for the production of particular substances.

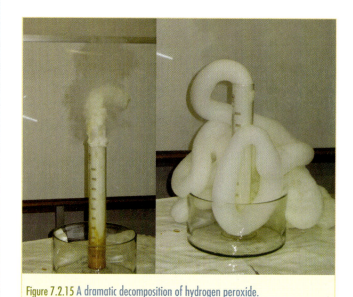

Figure 7.2.15 A dramatic decomposition of hydrogen peroxide.

| TABLE 7.2.1 INDUSTRIAL CATALYSTS | |
|---|---|
| **Product** | **Catalyst** |
| Petrol/diesel fuels | $Al_2O_3/SiO_2$ zeolite honeycomb |
| Margarine | Alloy of nickel and aluminium |
| Polyethene (HDPE) | Ziegler–Natta organometallic compound |
| Ammonia | Porous iron (can use osmium) |
| Nitric acid | Platinum/rhodium alloy wire mesh |
| Sulfuric acid | Vanadium(V) oxide pellets |

Living systems rely heavily on catalysts for the swift and orderly progress of chemical reactions within cells. Biological catalysts are better known as enzymes, and it is estimated that the human body utilizes more than 100 000 different enzymes. For example, the enzyme lysozyme catalyses the breakdown of the cell walls of bacteria and so helps to protect us from infection. The three major food groups of carbohydrates, proteins and lipids are broken down within the body in enzyme-catalysed hydrolysis reactions.

The key point to remember is that catalysts are able to increase the rate of a reaction by providing an alternative pathway that has a lower activation energy. The lower activation energy means that there will be a greater proportion of the particles present at any given temperature with sufficient kinetic energy to overcome the activation energy.

The enthalpy level diagram for a catalysed and an uncatalysed reation is shown in figure 7.2.16. The lower activation energy pathway provided by the catalyst enables more reactant particles to have sufficient energy to react. Similarly, the Maxwell–Boltzman curve in figure 7.2.17 shows that the lower activation energy in the presence of a catalyst enables a greater proportion of reactants to react.

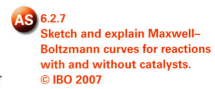

**6.2.7**
**Sketch and explain Maxwell–Boltzmann curves for reactions with and without catalysts.**
© IBO 2007

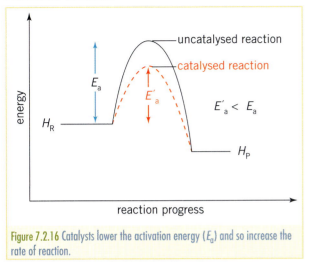

Figure 7.2.16 Catalysts lower the activation energy ($E_a$) and so increase the rate of reaction.

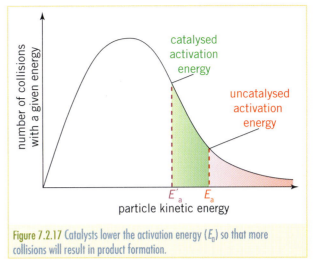

Figure 7.2.17 Catalysts lower the activation energy ($E_a$) so that more collisions will result in product formation.

## Section 7.2 Exercises

1. Use the kinetic molecular theory to explain why gases:
   a. are easily compressed
   b. completely fill any container
   c. are miscible (readily mix with one another)
   d. have lower densities than solids.

2. A sample of gas in a sealed fixed volume container is heated. Use the kinetic molecular theory to predict what will happen to the:
   a. average speed of the gas particles
   b. pressure of the gas on the container walls
   c. average space between the gas particles
   d. frequency of collisions between the particles.

3. a. Sketch a graph showing the distribution of particle speeds in a gas.
   b. State what happens to each of the following features of this graph when the gas is heated.
      i. Area under the graph
      ii. Spread of gas speeds
      iii. Velocity at which the graph peaks

4. According to collision theory, outline what must take place before a chemical reaction can occur.

5. Use collision theory to explain how changes in each of the following can affect the rate of a chemical reaction.
   a. Temperature     b. Concentration     c. Surface area

**6** Explain each of the following observations in terms of the factors that affect rate of reaction.

  **a** When milk is kept in the refrigerator it will remain fresh for up to a week, while it will become sour within 24 hours when left on the kitchen bench in summer.

  **b** Wheat silos have been known to explode when dust has been ignited by a stray spark or the flame of a cigarette lighter. A silo is far less likely to explode when completely full of wheat than when it is three-quarters empty.

  **c** Sulfur burns in air to produce sulfur dioxide. The rate of this reaction increases significantly when pure oxygen gas is used rather than air.

  **d** Small pieces of wood are far more effective for lighting a fire than large ones.

**7** Dinitrogen pentoxide decomposes over time according to the equation:
$2N_2O_5(g) \rightarrow 4NO_2(g) + O_2(g)$

A quantity of $N_2O_5$ was dissolved in tetrachloromethane at 60°C and its concentration was measured over a period of time. The results are shown in the table below.

| Reaction time (s) | Concentration of $N_2O_5$ (mol dm$^{-3}$) |
|---|---|
| 0 | 2.25 |
| 400 | 1.43 |
| 800 | 0.88 |
| 1200 | 0.46 |
| 1600 | 0.22 |
| 2000 | 0.13 |
| 2400 | 0.08 |

  **a** In terms of collision theory, why did the concentration of $N_2O_5$ decrease most quickly over the initial stages of the experiment?

  **b** The experiment was repeated using the same solvent and initial concentration of $N_2O_5$, but at a temperature of 10°C. Predict what differences might be expected in the results.

  **c** State two variations that could be introduced to the experiment that would result in a faster rate of decomposition of $N_2O_5$.

**8** In addition to the discovery of the preserved remains of Ötzi the Iceman, the remains of people have also been discovered in peat bogs in the United Kingdom and parts of Europe. These bogs tend to be rich in organic acids but very low in dissolved oxygen. Outline the factors that may contribute to the slow rate of decomposition of these bodies.

# Chapter 7 Summary

## Terms and definitions

**Activation energy**  The minimum amount of energy required to initiate a chemical reaction.

**Average rate of reaction**  The change in the concentration, mass or volume of the reactants or products over a period of time during the reaction.

**Catalyst**  A substance that changes the rate of reaction without itself being used up or permanently changed.

**Collision theory**  A theory which explains rates of reaction on a molecular level.

**Initial rate of reaction**  The change in the concentration, mass or volume of the reactants or products at the start of the reaction.

**Instantaneous rate of reaction**  The change in the concentration, mass or volume of the reactants or products at a particular point in time during the reaction.

**Kinetic molecular theory**  A model or a set of proposals, used to explain the properties and behaviour of gases.

**Maxwell–Boltzmann energy distribution curve**  A graph showing the distribution of particle velocities in a gas.

**Rate of reaction**  The change in the concentration, mass or volume of the reactants or products with time.

## Concepts

- Rate of reaction is defined as the 'change of concentration of reactants or products with time'.

$$\text{Rate} = \frac{\text{concentration of R at time } t_2 - \text{concentration of R at time } t_1}{t_2 - t_1}$$

$$= -\frac{\Delta[R]}{\Delta t}$$

- A range of methods can be used to experimentally measure rates of reaction including measuring mass loss, volume of gas produced and change in concentration, light absorption or colour of reactants or products.

- The kinetic molecular theory is used to explain these properties of gases. This theory states that the particles in gases:
  - are rapidly moving in random motion
  - have negligible volume compared to the gas volume
  - exert no forces on each other
  - collide elastically with each other and the container walls
  - have an average kinetic energy that is directly proportional to the absolute temperature of the gas.

- Increasing the temperature of a gas increases the average kinetic energy of the particles. It also increases the range of velocities of the particles.

- The activation energy ($E_a$) of a reaction is the minimum amount of energy required to initiate the reaction. This energy is used to break the bonds in the reactant molecules.

- At any given temperature, particles in a gas have a range of speeds. A graph of these particle speeds is called a Maxwell–Boltzmann distribution.

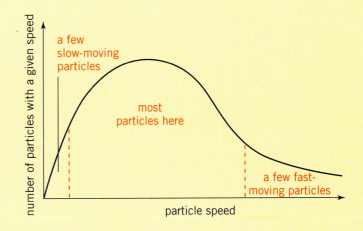

- Reaction rate depends on:
  - collision frequency
  - the number of particles with $E \geq E_a$
  - the appropriate orientation of the collision.

- Several factors can affect the rate at which a reaction proceeds:
  - a change in temperature
  - a change in concentration or pressure of reactants
  - the use of a catalyst
  - a change in the available surface area of a solid.
- A catalyst lowers the activation energy by providing an alternative pathway for a reaction to occur. By doing so, more particles have sufficient energy to initiate reaction.

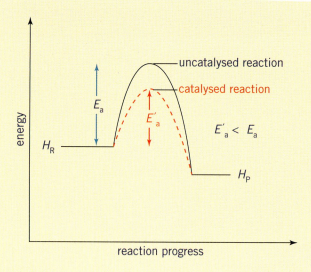

## Chapter 7 Review questions

1. Use the kinetic particle theory to explain the following.
   a. For virtually all substances (the exception being ice/water), the solid state of a particular substance is more dense than the liquid,
   b. If pressure is applied to a liquid, there is very little change in its volume, whereas a gas will compress readily.
   c. Cigarette lighters should never be disposed of in an incinerator.
   d. An increase in temperature almost invariably increases the rate of a chemical reaction.

2. Two students were having an argument about the motion of particles in the gaseous state. Both agreed that an increase in temperature causes particles to move more quickly, as they gain kinetic energy. One student had studied physics and knew that kinetic energy is directly related to the speed of the particles. This student insisted that all the particles at a particular temperature must be moving at the same speed as they have the same kinetic energy at that temperature. The second student disagreed and insisted that the particles had a range of speeds. Which student was correct? Explain your choice.

3. State the two criteria that, according to collision theory, must be satisfied before the collision of reactant particles will result in the formation of a new product.

4. On the same set of axes, draw graphs of the Boltzmann distributions of molecular speeds that would be expected for a sample of chlorine gas at room temperature and at 100°C.

5. Hydrogen peroxide is both an oxidizing agent and a reducing agent, and as a result will react with itself to produce water and oxygen gas according to the equation:
   $$2H_2O_2(aq) \to 2H_2O(l) + O_2(g)$$
   Solutions of hydrogen peroxide will remain fresh for several weeks if kept in the refrigerator. Explain this observation.

6. Explain why the activation energy of a chemical reaction is always endothermic.

7. Zinc metal dissolves in hydrochloric acid to produce hydrogen gas and zinc chloride, according to the equation:
   $$Zn(s) + 2HCl(aq) \to ZnCl_2(aq) + H_2(g)$$
   State three ways in which the rate of production of hydrogen gas could be increased in this reaction.

8. Ethene reacts with steam in the presence of a catalyst that consists of solid silicon dioxide coated with phosphoric(V) acid, to produce ethanol according to the equation:
   $$C_2H_4(g) + H_2O(g) \xrightarrow{H_3PO_4} C_2H_5OH(g)$$
   $$\Delta H = -45 \text{ kJ mol}^{-1}$$
   a. State what must happen to the bonds in the reactants before this reaction can take place.

**b** The activation energy for this catalysed reaction is 280 kJ mol$^{-1}$, under certain conditions of temperature and pressure. State whether the activation energy would be larger, smaller or the same if no catalyst was used? Explain your answer.

**c** Explain the role of the catalyst in this reaction.

**9** Explain each of the following observations in terms of the factors that affect rate of reaction.

**a** When setting up a camp fire, it is much easier to start the fire burning if you use small pieces of kindling rather than large pieces of wood.

**b** Large amounts of dust suspended in the air in a silo can result in an explosion.

**c** The body of a woolly mammoth (a type of extinct elephant) was recently discovered intact, frozen under the Russian tundra, having died some 20 000 years previously.

**d** Powdered zinc reacts violently with dilute hydrochloric acid, but large pieces of the same element react only slowly.

**e** Hydrogen peroxide decomposes to water and oxygen gas much more quickly when a small quantity of manganese(IV) oxide is added.

**10** An excess of calcium carbonate is reacted with a fixed volume of 1.0 mol dm$^{-3}$ hydrochloric acid at 25°C.

Predict the change (if any) to (i) the rate of reaction and (ii) the volume of CO$_2$ produced in the following conditions.

**a** The concentration of hydrochloric acid is increased to 2.0 mol dm$^{-3}$.

**b** The volume of 1.0 mol dm$^{-3}$ hydrochloric acid is halved.

**c** The size of the calcium carbonate pieces is decreased by crushing the calcium carbonate.

**d** The temperature of the 1.0 mol dm$^{-3}$ hydrochloric acid is decreased to 10°C by refrigeration.

www.pearsoned.com.au/schools

Weblinks are available on the Chemistry: For use with the IB Diploma Programme Standard Level Companion Website to support learning and research related to this chapter.

# Chapter 7 Test

## Part A: Multiple-choice questions

**1** Some collisions between reactant molecules do not form products. This is most likely because:

**A** the molecules do not collide in the proper ratio.

**B** the molecules do not have enough energy.

**C** the concentration is too low.

**D** the reaction is at equilibrium.

© IBO 2000, May P1 Q20

**2** As the temperature of a reaction between two gases is increased, the rate of the reaction increases. This is **mainly** because:

**A** the concentrations of the reactants increase.

**B** the molecules collide more frequently.

**C** the pressure exerted by the molecules increases.

**D** the fraction of molecules with the energy needed to react increases.

© IBO 2000, Nov P1 Q19

**3**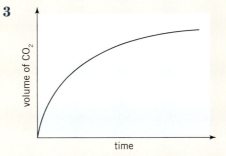

The curve above is obtained for the reaction of an excess of CaCO$_3$ with hydrochloric acid. How and why does the rate of reaction change with time?

| | Rate of reaction | Reason |
|---|---|---|
| A | Decreases | The HCl becomes more dilute. |
| B | Decreases | The pieces of CaCO$_3$ become smaller. |
| C | Increases | The temperature increases. |
| D | Increases | The CO$_2$ produced acts as a catalyst. |

© IBO 2000, Nov P1 Q20

4. The reaction between nitrogen and oxygen in the atmosphere under normal conditions is extremely slow. Which statement best explains this?

   A The concentration of oxygen is much lower than that of nitrogen.

   B The molar mass of nitrogen is less than that of oxygen.

   C The frequency of collisions between nitrogen and oxygen molecules is lower than that between nitrogen molecules themselves.

   D Very few nitrogen and oxygen molecules have sufficient energy to react.

   © IBO 2001, May P1 Q19

5. What are the units for the rate of a reaction?

   A $mol\ dm^{-3}$

   B $s^{-1}$

   C $mol\ dm^{-3}\ s^{-1}$

   D $dm^3\ mol^{-1}\ s^{-1}$

   © IBO 2001, Nov P1 Q19

6. $CaCO_3(s) + 2HCl(aq) \rightarrow CaCl_2(aq) + H_2O(l) + CO_2(g)$

   Which change will increase the rate of the reaction when 50 $cm^3$ of 1.0 $mol\ dm^{-3}$ HCl is added to 1.0 g of $CaCO_3$?

   A The volume of HCl is increased.

   B The concentration of HCl is decreased.

   C The size of the $CaCO_3$ solid particles is decreased.

   D The pressure of the $CO_2$ is increased.

   © IBO 2002, May P1 Q19

7. In general, the rate of a reaction can be increased by all of the following **except**:

   A increasing the temperature.

   B increasing the activation energy.

   C increasing the concentration of reactants.

   D increasing the surface area of the reactants.

   © IBO 2002, Nov P1 Q19

8. Which statement is correct for a collision between reactant particles leading to a reaction?

   A Colliding particles must have different energy.

   B All reactant particles must have the same energy.

   C Colliding particles must have a kinetic energy higher than the activation energy.

   D Colliding particles must have the same velocity.

   © IBO 2005, Nov P1 Q19

9. Which change of condition will decrease the rate of the reaction between excess zinc granules and dilute hydrochloric acid?

   A Increasing the amount of zinc

   B Increasing the concentration of the acid

   C Pulverizing the zinc granules into powder

   D Decreasing the temperature

   © IBO 2005, Nov P1 Q20

10. The table shows the concentrations of reactants and products during this reaction.

    $2A + B \rightarrow C + 2D$

    |  | at the start | after 1 min |
    | --- | --- | --- |
    | [A]/$mol\ dm^{-3}$ | 6 | 4 |
    | [B]/$mol\ dm^{-3}$ | 3 | 2 |
    | [C]/$mol\ dm^{-3}$ | 0 | 1 |
    | [D]/$mol\ dm^{-3}$ | 0 | 2 |

    The rate of the reaction can be measured by reference to any reactant or product. Which rates are correct for this reaction?

    I Rate = $-2\ mol\ dm^{-3}\ min^{-1}$ for A

    II Rate = $-1\ mol\ dm^{-3}\ min^{-1}$ for B

    III Rate = $-1\ mol\ dm^{-3}\ min^{-1}$ for C

    A I and II only

    B I and III only

    C II and III only

    D I, II and III

    © IBO 2006, May P1 Q19

    (10 marks)

 www.pearsoned.com.au/schools

For more multiple-choice test questions connect to the Chemistry: For use with the IB Diploma Programme Standard Level Companion Website and select Chapter 7 Review Questions.

## Part B: Short-answer questions

1. Consider the following reaction in the Contact process for the production of sulfuric acid.

   $2SO_2 + O_2 \rightleftharpoons 2SO_3$

   Use the collision theory to explain why increasing the temperature increases the rate of the reaction between sulfur dioxide and oxygen.

   (2 marks)

   © IBO 2006, May P2 Q6c

2. **a** Define the term *rate of reaction*.

   (1 mark)

   **b** The reaction between gases **C** and **D** is slow at room temperature.

   **i** Suggest **two** reasons why the reaction is slow at room temperature.

   (2 marks)

   **ii** A relatively small increase in temperature causes a relatively large increase in the rate of this reaction. State **two** reasons for this.

   (2 marks)

   **iii** Suggest **two** ways of increasing the rate of reaction between C and D other than increasing temperature.

   (2 marks)

   © IBO 2005, Nov P2 Q2

3. Lumps of magnesium carbonate react with dilute hydrochloric acid according to the equation:

   $MgCO_3(s) + 2HCl(aq)$
   $\rightarrow MgCl_2(aq) + CO_2(g) + H_2O(l)$

   **a** State **three** ways in which the rate of this reaction may be increased.

   (3 marks)

   **b** Sketch a graph of the Maxwell–Boltzmann distribution curve for this reaction.

   (1 mark)

   **c** On the same graph, sketch the curve if the reaction was carried out at a temperature that was 10°C greater than the original reaction.

   (2 marks)

## Part C: Data-based question

The graph below was obtained when copper(II) carbonate reacted with dilute hydrochloric acid under two different conditions, A and B.

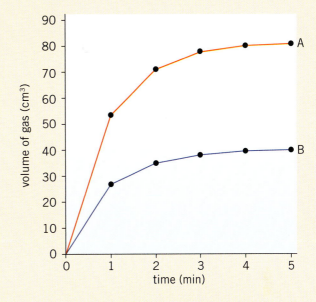

**a** **i** Name the gas produced in the reaction.

(1 mark)

**ii** Write a balanced chemical equation for the reaction occurring.

(2 marks)

**b** Identify the volume of gas produced and the time taken for the reaction under condition A to be complete.

(1 mark)

**c** **i** Explain the shape of curve A in terms of the collision theory.

(2 marks)

**ii** Suggest two possible reasons for the differences between curves A and B.

(2 marks)

**d** On the graph above draw a line, labelled C, for the curve if the reaction were repeated under condition A but in the presence of a catalyst.

(2 marks)

© IBO 2002, Nov P2 Q1

## Part D: Extended-response question

When dilute hydrochloric acid is added to a solution of sodium thiosulfate, a white precipitate of sulfur is formed.

**a** Describe the effect of the following changes on the rate of the reaction between hydrochloric acid and sodium thiosulfate and explain each of your answers using the kinetic molecular theory.

  **i** The concentration of the hydrochloric acid used is increased.
  (2 marks)

  **ii** The volume of the hydrochloric acid used is increased.
  (3 marks)

  **iii** The temperature is raised from 20°C to 30°C.
  (4 marks)

**b** Explain why a small increase in temperature has a much greater effect on the rate of reaction than a small increase in concentration.
(2 marks)

**c** A series of reactions is carried out where a fixed volume of 1.0 mol dm$^{-3}$ hydrochloric acid is added to a fixed volume of sodium thiosulfate solution of different concentrations. The time taken to form a visible precipitate of sulfur is noted.

  **i** Sketch a graph of sodium thiosulfate concentration (vertical axis) against time to form the visible precipitate (horizontal axis).
  (2 marks)

  **ii** How would you calculate the rate of the reaction at different times from the above graph?
  (2 marks)

© IBO 2000, Nov P2 Q7
Total marks 50

# 8 EQUILIBRIUM

## Chapter overview

This chapter covers the IB Chemistry syllabus Topic 7: Equilibrium.

By the end of this chapter, you should be able to:

- recognize that the principles of chemical equilibrium are essential to our understanding of how reactions proceed
- recall that equilibrium is a dynamic state in which the rates of the forward and reverse reactions are balanced
- deduce the equilibrium constant expression for a homogeneous reaction
- recall that the magnitude of the equilibrium constant indicates the extent of the reaction and that it is a constant value at a given temperature
- recognize the difference between the equilibrium constant and position of equilibrium
- recall that temperature is the only macroscopic property that can change the equilibrium constant for a particular reaction
- recall that equilibrium position can be moved by changes in the conditions under which the reaction proceeds: temperature, concentration and pressure all change the position of equilibrium
- use Le Chatelier's principle to explain movement of the equilibrium position by changes to reaction conditions
- state that a catalyst increases reaction rate, but has no effect on the position of equilibrium
- explain the effect of a catalyst on an equilibrium reaction
- apply the principles of chemical kinetics and of chemical equilibrium in explaining how rate of reaction and yield for two industrial processes can be maximized by changing conditions of temperature, pressure and concentration.

**IBO Assessment statements 7.1.1 to 7.2.5**

The notion of maintaining balance—a sense of equilibrium—is central to our very existence. Our relationships with each other and with the plants and animals we rely on and that rely on us are highly interdependent. Indeed, the health and viability of the Earth requires that its physical and biological systems work together to maintain a careful balance called equilibrium. Whenever there is a change, disturbance or fluctuation, the Earth's systems respond to these stresses in an attempt to maintain equilibrium. In many ways, chemical reactions are no different. This should come as no surprise to us, as chemistry is truly the 'stuff of life'. In this chapter, we will look at how chemical reactions maintain balance and how they react to changes in the conditions of concentration, temperature and pressure that affect them.

## 8.1 DYNAMIC EQUILIBRIUM

Reactions are based on collisions of particles, in which the particles must have the correct orientation and sufficient energy. (See chapter 7.) It follows that the products, once formed, can also undergo molecular collisions energetic enough to break them back down into the reactants. When the reaction has just begun, the amount of reactants available is relatively large, but there is very little product. Under these circumstances, the rate at which new products are formed is at its maximum.

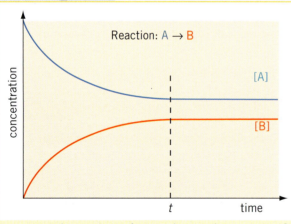

**Figure 8.1.1** As the concentration of reactants decreases, the concentration of products increases.

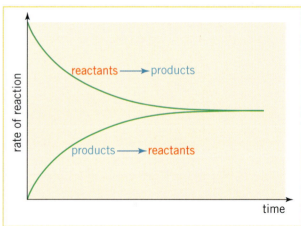

**Figure 8.1.2** The rates of the forward and reverse reactions are equal when equilibrium is reached.

With time, more product is formed and the rate of the forward reaction (left → right) decreases as the amount of reactant available decreases. At the same time, the amount of product formed is increasing, and so there is a greater likelihood that it will be broken down and reconverted into reactants. Hence, the rate of the forward reaction is decreasing while, at the same time, the rate of the reverse reaction is increasing.

In figure 8.1.1 it can be seen that at time $t$ the concentration of reactants and of products has stopped changing. The rate at which reactants are being converted into products is matched by the rate of the conversion of product back to reactant. The reaction will *appear* to have 'stopped', as no more changes will be visible to the eye or to conventional measurements: no bubbles of gas, no colour change, constant pH, no heat evolved or absorbed, and so on. The **macroscopic properties** of the reaction are constant and the reaction has reached a state of **dynamic equilibrium**.

Chemical equilibrium
Concentration versus time

## Characteristics of dynamic equilibrium

A chemical reaction is said to be in a state of dynamic equilibrium at a specified temperature when the rate of the forward reaction is exactly equal to the rate of the reverse reaction. Observation of macroscopic properties will suggest nothing is happening; everything will seem to have stopped. If, however, we could look at the reaction on a molecular level, the picture would be very different. Compounds would be constantly breaking apart into their constituent particles and rearranging themselves into new products. These products would similarly be breaking apart and rearranging themselves back into the reactants. The changes would be perfectly balanced, but nevertheless would both be proceeding.

**7.1.1
Outline the characteristics of chemical and physical systems in a state of equilibrium.
© IBO 2007**

This concept of equilibrium is not restricted to chemical reactions; it is also true for physical systems. Take for example the physical process of changing state. Imagine that you have left a half-filled, stoppered bottle of water on the kitchen bench on a warm day. The surface of the water is perfectly flat and nothing appears to be happening. In fact, water molecules are constantly in motion and some are escaping the surface of the water and entering the gaseous phase. At exactly the same rate, however, water molecules from the air above the surface are colliding with the water surface and forming bonds with the water molecules to return to the liquid state. A state of equilibrium exists between the liquid and gaseous states of water:

$$H_2O(l) \rightleftharpoons H_2O(g)$$

Note that the two-way arrow '$\rightleftharpoons$' is used to represent a reaction at equilibrium.

Figure 8.1.3 Water molecules in equilibrium between liquid and gaseous states.

It is at this point that the importance of temperature can be seen. If the water in the bottle is transferred to a small saucepan, a lid is placed and the saucepan is heated on the stove, its temperature will increase. As the water molecules gain kinetic energy, more and more of them will have sufficient energy to overcome the forces of attraction to other water molecules at the surface and so will enter the gaseous state. If, for example, the stove is providing just enough heat to maintain the temperature of the water at 70°C, the water will once more be in a state of equilibrium, but this will be a different equilibrium from that which existed when it was on the bench. Note that in both cases the containers must be sealed if equilibrium is to be reached, otherwise the water would simply evaporate and be lost. In an unsealed container, we would consider the water to be part of an **open system**, and equilibrium could not be reached. When the container is sealed, the liquid water and its vapour are in a constant state of flux with one another and the system is said to be a **closed system**. Chemical equilibrium can only be reached in a closed system.

Liquid–vapour equilibrium

What proof do we have that supports the hypothesis that systems enter a state of equilibrium if we cannot see this occur or measure it? The answer lies in the use of radioactive tracers. Iodine is only slightly soluble in water, so if a small quantity is placed in a beaker of water and stirred for a period of time, some of the iodine will dissolve, the rest will sink to the bottom—the solution has become saturated. No further changes will be noted in the beaker of purple-coloured water; the intensity of colour will remain constant, as will the quantity of crystals on the bottom of the beaker. A simple equation to represent this process is:

$$I_2(s) + aq \rightleftharpoons I_2(aq)$$

If we now add a small quantity of radioactive iodine crystals ($^{131}I_2$) to the beaker, they should simply sink to the bottom and have no visible effect—the solution is already saturated. According to the theory of dynamic equilibrium, after some time some radioactive iodine molecules should dissolve into the solution and other non-radioactive iodine molecules will crystallize out of solution. Measurement of the radioactive emissions of the $^{131}I_2$ with a sensitive Geiger counter proves that this is the case—the radioactive molecules are shown to be evenly distributed throughout the solution as well as in the solid crystals.

### Section 8.1 Exercises

1. Describe how the concentration of reactants and products changes as a reaction in a closed system proceeds.

2. Compare open and closed systems and explain why only a closed system can reach equilibrium.

3. Describe three experimental observations that you might make to identify that a system has reached equilibrium.

4. Explain the meaning of the term *dynamic equilibrium*.

## 8.2 THE POSITION OF EQUILIBRIUM

In our study of chemistry to date, we have always assumed that a reaction proceeds fully to the right. All the stoichiometric calculations we have performed rely on this premise. For example, the equation representing the formation of nitrogen(II) oxide from its constituent elements is:

$$N_2(g) + O_2(g) \rightleftharpoons 2NO(g)$$

This equation shows us that 1 mol of nitrogen gas will react with 1 mol of oxygen gas to produce 2 mol of nitrogen(II) oxide as product, which is, of course, true. However, if we filled a container with 1 mol of nitrogen and 1 mol of oxygen gas, when equilibrium was reached, there would not be precisely 2 mol of nitrogen(II) oxide formed. For this to be the case, the reaction would need to proceed completely to the right. In fact, at a temperature of 25°C the **extent of reaction** is so small as to be virtually non-existent. Even when the temperature is 1000°C, the reaction proceeds to only 0.04%. This should not really surprise us, because if nitrogen and oxygen gases (the principal components of our atmosphere) did readily react with one another, our atmosphere would swiftly be converted into poisonous nitrous oxide, and life on Earth as we know it would cease to exist!

The reaction between nitrogen and oxygen is an example of a **homogeneous** equilibrium. All reactants and products are in the gas phase. If reactants are in different phases, the equilibrium is described as **heterogeneous**. The reaction between steam and heated iron to form $Fe_3O_4$ and hydrogen:

$$3Fe(s) + 4H_2O(g) \rightleftharpoons Fe_3O_4(s) + 4H_2(g)$$

is an example of heterogeneous equilibrium.

It is important to recognize that it is not possible to determine the extent of a reaction on the basis of its equation alone. To do this, we need more information, which must be determined experimentally. To know how to use this information, we need to consider the equilibrium law.

## The equilibrium law

As early as 1799, the French chemist Claude Berthollet (1748–1822) proposed that reactions were reversible. Some years later, he showed that the concentration of reactants had a direct effect on the concentration of products formed. This work led two Norwegian chemists, Cato Maximilian Guldberg (1836–1902) and Peter Waage (1833–1900), to look more closely at the relationship between masses of reactants and products in reactions. After conducting numerous experiments, they were to propose in 1864 what came to be known as the *equilibrium law*.

Their calculations showed that there was a mathematical relationship between the concentrations of the products formed in a reaction once equilibrium had been reached and the concentrations of the reactants left behind. The coefficients of the species involved in the equation were important, as was the temperature at which the reaction proceeded. Guldberg and Waage determined that for the general reaction:

$$aA + bB + \ldots \rightleftharpoons pP + qQ + \ldots$$

At constant temperature, the concentrations of reactants and products are related by the ratio:

$$K_c = \frac{[P]^p [Q]^q \ldots}{[A]^a [B]^b \ldots}$$

**7.2.1 Deduce the equilibrium constant expression ($K_c$) from the equation for a homogeneous reaction.**
© IBO 2007

where $K_c$ is the **equilibrium constant** for the reaction at a specified temperature. This expression is therefore known as the **equilibrium constant expression**.

Note that it is a common convention to use square brackets when showing the concentration of a species in an equilibrium constant expression. Thus, the concentration of the reactant species P is written as [P].

The units of the equilibrium expression vary, depending on the reaction for which the expression is being written. As the expression is made up of concentration terms, the units will be related to concentration units, although some equilibrium expressions have no units at all due to cancelling within the expression. This can be best understood by examining some particular equilibrium expressions and their constants. See worked example 1 on the next page.

Equilibrium constant expression
Modifying chemical equilibrium expressions
Review of equilibrium constant expressions

Although these will not be considered, it is important to know that equilibrium constant expressions can also be written using pressures of gaseous species. In this case the symbol for the equilibrium constant would be $K_p$. The equilibrium constant expressions that we will be investigating will only be made up of concentrations, so the symbol is $K_c$.

### Worked example 1

Write an expression and state the units for the equilibrium constant, $K_c$, for each of the following reactions.

**a** $2NO_2(g) \rightleftharpoons N_2O_4(g)$    **b** $H_2(g) + I_2(g) \rightleftharpoons 2HI(g)$    **c** $2CO(g) + O_2(g) \rightleftharpoons 2CO_2(g)$

### Solution

**a** $K_c = \dfrac{[N_2O_4]}{[NO_2]^2}$

Units: $\dfrac{\text{mol dm}^{-3}}{\text{mol}^2 \text{ dm}^{-6}} = \text{mol}^{-1} \text{ dm}^3$

**b** $K_c = \dfrac{[HI]^2}{[H_2][I_2]}$

Units: $\dfrac{\text{mol}^2 \text{ dm}^{-6}}{\text{mol}^2 \text{ dm}^{-6}} = \text{no units}$

**c** $K_c = \dfrac{[CO_2]^2}{[CO]^2[O_2]}$

Units: $\dfrac{\text{mol}^2 \text{ dm}^{-6}}{\text{mol}^3 \text{ dm}^{-9}} = \text{mol}^{-1} \text{ dm}^3$

---

You will have noticed that the formulation of an equilibrium expression is based solely upon the way in which the equation for the equilibrium has been written. The coefficients in the equation are the ones that should be used in the equilibrium expression. With this in mind we can observe some interesting trends.

Consider the equation:

$$2NO_2(g) \rightleftharpoons N_2O_4(g)$$

The equilibrium expression for this equation is:

$$K_c = \dfrac{[N_2O_4]}{[NO_2]^2}$$

If the equation is reversed:

$$N_2O_4(g) \rightleftharpoons 2NO_2(g)$$

the new value of $K_c$, $K_c' = \dfrac{[NO_2]^2}{[N_2O_4]} = \dfrac{1}{K_c}$

If the coefficients are changed:

$$4NO_2(g) \rightleftharpoons 2N_2O_4(g)$$

the new value of $K_c$, $K_c'' = \dfrac{[N_2O_4]^2}{[NO_2]^4} = K_c^2$

or if we have $NO_2(g) \rightleftharpoons \dfrac{1}{2}N_2O_4(g)$

the new value of $K_c$, $K_c''' = \dfrac{[N_2O_4]^{\frac{1}{2}}}{[NO_2]} = K_c^{\frac{1}{2}} = \sqrt{K_c}$

The equilibrium constant indicates the extent of a chemical reaction; the larger the value, the further to the right the reaction proceeds and so the more product is formed. A chemical reaction with an equilibrium constant that is much greater than 1 will proceed to a very large extent, making large amounts of products and effectively proceeding to completion. A chemical reaction with an equilibrium constant that is much smaller than 1 will proceed to a very small extent and few products will be made. In fact the reaction hardly proceeds at all.

If $K_c \gg 1$, the reaction goes almost to completion.
If $K_c \ll 1$, the reaction hardly proceeds.

**7.2.2**
**Deduce the extent of a reaction from the magnitude of the equilibrium constant.**
© IBO 2007

## The position of equilibrium

The initial amounts of reactant (or product) in a reaction mixture have no influence on the value of the equilibrium constant for a particular reaction. Temperature does have an effect on the value of the constant, but concentration, gas pressure and the use of a **catalyst** do not. Nevertheless, it is possible to manipulate the *position of equilibrium*, and so the amount of product obtained, by varying the conditions under which a reaction is performed. This quite remarkable insight was first proposed by the French chemist Henri Le Chatelier in 1884.

Reaching equilibrium from differing starting concentration

### CHEM COMPLEMENT

#### Henri-Louis Le Chatelier

Henri-Louis Le Chatelier was born in Paris in 1850. His father was an engineer and helped to develop the fledgling aluminium industry there. Having completed his formal studies in Paris at the École Polytechnique and the École des Mines, he worked for two years as a mining engineer before being offered a position as a professor of chemistry at the Écoles des Mines. In 1887, he became a professor at the Sorbonne.

His initial research projects were centred on the chemistry of cement and how it set. He then diversified into an examination of thermochemical principles, from which arose his famous principle. His work on the often explosive mixture of gases found in mines, and how to deal with them effectively, undoubtedly led to saving many lives in coal mines around the world. Le Chatelier was to use this insight into thermochemical principles to devise an industrial process for the production of ammonia before the German chemists Fritz Haber and Karl Bosch developed it further in the early years of the 20th century. Le Chatelier died in 1936 at the age of 85, having made an enormous contribution to science in the industrial and metallurgical fields.

Figure 8.2.1 Henri Le Chatelier (1850–1936)

## Le Chatelier's principle

Le Chatelier investigated the thermodynamics (particle energy) of a number of chemical reactions at equilibrium to establish how reactions behaved when external conditions were changed. He looked at changes in temperature, and the concentrations of reactants and products, and, for gaseous systems, at the pressure applied to the system, and realized that he could make general observations about any such changes. If the reaction was forced out of equilibrium by some external change, it would swiftly return to it by changing the relative proportions of reactant and product so as to partly negate the effect of the

Le Chatelier's principle

**7.2.3 Apply Le Chatelier's principle to predict the qualitative effects of changes of temperature, pressure and concentration on the position of equilibrium and on the value of the equilibrium constant.** © IBO 2007

external influence. He summarized this simple but powerful insight in what came to be known as **Le Chatelier's principle**:

> **If a system is at equilibrium and any of the temperature, pressure or concentrations of the species are changed, the reaction will proceed in such a direction as to partially compensate for this change.**

To see how useful Le Chatelier's principle is in determining how reactions may be manipulated to our benefit, we will consider each of the factors in turn.

## Temperature

Temperature is the only factor that can affect both the equilibrium position and the equilibrium constant. All chemical reactions either release energy to their surroundings or absorb energy from them. Recall from chapter 6 that any reaction that releases energy is exothermic ($\Delta H < 0$), and any reaction that absorbs energy is endothermic ($\Delta H > 0$). Recall also that a reaction that gives off heat when generating products (an exothermic reaction) is endothermic in the reverse direction.

Let us consider an exothermic reaction, such as the production of sulfur trioxide. As heat energy is given out when two mole of $SO_2$ reacts with one mole of $O_2$, the equation can be written as

$$2SO_2(g) + O_2(g) \rightleftharpoons 2SO_3(g) \; (+ \; 197 \text{ kJ } \textbf{energy})$$

Heat can be considered in the same way that you would consider a product (although of course it is not actually a chemical product). If the temperature of this system is increased, this has the same effect as increasing the concentration of a product. The system acts to oppose the change and the position of equilibrium moves to the left, making more reactants. The value of the equilibrium constant decreases as the denominator increases. If the temperature of this system is decreased (the same effect as removing a product), the position of equilibrium moves to the right, making more products. The value of the equilibrium constant increases due to the larger numerator.

For an endothermic reaction, heat can be considered as a reactant. For example, consider the following reaction:

$$NH_4NO_3(s) + aq \; (+ \; 25 \text{ kJ } \textbf{energy}) \rightleftharpoons NH_4^+(aq) + NO_3^-(aq)$$

If the temperature is increased, it has the same effect as adding a reactant and the position of equilibrium moves to the right, making more products. The equilibrium constant increases. If the temperature is decreased, it has the same effect as removing a reactant, so the position of equilibrium moves to the left, making more reactants. The equilibrium constant decreases.

**TABLE 8.2.1 THE EFFECT OF TEMPERATURE CHANGE ON EQUILIBRIUM POSITION AND EQUILIBRIUM CONSTANT VALUE**

| Temperature change | Exothermic reaction | | Endothermic reaction | |
| --- | --- | --- | --- | --- |
| | Position of equilibrium | Equilibrium constant, $K_c$ | Position of equilibrium | Equilibrium constant, $K_c$ |
| Increase | Moves to the left | Decreases | Moves to the right | Increases |
| Decrease | Moves to the right | Increases | Moves to the left | Decreases |

A useful example of the effect of temperature change on both the equilibrium constant and position is provided by the reaction involving nitrogen dioxide and its dimer, dinitrogen tetraoxide. (When two identical molecules combine, the resulting molecule is called a *dimer*). Nitrogen dioxide is a brown, foul-smelling gas and is a major component of the brownish haze of smog that settles over major cities. Dinitrogen tetraoxide is colourless, and so the depth of colour of the equilibrium mixture of these gases provides a good indication of which of the two gases is the major component of any mixture.

Nitrogen dioxide and dinitrogen tetraoxide exist in equilibrium with one another according to the equation:

$N_2O_4(g) \rightleftharpoons 2NO_2(g)$  $\Delta H = +58$ kJ mol$^{-1}$, $K_c = 4.7 \times 10^{-3}$ mol dm$^{-3}$ at 25°C
colourless    brown

At room temperature, the gas mixture is a light-brown colour. If a container of gas is placed in hot water for a short period of time, the colour deepens significantly, indicating that the proportion of $NO_2$ has increased. When it is removed from the boiling water and the temperature drops, the colour gradually returns to the original light brown. If the same container is then plunged into ice-cold water, the colour lightens to a light-brown haze. In this case, the proportion of $N_2O_4$ is much greater than that of the $NO_2$. How are these observations explained in terms of Le Chatelier's principle?

As the conversion of $N_2O_4$ to $NO_2$ is an endothermic process ($\Delta H$ positive), an increase in temperature will favour the forward reaction. In accordance with Le Chatelier's principle, the reaction proceeds in the forward direction using the supplied energy. The concentration of $NO_2$ should increase and that of $N_2O_4$ decrease. The deeper colour of the gas mixture indicates that this has occurred. Accurate measurements of concentrations of both species at elevated temperatures indicate that the equilibrium constant increases significantly, from 0.0047 mol dm$^{-3}$ at 25°C to 0.147 mol dm$^{-3}$ at 77°C.

$$K_c = \frac{[NO_2]^2}{[N_2O_4]}$$

Figure 8.2.2 Effect of heat on $N_2O_4$/$NO_2$ equilibrium. The beaker on the left contains hot water, the beaker on the right contains iced water.

The effect of a temperature change

When the gas mixture was cooled in the ice water, the colour lightened. For this endothermic process, a decrease in temperature favoured the backwards reaction, and so $N_2O_4$ formed as $NO_2$ was consumed. As $N_2O_4$ is colourless, a lighter shade of brown resulted. As the position of equilibrium moved to the left, the concentration of $NO_2$ decreased, so the numerator of the equilibrium expression decreased while the denominator increased. The result of these quantitative changes was that the equilibrium constant decreased.

Changes to the position of equilibrium caused by modifying external conditions, such as temperature, can also be shown graphically. Figure 8.2.3(a) shows how the concentrations of $N_2O_4$ and $NO_2$ change with time as the reaction proceeds. The initial concentration of $N_2O_4$ is at its maximum value, while the product ($NO_2$) has an initial concentration of zero. As the reaction progresses, the concentration of reactant decreases and product increases. From the equation, it can be seen that 2 mol of $NO_2$ are produced for every 1 mol of $N_2O_4$ consumed, which is why the concentration change of $NO_2$ is twice that of $N_2O_4$. When the concentrations of both species are no longer changing (the lines have become horizontal), a state of chemical equilibrium has been reached.

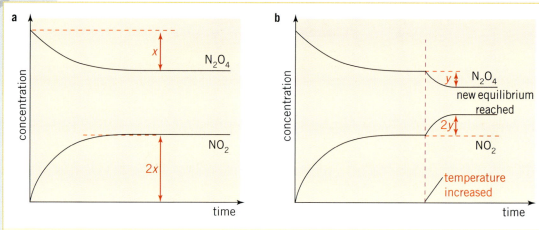

**Figure 8.2.3** (a) Graph of concentration versus time for $NO_2/N_2O_4$ mixture. (b) Graph of concentration versus time for $NO_2/N_2O_4$ mixture as temperature is increased.

Figure 8.2.3(b) shows what happens when the temperature is increased for this endothermic reaction. Le Chatelier's principle indicates that the reactions will proceed to the right, generating more product. The concentration of $N_2O_4$ decreases, while the concentration of $NO_2$ increases twice as much. When equilibrium has been re-established, measurements of actual concentrations would show that the equilibrium constant had increased.

### Worked example 2

Determine the direction in which the equilibrium systems listed below will proceed, and the effect on the equilibrium constant, following an increase in the external temperature:

**a** $N_2(g) + 3H_2(g) \rightleftharpoons 2NH_3(g)$     $\Delta H = -92.3$ kJ mol$^{-1}$

**b** $CO_2(g) + H_2(g) \rightleftharpoons CO(g) + H_2O(g)$     $\Delta H = +282$ kJ mol$^{-1}$

### Solution

**a** This is an exothermic reaction in the forward direction, so an increase in temperature will cause the reaction to proceed backwards (to the left), in favour of the endothermic (temperature-lowering) process. As a result, ammonia will dissociate to $N_2(g)$ and $H_2(g)$.

$K_c = \dfrac{[NH_3]^2}{[N_2][H_2]^3}$, so as the concentration of $NH_3$ decreases, the numerator will decrease and the equilibrium constant will decrease.

**b** This is an endothermic reaction in the forward direction, so an increase in temperature will cause the reaction to proceed forwards (to the right). As a result, more carbon monoxide and water vapour will form.

$K_c = \dfrac{[CO][H_2O]}{[CO_2][H_2]}$, so as the concentration of CO and $H_2O$ increase, the numerator of the equilibrium expression fraction will increase and the equilibrium constant will increase.

**PRAC 8.1**
The effect of temperature on the position of equilibrium

## Concentration

Consider the reaction system shown in the equation below. This reaction is very useful, as both $Fe^{3+}$ and $SCN^-$ are effectively colourless at low concentrations, while the $Fe(SCN)^{2+}$ produces a solution of intense blood-red colour. Even a very small change in the relative concentration of the $Fe(SCN)^{2+}$ species can be seen clearly:

$$Fe^{3+}(aq) + SCN^-(aq) \rightleftharpoons Fe(SCN)^{2+}(aq)$$
**colourless      colourless           blood-red**

A series of test tubes are partially filled with a dilute aqueous solution of $Fe(SCN)^{2+}$, which will contain all three ions in equilibrium. To one test tube is added a few drops of $FeCl_3$ solution, which results in an immediate increase in the intensity of the red colour. Why has this happened?

Adding the $FeCl_3$ instantaneously increases the $Fe^{3+}$ ion concentration. According to Le Chatelier's principle, an increase in the concentration of any reactant will cause the reaction to proceed in such a way as to partly compensate for this change. To overcome the increased concentration of $Fe^{3+}$ ions, the equilibrium position will move to the right. As this occurs, the concentration of the product species ($Fe(SCN)^{2+}$) will increase and so the intensity of colour will increase. A move of the equilibrium position to the right requires that reactants be consumed; both the $Fe^{3+}$ and $SCN^-$ ions will be used. When the new equilibrium position has been established, the concentration of $SCN^-$ will be less than it was before the addition of the $Fe^{3+}$.

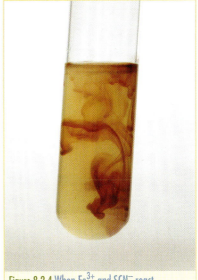

Figure 8.2.4 When $Fe^{3+}$ and $SCN^-$ react together, a solution of a deep red colour ($Fe(SCN)^{2+}$) forms.

How does the concentration of $Fe^{3+}$ at the final equilibrium position compare with its original equilibrium concentration? The instant the $Fe^{3+}$ was added, its concentration increased. The reaction then proceeded forward to partly compensate for the addition and so its concentration decreased, but once the new equilibrium is reached, it is still higher than it was originally. Why? Le Chatelier's principle provides the answer: the reaction proceeds in such a manner as to *partly* compensate for the increase in concentration of the $Fe^{3+}$.

Figure 8.2.6 shows graphically how the concentrations of the three species involved in this equilibrium change with the addition of $Fe^{3+}$. Note that the concentration of $Fe^{3+}$ increases instantaneously upon the addition of a few of drops of this reactant, but then its concentration decreases as the reaction proceeds back to equilibrium. The concentration of $Fe^{3+}$ is still greater at the new equilibrium position than it was at equilibrium before the addition.

Figure 8.2.5 The effect of adding $Fe^{3+}$ to an equilibrium mixture of $Fe^{3+}$, $SCN^-$ and $Fe(SCN)^{2+}$

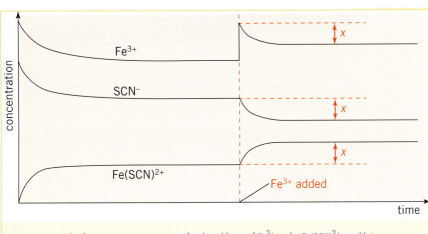

Figure 8.2.6 Graph of concentration versus time for the addition of $Fe^{3+}$ to the $Fe(SCN)^{2+}$ equilibrium system.

As the temperature has not changed, the equilibrium constant for this reaction remains the same, despite the fact that the concentrations of all three species involved have themselves changed. Mathematically, we can say that:

$$\frac{[Fe(SCN)^{2+}]}{[Fe^{3+}][SCN^-]} \text{ before the addition of } Fe^{3+} = \frac{[Fe(SCN)^{2+}]}{[Fe^{3+}][SCN^-]} \text{ once equilibrium is re-established.}$$

Addition of $SCN^-$ to another test tube will have precisely the same effect. Once again, the concentration of one of the reactants has increased and the reaction will proceed so as to compensate for the change.

**TABLE 8.2.2 THE EFFECT OF CONCENTRATION CHANGE ON EQUILIBRIUM POSITION AND EQUILIBRIUM CONSTANT VALUE**

| Change in concentration | of reactants | | of products | |
|---|---|---|---|---|
| | Position of equilibrium | Equilibrium constant, $K_c$ | Position of equilibrium | Equilibrium constant, $K_c$ |
| Increase | Moves to the right | Unchanged | Moves to the left | Unchanged |
| Decrease | Moves to the left | Unchanged | Moves to the right | Unchanged |

### Worked example 3

A test tube contains a small volume of the equilibrium system described below:

$Fe^{3+}(aq) + SCN^-(aq) \rightleftharpoons Fe(SCN)^{2+}(aq)$

A number of drops of silver nitrate solution ($AgNO_3$) is added and the $Ag^+$ ions react with the $SCN^-$ ions present to form an insoluble precipitate of AgSCN:

$Ag^+(aq) + SCN^-(aq) \rightleftharpoons AgSCN(s)$

**a** Predict what will happen to the colour of the equilibrium solution, given that the reactant species are effectively colourless and $Fe(SCN)^{2+}(aq)$ is an intense red colour.

**b** Sketch a graph of concentration versus time to show how the concentrations of $Fe^{3+}$, $SCN^-$ and $Fe(SCN)^{2+}$ would change with the addition of $AgNO_3$ and as the reaction proceeds to a new equilibrium position.

### Solution

**a** The thiocyanate ion ($SCN^-$) effectively has been removed from the equilibrium system, because it forms an insoluble precipitate. As a consequence, the reaction will proceed to the left (backwards) so as to partly compensate for this loss of reactant. The solution will become lighter in colour as the concentration of the $Fe(SCN)^{2+}$ decreases.

**b** See figure 8.2.7. Note that the decrease in concentration of $SCN^-$ caused by the reaction with $AgNO_3$ is only partly compensated for as the reaction proceeds to a new equilibrium position.

PRAC 8.2
The effect of concentration on the position of equilibrium

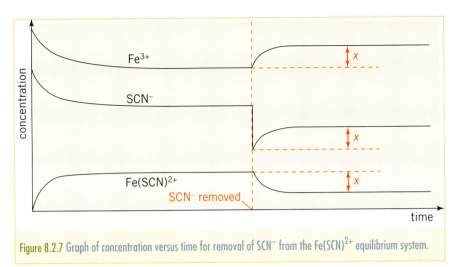

Figure 8.2.7 Graph of concentration versus time for removal of SCN$^-$ from the Fe(SCN)$^{2+}$ equilibrium system.

## Changes to physical equilibrium systems

The concept of concentration is difficult to achieve in physical equilibria since pure substances are involved. Consider two physical (heterogeneous) equilibria:

$$Br_2(l) \rightleftharpoons Br(g) \quad \text{and} \quad H_2O(s) \rightleftharpoons H_2O(l)$$

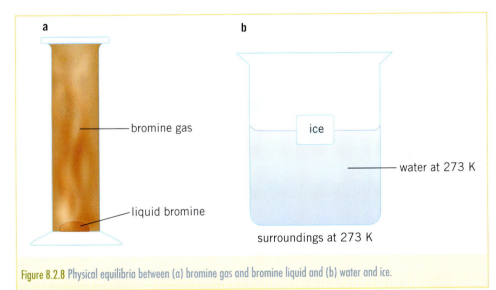

Figure 8.2.8 Physical equilibria between (a) bromine gas and bromine liquid and (b) water and ice.

When a small amount of liquid bromine is introduced into the gas jar containing bromine liquid and gas in equilibrium, the concentration of bromine does not change even though the quantity of Br$_2$ increases slightly. As a result there is no change to the position of equilibrium and no change in the depth of colour of the brown gas (Br$_2$(g)).

Similarly, if a small amount of water at 273 K is removed from the water/ice equilibrium system, although the level of the water will drop slightly, the concentration of H$_2$O(l) will not change, so there is no change to the position of equilibrium.

Physical equilibria obey Le Chatelier's principle in the same way as do chemical equilibria, however a change in concentration of pure substances cannot easily be achieved.

Note that a pure substance such as water is not included in the expression for a chemical equilibrium.

### CHEM COMPLEMENT

#### Dissolving gold

*Aqua regia*, or 'royal water', is the only substance capable of dissolving the exceedingly unreactive metals gold and platinum. It was named by the alchemists, who used it to etch gold, as it was able to mix with the 'noble' metals, which otherwise would react with nothing. Aqua regia is a corrosive, fuming, yellow liquid consisting of a mixture of one part concentrated nitric acid to three parts concentrated hydrochloric acid. While neither of these acids alone will dissolve gold, in combination the reaction is quite vigorous. Why? Nitric acid is a powerful oxidizing agent and oxidizes a small amount of gold to Au$^{3+}$—the equilibrium constant for the reaction is very small. Once formed, the Au$^{3+}$ reacts with chloride ions from the hydrochloric acid to produce the stable chloraurate complex ion (AuCl$_4^-$)—the equilibrium constant for the reaction is large. This second reaction consumes the available Au$^{3+}$ ions and forces the first oxidation reaction of gold to the right so as to compensate for the loss of a reactant. The combined effect is that the gold quickly dissolves. The relevant equations are:

$$Au(s) \rightleftharpoons Au^{3+}(aq) + 3e^-$$
$$Au^{3+}(aq) + 4Cl^-(aq) \rightleftharpoons AuCl_4^-(aq)$$

Saturated solution equilibrium

## Pressure (of gaseous systems)

The kinetic particle theory states that the particles that make up gases are moving randomly at high speed, independently of one another, and they completely fill any container into which they are placed. Gases exert pressure on the sides of the container as a result of their collisions with them. The more frequent the collisions and the greater their energy, the higher will be the gas pressure exerted. Gas pressure increases with an increase in temperature or a decrease in the volume of the container.

In a gaseous system at equilibrium, an increase in external pressure causes both the reactant and product gas particles to be compressed into a smaller volume. According to Le Chatelier's principle, the reaction will proceed so as to partly compensate for this increase in pressure. To achieve this, the reaction proceeds to the side of the reaction with the lower number of particles, since by reducing the overall number of gas particles the pressure will be reduced. Similarly, a decrease in external pressure will cause the reaction to proceed in such a manner as to build the pressure back up again; the reaction moves toward the side with the greater number of particles. If the equilibrium system has the same number of reactant gas particles as product gas particles, a change in pressure will have no effect whatsoever on the position of equilibrium.

**TABLE 8.2.3 THE EFFECT OF VOLUME/PRESSURE CHANGE ON EQUILIBRIUM POSITION AND EQUILIBRIUM CONSTANT VALUE**

| Change in pressure | Reactions where $n$(reactants) > $n$(products) | | Reactions where $n$(reactants) < $n$(products) | |
|---|---|---|---|---|
| | Position of equilibrium | Equilibrium constant, $K_c$ | Position of equilibrium | Equilibrium constant, $K_c$ |
| Increase | Moves forwards to the right | Unchanged | Moves backwards to the left | Unchanged |
| Decrease | Moves backwards to the left | Unchanged | Moves forwards to the right | Unchanged |

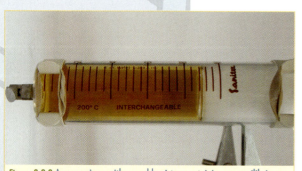

Figure 8.2.9 A gas syringe with movable piston containing an equilibrium mixture of $NO_2$ and $N_2O_4$.

QuickTime Video
The effect of a volume change

Consider again the gaseous system of $NO_2$ and $N_2O_4$ in equilibrium. $NO_2$ is a deep brown colour and $N_2O_4$ is colourless. The equation representing this reaction is:

$N_2O_4(g) \rightleftharpoons 2NO_2(g)$

colourless     brown

An equilibrium mixture of these two gases is held in a movable gas syringe. The movable piston within the cylinder of the syringe can be pushed in to compress the gas mixture, or drawn out to increase the volume and so reduce the pressure.

If the gas syringe is pushed in rapidly, the volume of the gas mixture is dramatically decreased, 'instantly' increasing the pressure. The colour of the gas mixture is darker, as the gas is occupying a smaller volume and so is more concentrated. The gas mixture is no longer in equilibrium. According to Le Chatelier's principle, the reaction will proceed to partially compensate for this change. In this case, the reaction will move in the direction of the lesser number of moles, as this will reduce the pressure.

According to the equation, 1 mol of $N_2O_4$ dissociates to 2 mol of $NO_2$, so the reaction will proceed backwards (to the left). The gas mixture will lighten in colour as a result of this change. Note that when the new equilibrium position has been established the gas mixture will still be darker than it was before the change, as the concentration of $NO_2$ has increased. This situation is represented graphically in figure 8.2.10.

What happens if the volume of the syringe is increased? With a volume increase, pressure decreases. The reaction proceeds to increase the gas pressure by proceeding to the side with the greater number of moles—it moves to the right. The gas mixture initially lightens with the increase in volume, and then darkens slightly as the reaction moves to a new equilibrium position. There is no change in the equilibrium constant associated with any of these volume changes.

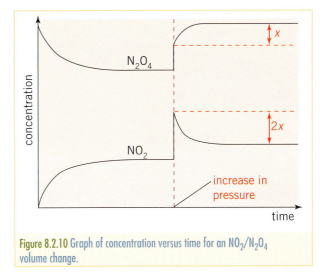

**Figure 8.2.10** Graph of concentration versus time for an $NO_2/N_2O_4$ volume change.

## Worked example 4

Engineers wish to increase the yield of product for each of the equilibrium systems described below. Ignoring all other factors, should the external pressure be increased or decreased to achieve this aim?

**a** $N_2(g) + 3H_2(g) \rightleftharpoons 2NH_3(g)$

**b** $2NH_3(g) + 4H_2O(g) \rightleftharpoons 2NO_2(g) + 7H_2(g)$

**c** $H_2(g) + Cl_2(g) \rightleftharpoons 2HCl(g)$

### Solution

**a** The mole ratio of reactants to products in this reaction is 4:2, so an increase in the external pressure will be required. According to Le Chatelier's principle, an increase in pressure will favour the side with fewer mole of particles (the products).

**b** The mole ratio of reactants to products in this reaction is 6:9, so a decrease in the external pressure will be required. According to Le Chatelier's principle, a decrease in pressure will favour the side with the greater number of particles (the products).

**c** In this reaction, the mole ratio of reactant to product is 2:2, so no advantage is to be gained by changing the external pressure.

## Dilution of aqueous solutions

The dilution of an aqueous equilibrium system by adding water is similar to the situation discussed in the increase of volume of a gas system. If water is added, the concentrations of all species will decrease. The position of equilibrium will shift in such a direction as to partly compensate for this change by proceeding to the side of the reaction with the greater number of mole of particles, as this will result in an overall increase in concentration.

## Worked example 5

For each of the following equilibrium systems, determine the direction in which the reaction will proceed if a volume of water is added to the reaction vessel.

**a** $Fe^{3+}(aq) + SCN^-(aq) \rightleftharpoons Fe(SCN)^{2+}(aq)$

**b** $Cu(NH_3)_4(H_2O)_2^{2+}(aq) \rightleftharpoons Cu(H_2O)_6^{2+}(aq) + 4NH_3(aq)$

**c** $KNO_3(aq) + NaCl(aq) \rightleftharpoons NaNO_3(aq) + KCl(aq)$

### Solution

**a** The mole ratio of reactants to products in this reaction is 2:1. According to Le Chatelier's principle, diluting with water will favour the side with the greater number of particles and so the reaction will proceed backwards (to the left).

**b** The mole ratio of reactants to products in this reaction is 1:5. Diluting with water in this case will favour the forward reaction (to the right).

**c** In this reaction, the mole ratio of reactant to product is 2:2, so the addition of water will have no effect on the equilibrium system.

**WORKSHEET 8.1**
Changing the position of equilibrium

**AS 7.2.4**
State and explain the effect of a catalyst on an equilibrium reaction. © IBO 2007

## The role of a catalyst

Recall that the role of a catalyst is to increase the rate of a reaction by providing an alternative pathway with lower activation energy. When a catalyst is used in a reaction, equilibrium is reached more quickly, but neither the equilibrium position nor the constant are affected in any way. Figure 8.2.11 illustrates how the decomposition reaction of nitrogen monoxide occurs more rapidly when a catalyst is used, even though the concentrations reached at equilibrium have not changed.

Rate and extent of reaction are two quite separate concepts. A reaction may proceed to a large extent, but proceed very slowly. An iron nail, or any piece of iron, is susceptible to reaction with oxygen from the air in the presence of water, so we say that the iron rusts. Under most circumstances, rusting is a quite slow process—it may take from weeks to years for an object made of iron to completely rust, but eventually it will. This combination of rate and extent of reaction requires special attention in industrial chemistry, as will be examined in the next section.

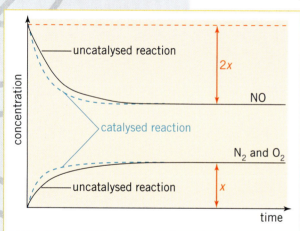

Figure 8.2.11 Graph of concentration versus time for the $N_2(g) + O_2(g) \rightleftharpoons 2NO(g)$ reaction.

## Section 8.2 Exercises

1. State the equilibrium constant expression for each of the following reactions.
   **a** $2Cu(NO_3)_2(s) \rightleftharpoons 2CuO(s) + 4NO_2(g) + O_2(g)$
   **b** $CO(g) + 2H_2(g) \rightleftharpoons CH_3OH(g)$
   **c** $H_2(g) + I_2(g) \rightleftharpoons 2HI(g)$
   **d** $I_3^-(aq) \rightleftharpoons I_2(aq) + I^-(aq)$

**2** Phosgene (carbonyl dichloride) decomposes to carbon monoxide and chlorine gas according to the equation:

$COCl_2(g) \rightleftharpoons CO(g) + Cl_2(g)$    $K_c = 6.37 \times 10^5$ mol dm$^{-3}$ at 50°C

State the equilibrium constant expression for the following reactions in terms of $K_c$.

**a** $2COCl_2(g) \rightleftharpoons 2CO(g) + 2Cl_2(g)$

**b** $\frac{1}{2}CO(g) + \frac{1}{2}Cl_2(g) \rightleftharpoons \frac{1}{2}COCl_2(g)$

**3 a** State Le Chatelier's principle.

**b** In which direction would an exothermic chemical reaction proceed if the temperature was increased?

**c** What would be the effect, if any, of the increase in temperature on the equilibrium constant for this reaction?

**4** For each of the following reactions, determine the effect of increasing the temperature on:

**i** the position of equilibrium

**ii** the equilibrium constant.

**a** $I_2(g) + Cl_2(g) \rightleftharpoons 2ICl(g)$    $\Delta H = +35$ kJ mol$^{-1}$

**b** $2SO_2(g) + O_2(g) \rightleftharpoons 2SO_3(g)$    $\Delta H = -92.3$ kJ mol$^{-1}$

**c** $CO_2(g) + C(s) \rightleftharpoons 2CO(g)$    $\Delta H = +173$ kJ mol$^{-1}$

**d** $Ag^+(aq) + 2NH_3(aq) \rightleftharpoons Ag(NH_3)_2^+(aq)$    $\Delta H = -111$ kJ mol$^{-1}$

**5** Consider the graph of concentration versus time provided in the figure below for the reaction:

$2H_2(g) + O_2(g) \rightleftharpoons 2H_2O(g)$    $\Delta H = -484$ kJ mol$^{-1}$

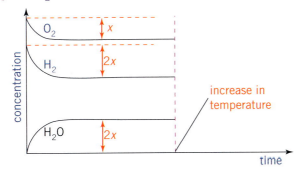

Copy the graph and show what changes would occur in the concentrations of all three species when the temperature is increased.

**6** State the direction in which the equilibrium position would move on the addition of chlorine gas for each of the following reactions.

**a** $Cl_2(g) + 3F_2(g) \rightleftharpoons 2ClF_3(g)$

**b** $4HCl(g) + O_2(g) \rightleftharpoons 2H_2O(l) + Cl_2(g)$

**c** $C_3H_8(g) + Cl_2(g) \rightleftharpoons C_3H_7Cl(g) + HCl(g)$

**d** $2HCl(g) \rightleftharpoons Cl_2(g) + H_2(g)$

7 Some liquid boron trichloride is added to a vessel containing hydrogen gas. The substances react and reach equilibrium according to the following equation:

$2BCl_3(l) + 6H_2(g) \rightleftharpoons B_2H_6(g) + 6HCl(g)$   $\Delta H = +315$ kJ mol$^{-1}$

What would be the effect on the position of equilibrium of each of the following changes?

a Extra hydrogen gas is added.

b The vessel is heated.

c HCl gas is removed.

8 Hypochlorous acid is employed as the active ingredient in many bleaches, where it reacts according to the equation:

$2HOCl(aq) + 2H_2O(l) \rightleftharpoons 2H_3O^+(aq) + 2Cl^-(aq) + O_2(g)$

State the effect on the position of equilibrium of the following changes.

a Water is added to dilute the mixture.

b Oxygen gas bubbles out of solution into the atmosphere.

c Concentrated hypochlorous acid is added.

d Some sodium hydroxide is added.

9 Iron pyrite (FeS$_2$) can be used as a source of sulfur dioxide in the production of sulfuric acid. The iron pyrite is 'roasted' in air according to the equation:

$4FeS_2(s) + 11O_2(g) \rightleftharpoons 8SO_2(g) + 2Fe_2O_3(s)$   $\Delta H = -3642$ kJ mol$^{-1}$

Discuss whether each of the following measures would ensure that the yield of sulfur dioxide from this process is increased.

a Use an excess amount of air (O$_2$).

b Employ a catalyst.

c Supply extra heat.

10 Consider the following graph of concentration versus time for the reaction:

$2SO_2(g) + O_2(g) \rightleftharpoons 2SO_3(g)$

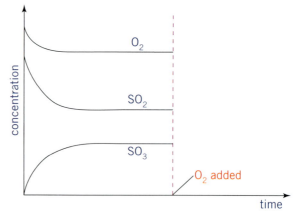

Copy the graph and show what changes would occur in the concentrations of all three species when some extra oxygen is added to the reaction vessel.

# 8.3 INDUSTRIAL PROCESSES

We rely on industrial processes to provide us with the chemicals we need for everyday life. Ideally these chemicals should be provided cheaply; however, chemical equilibrium sometimes stands in the way of simply being able to add two reactants together and produce 100% products. Maintaining a useful balance between rate and extent of reaction produces a challenge for industrial chemists who meet it by using Le Chatelier's principle (see section 8.2) together with the principles of collision theory (see chapter 7).

 7.2.5
**Apply the concepts of kinetics and equilibrium to industrial processes.** © IBO 2007

## The Haber process

Ammonia provides the characteristic pungent smell of household cleaning agents—but what else can it be used for? Ammonia has a range of applications and is one of the most widely produced inorganic chemicals in the world. In 2005, world production of this chemical was 130 million tonnes, with approximately 85% of this being used in fertilizer manufacture. A large proportion of the ammonia produced is first oxidized to nitric acid via the Ostwald process. About 500 million tonnes of fertilizer are produced using ammonia, including ammonia itself, urea, ammonium sulfate and ammonium nitrate.

Before World War I, much of the world's ammonia was produced by the distillation of plant and animal waste products, including camel dung! A more reliable source of raw material was the nitrate salts from the deserts of Chile. Potassium nitrate (Chilean saltpetre) was imported by Germany until the Allied forces cut off its supply leading up to World War I. As ammonia is a raw material for many explosives, there was a need to manufacture large quantities of ammonia from synthetic raw materials. The German chemists Fritz Haber (1868–1934) and Carl Bosch (1874–1940) developed such a process, which became known as the Haber–Bosch process or, more simply, the **Haber process**.

Ammonia is produced industrially by the Haber process (figure 8.3.1). Elevated temperatures are used to increase the rate of reaction and higher pressures are used to increase both the extent and rate of the reaction:

$$N_2(g) + 3H_2(g) \rightleftharpoons 2NH_3(g) \qquad \Delta H = -92.4 \text{ kJ mol}^{-1}$$

This is an example of a reaction for which both rate and equilibrium considerations are important. According to Le Chatelier's principle, an increased yield can be achieved by:

- increasing the concentration of one of the reactants
- removing product from the reaction mixture as it forms
- increasing the gas pressure (4 mol of reactant gases produce only 2 mol of product gases, so an increase in pressure would enhance the forward reaction)
- decreasing the temperature, as the forward reaction is exothermic.

Pressures of between 100 and 350 atm are generally used—a compromise between the optimal yield and the expense and safety considerations of using high-pressure equipment—as increasing the pressure enhances both the rate and extent of the reaction. While it would be cheaper to carry out the conversion at atmospheric pressure, the yield of product at this pressure is simply uneconomic, so in practice, most plants use a pressure of around 200 atm.

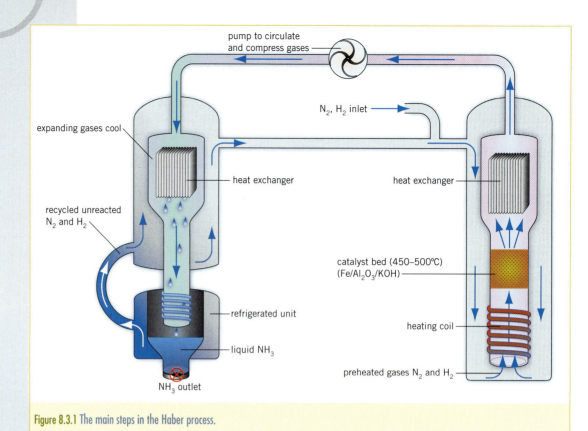

**Figure 8.3.1** The main steps in the Haber process.

As this is an exothermic reaction, low temperatures should be used to increase the value of the equilibrium constant and the extent of forward reaction (table 8.3.1). At room temperature, the reaction proceeds effectively to completion and has a large equilibrium constant. Why then is the industrial process carried out at elevated temperatures?

**TABLE 8.3.1 THE EFFECT OF VOLUME/PRESSURE CHANGE ON EQUILIBRIUM POSITION AND EQUILIBRIUM CONSTANT VALUE**

| Temperature (°C) | Equilibrium constant ($K_c$) |
| --- | --- |
| 25 | 640 |
| 200 | 0.44 |
| 300 | 0.0043 |
| 400 | 0.00016 |
| 500 | 0.000015 |

To understand this apparent contradiction, we must consider the rate of reaction. Recall that for most reactions, rate increases with temperature. Thus, low temperatures ensure that a significant yield of ammonia is formed, but very slowly, while at high temperatures, the equilibrium yield is poor, but it forms quickly. A compromise in temperature must be reached to achieve acceptable yields as quickly as possible. A temperature of around 400°C to 450°C for the incoming gases is appropriate.

To increase the rate of reaction further a cheap, but effective, catalyst, a porous form of iron, is used.

As the gaseous mixture is passed over the catalyst, about 15% of the mixture is converted to ammonia. The product is cooled to about −50°C, which liquefies the ammonia and allows it to be easily removed. The gas mixture is then recycled across the catalyst several more times until a yield of around 98% is achieved. Removing the ammonia shifts the equilibrium to the right to partly compensate for the loss.

Table 8.3.2 summarizes the conditions that are used in the Haber process. Notice how the concepts of kinetics and equilibrium are combined in manipulating the conditions to produce the maximum amount of product at the highest possible rate.

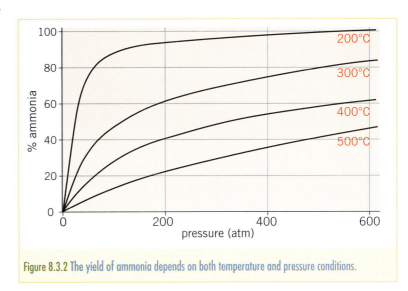

Figure 8.3.2 The yield of ammonia depends on both temperature and pressure conditions.

**TABLE 8.3.2 SUMMARY OF CONDITIONS FOR THE HABER PROCESS**

|  | Temperature | Concentration (gas pressure) | Catalyst |
|---|---|---|---|
| Rapid rate | High | High | Most effective available High surface area (finely divided) |
| High yield (large equilibrium constant) | Low | High for reactants Remove products as formed | No effect |
| Actual conditions used | Low | Low | Cheapest catalyst available |
| Actual conditions used | 400–450°C | 200 atm | Porous iron pellets with $Al_2O_3$ and KOH |

Figure 8.3.3 A Haber plant for producing ammonia.

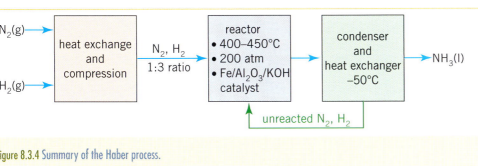

Figure 8.3.4 Summary of the Haber process.

**THEORY OF KNOWLEDGE**

In 1912 Fritz Haber was appointed Director of the Kaiser Wilhelm Institute of Physical Chemistry in Berlin. During World War I the institute was placed in the service of the German military and became involved in research into the use of poison gas during battle. Haber supervised the first deployment of chlorine gas on the Western Front at Ypres, Belgium, in 1915. The use of phosgene and mustard gas, both chlorine compounds soon followed. The gas gave the German's a tactical advantage and it wasn't long before the British starting using poisonous gas. After the war Haber continued as director and the institute was recognized for many outstanding scientific achievements. One achievement was the development of hydrogen cyanide gas, for use as an insecticide. It was later used in World War II to exterminate millions of Jews in the Nazi concentration camps. In 1933, with the rise of Hitler's Nazi party many Jewish staff at the institute lost their positions. Haber submitted his resignation and fled to Switzerland where he died in 1934.

Ethics is a study of how our values affect the choices we make. For example, what does it mean to be a good scientist, and why do scientists act the way they do in certain situations. Immoral choices are those we personally don't approve of, that we think are wrong, whereas moral choices are those we do approve of or agree with. In 1918 Haber was awarded the Nobel Prize in Chemistry for the synthesis of ammonia. At the time there was disagreement in the scientific community about whether Haber should have received the prize because many believed he used his knowledge in an immoral way.

- Discuss possible reasons why some members of the scientific community did not approve of the decision to award Haber the Nobel Prize in Chemistry?
- Fritz Haber is quoted as saying 'A scientist belongs to his country in times of war and to all mankind in times of peace'. How do you feel about Haber's claim?
- What questions does the following claim raise? 'Scientists should be held responsible for the applications of discoveries that decrease the quality of human health and wellbeing.'
- Make a list of scientific research you consider to be morally unacceptable and a list of research that is morally required. Can you justify your choices to others? What role did your emotions, personal and cultural identity, beliefs and religion play in the formation of your list?
- During World War I the political context of the time affected the type of scientific research undertaken at the Kaiser Wilhelm Institute. How does the social or political context of scientific work affect the type of scientific research undertaken today?

## The Contact process

Sulfuric acid is the most widely produced synthetic chemical in the world and, indeed, is generally accepted as a good indicator of the economic health of a nation. Cheap to produce and easily stored in steel drums, world production exceeds 165 million tonnes! Australia produces approximately 3.8 million tonnes per year, 70% of which is used in the manufacture of phosphate fertilizers such as superphosphate. Other uses include the removal of rust and scale from the surface of metals, production of paints and synthetic dyes, manufacture of explosives, soaps and detergents, and use in polymer industries.

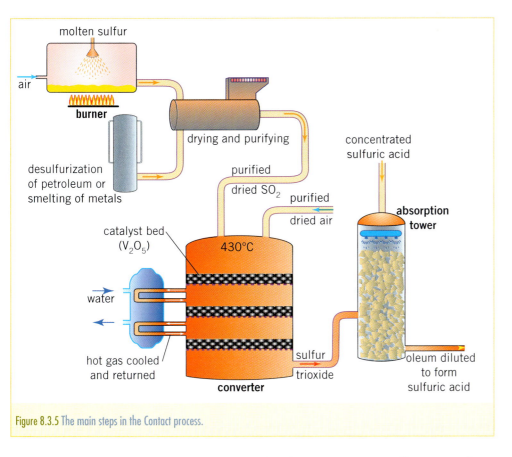

Figure 8.3.5 The main steps in the Contact process.

Sulfuric acid is produced commercially by the **Contact process** (figure 8.3.5). In the first step $SO_2$ is made when sulfur is burnt in air. The second stage:

$$2SO_2(g) + O_2(g) \rightleftharpoons 2SO_3(g) \qquad \Delta H = -197 \text{ kJ mol}^{-1}$$

provides another example of a reaction for which both rate and equilibrium considerations are important. According to Le Chatelier's principle, an increased yield can be achieved by:

- increasing the concentration of one of the reactants
- removing product from the reaction mixture as it forms
- increasing the gas pressure (3 mol of reactant gases produce only 2 mol of product gases, so an increase in pressure would enhance the forward reaction)
- decreasing the temperature, as the forward reaction is exothermic.

A moderate excess of air is used to enhance the forward reaction. At some plants, extra oxygen is added to increase the reactant concentration. The process is carried out at atmospheric pressures because the reaction can proceed almost to completion without the extra expense of building high-pressure containment vessels. As this is an exothermic reaction, low temperatures should increase the value of the equilibrium constant and the extent of forward reaction. At room temperature, the reaction proceeds effectively to completion and has a large equilibrium constant, but the rate of reaction is low. Thus, low temperatures ensure that a significant yield of sulfur trioxide is formed, but slowly, while at high temperatures the equilibrium yield is poor, but it forms quickly. A compromise in temperature must be reached to achieve acceptable yields quickly. A temperature of approximately 440°C for the incoming gases is appropriate.

The catalyst used in this equilibrium is vanadium(V) oxide ($V_2O_5$). It is laid out in a series of three or four trays in pellet form. By using trays of pellets, the catalyst surface area is increased, and so is its effectiveness. The procedure is called the 'contact' process because the reactant gases come into contact with the catalyst as they pass through. The length of time of contact is carefully controlled.

As the $SO_2$–air mixture makes its first pass through the catalyst beds, approximately 75% of the mixture is converted to $SO_3$. As the reaction is exothermic, the temperature of the gas mixture rises to approximately 600°C. At this temperature, the $SO_3$ would quickly decompose to $SO_2$ and $O_2$:

$$2SO_3(g) \rightleftharpoons 2SO_2(g) + O_2(g) \quad \Delta H = +197 \text{ kJ mol}^{-1}$$

To minimize this unwanted reaction, the gas mixture is rapidly cooled between passes through the catalyst beds using heat exchangers. The excess heat is used to heat the incoming gas mixture, thus reducing the overall cost of production. The gas mixture ($SO_3$ and unreacted $SO_2$ and $O_2$) is then cycled through the absorption tower, which is filled with previously generated 98% sulfuric acid. In the absorption tower, the $SO_3$ reacts to produce oleum, according to the equation:

$$SO_3(g) + H_2SO_4(l) \rightarrow H_2S_2O_7(l)$$

The oleum is then carefully reacted with water to produce 98% sulfuric acid:

$$H_2S_2O_7(l) + H_2O(l) \rightarrow 2H_2SO_4(l)$$

Why go to the extra trouble and expense of producing oleum and then diluting, when $SO_3$ readily undergoes a hydrolysis reaction?

$$SO_3(g) + H_2O(l) \rightarrow H_2SO_4(aq)$$

Although this process would be more straightforward and hence seem favourable, it is highly exothermic and would result in the sulfuric acid being vaporized by the evolved heat until its eventual condensation into tiny droplets.

Figure 8.3.6 A plant for producing sulfuric acid.

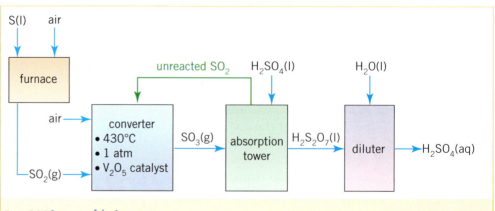

Figure 8.3.7 Summary of the Contact process.

Table 8.3.3 summarizes the conditions that are used in the Contact process. As in the case of the Haber process, the concepts of kinetics and equilibrium are combined in manipulating the conditions to produce the maximum amount of product at the highest possible rate.

**TABLE 8.3.3 SUMMARY OF CONDITIONS FOR THE CONTACT PROCESS**

|  | Temperature | Concentration (gas pressure) | Catalyst |
| --- | --- | --- | --- |
| Rapid rate | High | High | Most effective available High surface area |
| High yield (large equilibrium constant) | Low | High for reactants Remove products as formed | No effect |
| Minimal costs | Low | Low | Cheapest catalyst available |
| Actual conditions used | 400°C | Atmospheric pressure | Porous pellets of vanadium(V) oxide |

**WORKSHEET 8.2**
Industrial processes

## Section 8.3 Exercises

1 Describe the catalyst used in the Haber process. What characteristics does it possess that makes it suitable for this application?

2 Use Le Chatelier's principle to describe the theoretical conditions that should be used to maximize the yield of the reaction of nitrogen and hydrogen to produce ammonia. Are these the conditions actually used? If not, why not?

3 Describe the effect that the catalyst used in the Haber process has on
   a the position of equilibrium (the extent) of the reaction
   b the rate of the reaction (the kinetics).

4 Describe the catalyst used in the Contact process. What characteristics does it possess that makes it suitable for this application?

5 Use Le Chatelier's principle to describe the theoretical conditions that should be used to maximize the yield of the reaction of sulfur dioxide and oxygen gases to produce sulfur trioxide. Are these the conditions actually used? If not, why not?

## Chapter 8 Summary

**Terms and definitions**

**Catalyst**  A substance that increases the rate of reaction without being consumed in the reaction.

**Closed system**  A reaction mixture in a container that is sealed and does not allow reactants or products to escape to the surroundings.

**Contact process**  The industrial process in which sulfuric acid is made in a number of stages starting with the reaction between sulfur and oxygen.

**Dynamic equilibrium**  A state of balance in which the forward and the backward reactions are still progressing, but macroscopic properties are constant.

**Equilibrium constant, $K_c$**  The temperature-dependent value of the equilibrium constant expression.

**Equilibrium constant expression**  The fraction formed by the concentrations of products raised to the powers of their coefficients divided by the concentrations of reactants raised to the powers of their coefficients. For reaction $aA + bB + ... \rightleftharpoons pP + qQ + ...$,

$$K_c = \frac{[P]^p [Q]^q ...}{[A]^a [B]^b ...}$$

**Extent of reaction**  The degree to which products are made.

**Haber process**  The industrial process in which nitrogen and hydrogen react together to make ammonia.

**Heterogeneous**  In different physical states.

**Homogeneous**  In the same physical state.

**Le Chatelier's principle**  If a change is made to a system at equilibrium, the reaction will proceed in such a direction as to partially compensate for this change.

**Macroscopic properties**  The properties of a reaction that can be measured such as concentration, pressure, temperature, pH.

**Open system**  A reaction mixture in a container that allows products to escape to the surroundings.

**Concepts**

- All chemical reactions within a closed system involve a state of balance, or dynamic equilibrium, between the rate of forward and reverse reactions.

- At equilibrium, both reactions continue to proceed, even though macroscopic properties are constant and nothing may appear to be happening.

- The extent of reaction can be measured by the size of the equilibrium constant $(K_c)$, where:
  - if $K_c \gg 1$, the reaction goes almost to completion
  - if $K_c \ll 1$, the reaction hardly proceeds.

- Equilibrium constants are temperature dependent, but pressure and concentration changes have no effect on the constant.

- The position of equilibrium can be manipulated by changes in temperature, pressure (for gases) and concentration.

- Le Chatelier's principle is very useful in determining how the position of equilibrium can be changed to ensure more product is formed.

- If a system is at equilibrium and the temperature, pressure or concentrations of the species are changed, the reaction will proceed in such a direction as to partially compensate for this change.

- For both endothermic and exothermic reactions, a change in temperature alters both the value of the equilibrium constant and its position.

| Temperature change | Exothermic reaction | |
|---|---|---|
| | Position of equilibrium | Equilibrium constant, $K_c$ |
| Increase | Moves to the left | Decreases |
| Decrease | Moves to the right | Increases |

| Temperature change | Endothermic reaction | |
|---|---|---|
| | Position of equilibrium | Equilibrium constant, $K_c$ |
| Increase | Moves to the right | Increases |
| Decrease | Moves to the left | Decreases |

- Changes in concentration and changes in pressure for gaseous systems will alter the equilibrium position but not the value of the equilibrium constant.

- Addition of a reactant moves the position of equilibrium forward, producing more product. The value of the equilibrium constant is unchanged.

- Increase in pressure for a gaseous system moves the position of equilibrium to the side of the equation that has fewer moles of substance, and hence lower pressure. The value of the equilibrium constant is unchanged.

- The addition of a catalyst does not change the position of equilibrium or the value of the equilibrium constant. Equilibrium is simply reached more quickly.
- In both the Haber process and the Contact process, a compromise is reached between the optimum conditions for a fast rate and those for a high yield of products.

## Chapter 8 Review questions

1. Write equilibrium constant expressions for each of the following equations.
   a. $2SO_2(g) + O_2(g) \rightleftharpoons 2SO_3(g)$
   b. $P_4(s) + 6H_2(g) \rightleftharpoons 4PH_3(g)$
   c. $2LiOH(aq) + CO_2(g) \rightleftharpoons Li_2CO_3(aq) + H_2O(l)$
   d. $4NH_3(g) + 7O_2(g) \rightleftharpoons 4NO_2(g) + 6H_2O(g)$

2. For each of the following equations, state whether the formation of reactants or products would be most favoured. All values of $K_c$ are measured at 25°C.
   a. $N_2(g) + O_2(g) \rightleftharpoons 2NO(g)$
   $$K_c = 10^{-31}$$
   b. $Ag^+(aq) + 2NH_3(aq) \rightleftharpoons Ag(NH_3)_2^{2+}(aq)$
   $$K_c = 1.6 \times 10^7 \text{ mol}^{-2} \text{ dm}^6$$
   c. $PCl_5(g) \rightleftharpoons PCl_3(g) + Cl_2(g)$
   $$K_c = 6.4 \times 10^{-5} \text{ mol dm}^{-3}$$
   d. $2NO(g) + Cl_2(g) \rightleftharpoons 2NOCl(g)$
   $$K_c = 4.6 \times 10^4 \text{ mol}^{-1} \text{ dm}^3$$

3. Two experiments are performed which involve the oxidation of sulfur dioxide to sulfur trioxide according to the following equation:
   $$2SO_2(g) + O_2(g) \rightleftharpoons 2SO_3(g) \quad \Delta H = -197 \text{ kJ mol}^{-1}$$

   |  | Experiment 1 | Experiment 2 |
   | --- | --- | --- |
   | $[SO_2]_{eq}$ (mol dm$^{-3}$) | 0.0831 | 0.362 |
   | $[O_2]_{eq}$ (mol dm$^{-3}$) | 0.161 | 0.819 |
   | $[SO_3]_{eq}$ (mol dm$^{-3}$) | 0.855 | 0.433 |
   | $K_c$ | 657.5 | 1.746 |

   a. State the equilibrium constant expression for this reaction.
   b. Which experiment was conducted at the higher temperature? Explain your answer.

4. Consider the general equation:
   $2P + Q \rightleftharpoons R + 3S \quad K_c = x$
   If the magnitude of the equilibrium constant for this reaction is $x$, determine the constant for each of the following reactions in terms of $x$.
   a. $R + 3S \rightleftharpoons 2P + Q$
   b. $4P + 2Q \rightleftharpoons 2R + 6S$
   c. $3R + 9S \rightleftharpoons 6P + 3Q$

5. Ammonia and hydrogen react according to the equation:
   $N_2(g) + 3H_2(g) \rightleftharpoons 2NH_3(g)$
   $$K_c = 0.650 \text{ mol}^{-2} \text{ dm}^6 \text{ at } 200°C$$
   Would the value of the equilibrium constant increase, decrease or remain constant with an increase in temperature to 600°C, given that the forward reaction is exothermic?

6. For each of the following reactions, determine the effect of decreasing the temperature on:
   i. the position of equilibrium
   ii. the equilibrium constant.
   a. $2CO(g) + O_2(g) \rightleftharpoons 2CO_2(g)$
   $$\Delta H = -564 \text{ kJ mol}^{-1}$$
   b. $2BCl_3(l) + 6H_2(g) \rightleftharpoons B_2H_6(g) + 6HCl(g)$
   $$\Delta H = +315 \text{ kJ mol}^{-1}$$
   c. $N_2(g) \rightleftharpoons 2N(g)$
   $$\Delta H = +946 \text{ kJ mol}^{-1}$$
   d. $PCl_3(g) + Cl_2(g) \rightleftharpoons PCl_5(g)$
   $$\Delta H = -88 \text{ kJ mol}^{-1}$$

7. State the direction in which the equilibrium position would move on the removal of carbon dioxide gas for each of the following reactions:
   a. $CaCO_3(s) \rightleftharpoons CaO(s) + CO_2(g)$
   b. $2CO(g) + NH_3(g) \rightleftharpoons HCN(g) + CO_2(g) + H_2(g)$
   c. $Ca(OH)_2(aq) + CO_2(g) \rightleftharpoons CaCO_3(s) + H_2O(l)$
   d. $H_2(g) + CO_2(g) \rightleftharpoons H_2O(g) + CO(g)$

8. Ammonia is produced via the Haber process according to the following equation:
   $N_2(g) + 3H_2(g) \rightleftharpoons 2NH_3(g) \quad \Delta H = -92.3 \text{ kJ mol}^{-1}$
   Complete the table below to indicate the effect of the changes listed on the equilibrium position and constant. Use the words *increases*, *decreases*, *left* or *right* to indicate your answers.

   | Change | Equilibrium position | Equilibrium constant |
   | --- | --- | --- |
   | Add nitrogen gas | | |
   | Cool gases | | |
   | Increase pressure | | |
   | Remove hydrogen | | |
   | Employ a catalyst | | |

9. A sample of NO(g) is introduced into a vessel and allowed to come to equilibrium according to the equation:
$$3NO(g) \rightleftharpoons N_2O(g) + NO_2(g) \quad \Delta H = -155 \text{ kJ mol}^{-1}$$
Predict the effect on the equilibrium position of the following changes.
   a  Extra $N_2O$ is added to the container.
   b  The vessel is heated.
   c  Some unreactive argon gas is pumped into the container.
   d  Some $NO_2$ is removed from the reaction.
   e  A catalyst is employed.

10. Consider the graph below of concentration versus time for the reaction:
$$2NO(g) + O_2(g) \rightleftharpoons 2NO_2(g)$$

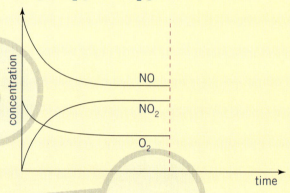

Copy the graph and show what changes would occur in the concentrations of all three species when some extra nitrogen dioxide ($NO_2$) is added to the reaction vessel.

11. What would be the effect on the position of equilibrium for the reaction in each of the following cases:
$$Zn^{2+}(aq) + 4NH_3(aq) \rightleftharpoons Zn(NH_3)_4^{2+}(aq) \quad \Delta H < 0$$
   a  A small volume of $NH_3$ is added.
   b  A small amount of $Zn^{2+}$ is removed.
   c  The system is diluted by the addition of deionized water.
   d  The system is heated on a water bath.

12. A gaseous mixture of chlorine and carbon monoxide is placed into an evacuated vessel and allowed to come to equilibrium according to the equation:
$$CO(g) + Cl_2(g) \rightleftharpoons COCl_2(g) \quad \Delta H > 0$$
A series of different events occurs, and the concentrations of the gases are measured and recorded on the graph below.

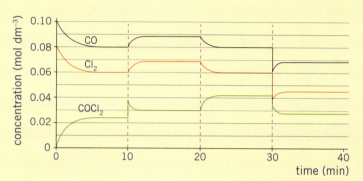

   a  How many times does the reaction system reach equilibrium over the period of time that the experiment is conducted?
   b  What event occurred at the 10-minute mark?
   c  In which direction was the reaction proceeding between 10 and 15 minutes?
   d  State the equilibrium expression for this reaction.
   e  State what event occurred at the 30-minute mark. In which direction did the reaction proceed? Explain your answer

13. a  State the conditions of temperature and pressure that favour a high yield in the Contact process:
$$2SO_2(g) + O_2(g) \rightleftharpoons 2SO_3(g) \quad \Delta H = -197 \text{ kJ mol}^{-1}$$
    b  Identify other factors that are used to raise the yield of $SO_3$ in the Contact process.

14. Explain why a compromise must be reached in order to produce $SO_3$ industrially.

15. State the chemical that is produced in the Haber process, and write the equation and the equilibrium expression for the reaction.

 www.pearsoned.com.au/schools

Weblinks are available on the Chemistry: For use with the IB Diploma Programme Standard Level Companion Website to support learning and research related to this chapter.

# Chapter 8 Test

## Part A: Multiple-choice questions

1. Chemical equilibrium is referred to as **dynamic** because, at equilibrium, the:
   - A equilibrium constant changes.
   - B reactants and products keep reacting.
   - C rates of the forward and backward reactions change.
   - D concentrations of the reactants and products continue to change.

   © IBO 1999, Nov P1 Q21

2. Which statement is true about chemical reactions at equilibrium?
   - A The forward and backward reactions proceed at equal rates.
   - B The forward and backward reactions have stopped.
   - C The concentrations of the reactants and products are equal.
   - D The forward reaction is exothermic.

   © IBO 2000, May P1 Q21

3. $2H_2O(l) \rightleftharpoons H_3O^+(aq) + OH^-(aq)$

   The equilibrium constant for the reaction above is $1.0 \times 10^{-14}$ at 25°C and $2.1 \times 10^{-14}$ at 35°C. What can be concluded from this information?
   - A $[H_3O^+]$ decreases as the temperature is raised.
   - B $[H_3O^+]$ is greater than $[OH^-]$ at 35°C.
   - C Water is a stronger electrolyte at 25°C.
   - D The ionization of water is endothermic.

   © IBO 2000, May P1 Q22

4. Which change will shift the position of equilibrium to the right in this reaction?

   $N_2(g) + 3H_2(g) \rightleftharpoons 2NH_3(g)$   $\Delta H = -92$ kJ
   - A Increasing the temperature
   - B Decreasing the pressure
   - C Adding a catalyst
   - D Removing ammonia from the equilibrium mixture

   © IBO 2001, May P1 Q20

5. Which statement(s) is(are) correct about the effect of adding a catalyst to a system at equilibrium?
   - I The rate of the forward reaction increases.
   - II The rate of the reverse reaction increases.
   - III The yield of the products increases.
   - A I only
   - B III only
   - C I and II only
   - D I, II and III

   © IBO 2001, Nov P1 Q22

6. The reaction
   $2NO_2(g) \rightleftharpoons N_2O_4(g)$
   is exothermic. Which of the following could be used to shift the equilibrium to the right?
   - I Increasing the pressure
   - II Increasing the temperature
   - A I only
   - B II only
   - C Both I and II
   - D Neither I nor II

   © IBO 2002, May P1 Q22

7. For a gaseous reaction, the equilibrium constant expression is:

   $K_c = \dfrac{[O_2]^5[NH_3]^4}{[NO]^4[H_2O]^6}$

   Which equation corresponds to this equilibrium expression?
   - A $4NH_3 + 5O_2 \rightleftharpoons 4NO + 6H_2O$
   - B $4NO + 6H_2O \rightleftharpoons 4NH_3 + 5O_2$
   - C $8NH_3 + 10O_2 \rightleftharpoons 8NO + 12H_2O$
   - D $2NO + 3H_2O \rightleftharpoons 2NH_3 + \dfrac{5}{2}O_2$

   © IBO 2002, May P1 Q21

8. Which changes will shift the position of equilibrium to the right in the following reaction?

   $2CO_2(g) \rightleftharpoons 2CO(g) + O_2(g)$
   - I Adding a catalyst
   - II Decreasing the oxygen concentration
   - III Increasing the volume of the container
   - A I and II only
   - B I and III only
   - C II and III only
   - D I, II and III

   © IBO 2005, Nov P1 Q21

9 The equation for a reaction used in the manufacture of nitric acid is

$$4NH_3(g) + 5O_2(g) \rightleftharpoons 4NO(g) + 6H_2O(g)$$
$$\Delta H^\ominus = -900 \text{ kJ}$$

Which changes occur when the temperature of the reaction is increased?

|   | Position of equilibrium | Value of $K_c$ |
|---|---|---|
| A | Shifts to the left | Increases |
| B | Shifts to the left | Decreases |
| C | Shifts to the right | Increases |
| D | Shifts to the right | Decreases |

© IBO 2006, May P1 Q22

10 What is the magnitude of the equilibrium constant, $K_c$, for a reversible reaction which goes almost to completion?

**A** $K_c = 1$
**B** $K_c = 0$
**C** $K_c \gg 1$
**D** $K_c \ll 1$

© IBO 2006, Nov P1 Q22

(10 marks)

  www.pearsoned.com.au/schools

For more multiple-choice test questions, connect to the Chemistry: For use with the IB Diploma Programme Standard Level Companion Website and select Chapter 8 Review Questions.

## Part B: Short-answer questions

1 Ammonia is made on a large scale by the Haber process. The main reaction occurring is

$$N_2(g) + 3H_2(g) \rightleftharpoons 2NH_3(g) \quad \Delta H = -92 \text{ kJ mol}^{-1}$$

a State **two** characteristics of a reversible reaction at equilibrium.

(2 marks)

b The reaction is described as homogeneous. State what is meant by the term *homogeneous*.

(1 mark)

c Write the equilibrium constant expression for the reaction.

(2 marks)

d Name the catalyst used in the Haber process. State and explain its effect on the value of the equilibrium constant.

(3 marks)

© IBO 2002, May P2 Q5a-c,e

2 For the reversible reaction:

$$H_2(g) + I_2(g) \rightleftharpoons 2HI(g) \quad \Delta H > 0$$

the equilibrium constant $K_c = 60$ at a particular temperature.

a Give the equilibrium expression and explain why the equilibrium constant has no units.

(2 marks)

b For this reaction, what information does the value of $K_c$ provide about the **relative** concentrations of the products and reactants at equilibrium?

(1 mark)

c What effect, if any, will an increase in pressure have on the **equilibrium position**?

(1 mark)

d Explain why an increase in temperature increases the value of the **equilibrium constant** for the above reaction.

(1 mark)

© IBO 2000, Nov P2 Q3

3 When nitrogen and hydrogen are mixed together at room temperature and atmospheric pressure the reaction is very slow. In industry, typical values of pressure and temperature used are 250 atmospheres and 450°C.

a State the effects on both the rate of reaction and the value of the equilibrium constant of increasing the temperature.

(2 marks)

b State the effects on both the rate of reaction and the value of the equilibrium constant of increasing the pressure.

(2 marks)

c Suggest why a pressure of 1000 atmospheres is not used.

(1 mark)

© IBO 2002, May P2 Q5d

## Part C: Data-based question

Information about some reactions used in industry is shown in the following table:

| Reaction | Equation | $\Delta H^\ominus$ (kJ$^{-1}$) |
|---|---|---|
| A | $H_2(g) + Cl_2(g) \rightarrow 2HCl(g)$ | −184 |
| B | $CH_4(g) + H_2O(g) \rightleftharpoons 3H_2(g) + CO(g)$ | +210 |
| C | $CO(g) + H_2O(g) \rightleftharpoons H_2(g) + CO_2(g)$ | −42 |
| D | $CaCO_3(s) \rightleftharpoons CaO(s) + CO_2(g)$ | +180 |
| E | $nC_2H_4(g) \rightarrow (-CH_2-CH_2-)_n(s)$ | −92 |

**a** Identify, with a reason, which of the reactions **A** to **E** is/are

  **i** the **two** in which an increase in temperature shifts the position of equilibrium to the right.
  (2 marks)

  **ii** the **two** in which an increase in pressure shifts the position of equilibrium to the left.
  (2 marks)

**b** Many reversible reactions in industry use a catalyst. State and explain the effect of a catalyst on the position of equilibrium and on the value of $K_c$.
(4 marks)

© IBO 2006, Nov P2 Q6ai,ii,b

## Part D: Extended-response question

Consider the following reaction in the Contact process for the production of sulfuric acid for parts **a**–**d** in this question.

$$2SO_2 + O_2 \rightleftharpoons 2SO_3$$

**a**  **i** State the catalyst used in this reaction of the Contact process.
(1 mark)

  **ii** State and explain the effect of the catalyst on the value of the equilibrium constant and on the rate of the reaction.
(4 marks)

**b** Use the collision theory to explain why increasing the temperature increases the rate of the reaction between sulfur dioxide and oxygen.
(2 marks)

**c** Using Le Chatelier's principle state and explain the effect on the position of equilibrium of:

  **i** increasing the pressure at constant temperature.
(2 marks)

  **ii** removing sulfur trioxide.
(2 marks)

  **iii** using a catalyst.
(1 mark)

**d** Using the following data, explain whether the above reaction is exothermic or endothermic.
(2 marks)

| Temperature (K) | Equilibrium constant $K_c$ (dm$^3$ mol$^{-1}$) |
|---|---|
| 298 | $9.77 \times 10^{25}$ |
| 500 | $8.61 \times 10^{11}$ |
| 700 | $1.75 \times 10^{6}$ |

© IBO 2006, May P2 Q6a–e
Total marks = 50

# 9 ACIDS AND BASES

## Chapter overview

This chapter covers the IB Chemistry syllabus Topic 8: Acids and Bases.

**By the end of this chapter, you should be able to:**

- name and give the formula and uses of some common acids and bases
- identify whether a substance is acting as a Brønsted–Lowry acid, base or neither in a given chemical reaction
- identify whether a substance is acting as a Lewis acid, base or neither in a given chemical reaction
- explain the meaning of the terms *conjugate pair*, *polyprotic* and *amphiprotic* when applied to acids
- identify the distinguishing properties of acids and bases
- write formulas and ionic equations for reactions involving acids and bases
- explain the meaning of the terms *strong*, *weak*, *concentrated* and *dilute* when applied to acidic and basic solutions
- use the pH scale as a measure of the acidity or basicity of a solution
- recognize that a change of 1 unit on the pH scale is equal to a 10-fold change in concentration of $H^+$ in the acid or base.

**IBO Assessment statements 8.1.1 to 8.4.4**

Many of the most familiar chemicals we meet every day are either acids or bases. They are important in many of the chemical reactions occurring around us and within our bodies. Acids and bases are used widely in agriculture, industry and the laboratory. An understanding of the properties of acids and bases is therefore necessary in our study of chemistry.

## 9.1 THEORIES OF ACIDS AND BASES

*Acids* (from the Latin *acidus* meaning 'sour') have been known for hundreds of years. Many of the foods we eat contain acids. An acid gives vinegar (containing ethanoic acid) its tang. The sour taste of certain fruits such as lemons and grapefruit is due to acids (citric acid and ascorbic acid), and soft drinks owe their 'fizz' to an acid (carbonic acid). The stomach produces gastric juice containing hydrochloric acid, which provides an acidic environment for the breakdown of certain foods.

Bases are the chemical opposite of acids. They are used in a variety of cleaning products such as oven, drain and window cleaners. Common household and industrial bases include sodium hydroxide (caustic soda), sodium carbonate, hydrated sodium carbonate and ammonia. Bases such as magnesium hydroxide and aluminium hydroxide are used in various antacid mixtures for the relief of upset stomachs.

Tables 9.1.1 and 9.1.2 give the names, formulas and uses of some common acids and bases.

Figure 9.1.1 Many common household substances contain acids.

Figure 9.1.2 Many common household substances contain bases.

### TABLE 9.1.1 NAMES, FORMULAS AND USES OF SOME COMMON ACIDS

| Name of acid | Chemical formula | Use or occurrence |
|---|---|---|
| **Inorganic acids** | | |
| Hydrochloric acid (spirits of salt) | HCl | Concrete cleaner and stomach acid |
| Nitric acid | $HNO_3$ | Fertilizer manufacture, dyes and explosives |
| Phosphoric acid (orthophosphoric acid) | $H_3PO_4$ | Fertilizer manufacture, food additive |
| Sulfuric acid (oil of vitriol) | $H_2SO_4$ | Car batteries and in the manufacture of fertilizers |
| **Organic acids** | | |
| Acetylsalicylic acid | $C_9H_8O_4$ | Aspirin |
| Ascorbic acid (vitamin C) | $C_6H_8O_6$ | Citrus fruits |
| Benzoic acid | $C_6H_5COOH$ | Food preservative |
| Carbonic acid | $H_2CO_3$ | Soft drinks and rainwater |
| Citric acid | $C_6H_8O_7$ | Lemons |
| Ethanoic (acetic) acid | $CH_3COOH$ | Vinegar |
| Lactic acid | $C_3H_6O_3$ | Milk and human muscle during strenuous exercise |
| Methanoic acid (formic) | HCOOH | Ant stings and stinging nettles |

CHEMISTRY: FOR USE WITH THE IB DIPLOMA PROGRAMME **STANDARD LEVEL**

Introduction to bases

WORKSHEET 9.1
Common acids and bases crossword

### TABLE 9.1.2 NAMES, FORMULAS AND USES OF SOME COMMON BASES

| Name of base | Chemical formula | Use or occurrence |
|---|---|---|
| **Inorganic bases** | | |
| Ammonium hydroxide | $NH_4OH$ | Window cleaner |
| Barium hydroxide | $Ba(OH)_2$ | Used as a strong base in organic syntheses |
| Calcium hydroxide (slaked lime) | $Ca(OH)_2$ | Plaster and cement, and to neutralize soil acidity |
| Calcium oxide (quicklime) | $CaO$ | Mortar for bricks, and to neutralize soil acidity |
| Hydrated sodium carbonate | $Na_2CO_3.10H_2O$ | Washing soda |
| Magnesium hydroxide | $Mg(OH)_2$ | Antacid |
| Sodium carbonate | $Na_2CO_3$ | Washing powders and in the manufacture of glass |
| Sodium hydrogen carbonate (sodium bicarbonate) | $NaHCO_3$ | Baking soda |
| Sodium hydroxide (caustic soda) | $NaOH$ | Drain and oven cleaners, and in soap production |
| **Organic bases** | | |
| Ammonia | $NH_3$ | Fish urine |
| Methylamine | $CH_3NH_2$ | Agricultural chemicals |

### CHEM COMPLEMENT

#### The Arrhenius theory of acids and bases

Before Brønsted, Lowry and Lewis, a definition of acids had been suggested by Svante Arrhenius, a Swedish chemist and one of the founders of physical chemistry. In his doctoral dissertation in 1883 he discussed ionic solutions and explained that charge was carried in salt solutions by ions that were present even in the absence of an electric current (ionic compounds). His work was groundbreaking but was given a low grade by the university examiners. Fortunately his work was regarded much more respectfully by scientists throughout Europe and he was able to use these contacts to further his career.

In 1884 Arrhenius extended his theory of ions to acids, and described acids as substances that contained hydrogen and yielded hydrogen ions in aqueous solution; bases contained the OH group and yielded hydroxide ions in aqueous solution.

In 1903 Svante August Arrhenius became the first Swede to be awarded the Nobel Prize in chemistry.

Figure 9.1.3 Svante Arrhenius (1859–1927), Nobel Prize winning Swedish chemist, at work in his laboratory.

### Brønsted–Lowry acids and bases

In 1923, Johannes Brønsted in Denmark and Thomas Lowry in England each independently developed a general model of acids and bases. In the Brønsted–Lowry theory an acid–base reaction is one that involves the transfer of a hydrogen ion ($H^+$) from one species to another. The substance that donates the hydrogen ion is an acid, while the hydrogen ion acceptor is the base. A substance cannot act as an acid unless a base is present.

**AS 8.1.1**
Define *acids* and *bases* according to the Brønsted–Lowry and Lewis theories.
© IBO 2007

**A Brønsted–Lowry acid is a hydrogen ion (proton) donor.**

**A Brønsted–Lowry base is a hydrogen ion (proton) acceptor.**

Most hydrogen atoms consist of one proton and one electron. Most hydrogen ions, $H^+$, therefore consist of one proton only. When describing the Brønsted–Lowry theory, the terms *hydrogen ion* and *proton* are both used. It is important to remember, however, that the 'proton' being referred to is just a hydrogen nucleus, not a proton plucked from the centre of another atom!

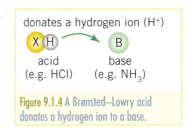

Figure 9.1.4 A Brønsted–Lowry acid donates a hydrogen ion to a base.

A hydrogen ion has no electrons and is therefore very small. Because the hydrogen ion's single positive charge is located in such a small volume, the attraction between a polar water molecule and a hydrogen ion is greater than for other ions. The hydrogen ion is in fact more likely to exist as a *hydronium* ($H_3O^+$) *ion*. Hydrogen ions produced in aqueous solution may be represented as $H^+$, $H_3O^+$ or $H^+(aq)$. Although $H_3O^+$ is a more accurate representation, $H^+$ is commonly used for convenience.

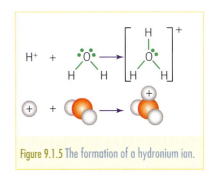

In the Brønsted–Lowry model, the reaction of HCl with water is represented by the equation:

$$HCl(g) + H_2O(l) \rightarrow H_3O^+(aq) + Cl^-(aq)$$

Figure 9.1.5 The formation of a hydronium ion.

This reaction is known as **dissociation** or **ionization** because ions are formed when the two molecular substances react, which then separate (dissociate). The reaction may also simply be termed *hydrolysis*, meaning 'reaction with water'. In this process the HCl donates a proton and is therefore acting as an acid. The $H_2O$, which accepts a proton, is classified as a base.

When ammonia dissolves in water it reacts as shown by the equation:

$$NH_3(aq) + H_2O(l) \rightleftharpoons NH_4^+(aq) + OH^-(aq)$$

In terms of the Brønsted–Lowry theory, the $NH_3$ accepts a proton and is acting as a base. The $H_2O$ donates a proton and is acting as an acid.

Aqueous acids and bases

## Lewis acidity

Also in the early 1920s, Gilbert Lewis (1875–1946), the same American chemist after whom the Lewis structures of molecules are named, hypothesized that acid–base behaviour is not tied to water or restricted to protons. Lewis' theory describes an acid as an electron-pair acceptor, and a base as an electron-pair donor. This requires a **Lewis base** to have a non-bonding pair of electrons and a **Lewis acid** to be electron deficient in some way. Unlike **Brønsted–Lowry acids**, Lewis acids do not have to have an H in their formula and include $BF_3$, $SO_3$, $H^+$ and $SbF_5$. Ammonia is both a **Brønsted–Lowry base** and a Lewis base due to its pair of non-bonding electrons. Other Lewis bases include $H_2O$, $OH^-$ and $F^-$.

DEMO 9.1
An everyday acid

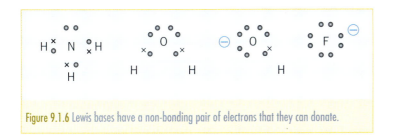

Figure 9.1.6 Lewis bases have a non-bonding pair of electrons that they can donate.

**A Lewis acid is an electron pair acceptor.**

**A Lewis base is an electron pair donor.**

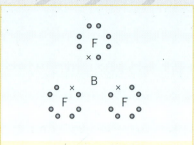

**Figure 9.1.7** BF$_3$ is electron deficient because there are only six electrons in the valence shell of boron, even when it has formed three covalent bonds.

Lewis acid–base theory

**AS 8.1.2**
Decide whether or not a species could act as a Brønsted–Lowry and/or a Lewis acid or base.
© IBO 2007

The donation of an electron pair by a Lewis base to a Lewis acid creates a covalent bond between the two species. This covalent bond is known as a **dative** or **coordinate bond** because the two electrons being shared in the bond have come from the same species. Lewis acids are often electron deficient. They do not have a full outer shell of electrons and they are very reactive towards any electron pair donor. Boron trifluoride, BF$_3$, is a good example of a Lewis acid. This compound is electron deficient as there are only three shared pairs of electrons between the boron atom and the fluorine atoms. There are no non-bonding pairs of electrons on the boron atom.

The equation for the reaction between the Lewis acid BF$_3$ and the Lewis base NH$_3$ can be written as

$$BF_3 + NH_3 \rightarrow BF_3NH_3$$

Lewis acid + Lewis base → Lewis acid–base complex

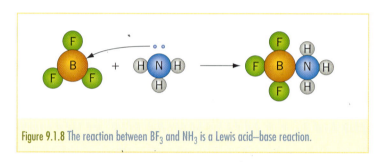

**Figure 9.1.8** The reaction between BF$_3$ and NH$_3$ is a Lewis acid–base reaction.

The Lewis definition allows for many reactions that would not otherwise be classified as acid–base reactions, as well as encompassing acid–base reactions under the Brønsted–Lowry definition.

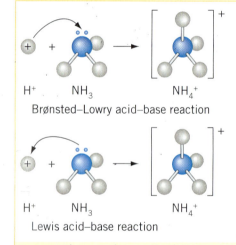

**Figure 9.1.9** The reaction between H$^+$ and NH$_3$ can be regarded as (a) Brønsted–Lowry or (b) Lewis acid–base reaction.

### Worked example 1

Classify each of the following substances as either Brønsted–Lowry acids or bases or Lewis acids or bases.

H$_2$O, BF$_3$, OH$^-$, H$_2$SO$_4$

Solution

| Substance | H$_2$O | BF$_3$ | OH$^-$ | H$_2$SO$_4$ |
|---|---|---|---|---|
| Brønsted–Lowry acid or base? | Can be both an acid and a base. | Is neither an acid nor a base. | Brønsted–Lowry base | Brønsted–Lowry acid |
| Reason | Can donate and accept H$^+$. | Has no H to donate and cannot accept H$^+$ due to lack of non-bonding pair of electrons. | Can accept H$^+$ due to lack of non-bonding pair of electrons. | Can donate H$^+$ as there are two Hs in the molecule which are bonded to O (polar bond). |
| Lewis acid or base? | Lewis base | Lewis acid | Lewis base | Neither |
| Reason | Non-bonding pair of electrons on O can be donated. | Boron has only 6 electrons in valence shell, so can accept an electron pair. | Non-bonding pair of electrons on O can be donated. | Cannot accept or donate electron pairs. (Already more than 4 electron pairs around S, but no non-bonding pairs.) |

## Amphiprotic substances

In the Brønsted–Lowry theory, many substances can act as acids or bases. In the examples given, water acts as a base in its reaction with HCl and as an acid in its reaction with NH$_3$.

$$HCl(g) + H_2O(l) \rightarrow H_3O^+(aq) + Cl^-(aq)$$
acid   base

$$NH_3(aq) + H_2O(l) \rightleftharpoons NH_4^+(aq) + OH^-(aq)$$
base   acid

This is also shown in the equation for the **self-ionization** of water.

$$H_2O(l) + H_2O(l) \rightleftharpoons H_3O^+(aq) + OH^-(aq)$$

In this reaction, one water molecule donates a proton to the other. The proton donor molecule is the acid and the proton acceptor is the base. In a similar way, the hydrogen carbonate ion (HCO$_3^-$) can act as an acid or a base.

$$HCO_3^-(aq) + OH^-(aq) \rightarrow CO_3^{2-}(aq) + H_2O(l)$$
acid

$$HCO_3^-(aq) + H^+(aq) \rightarrow H_2CO_3(aq)$$
base

In the first reaction, the hydrogen carbonate ion is acting as an acid; in the second it is acting as a base. Substances that can act as either a Brønsted–Lowry acid or a Brønsted–Lowry base are said to be **amphiprotic**; they can be proton acceptors or proton donors. (The term *amphoteric* is also used to describe a substance that can act as both acid and base.)

The hydrogen sulfate ion (HSO$_4^-$) is another amphiprotic substance. When it is added to a basic solution it behaves like an acid. In a highly **acidic solution** it can behave like a base.

$$HSO_4^-(aq) + OH^-(aq) \rightarrow SO_4^{2-}(aq) + H_2O(l)$$
acid

$$HSO_4^-(aq) + H^+(aq) \rightarrow H_2SO_4(aq)$$
base

PRAC 9.1
Acids, bases and amphiprotic substances

Self-ionization of water

How do $HCO_3^-$ and $HSO_4^-$ act when placed in water? We cannot predict this; we must find out by experiment. When $HCO_3^-$ is added to water the solution is basic. This indicates that it is acting as a base, accepting a proton from $H_2O$. When $HSO_4^-$ is added to water the solution is acidic. This indicates that it is acting as an acid, donating a proton to $H_2O$.

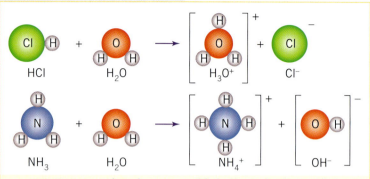

Figure 9.1.10 Water acts as a base when reacting with hydrochloric acid, and as an acid when reacting with ammonia.

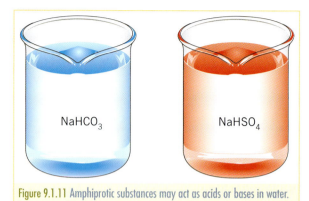

Figure 9.1.11 Amphiprotic substances may act as acids or bases in water. Litmus turns blue in aqueous $NaHCO_3$ and red in aqueous $NaHSO_4$.

### THEORY OF KNOWLEDGE

Arrhenius' theory was simple but somewhat limited because it could not explain why acids and bases not in an aqueous solution do not produce hydrogen and hydroxide ions.

The Brønsted–Lowry proton transfer theory was an extension of Arrhenius' theory and could account for the reactions that were not in solution. For example, the reaction between ammonia and hydrogen chloride was considered a Brønsted–Lowry acid–base reaction because ammonia acts as a base (a proton acceptor) and hydrogen chloride acts as an acid (a proton donor).

$NH_3(g) + HCl(g) \rightarrow NH_4^+(s) + Cl^-(s)$
base     acid

As useful as the Brønsted–Lowry definition was, it could not explain how or why these reactions occurred. This was soon changed by Lewis, who went one step further and proposed a mechanism or series of steps by which acids and bases react. Lewis' theory complemented the Brønsted–Lowry theory and extended the range of acids and bases to include those that did not have hydrogen in their formula.

In the Lewis theory, ammonia is a base because it has a non-bonding pair of electrons that it can donate to hydrogen, forming a dative covalent bond.

Lewis used curly arrows in the mechanism to show the movement of the pairs of electrons.

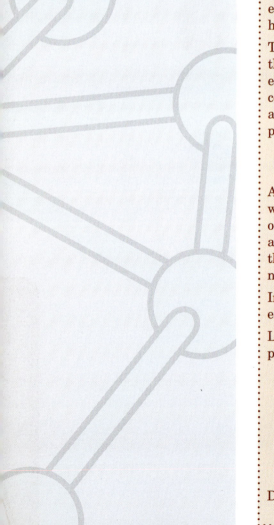

Discuss the value of having three different theories to explain acids and bases.

## Conjugate acid–base pairs

In the Brønsted–Lowry theory, a base, after it has received a proton, has the potential to react as an acid. Similarly, an acid that has donated a proton is a potential base. In the equation shown below, the $CH_3COOH$ is acting as an acid.

$$CH_3COOH(aq) + H_2O(l) \rightarrow CH_3COO^-(aq) + H_3O^+(aq)$$
acid          base          conjugate base          conjugate acid

The $CH_3COO^-$ can, under certain conditions, accept a proton. For example, in reaction with $H_3O^+$ the $CH_3COO^-$ accepts a proton and is therefore acting as a base.

$$CH_3COO^-(aq) + H_3O^+(aq) \rightarrow CH_3COOH(aq) + H_2O(l)$$

The $CH_3COOH$–$CH_3COO^-$ pair is described as a **conjugate acid–base** pair. Hydrogen chloride and the chloride ion, $HCl$–$Cl^-$, constitute another conjugate acid–base pair.

$$HCl(g) + H_2O(l) \rightarrow Cl^-(aq) + H_3O^+(aq)$$
acid          base          conjugate base          conjugate acid

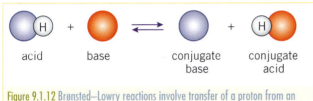

Figure 9.1.12 Brønsted–Lowry reactions involve transfer of a proton from an acid to a base, forming a conjugate acid–base pair.

Notice that an acid has one more proton than its conjugate base. Some conjugate acid–base pairs are shown in figure 9.1.13.

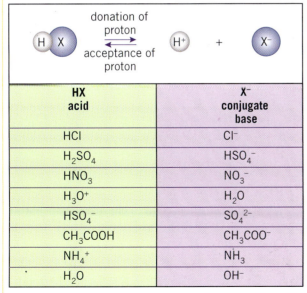

| HX acid | X⁻ conjugate base |
|---|---|
| $HCl$ | $Cl^-$ |
| $H_2SO_4$ | $HSO_4^-$ |
| $HNO_3$ | $NO_3^-$ |
| $H_3O^+$ | $H_2O$ |
| $HSO_4^-$ | $SO_4^{2-}$ |
| $CH_3COOH$ | $CH_3COO^-$ |
| $NH_4^+$ | $NH_3$ |
| $H_2O$ | $OH^-$ |

Figure 9.1.13 Conjugate acid–base pairs are molecules or ions that differ by one proton ($H^+$).

**AS 8.1.3**
**Deduce the formula of the conjugate acid (or base) of any Brønsted–Lowry base (or acid).**
© IBO 2007

Classifying acids and bases

**Section 9.1 Exercises**

1. Write equations for the reaction with water of:
   a. perchloric acid ($HClO_4$)
   b. the base methylamine ($CH_3NH_2$).

2. a. Identify the conjugate acids of $Br^-$, $SO_4^{2-}$, $NH_3$, $ClO_4^-$ and $HCO_3^-$.
   b. Identify the conjugate bases of $HI$, $HPO_4^{2-}$, $NH_4^+$, $HSO_4^-$ and $H_3O^+$.

3. Write an equation for the hydrogen sulfate ion ($HSO_4^-$) reacting with water as:
   a. a base
   b. an acid.

4. Write two equations showing the amphiprotic nature of the hydrogen carbonate ion ($HCO_3^-$).

5. For the following reactions identify the conjugate acid–base pairs.
   a. $HCO_3^-(aq) + Cl^-(aq) \rightarrow CO_3^{2-}(aq) + HCl(aq)$
   b. $HSO_4^-(aq) + NH_3(aq) \rightarrow SO_4^{2-}(aq) + NH_4^+(aq)$
   c. $HCl(aq) + H_2O(l) \rightarrow H_3O^+(aq) + Cl^-(aq)$

6. For each of the following reactions indicate whether the first reactant is acting as an acid, a base or neither.
   a. $CH_3COO^-(aq) + H_3O^+(aq) \rightarrow CH_3COOH(aq) + H_2O(l)$
   b. $NH_3(aq) + H_2O(l) \rightarrow NH_4^+(aq) + OH^-(aq)$
   c. $2H_2O(l) + 2Na(s) \rightarrow 2NaOH(aq) + H_2(g)$
   d. $HPO_4^{2-}(aq) + H_2O(l) \rightarrow H_2PO_4^-(aq) + OH^-(aq)$
   e. $HSO_4^-(aq) + NH_3(aq) \rightarrow SO_4^{2-}(aq) + NH_4^+(aq)$

7. Explain, with the aid of an equation, why solutions of sulfuric acid can conduct electricity.

8. A Venn diagram is a useful way of showing relationships. A Venn diagram for acids and bases is shown here.

   Copy this diagram and place the following in their correct locations on the diagram.

   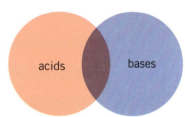

   $Cl^-$, $CO_3^{2-}$, $NH_3$, $SO_4^{2-}$, $HF$, $HSO_4^-$, $NH_4^+$, $HPO_4^{2-}$, $H_3O^+$

## 9.2 PROPERTIES OF ACIDS AND BASES

**AS 8.2.1**
Outline the characteristic properties of acids and bases in aqueous solution. ©IBO 2007

Acids may taste sour, but how do they behave chemically? We know that in aqueous solution a Brønsted–Lowry acid will donate an $H^+$ to water, but how does the reaction then proceed?

There are a number of characteristic reactions of acids with which you should become familiar. These are listed in table 9.2.1, together with the chemical properties of acids.

### TABLE 9.2.1 PROPERTIES AND REACTIONS OF ACIDS

**Properties of acids**

- Usually taste sour
- Turn blue litmus red
- Are corrosive
- Are molecular in structure
- Conduct electricity in aqueous solution

**Reactions of acids**

| | |
|---|---|
| 1 | React with most metals (not Cu, Ag or Hg) to produce a salt and hydrogen gas<br>e.g. $2HCl(aq) + Mg(s) \rightarrow MgCl_2(aq) + H_2(g)$ |
| 2 | React with metal hydroxides to produce a salt and water<br>e.g. $2HCl(aq) + Mg(OH)_2(s) \rightarrow MgCl_2(aq) + 2H_2O(l)$ |
| 3 | React with metal oxides to produce a salt and water<br>e.g. $2HCl(aq) + MgO(s) \rightarrow MgCl_2(aq) + H_2O(l)$ |
| 4 | React with metal carbonates to produce a salt, water and carbon dioxide<br>e.g. $2HCl(aq) + MgCO_3(s) \rightarrow MgCl_2(aq) + H_2O(l) + CO_2(g)$ |
| 5 | React with metal hydrogen carbonates to produce a salt, water and carbon dioxide<br>e.g. $2HCl(aq) + Mg(HCO_3)_2(s) \rightarrow MgCl_2(aq) + 2H_2O(l) + 2CO_2(g)$ |

Bases are commonly used in the home for cleaning purposes. One of the reasons for this is their ability to emulsify (break up), and therefore clean away fats and oils. Bases that dissolve in water are called **alkalis**. The term *caustic* is often used to describe a strongly **alkaline solution**. Caustic solutions are just as dangerous as strongly acidic solutions. They too can burn the skin and should be treated with extreme caution.

The chemical properties of bases are listed in table 9.2.2, together with some of their characteristic reations.

### TABLE 9.2.2 PROPERTIES AND REACTIONS OF BASES

**Properties of bases**

- Taste bitter
- Turn red litmus blue
- Feel slippery (Strong bases convert oils in your fingers to soap, hence they feel slippery.)
- Conduct electricity in aqueous solution

A solution of a base in water is usually called an **alkali**.

**Reactions of bases**

| | |
|---|---|
| 1 | Alkalis react with acids to produce a salt and water<br>e.g. $NaOH(aq) + HCl(aq) \rightarrow NaCl(aq) + H_2O(l)$ |
| 2 | Metal oxides react with acids to produce a salt and water<br>e.g. $MgO(s) + 2HCl(aq) \rightarrow MgCl_2(aq) + H_2O(l)$ |
| 3 | Metal carbonates react with acids to produce a salt, water and carbon dioxide<br>e.g. $Na_2CO_3(s) + 2HCl(aq) \rightarrow 2NaCl(aq) + H_2O(l) + CO_2(g)$ |
| 4 | Metal hydrogen carbonates react with acids to produce a salt, water and carbon dioxide<br>e.g. $NaHCO_3(s) + HCl(aq) \rightarrow NaCl(aq) + H_2O(l) + CO_2(g)$ |
| 5 | Ammonia reacts with acids to produce an ammonium salt<br>e.g. $NH_3(aq) + HCl(aq) \rightarrow NH_4Cl(aq)$ |

In chapter 3, section 3.3 we saw that some elements burn in air to produce oxides that then can react with water to make acidic or basic solutions. Alkali metals (group 1) react directly with water, producing the group 1 oxide and hydrogen gas. For example, recall the reaction of potassium with water:

$$2K(s) + H_2O(l) \rightarrow K_2O(aq) + H_2(g)$$

Further reaction of the oxide with water makes an alkaline solution:

$$K_2O(s) + H_2O(l) \rightarrow 2KOH(aq)$$

The reactions of the period 3 oxides with water are summarized in table 9.2.3.

PRAC 9.2
Reactions of acids

### TABLE 9.2.3 FORMATION OF ACIDIC OR ALKALINE SOLUTIONS BY PERIOD 3 OXIDES

| Element | Reaction | Nature of aqueous solution |
| --- | --- | --- |
| Sodium | $Na_2O(s) + H_2O(l) \rightarrow 2NaOH(aq)$ | Alkaline |
| Magnesium | $MgO(s) + H_2O(l) \rightarrow Mg(OH)_2(aq)$ | Alkaline |
| Aluminium | $Al_2O_3(s) + 6HCl(aq) \rightarrow 2AlCl_3(aq) + 3H_2O(l)$<br>$Al_2O_3(s) + 2NaOH(aq) + 3H_2O(l) \rightarrow 2NaAl(OH)_4(aq)$ | Amphoteric |
| Phosphorus | $P_4O_{10}(s) + 6H_2O(l) \rightarrow 4H_3PO_4(aq)$ | Acidic |
| Sulfur | $SO_3(g) + H_2O(l) \rightarrow H_2SO_4(l)$ | Acidic |

## Ionic equations

In many chemical reactions there are particles (usually ions) that do not participate in the actual reaction. Like spectators at a football match, who watch but do not participate in the match, these ions are called **spectator ions**. An ionic equation focuses on the part of the reaction that is actually undergoing chemical change: those atoms, molecules or ions whose bonds are breaking and which are forming new bonds. In addition these particles are changing state. Spectator ions are omitted from an ionic equation.

Ionic equations can be useful in accurately representing acid–base reactions. Consider the reaction between hydrochloric acid and sodium hydroxide.

Full equation: $HCl(aq) + NaOH(aq) \rightarrow NaCl(aq) + H_2O(l)$

In solution, HCl completely ionizes (it is a **strong acid**) and sodium hydroxide dissociates (it is a soluble salt). We can list all the species present.

Species present:
$H^+(aq) + Cl^-(aq) + Na^+(aq) + OH^-(aq) \rightarrow Na^+(aq) + Cl^-(aq) + H_2O(l)$

Neutralization reactions

The spectator ions (the ions that are in the same state on both sides of the equation) are $Na^+(aq)$ and $Cl^-(aq)$. Omitting these spectator ions gives the ionic equation.

Ionic equation: $H^+(aq) + OH^-(aq) \rightarrow H_2O(l)$

WORKSHEET 9.2
Reactions of acids and bases

### TABLE 9.2.4 IONIC EQUATIONS FOR SOME TYPICAL ACID–BASE REACTIONS

| Full equation | Ionic equation |
| --- | --- |
| $2HCl(aq) + Mg(s) \rightarrow MgCl_2(aq) + H_2(g)$ | $2H^+(aq) + Mg(s) \rightarrow Mg^{2+}(aq) + H_2(g)$ |
| $2HCl(aq) + Mg(OH)_2(s) \rightarrow MgCl_2(aq) + 2H_2O(l)$ | $H^+(aq) + OH^-(aq) \rightarrow H_2O(l)$ |
| $2HCl(aq) + MgO(s) \rightarrow MgCl_2(aq) + H_2O(l)$ | $2H^+(aq) + O^{2-}(aq) \rightarrow H_2O(l)$ |
| $2HCl(aq) + Na_2CO_3(aq) \rightarrow 2\rightarrow NaCl(aq) + H_2O(l) + CO_2(g)$ | $2H^+(aq) + CO_3^{2-}(aq) \rightarrow H_2O(l) + CO_2(g)$ |
| $HCl(aq) + NaHCO_3(aq) \rightarrow NaCl(aq) + H_2O(l) + CO_2(g)$ | $H^+(aq) + HCO_3^-(aq) \rightarrow H_2O(l) + CO_2(g)$ |

# Indicators

Acid–base indicators are substances that change colour to show the acidity or basicity of a solution. Many of the acid–base indicators used in laboratories are made from pigments extracted from the leaves and flowers of plants. These indicators are simply chemical substances that change colour depending on whether they are in acidic or basic solutions. They are often complex acidic organic molecules where the colour of the acid form is different from the colour of the base form. Litmus, a well known acid–base indicator, is a blue dye extracted from various species of lichens. Litmus is red in acidic solutions and blue in basic solutions. Indicators change colour over different acidity ranges; therefore, different indicators can be used in different situations. Some common indicators and their colour changes are shown in figure 9.2.2. Universal indicator is a mixture of several indicators. It undergoes a series of colour changes over a range of pH values.

A simple use of indicators is to classify substances as acidic, basic or **neutral**. The acidity or basicity of a substance can be determined by observing the colour change of an indicator as it is added to the substance being tested. Applications of indicators include their use in pool test kits to monitor the acidity levels in swimming pools, and by horticulturalists and farmers to test the acidity of soil so they can make adjustments to provide optimum conditions for plant growth. For example, if the soil is too acidic, chemicals such as lime (calcium oxide) are added to increase the soil pH.

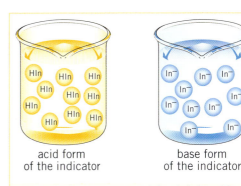

Figure 9.2.1 Indicators are compounds for which the acid form (HIn) is a different colour from the base form (In⁻). Bromothymol blue is blue in bases and yellow in acids.

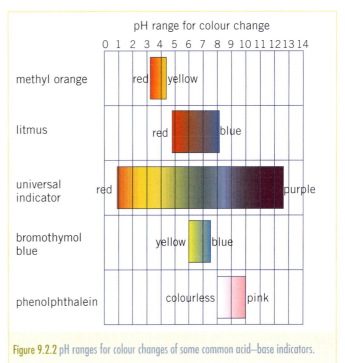

Figure 9.2.2 pH ranges for colour changes of some common acid–base indicators.

## CHEM COMPLEMENT

### Blue or pink? Acid or base?

Hydrangeas are common flowering bushes found in many gardens. When grown in soils that are naturally acidic they produce blue flowers; if the soil is basic the flowers are pink. By altering the acidity or basicity of the soil, gardeners can change the colour of these flowers. This occurs because the flowers of hydrangeas contain a chemical that changes colour in acidic or basic conditions.

Figure 9.2.3 The colour of hydrangeas depends on the acidity of the soil.

Natural indicators

**Section 9.2 Exercises**

1. Compare the terms *base* and *alkali*.

2. The word *acid* is often associated with danger and corrosiveness. Explain, using examples, whether all acids are dangerous and whether it is only concentrated acids that are dangerous.

3. State the balanced chemical equations for the reactions that occur when the following chemicals are mixed.
   a Aluminium and hydrofluoric acid (HF(aq))
   b Solid potassium hydrogen carbonate and sulfuric acid
   c Iron(III) oxide and nitric acid
   d Calcium hydroxide solution and hydrochloric acid
   e Ammonia and nitric acid

4. Describe two different tests you could perform on a dilute solution to determine whether it is acidic or basic.

5. Use figure 9.2.2 to answer the following questions.
   a Bromothymol blue, phenolphthalein and methyl orange are three common acid–base indicators used in school laboratories. State what colour each of these indicators would be in a solution with the following pH:
      i 3           ii 8           iii 11
   b A solution tested with bromothymol blue is yellow, while the same solution when tested with methyl orange is yellow. State the approximate pH range of the solution.

## 9.3 STRONG AND WEAK ACIDS AND BASES

**8.3.1**
Distinguish between *strong* and *weak* acids and bases in terms of the extent of dissociation, reaction with water and electrical conductivity.
© IBO 2007

Interactive Animation
Acid–base strength

QuickTime Video
Strong and weak acids
Strong and weak electrolytes

Acids vary in their strength. The *strength* of an acid is a measure of how readily it can dissociate in water and donate a proton. Consider two acid solutions of equal concentration.

$$HCl(aq) + H_2O(l) \rightarrow H_3O^+(aq) + Cl^-(aq) \qquad (1)$$

$$CH_3COOH(aq) + H_2O(l) \rightleftharpoons H_3O^+(aq) + CH_3COO^-(aq) \qquad (2)$$

Hydrochloric acid is a strong acid because the hydrogen chloride molecules completely dissociate to form hydrogen ions and chloride ions. This is represented by the one-way arrow '→'. Ethanoic acid is a **weak acid** as it only partially dissociates in water to form hydrogen ions and ethanoate ions. This reaction exists as an equilibrium, as shown by the two-way arrow (⇌). The extent of reaction (1) is almost 100%, whereas the extent of reaction (2) is very small (less than 1% complete).

If two acid solutions of equal concentration, say 1.0 mol dm$^{-3}$, are compared, the solution of a weak acid will be a poorer conductor of electricity than the solution of a strong acid due to its lower concentration of ions.

Bases also vary in strength. A **strong base** is one that readily accepts a proton. The oxide ion ($O^{2-}$) and hydroxide ion ($OH^-$) are examples of strong bases, although the hydroxide ion is the strongest base that can exist in aqueous solution. When mixed with water, hydroxide ions in compounds such as the group 1 hydroxides gain an $H^+$ ion, and the water loses an $H^+$ ion forming $OH^-(aq)$.

$$OH^-(s) + H_2O(l) \rightarrow H_2O(l) + OH^-(aq)$$

Another example of a strong base is $Ba(OH)_2$. The concentration of $OH^-$ ions in a solution of $Ba(OH)_2$ is twice that of $Ba(OH)_2$:

$$Ba(OH)_2(s) + aq \rightarrow Ba^{2+}(aq) + 2OH^-(aq)$$

**Weak bases** are substances in which only a small proportion of the molecules or ions react with water to form hydroxide ions in aqueous solution. The weak base ammonia reacts with water according to the equation:

$$NH_3(aq) + H_2O(l) \rightleftharpoons NH_4^+(aq) + OH^-(aq)$$

Only a small fraction of ammonia molecules react. Amines also react in this way (see chapter 11).

Many weak bases are anions such as carbonate, ethanoate, fluoride and phosphate. All these ions react with water to a small extent to produce hydroxide ions. For example, carbonate ions react with water according to the equation:

$$CO_3^{2-}(aq) + H_2O(l) \rightleftharpoons HCO_3^-(aq) + OH^-(aq)$$

Although this reaction occurs only to a very limited extent, some hydroxide ions are produced, so the carbonate ion is acting as a weak base.

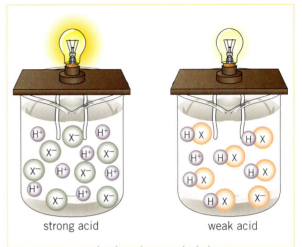

Figure 9.3.1 Strong acids HCl(aq) almost completely dissociate in water, while weak acids dissociate slightly. This can be shown by measuring the conductivity of the solutions.

| | Name of acid | Ionisation equation | Name of base | |
|---|---|---|---|---|
| very strong | Hydrochloric | $HCl(aq) + H_2O(l) \rightleftharpoons H_3O^+(aq) + Cl^-(aq)$ | Chloride ion | very weak |
| strong | Sulfuric | $H_2SO_4(aq) + H_2O(l) \rightleftharpoons H_3O^+(aq) + HSO_4^-(aq)$ | Hydrogen sulfate ion | |
| | Nitric | $HNO_3(aq) + H_2O(l) \rightleftharpoons H_3O^+(aq) + NO_3^-(aq)$ | Nitrate ion | |
| | Sulfurous | $H_2SO_3(aq) + H_2O(l) \rightleftharpoons H_3O^+(aq) + HSO_3^-(aq)$ | Hydrogen sulfite ion | |
| | Hydrogen sulfate ion | $HSO_4^-(aq) + H_2O(l) \rightleftharpoons H_3O^+(aq) + SO_4^{2-}(aq)$ | Sulfate ion | |
| | Phosphoric | $H_3PO_4(aq) + H_2O(l) \rightleftharpoons H_3O^+(aq) + H_2PO_4^-(aq)$ | Dihydrogen phosphate ion | |
| | Hydrofluoric | $HF(aq) + H_2O(l) \rightleftharpoons H_3O^+(aq) + F^-(aq)$ | Fluoride ion | |
| weak | Ethanoic | $CH_3COOH(aq) + H_2O(l) \rightleftharpoons H_3O^+(aq) + CH_3COO^-(aq)$ | Ethanoate ion | |
| | Carbonic | $H_2CO_3(aq) + H_2O(l) \rightleftharpoons H_3O^+(aq) + HCO_3^-(aq)$ | Hydrogen carbonate ion | |
| | Dihydrogen phosphate ion | $H_2PO_4^-(aq) + H_2O(l) \rightleftharpoons H_3O^+(aq) + HPO_4^{2-}(aq)$ | Hydrogen phosphate ion | |
| | Ammonium | $NH_4^+(aq) + H_2O(l) \rightleftharpoons H_3O^+(aq) + NH_3(aq)$ | Ammonia | weak |
| | Methylammonium | $CH_3NH_3^+(aq) + H_2O(l) \rightleftharpoons H_3O^+(aq) + CH_3NH_2(aq)$ | Methylamine | |
| | Water | $H_2O(l) + H_2O(l) \rightleftharpoons H_3O^+(aq) + OH^-(aq)$ | Hydroxide ion | strong |
| very weak | Hydroxide ion | $OH^-(aq) + H_2O(l) \rightleftharpoons H_3O^+(aq) + O^{2-}(aq)$ | Oxide ion | very strong |

decreasing acid strength ← / → decreasing base strength

Figure 9.3.2 Relative strengths of some common acids and their conjugate bases.

**AS 8.3.2**
State whether a given acid or base is strong or weak. © IBO 2007

**AS 8.3.3**
Distinguish between *strong* and *weak* acids and bases, and determine the relative strengths of acids and bases, using experimental data. © IBO 2007

**PRAC 9.3**
Conductivity of solutions

**WORKSHEET 9.3**
Ions in solution

Figure 9.3.2 shows acid strengths as well as base strengths. A strong acid has a weak conjugate base and a strong base has a weak conjugate acid. In the middle of the table are those conjugate pairs for which both acid and base are weak.

It should be noted that the terms *strong* and *weak* are often used in everyday language to describe the concentration of solutions. For example, you may say, 'I would like a strong cup of coffee' or 'the tea is very weak'. In Chemistry, the terms *strong* or *weak*, and *concentrated* or *dilute* have different and very specific meanings. Chemists use the terms *concentrated* and *dilute* to describe the concentrations of solutions, and the terms *strong* and *weak* to describe the degree to which the substance ionizes or dissociates. Therefore a chemist would say, 'I would like a concentrated cup of coffee' and 'the tea is very dilute'.

The relative strength of acids can be determined if their concentrations are the same and experimental data such as conductivity; rate of reaction with a metal, base or a carbonate; or pH is known. (pH will be discussed in section 9.4.)

Consider the following experiment. A student wishes to determine whether a solution of ethanoic acid is weaker than one of hydrochloric acid. The student is going to measure the conductivity of the two solutions and the rate of reaction with magnesium, but first the concentrations of the solutions must be the same. The student used 1.0 mol dm$^{-3}$ solutions for the experiment. The results for this experiment are shown below.

|  | 1.0 mol dm$^{-3}$ hydrochloric acid | 1.0 mol dm$^{-3}$ ethanoic acid |
|---|---|---|
| Conductivity of solution (mS) | 400 | 7.0 |
| Observations of rate of reaction with magnesium | Rapid effervescence, bubbles of hydrogen gas produced rapidly | Gradual effervescence, bubbles of hydrogen gas produced slowly |
| Temperature change when 100 cm$^3$ of acid reacted with 100 cm$^3$ of 1.0 mol dm$^{-3}$ NaOH (°C) | 6.9 | 6.7 |

These results show that HCl is a much stronger acid than ethanoic acid. When their concentrations are equal, the reaction of ethanoic acid with magnesium is much slower than that of HCl with magnesium. Similarly, the conductivity of 1.0 mol dm$^{-3}$ HCl is 50 times greater than that of ethanoic acid. The change in temperature for equal volumes (and numbers of mole) of HCl and NaOH is greater than for the same amounts of CH$_3$COOH and NaOH.

This can be explained in terms of the concentration of H$_3$O$^+$ ions in solution. The stronger acid, HCl, has a greater concentration of H$_3$O$^+$ ions. A greater frequency of collisions between H$_3$O$^+$ ions and Mg results in a more rapid reaction for HCl than for ethanoic acid. The higher conductivity of the HCl is due to the greater concentration of ions available to carry electrical charge, and the greater increase in temperature is due to the greater concentration of ions undergoing the reaction H$^+$(aq) + OH$^-$(aq) → H$_2$O(l).

Note that the standard enthalpy of neutralization ($\Delta H^\circ$) for HCl with NaOH is $-57.9$ kJ mol$^{-1}$, while that for CH$_3$COOH with NaOH is $-56.1$ kJ mol$^{-1}$.

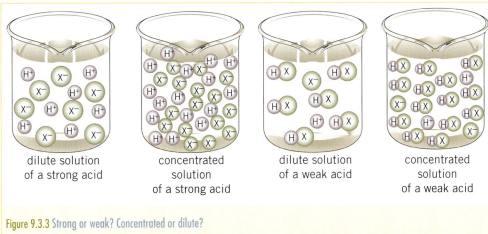

Figure 9.3.3 Strong or weak? Concentrated or dilute?

Acids such as hydrochloric (HCl), nitric (HNO$_3$) and ethanoic (CH$_3$COOH) acid contain only one acidic hydrogen atom per molecule which can be ionized in aqueous solution. They are called **monoprotic** acids. Other acids such as sulfuric (H$_2$SO$_4$) and carbonic (H$_2$CO$_3$) acid are **diprotic**, as they contain two ionizable hydrogen atoms per molecule of acid. In aqueous solutions of sulfuric acid the first proton is completely ionized, as shown by the equation:

$$H_2SO_4(aq) + H_2O(l) \rightarrow HSO_4^-(aq) + H_3O^+(aq)$$

The hydrogen sulfate ion is a weaker acid, so only some of these ions ionize further into hydrogen ions and sulfate ions according to the equation:

$$HSO_4^-(aq) + H_2O(l) \rightleftharpoons SO_4^{2-}(aq) + H_3O^+(aq)$$

Phosphoric acid (H$_3$PO$_4$), in addition to other substances, is added to cola beverages as a flavouring agent. It is a **triprotic acid** containing three ionizable hydrogen atoms. Phosphoric acid is a weak acid, so ionization occurs to only a small extent. Equations for the successive ionizations, which occur to progressively smaller extents, are:

$$H_3PO_4(aq) + H_2O(l) \rightleftharpoons H_2PO_4^-(aq) + H_3O^+(aq)$$

$$H_2PO_4^-(aq) + H_2O(l) \rightleftharpoons HPO_4^{2-}(aq) + H_3O^+(aq)$$

$$HPO_4^{2-}(aq) + H_2O(l) \rightleftharpoons PO_4^{3-}(aq) + H_3O^+(aq)$$

Some acids have more hydrogen atoms in their molecular formula than they are able to donate. These are usually organic acids such as carboxylic acids (see chapter 11). These acids include ethanoic acid, CH$_3$COO**H**, and lactic acid, CH$_3$CHOHCOO**H**. The hydrogen atom that is bolded in their molecular formula is the one that can be donated in acid–base reactions.

## CHEM COMPLEMENT

### Superacids

The most acidic substance that you will probably use at school is sulfuric acid, H$_2$SO$_4$. A 0.50 mol dm$^{-3}$ solution of sulfuric acid has a pH that is close to zero. Is there anything that is more acidic than this? Actually, there is a whole class of acids that are more acidic than H$_2$SO$_4$ and they are called superacids. In 1972, Ronald Gillespie defined superacids as acids that are stronger than 100% sulfuric acid.

Superacids include fluorosulfonic acid (HSO$_3$F), trifluoromethane-sulfonic acid (HSO$_3$CF$_3$, also known as triflic acid) and anhydrous hydrogen fluoride (HF). These acids are about a thousand times stronger than sulfuric acid. The very strongest superacids are prepared by the combination of two components, a strong Lewis acid and a strong Brønsted acid.

The strongest superacid system, fluoroantimonic acid, is made when the Lewis acid SbF$_5$ is mixed with anhydrous hydrogen fluoride. This superacid has an acidity that is 10$^{19}$ times stronger than that of sulfuric acid.

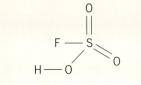

Figure 9.3.4 The structural formula of fluorosulfonic acid, HSO$_3$F.

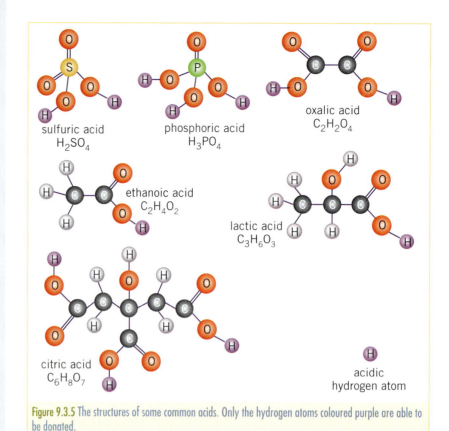

Figure 9.3.5 The structures of some common acids. Only the hydrogen atoms coloured purple are able to be donated.

### Acid rain

The atmosphere contains low concentrations of a number of non-metal oxides, including carbon dioxide, nitrogen dioxide, sulfur dioxide and sulfur trioxide. These oxides exist in the atmosphere from natural sources, but their concentrations have increased as a result of human activity, particularly during the last 100 years. Some of these acidic oxides are responsible for **acid rain**, a serious form of air pollution.

All rain contains dissolved carbon dioxide, so even unpolluted rain is slightly acidic. The carbon dioxide reacts with water to form carbonic acid.

$$CO_2(g) + H_2O(l) \rightleftharpoons H_2CO_3(aq)$$

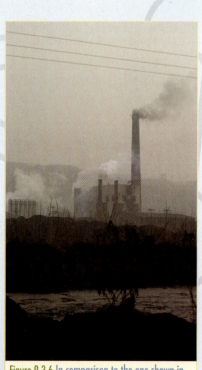

Figure 9.3.6 In comparison to the one shown in figure 9.3.7, this coal-fired power station pollutes heavily.

Figure 9.3.7 Coal-burning power stations have been major contributors to the amount of sulfur dioxide released into the atmosphere. The stacks seen here are releasing mostly steam, as strict controls on the release of sulfur dioxide are now in place.

Rain is usually described as *acid* rain if it has a pH lower than 5. The worst examples of acid rain occur around densely populated and highly industrialized regions. Acid rain is usually the result of rain dissolving sulfur oxides ($SO_2$ and $SO_3$) and nitrogen dioxide ($NO_2$) to produce a dilute solution of acids such as sulfurous ($H_2SO_3$), sulfuric ($H_2SO_4$) and nitric ($HNO_3$).

In addition to the natural sources of sulfur oxides such as volcanic eruptions, emission of sulfur oxides into the atmosphere is often the result of burning fossil fuels in electricity-generating power stations, and the smelting of sulfide ores in metal-smelting plants. The sulfur dioxide produced is oxidized in the air to form sulfur trioxide. Nitrogen oxide emissions are usually the result of high-temperature combustion processes such as those in motor vehicles and airplane emissions. Nitrogen oxides can also be formed naturally in the vicinity of lightning, which provides enough energy to allow nitrogen to be oxidized to NO and further to $NO_2$. When dissolved in rain, these sulfur and nitrogen oxides produce acids and, subsequently, acid rain.

$SO_2(g) + H_2O(l) \rightarrow H_2SO_3(aq)$

$SO_3(g) + H_2O(l) \rightarrow H_2SO_4(aq)$

$4NO_2(g) + 2H_2O(l) + O_2(g) \rightarrow 4HNO_3(aq)$

Although a solution with a pH of 4 or 5 would be described as only weakly acidic, these solutions can cause a lot of damage over long periods of time. Acid rain can cause several problems. Surface waters and lakes can become acidic, decreasing the populations of aquatic species that cannot tolerate these acidic conditions. Many lakes in Europe, Scandinavia and North America are now too acidic to support fish life. Acid rain can also cause damage to plants, including crops and forests. This occurs through the direct effect of the acid rain, but also indirectly because of its effects on soil. Another problem is the damage caused to metal and stone buildings and structures. In particular, many historical statues are slowly dissolving due to the effect of acid rain on the limestone of which they are composed. The reaction involved is:

$CaCO_3(s) + 2H^+(aq) \rightarrow Ca^{2+}(aq) + H_2O(l)$

Figure 9.3.8 The damage seen on this statue was caused by acid rain.

## Section 9.3 Exercises

1. With reference to two acid solutions, distinguish between the terms *concentrated* and *dilute*.

2. With reference to two acid solutions, distinguish between the terms *strong* and *weak*.

3. Describe how a strong acid differs from a weak acid in terms of extent of dissociation.

4. For each of the following pairs of solutions (of equal concentration), predict which solution:
   i   has the higher concentration of $H^+$
   ii  is the better electrical conductor.
   
   a  HCl and $CH_3COOH$  
   b  $H_2SO_4$ and $HNO_3$

## CHEM COMPLEMENT

### What does pH stand for?

The term *pH* was invented by Danish chemist Søren Sørenson in 1909, and stands for *pondus hydrogenii*, or 'potential hydrogen'. The phrase refers to the potential of a solution to produce hydrogen ions, as Sørenson recognized that the hydrogen ions were critical to the behaviour of enzymes that he was investigating. Strictly speaking, pH is a measure of what is known as the 'activity' of hydrogen ions, though for the types of problems that we will encounter, the concentration of hydrogen ions is quite sufficient. When Sørenson was recording the results of his experiments, he needed to record the exact hydrogen ion concentration every time. He soon tired of writing, for example, $10^{-6.4}$ mol dm$^{-3}$, and slipped into the habit of simply writing the exponent of the concentration, which in the example given is 6.4. His colleagues quickly followed suit and, as a result, a convenient and practical way of recording the relative acidity of solutions came into widespread use and acceptance.

**AS 8.4.1**
Distinguish between aqueous solutions that are *acidic, neutral* or *alkaline* using the pH scale.
© IBO 2007

5. For each of the following pairs of solutions (of equal concentration), predict which solution:
   i has the higher concentration of OH$^-$
   ii is the better electrical conductor.
   a  NaOH and NH$_3$
   b  KOH and Na$_2$CO$_3$

6. State the equation for the dissociation of each of the following.
   a  Strong acid HNO$_3$
   b  Weak acid HCN
   c  Strong base KOH
   d  Weak base KF

7. State whether each of the dilute solutions listed below is a good or poor electrical conductor. Explain your answer (including an equation where relevant) for each solution.
   a  KOH
   b  HCl
   c  NH$_3$
   d  H$_2$CO$_3$

8. Define the term *amphiprotic*.

9. Describe an experiment you would perform to determine which is the stronger base: an ammonia solution or a solution of sodium hydroxide.

10. Describe three experimental methods by which the strengths of two different acids could be compared.

11. Explain why acid rain occurs mainly in areas of high population density and high industrialized areas.

## 9.4 THE PH SCALE

In everyday substances, we encounter varying levels of acidity and basicity. The acidity of a solution is related to the concentration of hydrogen ions, H$^+$, or H$_3$O$^+$, in the solution. This concentration varies widely. For example, in battery acid it is $10^{0.9}$ mol dm$^{-3}$, in pure water $10^{-7}$ mol dm$^{-3}$, and in oven cleaner $10^{-13.5}$ mol dm$^{-3}$. For convenience, we express the acidity of a solution using the pH scale developed by Sørensen.

$$pH = -\log_{10}[H_3O^+]$$

where [H$_3$O$^+$] denotes the concentration of the hydronium ion expressed in mol dm$^{-3}$. (Note that either [H$_3$O$^+$] or [H$^+$] may be used in the expression for pH.)

The pH (power of the hydrogen ion) scale is usually applied over the range 0 to 14 and provides a convenient scale for describing whether an aqueous solution is acidic, neutral or basic. At 25°C, solutions with a pH less than 7 are described as acidic, those with a pH greater than 7 are basic, and solutions with a pH of exactly 7 are neutral. Figure 9.4.1 shows the pH values of a number of commonly used substances.

For an acidic solution:     pH < 7

For a neutral solution:     pH = 7

For a basic solution:       pH > 7

## Worked example 1

Classify each of the following solutions as acidic, basic or neutral.

a  Gastric juice of pH 1.0
b  Antacid emulsion of pH 10.5
c  Tomato juice of pH 4.2
d  Milk of pH 6.6
e  Pure water of pH 7.0

### Solution

| Substance | pH | Classification |
|---|---|---|
| a Gastric juice | 1.0 | Acidic |
| b Antacid emulsion | 10.5 | Basic |
| c Tomato juice | 4.2 | Acidic |
| d Milk | 6.6 | Acidic |
| e Pure water | 7.0 | Neutral |

> **THEORY OF KNOWLEDGE**
> - Advertisers often used science as a stamp of approval or guarantee of quality to sell their products. How could you determine if a claim made about the pH of a product is a scientific one?
> - Do you think a numerical scale like the pH scale affects the way knowledge claims in Chemistry are valued?

As indicated by the arrows to the left of figure 9.4.1, acidity increases with decreasing pH; that is the lower the pH, the more acidic the solution. Similarly, alkalinity increases with increasing pH, so the higher the pH the more alkaline the solution.

The pH scale is continuous from 0 to 14, so the relative acidity of two solutions can be determined by comparing their pH values.

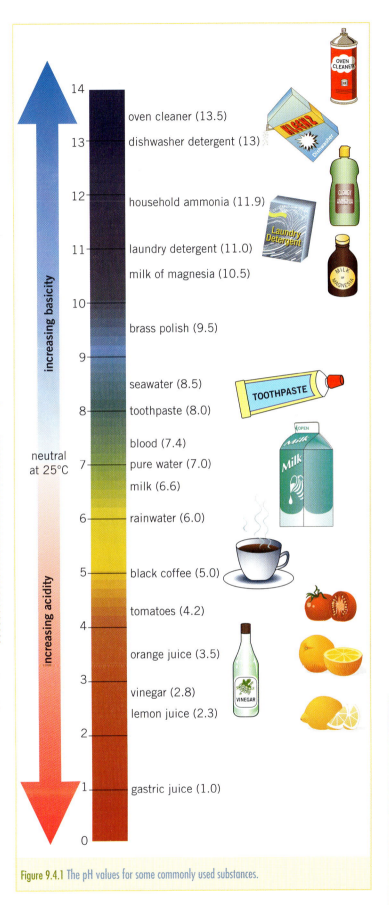

Figure 9.4.1 The pH values for some commonly used substances.

CHEMISTRY: FOR USE WITH THE IB DIPLOMA PROGRAMME **STANDARD LEVEL**

**AS 8.4.2**
Identify which of two or more aqueous solutions is more acidic or alkaline using pH values. © IBO 2007

**AS 8.4.3**
State that each change of one pH unit represents a 10-fold change in the hydrogen ion concentration [H⁺(aq)]. © IBO 2007

**AS 8.4.4**
Deduce changes in [H⁺(aq)] when the pH of a solution changes by more than one pH unit. © IBO 2007

WORKSHEET 9.4
Strenghts of acids and pH

### Worked example 2

For each of the following pairs of solutions, determine which is the more acidic.

**a** Blood of pH = 7.4 and brass polish of pH = 9.5

**b** Black coffee of pH = 5.0 and vinegar of pH = 2.8

### Solution

**a** The more acidic solution is blood because its pH (7.4) is lower than that of brass polish (9.5).

**b** The more acidic solution is vinegar because its pH (2.8) is lower than that of black coffee (5.0)

### Worked example 3

For each of the following pairs of solutions, determine which is the more alkaline.

**a** Toothpaste of pH = 8.0 and milk of magnesia of pH = 10.5

**b** Orange juice of pH = 3.5 and lemon juice of pH = 2.3

### Solution

**a** The more alkaline solution is milk of magnesia because its pH (10.5) is higher than that of toothpaste (8.0).

**b** The more alkaline solution is orange juice because its pH (3.5) is higher than that of lemon juice (2.3)

The definition of pH (pH = $-\log_{10}[H_3O^+]$) indicates that the pH scale is logarithmic:

**A difference of one pH unit represents a 10-fold change in the H⁺ ion concentration.**

For example, a solution with a pH of 2 is 10 times *more* acidic than a solution with a pH of 3, and a sample of brass polish with a pH of 9.5 is 10 times *less* acidic than a sample of seawater with a pH of 8.5.

### Worked example 4

Describe the difference in [H⁺(aq)] for each of the following pairs of solutions.

**a** Orange juice with a pH of 3.5 and lemon juice with a pH of 2.5

**b** Soft drink with a pH of 3 and coffee with a pH of 5

**c** A solution of NaOH with a pH of 13 and a solution of sodium carbonate with a pH of 10

### Solution

**a** The pH of the lemon juice (2.5) is one unit lower than the pH of the orange juice (3.5); therefore, the [H⁺(aq)] of the lemon juice is 10 times greater than that of the orange juice.

**b** The pH of the coffee (5) is two units higher than the pH of the soft drink (3); therefore, the [H⁺(aq)] of the coffee is 10 × 10 = 100 times less than that of the soft drink.

**c** The pH of the sodium carbonate solution (10) is three units lower than the pH of the NaOH solution (13); therefore, the [H⁺(aq)] of the sodium carbonate solution is 10 × 10 × 10 = 1000 times greater than that of the NaOH solution.

## CHEM COMPLEMENT

### Why is 7 a neutral pH?

To understand why a pH of 7 represents the neutral point, consider what happens in pure water. Recall that water undergoes self-ionization. The extent of this reaction is very slight and a small number of $H_3O^+$ and $OH^-$ ions exist in any solution in equilibrium with the water molecules.

$$H_2O(l) + H_2O(l) \rightleftharpoons H_3O^+(aq) + OH^-(aq)$$

The equilibrium constant for this reaction is $\frac{[H_3O^+][OH^-]}{[H_2O]^2}$.

Because the concentration of water is so great (pure water has a concentration of 55.6 mol dm$^{-3}$) compared to all other concentrations, $[H_2O]$ can be regarded as a constant and this equilibrium constant simplifies to $[H_3O^+][OH^-] = K_w$.

Experimentally it has been found that for dilute aqueous solutions at 25°C, $K_w = 10^{-14}$ mol$^2$ dm$^{-6}$.

$K_w$ is known as the *ionization constant of water* and allows us to calculate the pH values of neutral, acidic and basic solutions.

For an acidic solution:
$[H_3O^+] > [OH^-]$ hence $[H_3O^+] > 10^{-7}$ hence pH < 7

For a neutral solution:
$[H_3O^+] = [OH^-]$ hence $[H_3O^+] = 10^{-7}$ hence pH = 7

For a basic solution:
$[H_3O^+] < [OH^-]$ hence $[H_3O^+] < 10^{-7}$ hence pH > 7

Using the combination of pH = $-\log_{10}[H_3O^+]$ and $[H_3O^+][OH^-] = 10^{-14}$ mol$^2$ dm$^{-6}$, we can also calculate the pH of solutions of acids and bases of given concentrations.

## Section 9.4 Exercises

1. By making reference to the pH scale, distinguish between a neutral solution and an acidic solution.

2. For each of the following pairs of solutions, state which one is the more acidic.
   a. Rainwater of pH = 6.0 and tomato juice of pH = 4.2
   b. Milk of pH = 6.6 and seawater of pH = 8.5
   c. Vinegar of pH = 2.8 and orange juice of pH = 3.5

3. For each of the following pairs of solutions, state which one is the more alkaline.
   a. Rainwater of pH = 6.0 and seawater of pH = 8.5
   b. Orange juice of pH = 3.5 and black coffee of pH = 5.0
   c. Household ammonia of pH = 11.9 and brass polish of pH = 9.5

4. Solution A is 1000 times more acidic than solution B. If solution A has pH = 2, state the pH of solution B.

5. If a sodium hydroxide solution has pH = 12, what is the pH of a solution that is 100 times more acidic than the sodium hydroxide solution?

6. Blood has pH = 7.4. A urine sample has pH = 5.4. How many times more acidic is urine than blood?

7. If the [H$^+$(aq)] of an acidic solution with pH = 3 is $10^{-3}$ mol dm$^{-3}$, state the [H$^+$(aq)] of a second solution with pH = 2.

8. If the [H$^+$(aq)] of an acidic solution is $10^{-5}$ mol dm$^{-3}$, state the pH of the solution.

9. Explain why a 1.0 mol dm$^{-3}$ solution of ammonia has pH = 12 whereas a 1.0 mol dm$^{-3}$ solution of sodium hydroxide has pH = 14. Write an equation for the dissociation of ammonia in water and a second equation for the dissociation of sodium hydroxide in water to help explain your answer.

# Chapter 9 Summary

## Terms and definitions

**Acid rain**   Rain with pH < 5.

**Acidic solution**   A solution with pH < 7 at 25°C.

**Alkali**   A base that dissolves in water.

**Alkaline solution**   A solution with pH > 7 at 25°C.

**Amphiprotic**   A substance that can be a proton acceptor or a proton donor.

**Amphoteric**   A substance that can act as an acid or as a base, e.g. $Al_2O_3$.

**Brønsted–Lowry acid**   A hydrogen ion ($H^+$) donor.

**Brønsted–Lowry base**   A hydrogen ion ($H^+$) acceptor.

**Conjugate acid**   The acid formed when a base gains a hydrogen ion.

**Conjugate base**   The base formed when an acid donates a hydrogen ion.

**Dative (coordinate) bond**   A covalent bond formed between two atoms, only one of which has provided electrons for the bond.

**Diprotic**   An acid contains two acidic hydrogen atoms per molecule which can be ionized in aqueous solution.

**Dissociation (ionization)**   The process by which ions are formed and separate when a compound dissolves in and reacts with water.

**Lewis acid**   An electron pair acceptor.

**Lewis base**   An electron pair donor.

**Monoprotic**   An acid that contains only one acidic hydrogen atom per molecule which can be ionized in aqueous solution.

**Neutral solution**   A solution with pH = 7 at 25°C.

**Self-ionization**   The reaction between two molecules of the same type resulting in two ions, e.g. the self-ionization of:
$H_2O(l) + H_2O(l) \rightleftharpoons H_3O^+(aq) + OH^-(aq)$.

**Spectator ions**   Ions that do not participate in an aqueous reaction, but are present in the solution.

**Strong acid**   An acid that dissociates almost completely in aqueous solution.

**Strong base**   A base that readily accepts a hydrogen ion from water.

**Triprotic acid**   An acid that contains three acidic hydrogen atoms per molecule which can be ionized in aqueous solution.

**Weak acid**   An acid that dissociates to a small extent in aqueous solution.

**Weak base**   A base that accepts a hydrogen ion from water with difficulty.

## Concepts

- A Brønsted–Lowry acid is a species that donates a hydrogen ion (a proton) to another species, a Brønsted–Lowry base.

  an acid donates a proton ...
  $$HA + B \longrightarrow A^- + HB^+$$
  ... to a base

- A Lewis acid is a species that accepts an electron pair from another species, a Lewis base.

  $$BF_3 + :NH_3 \longrightarrow BF_3NH_3$$
  acid   base

- A hydrogen ion in water ($H^+(aq)$) exists as a hydronium ion ($H_3O^+(aq)$), but is often represented simply as $H^+$ for convenience.

- Amphiprotic species (such as water) can act as both acids and bases, donating and accepting a proton.

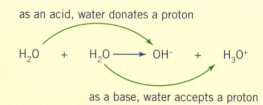

  as an acid, water donates a proton
  $$H_2O + H_2O \longrightarrow OH^- + H_3O^+$$
  as a base, water accepts a proton

- Brønsted–Lowry reactions involve the transfer of a proton from an acid to a base to form a conjugate base and a conjugate acid.

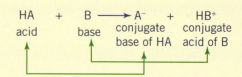

  $$HA + B \longrightarrow A^- + HB^+$$
  acid   base   conjugate   conjugate
                base of HA   acid of B

- Conjugate acid–base pairs are two species that differ by one proton. Examples include:

  HCl/Cl$^-$, H$_2$O/OH$^-$, H$_3$O$^+$/H$_2$O

- Properties of acids include: taste sour, are corrosive, turn blue litmus red, and conduct electricity in the aqueous state.

- Properties of bases include: taste bitter, feel slippery, turn red litmus blue, and conduct electricity in the aqueous state.

- Reactions of acids include:

  acid + most metals → salt + hydrogen gas
  acid + metal hydroxide → salt + water
  acid + metal carbonate → salt + water + carbon dioxide

- Reactions of bases include:

  metal hydroxide + acid → salt + water
  metal carbonate + acid → salt + water + carbon dioxide
  ammonia + acid → ammonium salt

- Strong acids dissociate completely in water. Weak acids only partially dissociate in water.

- The pH scale is used to describe the acidity or basicity of a solution.

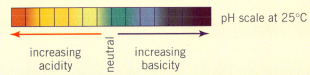

- A difference of one unit on the pH scale means a 10-fold difference in the hydrogen ion concentration.

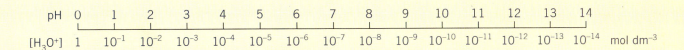

## Chapter 9 Review questions

1. Distinguish between a Brønsted–Lowry acid and a Lewis acid, using examples to support your answer.

2. Identify each of the following substances as Brønsted–Lowry acids, Brønsted–Lowry bases, Lewis acids or Lewis bases.
   a H$_2$O
   b NH$_3$
   c HCl
   d F$^-$
   e BF$_3$
   f OH$^-$

3. State the formula of:
   a the conjugate bases of HClO$_3$, HS$^-$, NH$_4^+$ and H$_2$O
   b the conjugate acids of HCO$_3^-$, HS$^-$, ClO$_4^-$ and H$_2$PO$_4^-$.

4. For each of the following reactions, identify the Brønsted–Lowry conjugate acid–base pairs.
   a H$_2$C$_2$O$_4$(aq) + H$_2$O(l) → H$_3$O$^+$(aq) + HC$_2$O$_4^-$(aq)
   b H$_2$O(l) + CN$^-$(aq) → OH$^-$(aq) + HCN(aq)
   c CH$_3$COOH(aq) + S$^{2-}$(aq) → CH$_3$COO$^-$(aq) + HS$^-$(aq)
   d HSO$_4^-$(aq) + F$^-$(aq) → SO$_4^{2-}$(aq) + HF(aq)

5. Identify each of the following as properties of Brønsted–Lowry (A) acids, (B) bases, (AB) acids and bases, or (N) neither acids nor bases.
   a Turns litmus red
   b Feels slippery
   c Used to clean drains
   d Corrosive
   e Produced in the stomach to aid digestion
   f Conducts electricity in aqueous solution
   g Has a sour taste
   h Dissolves limestone
   i Reacts by donating a single electron

6. From the list given, determine a use for each of the following acids and bases.

| Acid or base | Use |
|---|---|
| Sulfuric acid | Antacid powders |
| Sodium carbonate | Fertilizer manufacture |
| Magnesium hydroxide | Car batteries |
| Ammonia | Watch batteries |
| Phosphoric acid | Washing powder |
| Zinc hydroxide | Window cleaner |

7  Write balanced equations to illustrate the following reactions.
   a  Dissociation of HCl in aqueous solution
   b  Dissociation of $Ca(OH)_2$ in aqueous solution
   c  Successive dissociations of $H_2SO_4$
   d  Neutralization of $H_2SO_4$ solution with NaOH solution
   e  Neutralization of $Ba(OH)_2$ solution with HCl solution

8  Dilute acids react with most metals to produce hydrogen gas and a salt.
   a  State the balanced formula equation for the reaction of HCl with zinc (Zn).
   b  Explain why this reaction is not a acid–base reaction.

9  a  Define the following terms.
      i  Weak acid        ii  Amphiprotic
   b  Using chemical equations, show that the hydrogenchromate ion ($HCrO_4^-$) ion is amphiprotic.

10  Draw diagrams to represent these solutions.
    a  Concentrated solution of a strong acid
    b  Dilute solution of a strong acid
    c  Concentrated solution of a weak acid
    d  Dilute solution of a weak acid

11  Describe an experiment that could be used to determine whether a solution of unknown identity is more or less acidic than a solution of the same concentration of ethanoic acid.

12  a  List the ions and molecules present in an aqueous solution of $H_2SO_4$.
    b  State which ion or molecule is present in the highest concentration.
    c  State which ion or molecule is present in the lowest concentration.

13  Solutions of each of the following chemicals are prepared so that each has the same concentration. List the solutions in order of decreasing pH.
    NaOH, HCl, $HSO_4^-$, $H_2CO_3$, KCl, $NH_3$

14  Three acid solutions of equal concentration have the pH values listed here.
    HCl (pH = 2), $H_2SO_4$ (pH = 1.7), $CH_3COOH$ (pH = 3.5)
    With the aid of relevant equations, explain why these pH values differ.

15  Solutions of each of the following chemicals are prepared so that each has the same concentration.
    NaOH, $Ba(OH)_2$, $NH_3$, HCl
    a  State which solution would contain the greatest number of ions.
    b  State which solution would contain the smallest number of ions. Explain your choice.
    c  State which two solutions would contain the same number of ions.

16  For each of the following groups of solutions, state which one is the most acidic.
    a  Rainwater of pH = 6.0, seawater of pH = 8.5 and tomato juice of pH = 4.2
    b  Milk of pH = 6.6, brass polish of pH = 9.5 and laundry detergent of pH = 11

17  For each of the following groups of solutions, state which one is the most alkaline
    a  Rainwater of pH = 6.0, seawater of pH = 8.5 and black coffee of pH = 5.0
    b  Orange juice of pH = 3.5, household ammonia of pH = 11.9 and brass polish of pH = 9.5

18  Predict the pH of a solution that has an [$H^+$(aq)] of $10^{-4}$ mol dm$^{-3}$.

19  Solution X has pH = 7. The concentration of $H^+$ ions in solution X is 100 times smaller than the concentration of $H^+$ ions in solution Y.
    a  Determine the pH of solution Y.
    b  State whether solution Y is acidic, basic or neutral.

20  Solution A has pH = 8, solution B has a pH = 11. State how many times more alkaline solution B is than solution A.

 www.pearsoned.com.au/schools

**Weblinks are available on the Chemistry: For use with the IB Diploma Programme Standard Level Companion Website to support learning and research related to this chapter.**

# Chapter 9 Test

## Part A: Multiple-choice questions

1. Which of the following statements about aqueous solutions of most weak acids is/are correct?
   - I They react with carbonates to produce carbon dioxide.
   - II They conduct electricity better than strong acids.
   - A  I only
   - B  II only
   - C  Both I and II
   - D  Neither I nor II

   © IBO 2000, May P1 Q23

2. 10 cm³ of an HCl solution with a pH value of 2 was mixed with 90 cm³ of water. What will be the pH of the resulting solution?
   - A  1
   - B  3
   - C  5
   - D  7

   © IBO 2000, May P1 Q24

3. When the pH of a solution changes from 2.0 to 4.0, the hydrogen ion concentration
   - A  increases by a factor of 100.
   - B  increases by a factor of 2.
   - C  decreases by a factor of 2.
   - D  decreases by a factor of 100.

   © IBO 2000, Nov P1 Q23

4. Which will be the same for separate 1 mol dm⁻³ solutions of a strong acid and a weak acid?
   - I  Electrical conductivity
   - II Concentration of H⁺ ions
   - A  I only
   - B  II only
   - C  Both I and II
   - D  Neither I nor II

   © IBO 2000, Nov P1 Q24

5. Which statement describes the Brønsted–Lowry behaviour of H₂O molecules in aqueous solutions?
   - A  They cannot act as either acids or bases.
   - B  They can act as acids but not bases.
   - C  They can act as acids or bases when reacting with each other.
   - D  They can act as acids when reacting with HCl molecules.

   © IBO 2001, May P1 Q21

6. Aqueous solutions of each of the following have a concentration of 0.100 mol dm⁻³. Which has the highest pH?
   - A  HCl
   - B  CH₃COOH
   - C  NaOH
   - D  NH₃

   © IBO 2001, May P1 Q22

7. A Brønsted–Lowry base is defined as a substance which:
   - A  accepts H⁺ ions.
   - B  produces OH⁻ ions.
   - C  conducts electricity.
   - D  donates protons.

   © IBO 2001, Nov P1 Q23

8. Which statement best describes the difference between solutions of strong and weak acids of equal concentration?
   - A  Weak acids have lower pH values than strong acids.
   - B  Weak acid solutions react more slowly with sodium carbonate than strong acids.
   - C  Weak acid solutions require fewer moles of base for neutralization than strong acids.
   - D  Weak acid solutions do not react with magnesium while strong acids do.

   © IBO 2001, Nov P1 Q24

9. Solutions **P**, **Q**, **R** and **S** have the following properties:
   **P**: pH = 8   **Q**: [H⁺] = 1 × 10⁻³ mol dm⁻³   **R**: pH = 5
   **S**: [H⁺] = 2 × 10⁻⁷ mol dm⁻³

   When these solutions are arranged in order of increasing acidity (least acidic first), the correct order is:
   - A  P, S, R, Q.
   - B  Q, R, S, P.
   - C  S, R, P, Q.
   - D  R, P, Q, S.

   © IBO 2002, May P1 Q23

10. The ionization of sulfuric acid is represented by the equations below:

    $H_2SO_4(aq) + H_2O(l) \rightleftharpoons H_3O^+(aq) + HSO_4^-(aq)$
    $HSO_4^-(aq) + H_2O(l) \rightleftharpoons H_3O^+(aq) + SO_4^{2-}(aq)$

    What is the conjugate base of $HSO_4^-(aq)$?
    - A  $H_2O(l)$
    - B  $H_3O^+(aq)$
    - C  $H_2SO_4(aq)$
    - D  $SO_4^{2-}(aq)$

    © IBO 2002, May P1 Q24

    (10 marks)

www.pearsoned.com.au/schools

For more multiple-choice test questions, connect to the Chemistry: For use with the IB Diploma Programme Standard Level Companion Website and select Chapter 9 Review Questions.

## Part B: Short answer questions

1. In aqueous solutions, sodium hydroxide is a strong base and ammonia is a weak base.
   a Use the Brønsted–Lowry theory to state why both substances are classified as bases.

   (1 mark)

**b** Solutions of 0.1 mol dm$^{-3}$ sodium hydroxide and 0.1 mol dm$^{-3}$ ammonia have different electrical conductivities.

  **i** State and explain which solution has the greater electrical conductivity.
  (1 mark)

  **ii** The pH value of 0.1 mol dm$^{-3}$ ammonia solution is approximately 11. State and explain how the pH value of the 0.1 mol dm$^{-3}$ sodium hydroxide solution would compare.
  (2 marks)

  © IBO 2002, Nov P2 Q2

**2** Sodium hydrogencarbonate dissolves in water forming an alkaline solution according to the following equilibrium:

$HCO_3^-(aq) + H_2O(l) \rightleftharpoons H_2CO_3(aq) + OH^-(aq)$

  **a** Why is the solution alkaline?
  (1 mark)

  **b** Using the Brønsted–Lowry theory, state, with a brief explanation, whether the $HCO_3^-$ ion is behaving as an acid or as a base.
  (2 marks)

  © IBO 2001, May P2 Q2

**3** Carbonic acid ($H_2CO_3$) is described as a weak acid and hydrochloric acid (HCl) is described as a strong acid.

  **a** Explain, with the help of equations, what is meant by *strong* and *weak* acid, using the above acids as examples.
  (4 marks)

  **b** A solution of hydrochloric acid, HCl(aq), has a pH of 1 and a solution of carbonic acid, $H_2CO_3$(aq), has a pH of 5. Determine the ratio of the hydrogen ion concentrations in these solutions.
  (2 marks)

  © IBO 2002, Nov P2 Q5ac

## Part C: Data-based question

The graph below shows how the conductivity of a strong and a weak monoprotic acid change as the concentration changes.

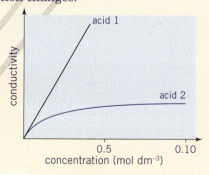

**a** Identify the strong acid and the weak acid from the above data. Give reasons for your choices.
(6 marks)

**b** Describe how magnesium metal can be used to distinguish between solutions of the strong acid and the weak acid of the same concentration.
(1 mark)

© IBO 2001, Nov P2 Q7d

## Part D: Extended-response question

**a** By means of balanced equations, give **three** different types of chemical reaction of an acid, such as aqueous sulfuric acid.
(3 marks)

**b** **i** Define an acid and a base according to the Brønsted–Lowry theory.
(1 mark)

  **ii** Ammonia acts as a base in water. Write a balanced equation for this reaction and state what would be observed if the final solution were tested with pH paper.
(3 marks)

**c** In the following reactions identify clearly the acid, conjugate base, base and conjugate acid:

  **i** $HNO_3 + H_2SO_4 \rightleftharpoons H_2NO_3^+ + HSO_4^-$
  (1 mark)

  **ii** $CH_3CH_2NH_2 + H_2O \rightleftharpoons OH^- + CH_3CH_2NH_3^+$
  (1 mark)

**d** Using equation (i) in part **c**, state and explain the relative strengths of nitric and sulfuric acid.
(2 marks)

**e** What is the difference between a strong acid and a weak acid? How could you distinguish between them experimentally?
(4 marks)

**f** Two acidic solutions, **A** and **B**, of equal concentration, have pH values of 2 and 6 respectively.

  **i** Indicate which acid is stronger and calculate how many times more acidic it is.
  (3 marks)

  **ii** State **two** ways in which solution A could be treated to produce a solution of pH 6.
  (2 marks)

© IBO 1999, May P2 Q6

Total marks 50

# 10 OXIDATION AND REDUCTION

## Chapter overview

This chapter covers the IB Chemistry syllabus Topic 9: Oxidation and Reduction.

**By the end of this chapter, you should be able to:**

- explain the meaning of the terms *reduction*, *oxidation*, *reducing agent* and *oxidizing agent*
- deduce the oxidation number of an element in a compound
- use oxidation numbers to identify whether a given reaction involves reduction or oxidation
- use oxidation numbers to identify the oxidizing agent and reducing agent in a redox reaction
- write balanced equations for redox reactions
- design and conduct experiments to establish an order of reactivity for a set of metals
- describe the essential features and operation of a voltaic cell
- describe how the electrochemical series was generated
- use the electrochemical series to write balanced equations for redox reactions
- describe the principal features and operation of an electrolytic cell
- use the electrochemical series to predict the products in an electrolytic cell
- explain how an electrolytic cell differs from a voltaic cell.

IBO Assessment statements 9.1.1 to 9.5.4

Redox (**red**uction and **ox**idation) reactions form one of the largest groups of chemical reactions, and include the reactions involved in rusting, combustion, the provision of energy by batteries, respiration, photosynthesis and, at least in part, the ageing of the human body. Many natural and synthetic chemical reactions involve the same fundamental redox mechanism—the transfer of electrons from one substance to another.

## 10.1 OXIDATION AND REDUCTION

Magnesium burns brightly in oxygen to produce the white powder magnesium oxide.

$$2Mg(s) + O_2(g) \rightarrow 2MgO(s)$$

For some time, chemists classified reactions such as this, in which elements gained oxygen, as **oxidation**. The opposite, removal of oxygen from a compound, was called **reduction**.

$$CuO(s) + H_2(g) \rightarrow Cu(s) + H_2O(g)$$

The term *reduction* was in use long before the term *oxidation*. Reduction was used to describe the process of 'restoring' a metal from its compound, such as the copper from the copper oxide shown in the equation above. During this 'restoration', the mass of the compound was reduced, hence the term *reduction*. The definitions of oxidation and reduction were later extended to include the removal and addition respectively of hydrogen.

Figure 10.1.1 Magnesium burns in oxygen to form magnesium oxide.

Figure 10.1.2 Copper oxide can be reduced to form copper in the laboratory.

QuickTime Video
Reduction of CuO

These definitions, however, did not explain how the reactions occurred. Consider again the reaction of magnesium with oxygen to produce the ionic compound magnesium oxide. The magnesium atom has become a magnesium ion by the removal of two electrons.

$$Mg \rightarrow Mg^{2+} + 2e^-$$

Each oxygen atom in the oxygen molecule has become an oxide ion by the addition of two electrons.

$$O_2 + 4e^- \rightarrow 2O^{2-}$$

When written in this form, known as a **half-equation**, the reaction can be seen clearly to involve the transfer of electrons. The loss of electrons by the magnesium atom is oxidation. The gain of electrons by the atoms in the oxygen molecule is reduction. Notice that two magnesium atoms are required to donate the total of four electrons required by the oxygen molecule.

**Oxidation is the loss of electrons.**

**Reduction is the gain of electrons.**

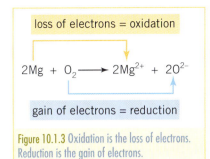

Figure 10.1.3 Oxidation is the loss of electrons. Reduction is the gain of electrons.

9.1.1
Define *oxidation* and *reduction* in terms of electron loss and gain. © IBO 2007

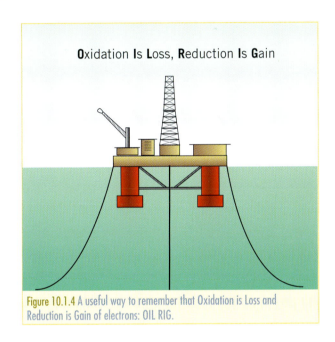

Figure 10.1.4 A useful way to remember that Oxidation is Loss and Reduction is Gain of electrons: OIL RIG.

Oxidation–reduction reactions: part 1
Oxidation–reduction reactions: part 2

## Oxidation numbers

In the redox reaction between magnesium and oxygen, the transfer of electrons was easily seen because the atoms or molecules formed ions. However, **electron transfers** are not always obvious, as many redox reactions do not involve metals and non-metals in elemental state or as simple ions. Consider, for example, the following redox reaction:

$$C(s) + O_2(g) \rightarrow CO_2(g)$$

The carbon has gained oxygen and so has been oxidized. The oxygen has been reduced. However, the transfer of electrons between carbon and oxygen is not complete, because the product, carbon dioxide, is molecular. We may say that a partial electron transfer occurs, in that the bonds in the carbon dioxide molecule are polar, with the oxygen atom having a greater share of the electrons. In order to classify reactions like this one, in which a complete electron transfer does not occur, a broader definition has been developed using a system of **oxidation numbers**.

Oxidation numbers are somewhat artificial; they are numbers assigned to atoms by a set of rules. They show the charge an atom would have if the electron transfer occurred completely. Their purpose is to enable us to determine whether a redox reaction has occurred and, if so, to identify the

Figure 10.1.5 When magnesium is added to hydrochloric acid a redox reaction occurs.

**THEORY OF KNOWLEDGE**
If oxidation numbers are 'artificial' or 'not real' why are they used?

**9.1.2 Deduce the oxidation number of an element in a compound.**
© IBO 2007

oxidizing agent and reducing agent. Table 10.1.1 shows the rules used for assigning oxidation numbers, and examples of the application of each rule.

Notice that when we write oxidation numbers, the number is preceded by a plus or minus sign. Also note that in the examples given, oxidation numbers are whole numbers. This is usually, but not always, the case.

**TABLE 10.1.1 ASSIGNING OXIDATION NUMBERS**

| Rule for assigning oxidation numbers | Examples of application of the rule |
|---|---|
| 1  The oxidation number of each atom in a pure element is zero. | Zn, O in $O_2$, and P in $P_4$ all have an oxidation number of zero. |
| 2  The oxidation number of an atom in a monatomic ion is equal to the charge on the ion. | $Na^+$ has an oxidation number of +1.<br>$S^{2-}$ has an oxidation number of −2. |
| 3  In compounds containing oxygen, each oxygen atom has an oxidation number of −2 (exceptions are $OF_2$, in which the oxidation number of oxygen is +2, and peroxides ($O_2^{2-}$), in which it is −1). | In $H_2O$ and $CO_2$ each oxygen atom has an oxidation number of −2.<br>In $H_2O_2$ each oxygen atom has an oxidation number of −1. |
| 4  In compounds containing hydrogen, each hydrogen atom has an oxidation number of +1 (exceptions are metal hydrides such as NaH, in which the oxidation number of hydrogen is −1). | In $NH_3$ and $H_2O$ each hydrogen atom has an oxidation number of +1.<br>In KH the hydrogen atom has an oxidation number of −1. |
| 5  For a molecule, the sum of the oxidation numbers of the atoms equals zero. | The sum of the oxidation numbers of the atoms in $CH_4$ is zero. As each hydrogen atom has an oxidation number of +1, the oxidation number of the carbon atom is −4:<br>$(x + (4 \times +1) = 0, x = -4)$ |
| 6  For a polyatomic ion, the sum of the oxidation numbers of the atoms equals the charge on the ion. | The sum of the oxidation numbers of the atoms in $PO_4^{3-}$ is −3. As each oxygen atom has an oxidation number of −2, the oxidation number of the phosphorus atom is +5:<br>$(x + (4 \times -2) = -3, x = +5)$ |
| 7  In a compound, the most electronegative atom is assigned the negative oxidation number. | In $SF_6$, the oxidation number of each fluorine atom is −1. The oxidation number of the sulfur atom is +6:<br>$(x + (6 \times -1) = 0, x = +6)$ |

### Worked example 1

Determine the oxidation number of each element in each of the following species:

**a** $H_2S$  **b** $H_2SO_4$  **c** $KClO_4$

### Solution

**a** Each hydrogen atom has an oxidation number of +1 (rule 4)—which gives a total of +2. The sum of the oxidation numbers is 0 for a molecule (rule 5).

$+2 + x = 0$
$x = -2$

Therefore the oxidation number of S is −2.

**b** Each hydrogen atom has an oxidation number of +1 (rule 4)—a total of +2.

Each oxygen atom has an oxidation number of −2 (rule 3)—a total of −8.

The sum of the oxidation numbers is 0 for a molecule (rule 5).

$+2 + x − 8 = 0$
$x = +6$

Therefore the oxidation number of S is +6.

**c** $KClO_4$ is ionic, with $K^+$ and $ClO_4^-$ ions.

The oxidation number of K is +1 (rule 2).

Each oxygen atom has an oxidation number of –2 (rule 3)—a total of –8.

The sum of the oxidation numbers of the Cl atom and the four O atoms is –1, the charge of the ion (rule 6).

$x - 8 = -1$
$x = +7$

Therefore the oxidation number of Cl is +7.

---

Recall that oxidation numbers are artificial; they have no physical meaning. When we say that carbon has an oxidation number of +4 in $CO_2$, we do not mean that the species $C^{+4}$ actually exists. The +4 is assigned for convenience. Consider again the reaction of carbon with oxygen to produce carbon dioxide, and assign oxidation numbers to all atoms involved. Note that oxidation numbers are written above each element.

$$\overset{0}{C}(s) + \overset{0}{O_2}(g) \rightarrow \overset{+4\ -2}{CO_2}(g)$$

The oxidation number of carbon has increased from 0 to +4. This increase in oxidation number shows that oxidation has occurred. For oxygen, the oxidation number has decreased from 0 to –2. This decrease in oxidation number shows that reduction has occurred. Notice that we now have four definitions of oxidation: gain of oxygen, loss of hydrogen, loss of electrons and increase in oxidation number. While any of these definitions may be used to recognize a redox reaction, in practice it is the one based on oxidation numbers that is usually the most convenient.

**Oxidation is an increase in oxidation number.**

**Reduction is a decrease in oxidation number.**

**AS** 9.1.4
**Deduce whether an element undergoes oxidation or reduction in reactions using oxidation numbers.** © IBO 2007

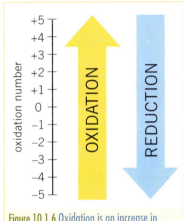

Figure 10.1.6 Oxidation is an increase in oxidation number. Reduction is a decrease in oxidation number.

## Worked example 2

Use oxidation numbers to determine which reactant has been oxidized and which has been reduced.

$SO_2(g) + NO_2(g) \rightarrow SO_3(g) + NO(g)$

### Solution

Assign oxidation numbers to all atoms using the set of rules provided in table 10.1.1.

$$\overset{+4\ -2}{SO_2}(g) + \overset{+4\ -2}{NO_2}(g) \rightarrow \overset{+6\ -2}{SO_3}(g) + \overset{+2\ -2}{NO}(s)$$

The oxidation number of sulfur has changed from +4 in $SO_2$ to +6 in $SO_3$. An increase in oxidation number shows that oxidation has occurred.

The oxidation number of nitrogen has changed from +4 in $NO_2$ to +2 in NO. A decrease in oxidation number shows that reduction has occurred.

Oxidation numbers are also used in the systematic naming of compounds. Transition metal ions can occur in a range of oxidation states. The oxidation number of the transition metal ion is indicated in the name of the compound. For example, iron(II) oxide has the formula FeO, as the oxidation number of the iron is +2. This is indicated in the name by the Roman numerals (II). Iron(III) oxide, $Fe_2O_3$, contains iron with an oxidation number of +3.

Other elements, such as nitrogen, can occur in a range of oxidation states, so for these elements the oxidation state (oxidation number) of the variable element can also be indicated in the name using Roman numerals. For example the brown, noxious gas $NO_2$ is known by the name nitrogen dioxide, but systematically it would be nitrogen(IV) oxide, while nitrous oxide, $N_2O$, also known as laughing gas and sometimes used as an anaesthetic, is systematically named nitrogen(I) oxide.

## Section 10.1 Exercises

1. Which of the following reactions would be classed as redox reactions? Explain your choices.
   a. Treatment of copper(II) sulfide to produce copper
   b. Melting of ice to form liquid water
   c. Corrosion of iron to iron(III) oxide ($Fe_2O_3$)
   d. Burning of magnesium in air

2. Determine the oxidation number of each element in each of the following species.
   a. MgO
   b. $H_2O$
   c. $HPO_4^-$
   d. $C_2H_4$
   e. $N_2O_4$
   f. $KMnO_4$

3. Determine the oxidation number of the metal in each of the following metal oxides and hence write the name of the compound using Roman numerals to show the oxidation number of the metal.
   a. CuO
   b. $Cu_2O$
   c. $MnO_2$
   d. $V_2O_5$

4. Complete the following table for oxides of nitrogen.

   TABLE 10.1.2 OXIDES OF NITROGEN AND THEIR NAMES

   | Oxide of nitrogen | Systematic name |
   |---|---|
   | $N_2O$ | |
   | NO | |
   | $NO_2$ | |
   | $N_2O_3$ | |
   | $N_2O_5$ | |

5. Use examples to show that the oxidation numbers of carbon atoms in carbon compounds range from −4 to +4.

6. For each of the questions given below, use oxidation numbers to determine whether the bolded chemical is being oxidized (O), reduced (R), neither oxidized nor reduced (N), or both oxidized and reduced (B).
   a. $Cl_2(g) + 2e^- \rightarrow 2Cl^-(aq)$
   b. $Ni(s) \rightarrow Ni^{2+}(aq) + 2e^-$
   c. $Ba^{2+}(aq) + SO_4^{2-}(aq) \rightarrow BaSO_4(s)$
   d. $2Fe^{2+}(aq) \rightarrow Fe(s) + Fe^{3+}(aq)$

7. Use oxidation numbers to show that the respiration reaction shown in the equation below is a redox reaction.

   $C_6H_{12}O_6(aq) + 6O_2(g) \rightarrow 6CO_2(g) + 6H_2O(l)$

8. Use oxidation numbers to identify which reactant has been oxidized in each of the following redox reactions.

   **a** $CuO(s) + H_2(g) \rightarrow Cu(s) + H_2O(l)$
   **b** $Cr_2O_7^{2-}(aq) + 14H^+(aq) + 6I^-(aq) \rightarrow 2Cr^{3+}(aq) + 3I_2(l) + 7H_2O(l)$
   **c** $MnO_2(s) + SO_2(aq) \rightarrow Mn^{2+}(aq) + SO_4^{2-}(aq)$

## 10.2 REDOX EQUATIONS

Consider the reaction between copper oxide and zinc:

$CuO(s) + Zn(s) \rightarrow ZnO(s) + Cu(s)$

This can be written as two half-equations:

Oxidation of zinc:   $Zn \rightarrow Zn^{2+} + 2e^-$

Reduction of copper ions:   $Cu^{2+} + 2e^- \rightarrow Cu$

During this reaction, two electrons are transferred from the zinc to the copper ions. The oxygen does not change; it is a spectator ion.

### Oxidizing agents and reducing agents

The zinc is an electron donor, which allows the copper oxide to be reduced. The zinc is therefore called a **reducing agent** or **reductant**. Similarly, the copper ion is an electron acceptor, which allows the zinc to be oxidized. The copper is therefore called an **oxidizing agent** or **oxidant**.

**A reducing agent causes another reactant to be reduced.**

**An oxidizing agent causes another reactant to be oxidized.**

Although it is easier to remember the definitions of reducing agent and oxidizing agent as given above, it is important to realize that, in a redox reaction, the reducing agent will always be oxidized and the oxidizing agent will always be reduced.

When copper turnings are placed in silver nitrate solution, $AgNO_3$, the solution will gradually turn blue and the solid will be a dark grey colour.

The equation for this reaction is:

$Cu(s) + 2AgNO_3(aq) \rightarrow 2Ag(s) + Cu(NO_3)_2(aq)$

Examination of the oxidation numbers shows that the oxidation number of copper increases from 0 to +2, while that of silver decreases from +1 to 0. The copper is being oxidized and the silver is being reduced.

The reactant that must be causing copper to lose electrons (be oxidized) is $Ag^+$, so it is the oxidizing agent. It is, itself, being reduced and so gains electrons to form Ag.

The reactant that is causing silver ions to gain electrons (be reduced) is Cu, so it is the reducing agent. It is, itself, being oxidized and so loses electrons to form $Cu^{2+}$.

Figure 10.2.1 Zinc reduces copper oxide in a spectacular fashion.

Redox chemistry of iron and copper
Redox chemistry of tin and zinc
Formation of silver crystals

 9.2.3
**Define the terms *oxidizing agent* and *reducing agent*.** © IBO 2007

DEMO 10.1
Crystal trees

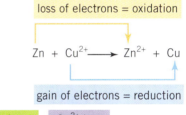

Figure 10.2.2 Zn atoms donate electrons and are therefore oxidized. $Cu^{2+}$ ions accept electrons and are therefore reduced. Zn acts as a reducing agent, $Cu^{2+}$ acts as an oxidizing agent.

 9.2.1
**Deduce simple oxidation and reduction half-equations given the species involved in a redox reaction.** © IBO 2007

 9.2.4
**Identify the oxidizing and reducing agents in redox equations.** © IBO 2007

### Worked example 1

When magnesium is burnt in air, white magnesium oxide is produced. The equation for the reaction is $2Mg(s) + O_2(g) \rightarrow 2MgO(s)$.

For this reaction:

**a** use oxidation numbers to identify which reactant is oxidized and which is reduced

**b** write half-equations to show the oxidation and reduction reactions

**c** identify the oxidizing agent and the reducing agent.

#### Solution

**a** Oxidation numbers:

$$\overset{0}{2Mg(s)} + \overset{0}{O_2(g)} \rightarrow \overset{+2\ -2}{2MgO(s)}$$

Magnesium has increased in oxidation number (from 0 to +2)

∴ it has been oxidized.

Oxygen has decreased in oxidation number (from 0 to –2)

∴ it has been reduced.

**b** Oxidation half-equation: $Mg \rightarrow Mg^{2+} + 2e^-$

Reduction half-equation: $O_2 + 4e^- \rightarrow 2O^{2-}$

**c** Magnesium causes the oxygen to be reduced, ∴ Mg is the reducing agent.

Oxygen causes the magnesium to be oxidized ∴ $O_2$ is the oxidizing agent.

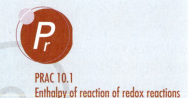

PRAC 10.1
Enthalpy of reaction of redox reactions

### Worked example 2

When a strip of nickel is placed in a silver nitrate solution, the nickel dissolves and a deposit of silver forms.

**a** Write half-equations and an overall ionic equation for this reaction.

**b** Identify the oxidizing agent and the reducing agent.

#### Solution

**a**

| | | |
|---|---|---|
| Nickel dissolves: | $Ni(s) \rightarrow Ni^{2+}(aq) + 2e^-$ | oxidation |
| Silver forms: | $(Ag^+(aq) + e^- \rightarrow Ag(s)) \times 2$ | reduction |
| Overall ionic equation: | $Ni(s) + 2Ag^+(aq) \rightarrow Ni^{2+}(aq) + 2Ag(s)$ | redox |

Notice that the $Ag^+/Ag$ half-equation must be multiplied by two in order to provide the two electrons needed for the $Ni/Ni^{2+}$ reaction.

**b** Ni has undergone oxidation. It is therefore an electron donor, the reducing agent. $Ag^+$ has undergone reduction. It is therefore an electron acceptor, the oxidizing agent.

## THEORY OF KNOWLEDGE

### Shields Silver Polish

Shields Silver Polish with tarnish control is formulated to gently clean and safely remove even the heaviest tarnish from silver. The quick rinse formula lets you finish the job in less time. Your silver is left with a sparkling shine and an anti-tarnish shield that controls tarnish build-up for months.

Advertisers often use the science to promote the effectiveness of their product and make its claims more believable.

- How does the language in this advertisement use the authority of science to sell silver polish?
- What limitations are there in relying on second-hand sources of knowledge such as this?
- Are the claims made by the advertiser scientifically testable?

## CHEM COMPLEMENT

### How best to clean the silverware?

The dark tarnish that forms on silver objects is a silver sulfide ($Ag_2S$). As polishing may remove silver, a better method of cleaning is to use a redox reaction. The tarnished object is placed in an aluminium dish or on aluminium foil in a non-metallic dish. A hot solution of baking soda is added. The redox reaction below occurs, safely removing the tarnish.

$Ag_2S(s) + Al(s) \rightarrow Al_2S(s) + Ag(s)$

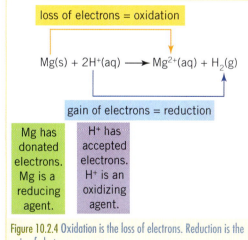

Figure 10.2.3 A redox reaction safely removes tarnish from silverware.

## Writing equations for redox reactions

Half-equations clearly show the electron transfer occurring in redox reactions. For cases such as the $Cu^{2+}/Zn$ displacement reaction, the writing of half-equations, given an overall equation, is relatively straightforward. The conjugate pairs $Cu^{2+}/Cu$ and $Zn^{2+}/Zn$ are easily identified, and the half-equations are simply balanced by adding the appropriate numbers of electrons.

Consider the reaction between hydrochloric acid and magnesium, which produces hydrogen gas. The overall equation for the reaction is:

$$2HCl(aq) + Mg(s) \rightarrow H_2(g) + MgCl_2(aq)$$

In the ionic equation, we can see that the reaction occurs between magnesium atoms and hydrogen ions, while chloride ions act as spectator ions (they are not involved, but merely 'watch' the reaction).

$$2H^+(aq) + Mg(s) \rightarrow H_2(g) + Mg^{2+}(aq)$$

loss of electrons = oxidation

$Mg(s) + 2H^+(aq) \longrightarrow Mg^{2+}(aq) + H_2(g)$

gain of electrons = reduction

Mg has donated electrons. Mg is a reducing agent.

$H^+$ has accepted electrons. $H^+$ is an oxidizing agent.

Figure 10.2.4 Oxidation is the loss of electrons. Reduction is the gain of electrons.

**AS** 9.2.2
Deduce redox equations using half-equations. © IBO 2007

Balancing redox equations

By separating the ionic equation into two half-equations we can focus on what is happening to each element. The conjugate pairs are $H^+/H_2$ and $Mg/Mg^{2+}$. Reactants are still on the left and products on the right and the equations are balanced both for numbers of atoms and for charge.

$$2H^+(aq) + 2e^- \rightarrow H_2(g)$$

$$Mg(s) \rightarrow Mg^{2+}(aq) + 2e^-$$

Writing half-equations for redox reactions involving polyatomic molecules such as $H_2O_2$ and polyatomic ions such as $MnO_4^-$ is a little more complex. Here we need a series of steps involving $H^+(aq)$ ions and $H_2O(l)$ molecules to produce a balanced half-equation. The steps involved are outlined in table 10.2.1.

### TABLE 10.2.1 BALANCING REDOX HALF-EQUATIONS

| Procedure | Examples | |
|---|---|---|
| | Oxidation of $SO_2$ to $SO_4^{2-}$ in acidic solution | Reduction of $Cr_2O_7^{2-}$ to $Cr^{3+}$ in acidic solution |
| 1 Write the formula of the species being oxidized/reduced and its conjugate pair. | $SO_2 \rightarrow SO_4^{2-}$ (oxidation number of S increases from +4 to +6) | $Cr_2O_7^{2-} \rightarrow Cr^{3+}$ (oxidation number of Cr decreases from +6 to +3) |
| 2 Balance the number of atoms of the species being oxidized/reduced. | $SO_2 \rightarrow SO_4^{2-}$ (S is already balanced) | $Cr_2O_7^{2-} \rightarrow 2Cr^{3+}$ |
| 3 Balance the number of oxygen atoms by adding $H_2O$ to the equation. | $SO_2 + 2H_2O \rightarrow SO_4^{2-}$ | $Cr_2O_7^{2-} \rightarrow 2Cr^{3+} + 7H_2O$ |
| 4 Balance the number of hydrogen atoms by adding $H^+$ to the equation ($H^+$ comes from the acidified solution). | $SO_2 + 2H_2O \rightarrow SO_4^{2-} + 4H^+$ | $Cr_2O_7^{2-} + 14H^+ \rightarrow 2Cr^{3+} + 7H_2O$ |
| 5 Balance the charges by adding electrons. | $SO_2 + 2H_2O \rightarrow SO_4^{2-} + 4H^+ + 2e^-$ (charge on both sides is zero) | $Cr_2O_7^{2-} + 14H^+ + 6e^- \rightarrow 2Cr^{3+} + 7H_2O$ (charge on both sides is +6) |
| 6 Add states for all species. | $SO_2(g) + 2H_2O(l) \rightarrow SO_4^{2-}(aq) + 4H^+(aq) + 2e^-$ | $Cr_2O_7^{2-}(aq) + 14H^+(aq) + 6e^- \rightarrow 2Cr^{3+}(aq) + 7H_2O(l)$ |

WORKSHEET 10.1
Oxidation numbers and redox equations

Notice in table 10.2.1 that step 4 requires the presence of $H^+$ ions. Many redox reactions involving polyatomic species only occur in acidified solutions; that is, solutions where $H^+$ ions are available. If the solution is not acidified the redox reaction will not occur. For simpler redox reactions, for example, the $Zn/Cu^{2+}$ reaction, the same steps may be followed, but steps 3 and 4 will not be required.

Using the steps shown in table 10.2.1 allows half-equations to be written for both oxidation and reduction processes. Once these half-equations are written, one further step is required to obtain the balanced overall redox equation. In a redox reaction the number of electrons released during oxidation must equal the number of electrons gained during reduction. Before half-equations can be combined they must involve the same number of electrons. Consider the example of the oxidation of $SO_2$ by $Cr_2O_7^{2-}$ to form $SO_4^{2-}$ and $Cr^{3+}$. From table 10.2.1 we have:

$$SO_2(g) + 2H_2O(l) \rightarrow SO_4^{2-}(aq) + 4H^+(aq) + 2e^- \quad \text{oxidation}$$

$$Cr_2O_7^{2-}(aq) + 14H^+(aq) + 6e^- \rightarrow 2Cr^{3+}(aq) + 7H_2O(l) \quad \text{reduction}$$

The oxidation produces 2 electrons, but the reduction requires 6. The oxidation half-equation must be multiplied by three in order to achieve electron balance.

$$3SO_2(g) + 6H_2O(l) \rightarrow 3SO_4^{2-}(aq) + 12H^+(aq) + 6e^- \quad \text{oxidation}$$
$$Cr_2O_7^{2-}(aq) + 14H^+(aq) + 6e^- \rightarrow 2Cr^{3+}(aq) + 7H_2O(l) \quad \text{reduction}$$
$$Cr_2O_7^{2-}(aq) + 2H^+(aq) + 3SO_2(g) \rightarrow 2Cr^{3+}(aq) + H_2O(l) + 3SO_4^{2-}(aq) \quad \text{redox}$$

When adding the two half-equations, the electrons and some $H^+$ ions and $H_2O$ molecules 'cancel' one another. Although redox reactions such as this one may appear complex, the writing of such equations can be easily mastered by a careful, step-by-step approach (and plenty of practice!).

## Section 10.2 Exercises

**1** Distinguish between the following terms.
  **a** Oxidation and reduction
  **b** Oxidizing agent and reducing agent

**2** For each of the following redox reactions, identify the oxidizing agent and the reducing agent.
  **a** $Cu(s) + Fe^{2+}(aq) \rightarrow Cu^{2+}(aq) + Fe(s)$
  **b** $2Ag^+(aq) + Sn^{2+}(aq) \rightarrow 2Ag(s) + Sn^{4+}(aq)$
  **c** $Ca(s) + 2H_2O(l) \rightarrow Ca(OH)_2(aq) + H_2(g)$
  **d** $2Mg(s) + O_2(g) \rightarrow 2MgO(s)$
  **e** $Cu^{2+}(s) + Fe(s) \rightarrow Cu(s) + Fe^{2+}(aq)$
  **f** $F_2(g) + Ni(s) \rightarrow NiF_2(s)$
  **g** $2Fe^{3+}(aq) + 2I^-(aq) \rightarrow 2Fe^{2+}(aq) + I_2(l)$

**3** Rewrite each of the following as two half-equations, indicating which represents oxidation and which represents reduction.
  **a** $Zn(s) + 2H^+(aq) \rightarrow Zn^{2+}(aq) + H_2(g)$
  **b** $2Mg(s) + O_2(g) \rightarrow 2MgO(s)$
  **c** $F_2(g) + Ni(s) \rightarrow NiF_2(aq)$
  **d** $Cu^{2+}(aq) + 2Fe^{2+}(aq) \rightarrow Cu(s) + 2Fe^{3+}(aq)$
  **e** $2Ag^+(aq) + Sn^{2+}(aq) \rightarrow 2Ag(s) + Sn^{4+}(aq)$
  **f** $2Na(s) + 2H_2O(l) \rightarrow 2NaOH(aq) + H_2(g)$

**4** Use oxidation numbers to identify the oxidizing agent and reducing agent in each of the following redox reactions:
  **a** $CuO(s) + H_2(g) \rightarrow Cu(s) + H_2O(l)$
  **b** $Cr_2O_7^{2-}(aq) + 14H^+(aq) + 6I^-(aq) \rightarrow 2Cr^{3+}(aq) + 3I_2(l) + 7H_2O(l)$
  **c** $MnO_2(s) + SO_2(aq) \rightarrow Mn^{2+}(aq) + SO_4^{2-}(aq)$

**5** Use oxidation numbers to decide whether or not any of the following equations represents a redox reaction. For each equation that does represent a redox reaction, identify the oxidizing agent and the reducing agent.
  **a** $2H_2(g) + O_2(g) \rightarrow 2H_2O(l)$
  **b** $SO_2(g) + H_2O(l) \rightarrow H_2SO_3(aq)$
  **c** $MgO(s) + H_2SO_4(aq) \rightarrow MgSO_4(aq) + H_2O(l)$
  **d** $2KMnO_4(aq) + 5H_2S(aq) + 6HCl(aq)$
  $\rightarrow 2MnCl_2(aq) + 5S(s) + 2KCl(aq) + 8H_2O$

6 Write balanced half-equations for the following processes:
   a  $O_2(g)$ to $H_2O(l)$
   b  $NO_2(g)$ to $NO_3^-(aq)$
   c  $HOCl(aq)$ to $Cl_2(g)$
   d  $MnO_4^-(aq)$ to $Mn^{2+}(aq)$
   e  $H_2S(g)$ to $SO_2(g)$
   f  $H_2O_2(aq)$ to $O_2(g)$
   g  $H_2O_2(aq)$ to $H_2(g)$
   h  $C_2O_4^{2-}(aq)$ to $CO_2(g)$

7 Shown below are four skeleton equations for redox reactions. Write a balanced equation for each reaction by first writing oxidation and reduction half-equations.
   a  $S_2O_3^{2-}(aq) + I_2(aq) \rightarrow S_4O_6^{2-}(aq) + I^-(aq)$
   b  $Zn(s) + NO_3^-(aq) + H^+(aq) \rightarrow Zn^{2+}(aq) + NO_2(g) + H_2O(l)$
   c  $ClO^-(aq) \rightarrow Cl^-(aq) + ClO_3^-(aq)$
   d  $Cr_2O_7^{2-}(aq) + CH_3CH_2OH(aq) + H^+(aq) \rightarrow Cr^{3+}(aq) + CH_3COOH(aq) + H_2O(l)$

## 10.3 VOLTAIC CELLS

The earliest electrochemical cell was the **voltaic cell**, constructed by Alessandro Volta (1745–1827). This cell consisted of a sheet of copper metal and a sheet of zinc metal separated by a sheet of paper soaked in concentrated salt solution. It produced very little electricity, so Volta came up with a better design by joining several cells together to form a battery. The correct name for a single chemical unit producing electricity is a *cell*. Several cells joined together in series is termed a *battery*. Volta's battery, known as the Volta pile, produced a steady electric current, and enabled scientists to study the effects of electric current. The unit for potential difference, the volt, is named in honour of Volta.

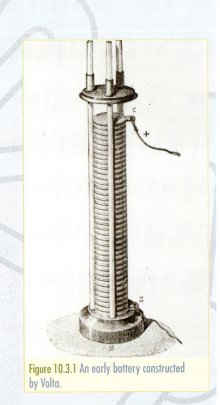

Figure 10.3.1 An early battery constructed by Volta.

### CHEM COMPLEMENT

#### A shocking experience!

While Volta is known and remembered for his contribution to the study of electrochemistry, not all his experimental methods bear repeating!

> I introduced into my ears two metal rods with rounded ends and joined them to the terminals of the apparatus. At the moment the circuit was completed I received a shock in the head—and began to hear a noise—a crackling and boiling. This disagreeable sensation, which I feared might be dangerous, has deterred me so that I have not repeated the experiment.  *Alessandro Volta*

A single electrochemical cell is sometimes named a voltaic cell, after Volta. The other frequently used name, a galvanic cell, is in honour of the Italian physiologist Luigi Galvani (1737–1798). Galvani discovered that frogs' legs could be made to twitch by connecting the nerve and muscle tissues to different metals such as copper and iron. He thought (incorrectly) that he had discovered a special form of 'animal electricity'.

The first voltaic cell produced on a large scale for practical commercial use was the Daniell cell. The first version of this Zn–Cu cell was constructed in 1836. Cells were placed in series to produce a battery. Large numbers of these were used in telegraph stations and railway signals until the beginning of the 20th century.

Today, we are heavily dependent on cells and batteries to provide a portable power source. These cells and batteries do not contain complex circuitry, but are simply voltaic cells arranged in a form that is rather more convenient than two beakers joined by a wet strip of filter paper! The design and operation of one commonly used cell is shown in figure 10.3.3.

Batteries have come a long way since Volta's pile! The technology of batteries is not static, and scientists are constantly researching new materials to make batteries cheaper or more efficient. New appliances are continually developed, laptops, mobile phones and iPods being typical examples, and many of these appliances require batteries with superior performance to those of the past. The use of electric vehicles and solar energy are both dependent on the availability of efficient, low-cost cells. For these reasons, modern battery manufacturers are continuing to research and develop more efficient and cheaper commercial cells.

Figure 10.3.2 Alessandro Volta (1745–1827), inventor of the first voltaic cell.

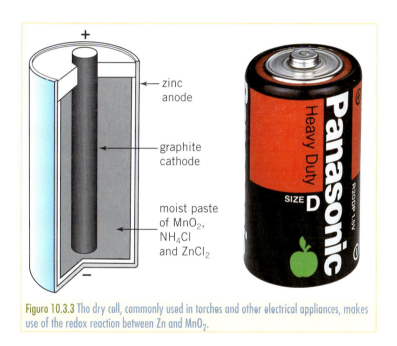

Figure 10.3.3 The dry cell, commonly used in torches and other electrical appliances, makes use of the redox reaction between Zn and $MnO_2$.

**DEMO 10.2**
A simple voltaic cell

**DEMO 10.3**
Constructing a voltaic pile

The reaction between zinc and copper ions can be described in terms of the transfer of electrons from the reducing agent, Zn, to the oxidizing agent, $Cu^{2+}$ ions. Can we actually detect this transfer of electrons? When the $Cu^{2+}$ ions and Zn atoms are in contact with each other, electron transfer is rapid and direct. A significant amount of heat energy is released. To detect the transfer of electrons, we must separate the $Cu^{2+}$ ions and Zn atoms and force the electrons to flow along a wire where the electron flow may be registered by a galvanometer (a meter for detecting current flow) or by a sensitive globe. Steps used to set up this current-producing arrangement are explained in table 10.3.1.

**9.4.1**
Explain how a redox reaction is used to produce electricity in a voltaic cell. © IBO 2007

### TABLE 10.3.1 CONSTRUCTING A VOLTAIC CELL

| Step | Explanation |
|---|---|
| 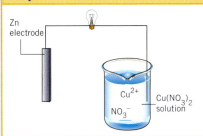 | The Zn and $Cu^{2+}$ ions are separated and connected by a wire. No current can yet be detected. |
| 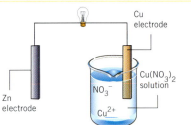 | A solid conducting surface is used as the site for each half-reaction. This solid is called the electrode. The electrode must not react with other species present, so a Cu strip is chosen for the $Cu^{2+}$ solution. Zn serves as the other electrode. |
| 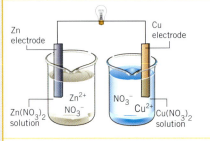 | A conducting solution is used to allow movement of reactant or product ions for each half-reaction. This solution is the electrolyte. The electrolyte must not react with other species present, so a $Zn(NO_3)_2$ solution is chosen for the Zn; $Cu(NO_3)_2$ serves as the other electrolyte. No current can yet be detected. The electrical circuit is not complete. |
| 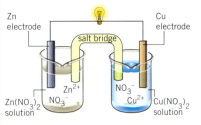 | The beakers containing the $Zn^{2+}/Zn$ and $Cu^{2+}/Cu$ are connected by an electrolyte that allows the flow of ions to and from each beaker. A convenient form is a piece of filter paper soaked in a salt solution. This electrolytic conductor connecting the two beakers is called a salt bridge. A current can now be detected. |
| 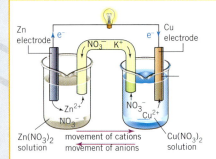 | Electrons are generated at the Zn electrode, giving this electrode a negative charge.<br>$$Zn(s) \rightarrow Zn^{2+}(aq) + 2e^-$$<br>Electrons flow towards the Cu electrode where they are used. This gives the Cu electrode a positive charge.<br>$$Cu^{2+}(aq) + 2e^- \rightarrow Cu(s)$$<br>Ions flow through the salt bridge ($NO_3^-$ towards the $Zn^{2+}/Zn$ beaker, $K^+$ towards the $Cu^{2+}/Cu$ beaker) to complete the circuit. |

**PRAC 10.2**
Voltaic cells

This completed current-producing arrangement is known as an electrochemical, voltaic or galvanic cell. Each part of the cell, in this case the $Zn^{2+}/Zn$ beaker and $Cu^{2+}/Cu$ beaker, is known as a half-cell. The **half-cell** contains a conjugate redox pair: an electrode and an electrolyte. An external circuit, consisting of conducting wires and a current-detecting device, connects the half-cells. Half-cells are also connected by an internal circuit (a salt bridge). This consists of an electrolytic conductor that allows ions to flow between the half-cells.

## Components of a voltaic cell

### Electrodes

Electrodes serve as the site for the oxidation and reduction reactions. At the surface of each electrode, electrons are either donated or accepted. The electrode at which oxidation occurs is called the **anode**. It is the *negative electrode*, as electrons are generated here (the electron source). The electrode at which reduction occurs is called the **cathode**. It is the *positive electrode* (the electron 'sink'), as electrons are consumed by the oxidizing agent in the solution at the positive electrode.

 9.4.2
**State that oxidation occurs at the negative electrode (anode) and reduction occurs at the positive electrode (cathode).**
© IBO 2007

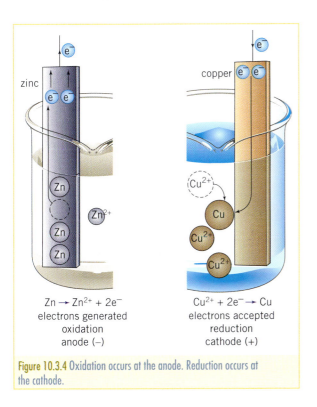

Figure 10.3.4 Oxidation occurs at the anode. Reduction occurs at the cathode.

Voltaic cells I
Voltaic cells II

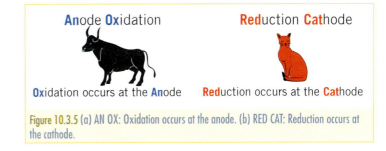

Figure 10.3.5 (a) AN OX: Oxidation occurs at the anode. (b) RED CAT: Reduction occurs at the cathode.

Electrodes are electronic conductors—they conduct current by the flow of electrons. They may be made of metal or graphite. Where a half-cell contains a metal reducing agent, the metal serves as the electrode. For example, zinc is an electrode in the $Zn^{2+}/Zn$ half cell. Where the reducing agent is not a metal, the electrode is either graphite or a non-reactive metal such as platinum. Examples are the graphite electrode in a $Fe^{3+}/Fe^{2+}$ half-cell, or a platinum electrode in a $Cl_2/Cl^-$ half-cell.

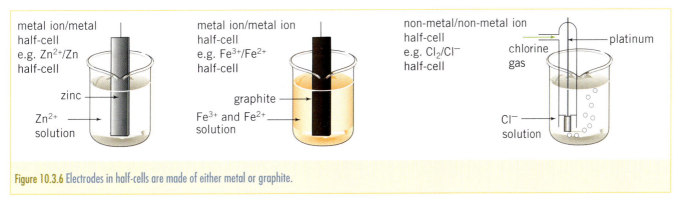

Figure 10.3.6 Electrodes in half-cells are made of either metal or graphite.

### The salt bridge

The salt bridge completes the circuit in a voltaic cell. It is an electrolytic conductor, conducting current by the flow of ions. These mobile ions migrate into each half-cell during cell operation to prevent the build-up of charges. Consider the $Zn^{2+}/Zn$ half-cell. As the Zn atoms are oxidized, $Zn^{2+}$ ions enter the half-cell solution. There are now more positive ions ($Zn^{2+}$) than negative ions ($NO_3^-$). This build-up of positive charge would soon stop the flow of negatively charged electrons away from the half-cell. Build-up is prevented by negative ions migrating into the half-cell from the salt bridge. Similarly, in the $Cu^{2+}/Cu$ half-cell, $Cu^{2+}$ ions are used up, leading to a build-up of negative charge in this half-cell. Positive ions from the salt bridge migrate into the half-cell to prevent charge build-up. In this way, the salt bridge maintains the electrical neutrality of each half-cell.

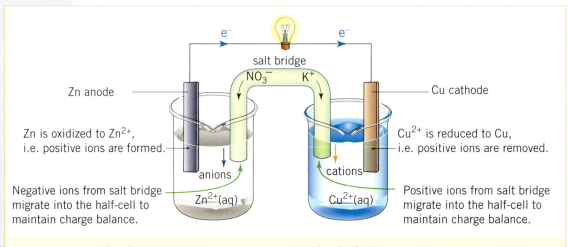

Figure 10.3.7 Movement of ions from the salt bridge maintains the electrical neutrality of half-cells.

The salt chosen for the salt bridge must meet two criteria: it must be soluble and it must not react with any of the species present in either half-cell. A suitable substance to use in many voltaic cells is aqueous potassium nitrate ($KNO_3(aq)$).

### Designing voltaic cells

When provided with a combination of two half-cells and the information necessary to establish which species is most reactive, we need to be able to determine the following:

- half-equations representing reactions in each half-cell
- overall ionic equation
- polarity of electrodes and their nature (anode and cathode)
- oxidizing agent and reducing agent
- direction of flow of electrons through the conducting wires and of ions through the salt bridge
- physical changes occurring at the electrodes or in their vicinity (metal deposition, electrode dissolution, changes in pH and so on).

WORKSHEET 10.2
Voltaic cells

## Worked example

Design a voltaic cell from iron and magnesium and their respective ions, given that magnesium is a stronger reducing agent than iron. Draw a diagram of the cell and fully label it, showing relevant half-equations, the polarity of the electrodes and the direction of flow of both the electrons through the wires and the ions through the potassium nitrate salt bridge. Label the anode and cathode and the oxidizing agent and reducing agent.

### Solution

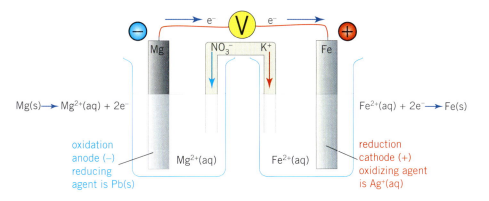

Magnesium is a stronger reducing agent than iron and so will be preferentially oxidized. Magnesium metal will be oxidized to magnesium ions and the iron(II) ions will be reduced to iron metal. The electrons released by the magnesium travel through the wires to the iron, making the magnesium electrode negative and the iron electrode positive. As oxidation occurs at the anode, the magnesium electrode must be the anode and iron the cathode. The strongest reducing agent (Mg) undergoes oxidation and the strongest oxidizing agent ($Fe^{2+}$) undergoes reduction. The overall equation for the reaction is derived by adding the oxidation and reduction half-equations in such a way that the electrons are balanced, and so cancel:

$Fe^{2+}(aq) + 2e^- \rightarrow Fe(s)$

$Mg(s) \rightarrow Mg(aq) + 2e^-$

Overall equation:

$Fe^{2+}(aq) + Mg(s) \rightarrow Fe(s) + Mg^{2+}(aq)$

The salt bridge consists of a saturated solution of potassium nitrate ($KNO_3$). The negative $NO_3^-$ ions in the salt bridge move into the magnesium half-cell solution to balance the extra $Mg^{2+}$ ions formed, while the positive ions move into the iron half-cell to replace the $Fe^{2+}$ ions being consumed.

## Section 10.3 Exercises

1. Explain the meaning of each of the following terms.
   a. Electrode
   b. Electrolyte
   c. Anode
   d. Cathode
   e. Half-cell

2. State two properties a salt must have if it is to be used in the salt bridge of a voltaic cell.

3. Complete the labelling of the voltaic cell shown in the diagram below, which uses the reaction between Ni(s) and $Pb^{2+}$(aq) to produce electrical energy.

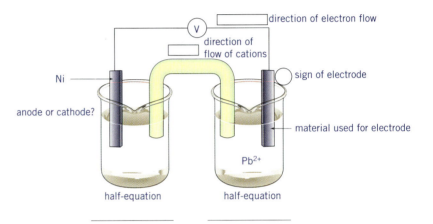

4. Suggest two reasons why the voltaic cell shown in question 3 might stop producing a current after several hours of operation.

5. A voltaic cell is set up by combining the $A^{2+}$(aq)/A(s) and $B^{2+}$(aq)/B(s) half-cells via a conducting wire and a salt bridge. After some time, the A and B electrodes are taken out of the half-cells and reweighed. The mass of the A electrode has increased, but the mass of the B electrode has decreased. Complete the labelling of the voltaic cell shown in the following diagram, including the overall redox equation for the reaction occurring in the cell.

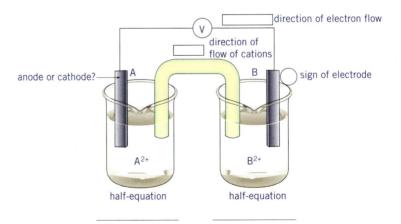

6  Copy and complete the table below.

| Reaction at anode (−) | Reaction at cathode (+) | Overall reaction | Oxidizing agent | Electrode material at cathode |
|---|---|---|---|---|
| $Cu(s) \rightarrow Cu^{2+}(aq) + 2e^-$ | $Ag^+(aq) + e^- \rightarrow Ag(s)$ | | | |
| $Mg(s) \rightarrow Mg^{2+}(aq) + 2e^-$ | | | $Pb^{2+}(aq)$ | |
| | | $2Fe^{3+}(aq) + 2I^-(aq) \rightarrow I_2(aq) + 2Fe^{2+}(aq)$ | | Pt (inert) |
| $Al(s) \rightarrow Al^{3+}(aq) + 3e^-$ | | | | Ni |

7  What is the role of a salt bridge in a voltaic cell?

8  On the diagram of a typical voltaic cell shown, indicate the direction of flow of:
   a  electrons through the conducting wires
   b  ions through the potassium nitrate salt bridge.

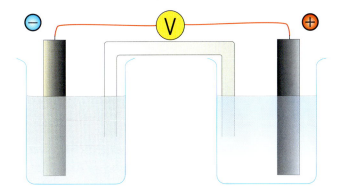

9  Sketch the half-cells in which each of the following reactions could occur.
   a  $Fe^{2+}(aq) + 2e^- \rightarrow Fe(s)$
   b  $Fe^{3+}(aq) + e^- \rightarrow Fe^{2+}(aq)$
   c  $Cl_2(aq) + 2e^- \rightarrow 2Cl^-(aq)$

10  A voltaic cell consists of half-cells of $Fe^{3+}/Fe^{2+}$ and $I_2/I^-$, both using an inert graphite electrode. $Fe^{3+}(aq)$ ion is a stronger oxidizing agent than $I_2(aq)$. Complete the labelling of the diagram of this cell shown below.

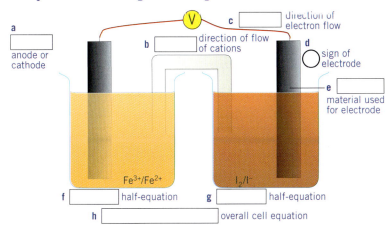

**AS** 9.3.1
**Deduce a reactivity series based on the chemical behaviour of a group of oxidizing and reducing agents. © IBO 2007**

## 10.4 REACTIVITY

Redox reactions involving metals placed in a metal ion solution such as the $Cu/Ag^+$ reaction, are also known as metal **displacement reactions**. One metal (Cu) has pushed or displaced another metal (Ag) from its salt solution. During a displacement reaction the solid metal acts as a reducing agent. Metals generally act as reducing agents (electron donors) because a relatively small amount of energy is needed to remove the small number of outer-shell electrons in metals. Metals clearly differ in their ability to act as reducing agents. Some, such as sodium, are so readily oxidized they must be stored under paraffin oil. Others, such as iron, corrode readily in moist air, while yet others, such as gold, are so hard to oxidize that they are found pure in nature.

By observing how readily metals react with oxygen, water, dilute acids and other metal salts, it is possible to determine an order of reactivity of metals. Metals may be ranked according to their ability to act as reducing agents. Table 10.4.1 shows such a ranking, or **reactivity series**, listing metals from the most reactive (strongest reducing agent) to the least reactive.

| TABLE 10.4.1 AN ACTIVITY SERIES FOR METALS ||| 
| Reactivity | Metals | Properties |
| --- | --- | --- |
| ↑ Increasing reactivity | K<br>Na<br>Ca<br>Mg | Burn readily in oxygen to form oxides<br>React with water to form hydrogen and hydroxides<br>React with acids to form hydrogen and salts |
| | Al<br>Zn<br>Cr<br>Fe<br>Cd<br>Co<br>Ni | Burn to form oxides if finely divided<br>React with steam to form hydrogen and oxides<br>React with cold acids to form hydrogen and salts |
| | Sn<br>Pb<br>Cu<br>Hg | Oxidize if heated in air or pure oxygen<br>No reaction with steam<br>Sn and Pb react with warm acids to form hydrogen and salts<br>Cu and Hg do not react with acids |
| | Ag<br>Pt<br>Au | Do not react with oxygen<br>No reaction with steam<br>No reaction with acids |

It can be noted from the activity series that the most reactive metals are found on the far left-hand side of the periodic table in groups 1 and 2. Transition metals and group 3 and group 4 metals are less reactive. Non-metals also vary in their reactivity. The most reactive, the strongest oxidizing agents, are those on the upper right-hand side of the table. Fluorine and oxygen are both very reactive as oxidizing agents (electron acceptors).

The metal activity series can be used to predict displacement reactions. Any metal will be oxidized by the ions of a less-reactive metal. For example, zinc is more reactive than copper and will therefore be oxidized by copper(II) ions. Chromium is more reactive than lead and will therefore be oxidized by lead(II) ions.

**WORKSHEET 10.3**
Metal reactivity and redox equations

## CHEM COMPLEMENT

### The Thermit reaction

A metal oxide can be reduced by heating it with another metal that is higher in the activity series. In this reaction, there is competition between the metals for oxygen; the more reactive metal, the one that holds oxygen more tightly, wins. Thermit, a powdered mixture of aluminium and iron(III) oxide, uses this reaction to produce incendiary bombs and for welding and foundry work. When Thermit is ignited, the reaction is rapid and very exothermic, and the iron formed is molten. In apparatus widely used in the past for welding railway tracks, the Thermit reaction occurred in its upper chamber and the molten iron flowed down through the bottom of the apparatus into the gap between the ends of two rails, welding the tracks together on cooling.

$Fe_2O_3(s) + 2Al(s) \rightarrow 2Fe(l) + Al_2O_3(s) + heat$

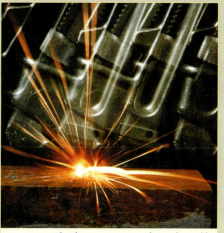

Figure 10.4.1 The Thermit reaction can be used to weld railway lines.

There is also, however, a wide range of other oxidizing agents and reducing agents, including non-metals. How are we to predict their reactivity with respect to one another and the metals?

The electrochemical series, as shown on page 323, is the most common data table used for comparing the reactivity of oxidizing agents and reducing agents. It is generated by preparing half-cells consisting of a conjugate redox pair and comparing them to a standard cell to determine their relative reactivity. To ensure consistency of results, all half-cells are prepared under standard conditions of concentration, temperature and pressure. These conditions are:

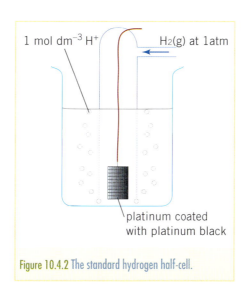

Figure 10.4.2 The standard hydrogen half-cell.

Thermit reaction

- 1 mol dm$^{-3}$ concentration of aqueous solutions
- 25°C temperature
- 1 atm pressure, if gases are used.

The cell to which all other half-cells are compared is called the standard hydrogen half-cell. It consists of a solution of 1 mol dm$^{-3}$ hydrochloric acid ([H$^+$] = 1 mol dm$^{-3}$) and hydrogen gas at 1 atm pressure (measured at 25°C) delivered through a glass tube to the half-cell. The hydrogen ions and the hydrogen gas are both in contact with a platinum electrode (coated with platinum black). The half-equation for the reaction occurring within this cell is:

$2H^+(aq) + 2e^- \rightarrow H_2(g)$

Figure 10.4.3 (a) Zinc is a more reactive metal than (b) copper.

Standard reduction potentials

## Potential difference

Potential difference is a measure of the energy difference between the reactants in an electrochemical cell, and is measured in units of volts. For a spontaneous reaction to occur, the oxidizing agent species must be sufficiently reactive to be able to oxidize the reducing agent. Thus, a strip of zinc metal will react with a solution of copper(II) ions, but copper metal will not react when placed into a solution of zinc ions.

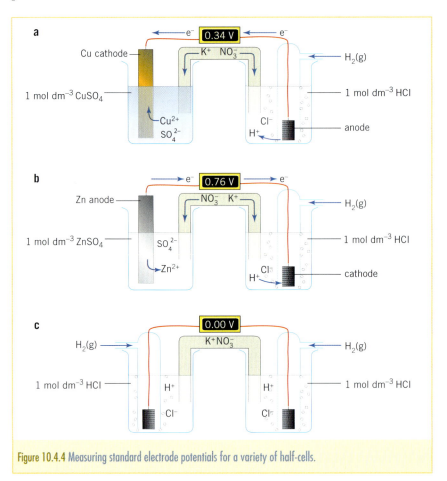

Figure 10.4.4 Measuring standard electrode potentials for a variety of half-cells.

Figure 10.4.4 shows the $Cu^{2+}/Cu$ and $Zn^{2+}/Zn$ half-cells connected to the standard hydrogen half-cell. $Cu^{2+}$ is a stronger oxidizing agent than $H^+$ and so it is preferentially reduced to copper metal. Reduction occurs at the copper electrode, and so it is acting as the cathode (+). In the $Zn^{2+}/Zn$ cell, the $H^+$ ions are stronger oxidizing agents than $Zn^{2+}$ and so are reduced in preference to the zinc ions. The zinc electrode is oxidized and so it is the anode (−).

The *potential difference* generated by a voltaic cell in which a particular half-cell is coupled to a hydrogen half-cell under standard conditions is known as the **standard electrode potential** ($E^\ominus$). If the $E^\ominus$ value for a particular half-cell is positive, it means that the oxidizing agent member of the conjugate pair is a stronger oxidizing agent than $H^+$ (the $E^\ominus$ for the $Cu^{2+}/Cu$ half-cell is + 0.34 V). A negative $E^\ominus$ value indicates a weaker oxidizing agent than $H^+$ ( the $E^\ominus$ for the $Zn^{2+}/Zn$ half-cell is −0.76 V). If the hydrogen half-cell is coupled with itself, there is no potential difference (as the cells are identical) and the standard electrode potential for the hydrogen half-cell is, by definition, 0.00 V. By convention, standard electrode potentials are always provided for reduction reactions, and together they make up the **electrochemical series**, or reduction potential table.

| Oxidized species ⇌ Reduced species | $E^⦵$ volts |
|---|---|
| $Li^+(aq) + e^- \rightleftharpoons Li(s)$ | −3.03 |
| $K^+(aq) + e^- \rightleftharpoons K(s)$ | −2.92 |
| $Ca^{2+}(aq) + 2e^- \rightleftharpoons Ca(s)$ | −2.87 |
| $Na^+(aq) + e^- \rightleftharpoons Na(s)$ | −2.71 |
| $Mg^{2+}(aq) + 2e^- \rightleftharpoons Mg(s)$ | −2.36 |
| $Al^{3+}(aq) + 3e^- \rightleftharpoons Al(s)$ | −1.66 |
| $Mn^{2+}(aq) + 2e^- \rightleftharpoons Mn(s)$ | −1.18 |
| $H_2O(l) + e^- \rightleftharpoons \frac{1}{2}H_2(g) + OH^-(aq)$ | −0.83 |
| $Zn^{2+}(aq) + 2e^- \rightleftharpoons Zn(s)$ | −0.76 |
| $Fe^{2+}(aq) + 2e^- \rightleftharpoons Fe(s)$ | −0.44 |
| $Ni^{2+}(aq) + 2e^- \rightleftharpoons Ni(s)$ | −0.23 |
| $Sn^{2+}(aq) + 2e^- \rightleftharpoons Sn(s)$ | −0.14 |
| $Pb^{2+}(aq) + 2e^- \rightleftharpoons Pb(s)$ | −0.13 |
| $H^+(aq) + e^- \rightleftharpoons \frac{1}{2}H_2(g)$ | 0.00 |
| $Cu^{2+}(aq) + e^- \rightleftharpoons Cu^+(aq)$ | +0.15 |
| $SO_4^{2-}(aq) + 4H^+(aq) + 2e^- \rightleftharpoons H_2SO_3(aq) + H_2O(l)$ | +0.17 |
| $Cu^{2+}(aq) + 2e^- \rightleftharpoons Cu(s)$ | +0.34 |
| $\frac{1}{2}O_2(g) + H_2O(l) + 2e^- \rightleftharpoons 2OH^-(aq)$ | +0.40 |
| $Cu^+(aq) + e^- \rightleftharpoons Cu(s)$ | +0.52 |
| $\frac{1}{2}I_2(s) + e^- \rightleftharpoons I^-(aq)$ | +0.54 |
| $Fe^{3+}(aq) + e^- \rightleftharpoons Fe^{2+}(aq)$ | +0.77 |
| $Ag^+(aq) + e^- \rightleftharpoons Ag(s)$ | +0.80 |
| $\frac{1}{2}Br_2(l) + e^- \rightleftharpoons Br^-(aq)$ | +1.09 |
| $\frac{1}{2}O_2(g) + 2H^+(aq) + 2e^- \rightleftharpoons H_2O(l)$ | +1.23 |
| $Cr_2O_7^{2-}(aq) + 14H^+(aq) + 6e^- \rightleftharpoons 2Cr^{3+}(aq) + 7H_2O(l)$ | +1.33 |
| $\frac{1}{2}Cl_2(g) + e^- \rightleftharpoons Cl^-(aq)$ | +1.36 |
| $MnO_4^-(aq) + 8H^+(aq) + 5e^- \rightleftharpoons Mn^{2+}(aq) + 4H_2O(l)$ | +1.51 |
| $\frac{1}{2}F_2(g) + e^- \rightleftharpoons F^-(aq)$ | +2.87 |

Figure 10.4.5 The electrochemical series. This can also be found in table 14 of the IB data booklet.

The electrochemical series shows the values obtained by connecting many half-cells to the standard hydrogen electrode. As all half-equations are written as reduction potentials, the species on the left-hand side of the series are all oxidizing agents, and the species on the right-hand side are reducing agents. The top right-hand corner shows the strongest reducing agent, lithium, and reducing agent strength decreases down the series. The bottom left-hand corner shows the strongest oxidizing agent, fluorine. Fluorine, the most electronegative element, has a strong tendency to accept electrons. Just as fluorine gas ($F_2$) is a powerful oxidizing agent, its conjugate, the fluoride ion ($F^-$), is a very weak reducing agent. The fluoride ion has a filled outer shell of eight electrons and has very little tendency to give away electrons.

At the top of the electrochemical series is the half-equation for the reduction of the lithium ion ($Li^+$). $Li^+$ is the weakest oxidized species (oxidizing agent) as it has the least tendency to accept electrons. Conversely, lithium metal is the strongest reduced species (reducing agent). In summary, oxidizing agent strength increases down the left-hand side of the electrochemical series and reducing agent strength increases up the right-hand side of the series.

### Predicting redox reactions using the electrochemical series

**AS 9.3.2 Deduce the feasibility of a redox reaction from a given reactivity series. © IBO 2007**

The electrochemical series allows the prediction of whether a redox reaction between two species will occur. To determine whether or not a spontaneous redox reaction will occur, we use the following steps:

1  Identify all possible reactants. Locate their half-equations on the electrochemical series. Underline the species present. (One reactant must be oxidized, while the other is reduced. This will be the case if you have underlined reactants on opposite sides of the series.)

2  The strongest oxidizing agent will undergo reduction. For this reason, the underlined oxidizing agent must be lower down on the left-hand side of the series than the underlined reducing agent, which will be on the right-hand side. Visually, a line joining the oxidizing agent and the reducing agent should diagonally cross the series going 'uphill' from left to right.

3  Write the reducing agent half-equation in reverse. The two half-equations can now be balanced and then added together.

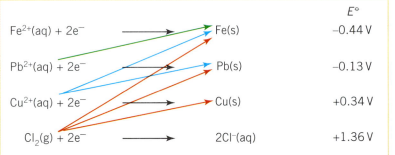

$Cl_2$, the strongest oxidizing agent, can oxidize Cu, Pb and Fe. $Cu^{2+}$ can oxidize Pb and Fe, and $Pb^{2+}$ can only oxidize Fe.

**Figure 10.4.6** Predicting reactions for selected conjugate redox pairs.

In chapter 3, section 3.3 the reactions between halogens and group 1 metals were discussed, as were reactions between halide ions and halogens. The electrochemical series can be used to predict the products of these reactions. The alkali metals are all strong reducing agents. They are found at the top on the right-hand side of the electrochemical series (figure 10.4.5). The halogens are all quite strong oxidizing agents. They are found on the left-hand side of the electrochemical series, at the bottom of the table. Consequently all the halogens will react with any of the alkali metals.

The displacement reactions between halide ions in solution and halogens can also be predicted from the electrochemical series. Of the halogens, fluorine, $F_2(g)$, is the strongest oxidizing agent, so it can displace any of the halide ions from solution. Chlorine, $Cl_2(g)$, is able to displace bromide and iodide ions from solution, and bromine, $Br_2$, can only displace iodide ions from solution.

**PRAC 10.3** Redox reactions

## Worked example

For each of the following combinations, determine if a reaction will occur. For each reaction that occurs, write:

 i half-equations for the oxidation and reduction reactions

 ii an overall ionic equation.

a Nickel metal and silver nitrate solution

b Iodine crystals in sodium chloride solution

c Chlorine gas bubbled through potassium iodide solution

### Solutions

a Reactants present:   Ni(s) and $Ag^+$(aq)

$Ni^{2+}$(aq) + 2e⁻ ⇌ **Ni(s)**          $E^\ominus = -0.23$ V

**$Ag^+$(aq)** + e⁻ ⇌ Ag(s)          $E^\ominus = +0.80$ V

The bolded oxidizing agent has a more positive $E^\ominus$ value than the bolded reducing agent, so a reaction can occur.

Note that the oxidation reaction involving Ni(s) must be reversed before the half-equation and the overall equation can be written.

Ni(s) → $Ni^{2+}$(aq) + 2e⁻

2 × [$Ag^+$(aq) + e⁻ → Ag(s)]

$2Ag^+$(aq) + Ni(s) → 2Ag(s) + $Ni^{2+}$(aq)

b Reactants present:   $I_2$(s), $Na^+$(aq), $Cl^-$(aq)

**$Na^+$(aq)** + e⁻ ⇌ Na(s)          $E^\ominus = -2.71$ V

$\frac{1}{2}I_2$(s) + e⁻ ⇌ $I^-$(aq)          $E^\ominus = +0.54$ V

$\frac{1}{2}Cl_2$(g) + e⁻ ⇌ **$Cl^-$(aq)**          $E^\ominus = +1.36$ V

No reaction will occur in this case. The oxidizing agents ($Na^+$ and $I_2$) are not strong enough (their $E^\ominus$ values are more negative than that of the reducing agents) to oxidize the reducing agent ($Cl^-$).

c Reactants present:   $Cl_2$(g); $K^+$(aq), $I^-$(aq)

**$K^+$(aq)** + e⁻ ⇌ K(s)          $E^\ominus = -2.71$ V

$\frac{1}{2}I_2$(s) + e⁻ ⇌ **$I^-$(aq)**          $E^\ominus = +0.54$ V

$\frac{1}{2}Cl_2$(g) + e⁻ ⇌ **$Cl^-$(aq)**          $E^\ominus = +1.36$ V

The second bolded oxidizing agent ($Cl_2$) has a more positive $E^\ominus$ value than the bolded reducing agent ($I^-$), so a reaction can occur. The other oxidizing agent ($K^+$) has a more negative $E^\ominus$ value than $I^-$(aq), so it does not react. It is a spectator ion.

Note that the oxidation reaction involving $I^-$(aq) must be reversed before the half-equation and the overall equation can be written.

$I^-$(aq) → $\frac{1}{2}I_2$(s) + e⁻

$\frac{1}{2}Cl_2$(g) + e⁻ → $Cl^-$(aq)

$I^-$(aq) + $\frac{1}{2}Cl_2$(g) → $\frac{1}{2}I_2$(s) + $Cl^-$(aq)

## Section 10.4 Exercises

1. Explain the following:
   a. Metals such as potassium and calcium are strong reducing agents.
   b. Non-metals such as oxygen and chlorine are strong oxidizing agents.
   c. Most metals in the Earth's crust are found combined with other elements.

2. W, X, Y and Z are metals.
   
   W reacts with dilute hydrochloric acid and steam, but not with hot or cold water.
   
   X reacts with hot water.
   
   Y does not react with water, steam or dilute hydrochloric acid.
   
   Z reacts vigorously with water.
   
   Arrange metals W, X, Y and Z in order of decreasing reactivity, that is, from the most reactive to the least reactive.

3. Using the activity series, write ionic equations for any reactions that would occur when:
   a. Fe is added to a solution containing $Ni^{2+}$ ions
   b. Ag is added to a solution containing $Pb^{2+}$ ions
   c. Mg is added to a solution containing $Sn^{2+}$ ions.

4. For each of the following situations, use the electrochemical series to determine whether or not a reaction will occur. For any reaction that occurs, write an overall equation for the process.
   a. Zinc chloride solution is stored in an iron container.
   b. Magnesium ribbon is added to a zinc iodide solution.
   c. Bromine liquid is poured into magnesium chloride solution.
   d. Sodium metal is dropped into a beaker of water.

5. Use the $E^\ominus$ values given for the following conjugate redox pairs to answer the questions below:

   | | | |
   |---|---|---|
   | I | $Sc^{2+}/Sc$ | $E^\ominus = -2.12$ |
   | II | $Zn^{2+}/Zn$ | $E^\ominus = -0.76$ |
   | III | $Sn^{4+}/Sn^{2+}$ | $E^\ominus = +0.15$ |
   | IV | $Ag^+/Ag$ | $E^\ominus = +0.80$ |
   | V | $V^{2+}/V$ | $E^\ominus = -1.18$ |
   | VI | $Na^+/Na$ | $E^\ominus = -2.71$ |

   Which species is the:
   a. strongest oxidizing agent?
   b. weakest oxidizing agent?
   c. strongest reducing agent?
   d. weakest reducing agent?

6. Iron nails are placed into separate beakers containing solutions of $CuSO_4$, $SnCl_2$, $Sn(NO_3)_4$ and $PbCl_2$. In which beaker(s) will a metal coating form on the iron nail?

7 Strips of an unknown elemental metal are placed in three beakers containing 1 mol dm$^{-3}$ solutions of hydrochloric acid, zinc chloride and nickel nitrate solutions. The beakers were observed over a period of one hour, and the following observations were noted.

| Solution | Observations |
| --- | --- |
| HCl | Metal dissolved; bubble of gas evolved |
| ZnCl$_2$ | No reaction observed |
| Ni(NO$_3$)$_2$ | Crystals form over surface of metal |

Use the electrochemical series to determine the possible identity of the metal.

8 A voltaic cell is constructed using half-cells of Sn$^{2+}$/Sn and Ag$^+$/Ag. Draw a diagram of this voltaic cell and clearly indicate the:
   a  two half-cell equations
   b  anode and cathode
   c  polarity of the electrodes
   d  overall ionic equation
   e  oxidizing agent and reducing agent
   f  direction of flow of electrons through the connecting wires
   g  direction of flow of ions through the salt bridge.

9 A strip of aluminium foil is placed into a beaker containing a solution of nickel nitrate. According to the electrochemical series, a reaction should occur; however, no reaction is observed. Explain this observation.

10 a  Give two reasons why an electrochemical half-cell may require the use of an inert electrode such as graphite or platinum.
   b  Write half-equations for three cells that use an inert electrode.
   c  What is the main advantage of using graphite rather than platinum as an electrode material?

11 List the halogens that will displace bromine from a solution of sodium bromide and write an overall equation for each displacement reaction that occurs.

## 10.5 ELECTROLYTIC CELLS

Figure 10.5.1 shows two metallic objects. Which is more valuable: the gold bracelet or the aluminium saucepan? The obvious answer is the gold bracelet, but this answer would not always have been correct! How could aluminium possibly be more expensive than gold? To answer this intriguing question we need to go back to the times of Napoleon III (1808–1873), the French emperor from 1852 to 1870 and nephew of the famous Napoleon I, who waged war against much of Europe in the Napoleonic wars. Napoleon III was so enamoured of what was then the rare and valuable aluminium that he would serve dinner to his lesser guests on plates of gold, and reserve his finest aluminium tableware for himself and his most honoured guests!

Figure 10.5.1 Which object is made of the more expensive metal?

It is relatively easy to picture the history of gold. Gold is an inert metal, found in its native state in nature. The mining of gold is a straightforward process of accumulating enough gold from the very small pieces available. The story of aluminium is more complex. Aluminium is a very reactive metal that is never found in its native state in nature, but is found in ionic compounds such as aluminium oxide and aluminosilicates. These compounds are so stable that the existence of aluminium as a separate element was not known until it was isolated in pure form in 1827 by German chemist Friedrich Wöhler (1800–1882). The separation of aluminium from its compounds is highly endothermic. Scientists puzzled for many years as to how to commercially reduce aluminium ions to aluminium metal. This problem also applied to other reactive metals, such as sodium and potassium. To see how this problem was solved, we must turn to the technique of **electrolysis**, in which electricity is used to pry elements apart from their compounds.

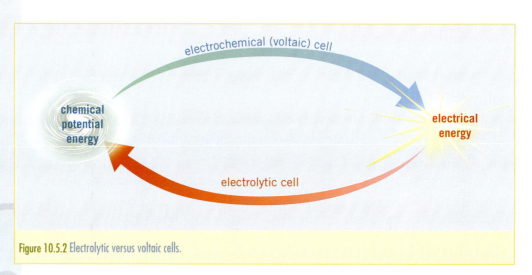

Figure 10.5.2 Electrolytic versus voltaic cells.

In electrolysis, reactions that would not normally occur are forced to proceed by supplying an external source of electrical energy to the reactants. Electrical energy from the power source is converted to chemical potential energy in the products. The earliest electrolysis experiments were performed by Sir Humphry Davy (1778–1829). In 1800, Davy read of Volta's invention of the first battery and was fascinated by the potential of this new invention. He immediately set to work to manufacture his own replica, which he used to electrolyse water into its constituent elements. In 1807, Davy passed an electric current through molten potash (mainly potassium carbonate) to produce small amounts of a silvery, light and enormously reactive metal. Davy had discovered potassium. He was astonished to observe its behaviour when added to water. Not only did it float (and so must be less dense than water), it immediately burst into flame and disintegrated in a shower of sparks while producing a magnificent violet flame. Davy was so entranced by his discovery that his laboratory assistant recorded at the time that 'he danced with joy around the lab'. Later in the same week, he used the same method to discover sodium from molten soda ash (sodium carbonate), and within 12 months he had discovered magnesium, calcium, strontium and barium. Today, reactive metals such as sodium, calcium and magnesium are produced industrially by electrolysis, along with other important chemicals, including chlorine and sodium hydroxide.

## CHEM COMPLEMENT

### Davy's other discoveries

Among his extraordinary list of discoveries and achievements, Davy invented the first effective miner's safety lamp (the Davy lamp) in 1815, which alerted coal miners to the presence of methane in underground mines. His invention saved countless thousands of lives, as methane explosions in mines were frequent and deadly until then. In 1814, Davy hired the 22-year-old Michael Faraday to be his assistant, and a lifelong friendship and collaboration came into being. Davy died in 1829 in Rome of a heart attack, possibly brought on by his many years of experimentation and the resultant inhalation of toxic gases such as chlorine.

Figure 10.5.3 Sir Humphry Davy (1778–1829).

## Essential components of an electrolytic cell

Electrolysis occurs in an **electrolytic cell** (figure 10.5.4). There are a number of important structural differences between voltaic and electrolytic cells. Key features to note are the presence of a power source and the fact that there is only one container, in which both oxidation and reduction reactions proceed.

 **9.5.1 Describe, using a diagram, the essential components of an electrolytic cell.** © IBO 2007

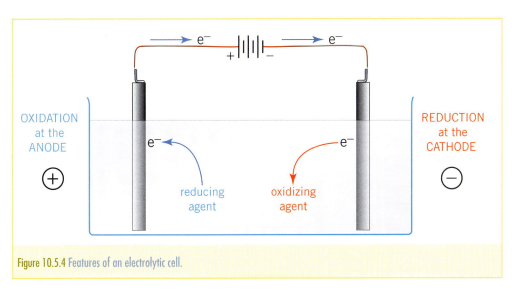

Figure 10.5.4 Features of an electrolytic cell.

To understand how electrolytic cells function, it is necessary to recognize that the power source defines the polarity of each electrode, not the reaction occurring there (as is the case in voltaic cells). The electrode connected to the negative terminal of the power source has an excess of electrons forced onto it and so a reduction reaction occurs (as reduction reactions involve electrons being accepted). An oxidizing agent present in the electrolyte gains the excess electrons. Which oxidizing agent? The strongest oxidizing agent: the one most capable of accepting electrons.

Similarly, the electrode connected to the positive terminal has electrons taken away from it and so a reaction occurs that can supply electrons. The strongest reducing agent undergoes oxidation at this electrode, as oxidation produces electrons. The anode is the electrode at which oxidation takes place, and the cathode the electrode at which reduction takes place.

**AS 9.5.2**
State that oxidation occurs at the positive electrode (anode) and reduction occurs at the negative electrode (cathode).
© IBO 2007

Thus, for an electrolytic cell, reduction of the strongest oxidizing agent occurs at the negatively charged cathode, and oxidation of the strongest reducing agent occurs at the positively charged anode. The polarity of the anode and cathode has reversed from that of voltaic cells.

As both oxidation and reduction reactions occur in the same reaction chamber, there is no need for a salt bridge in an electrolytic cell. It is not necessary to separate the reactants, as they are unable to react spontaneously with each other. However, there is a chance that the products may react spontaneously, so in commercial electrolysis cells care is taken to keep the products away from each other.

Another difference between the two types of cells is that in electrolytic cells it is common to use inert, or unreactive, electrodes.

### The electrolysis of molten salts

Sodium and chlorine react spontaneously and vigorously. Does $Na^+$ react with $Cl^-$? The existence of the stable salt NaCl suggests not. If we consider the relevant $E^\ominus$ values from the electrochemical series, it becomes obvious why no reaction is observed:

$$\mathbf{Na^+}(l) + e^- \rightarrow Na(l) \qquad E^\ominus = -2.71 \text{ V}$$
$$Cl_2(g) + 2e^- \rightarrow 2\mathbf{Cl^-}(l) \qquad E^\ominus = +1.36 \text{ V}$$

The reaction between sodium ions and chloride ions is non-spontaneous—it requires energy to proceed and so is endothermic. As the $E^\ominus$ values have a difference of $(1.36 - (-2.71)) = 4.07$ V, this indicates that a minimum of 4.07 V would need to be supplied to initiate the production of sodium metal and chlorine gas. The cell shown in figure 10.5.5 can be used to drive this endothermic reaction, and is an example of an electrolytic cell.

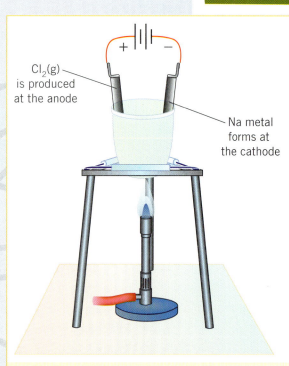

Figure 10.5.5 The electrolysis of molten NaCl.

The external power supply determines the polarity of the electrodes. The negative terminal of the power supply forces electrons into the electrode to which it is attached. This negative electrode now has an excess of electrons, and positive ions are attracted to it. The positive ions gain electrons and are reduced, so the negative electrode is the cathode. The only positive ion present is $Na^+$, and so the reduction at the cathode (−) is:

$$Na^+(l) + e^- \rightarrow Na(l)$$

**AS 9.5.3**
Describe how current is conducted in an electrolytic cell.
© IBO 2007

As $Na^+$ ions react at the cathode, more are attracted from the molten NaCl electrolyte, creating a flow of positive ions through the electrolyte towards the cathode.

**AS 9.5.4**
Deduce the products of the electrolysis of a molten salt.
© IBO 2007

Similarly, the positive terminal of the battery withdraws electrons from the electrode to which it is attached, making it positive. The positive electrode has a deficit of electrons, and negative ions are attracted to it. The negative ions lose electrons and are oxidized, so the positive electrode is the anode. The only negative ion present is $Cl^-$, and so the oxidation at the anode (+) is:

$$2Cl^-(l) \rightarrow Cl_2(g) + 2e^-$$

As the Cl⁻ ions react at the anode, more are attracted from the molten NaCl electrolyte, creating a flow of negative ions through the electrolyte. In this way, the current flowing through the electrolyte is made up of positive and negative ions, while electron current still flows through the electrodes and the wires connecting them to the power supply.

DEMO 10.4
Electrolysis of a molten salt

## Worked example

Draw an electrolytic cell representing the electrolysis of molten potassium bromide using inert electrodes. On or below your diagram you should clearly indicate the:

a polarity of the electrodes

b anode and the cathode

c oxidation and reduction reactions

d oxidizing agent and reducing agent species

e overall equation representing the cell reaction.

## Solution

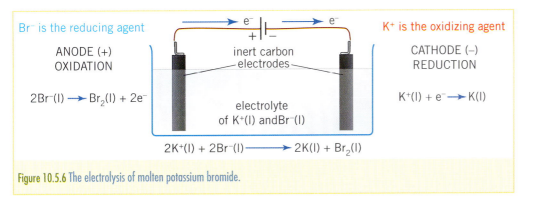

Figure 10.5.6 The electrolysis of molten potassium bromide.

The only species present are $K^+(l)$ and $Br^-(l)$ and so the relevant half-equations from the electrochemical series are:

$\mathbf{K}^+(l) + e^- \rightarrow K(l)$     $E^\ominus = -2.93$ V

$Br_2(l) + 2e^- \rightarrow 2\mathbf{Br}^-(l)$     $E^\ominus = +1.08$ V

At the cathode (−), reduction of the strongest oxidizing agent occurs. As there is only one oxidizing agent ($K^+$) present, the reaction is:

$K^+(l) + e^- \rightarrow K(l)$

The potassium cations are attracted to the negative electrode, where they accept electrons from the power supply to produce potassium metal.

At the anode (+), oxidation of the strongest reducing agent occurs. As there is only one reducing agent ($Br^-$) present, the reaction is:

$2Br^-(l) \rightarrow Br_2(l) + 2e^-$

The bromide anions are attracted to the positively charged anode, where they give up their electrons to produce molecular bromine.

Note that the half-equations from the electrochemical series for the reducing agent species must be reversed, as this species will undergo oxidation. As the $E^\circ$ values have a difference of $(1.08 - (-2.93)) = 4.01$ V, this indicates that a minimum 4.01 V of energy would need to be supplied to ensure the overall reaction:

$$2K^+(l) + 2Br^-(l) \rightarrow 2K(l) + Br_2(l)$$

## Section 10.5 Exercises

1. What is the role of the power source in an electrolytic cell?

2. Explain why reduction occurs at the negative electrode in an electrolytic cell.

3. In an electrolytic cell molten lithium chloride is decomposed to lithium and chlorine by the passage of an electric current through the liquid.
   a. Write the equation for the reaction occurring at the anode, and state the polarity of this electrode.
   b. Name the products of this electrolysis.
   c. Write an overall equation to represent the reaction in the electrolytic cell.

4. Molten magnesium sulfide was electrolysed using inert graphite electrodes. Complete the labelling of the diagram of the electrolytic cell shown in the diagram below.

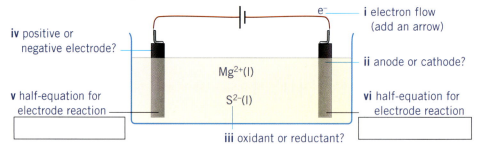

5. An electrolytic cell with graphite electrodes was used to electrolyse molten aluminium fluoride.
   a. Sketch the cell for this electrolytic process, showing the direction of ion flow and electron flow.
   b. Write half-equations for the reactions occurring at the electrodes.
   c. Label the anode and the cathode.
   d. Determine the oxidizing agent and reducing agent.
   e. Write an overall equation for the reaction.

6. Explain how current is conducted in an electrolytic cell.

# Chapter 10 Summary

## Terms and definitions

**Anode**  Electrode at which oxidation takes place.

**Cathode**  Electrode at which reduction takes place.

**Displacement reaction**  A reaction in which one substance (a solid) replaces another (aqueous ions). Often involves the reaction of metal atoms with the ions of another metal.

**Electrochemical series**  A list of oxidizing agents and their conjugate reducing agents written as half-equations and listed in order of increasing strength of the oxidizing agent.

**Electrolysis**  The decomposition of a compound into its constituent elements using electricity to drive the non-spontaneous reactions.

**Electrolytic cell**  A single chamber in which two electrodes are dipped into an electrolyte that is to be decomposed.

**Electron transfer**  The movement of electrons from a reducing agent to an oxidizing agent.

**Half-cell**  Half a voltaic cell. The half-cell contains a conjugate redox pair: an electrode and an electrolyte.

**Half-equation**  An equation that shows the loss or gain of electrons by an individual reactant.

**Molten salt**  An ionic compound that has been heated until it is a liquid.

**Oxidation**  The loss of electrons (or gain of oxygen or loss of hydrogen).

**Oxidation number**  A number that is assigned to an element in a compound and indicates the degree to which the atom has lost electrons.

**Oxidizing agent (oxidant)**  A reactant, often a non-metal, that causes another reactant to lose electrons. The oxidizing agent gains electrons.

**Reactivity series**  A list of metals in order of decreasing reactivity.

**Reduction**  The gain of electrons (or loss of oxygen or gain of hydrogen).

**Reducing agent (reductant)**  A reactant, usually a metal, that causes another reactant to gain electrons. The reducing agent loses electrons.

**Standard electrode potential ($E^\ominus$)**  The potential difference generated by a voltaic cell in which a particular half-cell under standard conditions is connected to a standard hydrogen half-cell.

**Voltaic cell (also known as an electrochemical cell or galvanic cell)**  An arrangement consisting of two half-cells connected internally by a salt bridge and externally by wires. It is used to generate electricity using a redox reaction.

## Concepts

- Redox reactions involve the transfer of electrons from one species to another.

- Oxidation is defined as the loss of electrons. Reduction is the gain of electrons. Reduction and oxidation occur simultaneously.

- An oxidizing agent is an electron acceptor. During a redox reaction the oxidizing agent is reduced.

- A reducing agent is an electron donor. It is oxidized during a reaction.

- Oxidation numbers are artificial numbers assigned to atoms according to a set of rules. They are the charge an atom would have if electrons were transferred during a reaction. Rules for assigning oxidation numbers are:

  1. Pure elements       ON = 0
  2. Ions               ON = charge on the ion
  3. Combined oxygen    ON = –2 (usually)
  4. Combined hydrogen  ON = +1 (usually)
  5. For polyatomic ions or molecules, the sum of the oxidation numbers equals the charge on the ion or molecule.

- A redox reaction may be identified by changes in oxidation numbers. Oxidation is an increase in oxidation number. Reduction is a decrease in oxidation number.

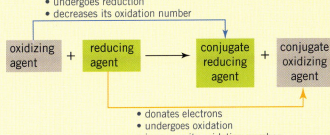

- Equations for redox processes may be written using a series of steps.
  1. Write conjugate pairs.
  2. Balance the number of atoms undergoing oxidation or reduction.
  3. Balance the oxygen atoms by adding water.
  4. Balance the hydrogen atoms by adding $H^+$ ions.
  5. Balance the charges by adding electrons.
  6. Balance the number of electrons produced in the oxidation with those used in the reduction.
  7. Add the half-equations together.
  8. Add states to all species in the overall balanced equation.

- Voltaic cells use the energy released in spontaneous redox reactions to generate electricity.

- Voltaic cells are composed of two half-cells, each containing a conjugate redox pair. Each half-cell contains an electrode (conducting solid) and an electrolyte (conducting liquid). Half-cells are connected externally by wiring and internally by a salt bridge.

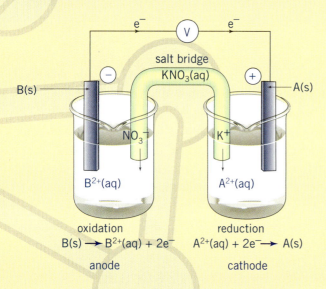

- The relative reactivities of chemical species is summarized in the electrochemical series. The standard electrochemical series is prepared by comparing each half-cell with the standard hydrogen half-cell. The standard hydrogen half-cell is assigned a value of 0.00 V. Standard conditions are 25°C, 1 mol dm$^{-3}$ and 1 atm pressure for gases.

- The electrochemical series can be used to predict whether a particular combination of chemicals will react.

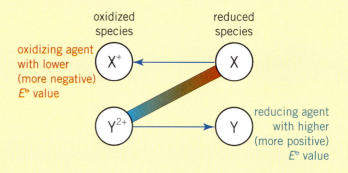

- The strongest reducing agents are found in groups 1 and 2 of the periodic table. The strongest oxidizing agents are found at the far upper-right of the periodic table.

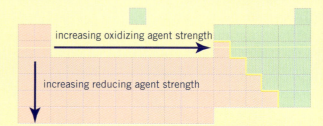

- The electrochemical series is only a predictor of reactions. An expected reaction may not occur, due to such factors as a slow rate of reaction, or non-standard conditions being used.

- Electrolysis is the process of using an external power supply to cause endothermic redox reactions to occur. Electrical energy is converted to chemical potential energy.

- The products of an electrolysis reaction can be predicted from the electrochemical series. In general, the strongest oxidizing agent is reduced at the cathode (negatively charged) and the strongest reducing agent is oxidized at the anode (positively charged).

- Voltaic and electrolytic cells both involve redox reactions and the inter-conversion of chemical and electrical energy.

| Voltaic cells | Electrolytic cells |
|---|---|
| 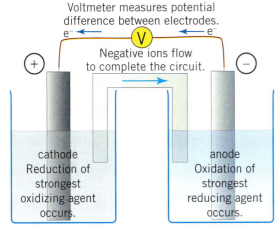<br>Two half-cells, each contains a redox pair, e.g. $X^{2+}(aq)/X(s)$. | 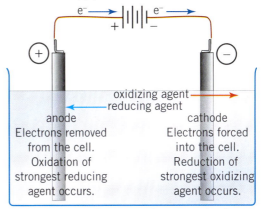<br>One cell only. Reactants do not need to be separated since reactions are not spontaneous. |
| • Spontaneous exothermic reactions convert chemical energy into electrical energy. | • Non-spontaneous endothermic reactions require electrical energy, which is converted into chemical energy. |
| • Can be used as a power source. | • Requires an external power supply. |
| • Comprises two separate half-cells connected by a salt bridge and connecting wires. | • No need to separate half-cells and so no salt bridge is required. |
| • Electrons flow from negative electrode to positive electrode. | • Power supply forces electrons onto negative electrode and takes them from the positive electrode. |
| • Oxidation occurs at the anode. | • Oxidation occurs at the anode. |
| • Reduction occurs at the cathode. | • Reduction occurs at the cathode. |
| • Anode is negative. | • Anode is positive. |
| • Cathode is positive. | • Cathode is negative. |
| • Polarity of the electrodes is determined by the reactions occurring there. | • Polarity of the electrodes is determined by the power source. |
| • Reducing agent donates electrons to oxidizing agent via the external circuit. | • Power source supplies electrons to oxidizing agent and accepts electrons from reducing agent. |

# Chapter 10 Review questions

1. Match each term to an appropriate definition.

   | | |
   |---|---|
   | Oxidizing agent | Involves the gain of electrons |
   | Reducing agent | An electron donor |
   | Oxidation | Undergoes reduction during a redox reaction |
   | Reduction | This happens to a reducing agent during a reaction |

2. Give three definitions of the process of oxidation.

3. Determine the oxidation number of each element in the following species.
   a  $NO_2$
   b  $K_2Cr_2O_7$
   c  $HSO_4^-$
   d  $HNO_2$
   e  $H_3PO_4$
   f  $S_2O_3^{2-}$
   g  $Ga_2(CO_3)_3$
   h  $Zn(BrO_3)_2$

4. The addition of potassium permanganate ($KMnO_4$) solution to a solution of manganese(II) sulfate ($MnSO_4$) results in the precipitation of manganese dioxide ($MnO_2$).
   a  State the oxidation number of manganese in each of the three compounds.
   b  Describe the reaction in terms of half-equations.

5. It is sometimes possible for an ion to undergo simultaneous oxidation and reduction. This is called disproportionation. Show that this occurs by writing half-equations for the conversion of hypochlorite ion ($ClO^-$) to chloride ($Cl^-$) and chlorate ($ClO_3^-$) ions.

6. Write balanced half-equations for each of the following conversions.
   a  Glucose ($C_6H_{12}O_6$) to sorbitol ($C_6H_8(OH)_6$)
   b  Sorbitol ($C_6H_8(OH)_6$) to ascorbic acid ($C_6H_8O_6$)
   c  Oxalate ion ($C_2O_4^{2-}$) to carbon dioxide ($CO_2$)
   d  Ethanol ($CH_3CH_2OH$) to ethanoic acid ($CH_3COOH$)
   e  Sulfite ions ($SO_3^{2-}$) to hydrogen sulfide ($H_2S$)
   f  Nitric acid ($HNO_3$) to nitrogen dioxide ($NO_2$)

7. Write a balanced equation for each of the processes below. Use oxidation numbers to show that each process is a redox reaction.
   a  Combustion of ethanol
   b  Decomposition of hydrogen peroxide to water and oxygen

8. Balance each of the following equations by first writing half-equations.
   a  $Al(s) + H^+(aq) \rightarrow Al^{3+}(aq) + H_2(g)$
   b  $Cu(s) + NO_3^-(aq) \rightarrow NO(g) + Cu^{2+}(aq)$
   c  $SO_2(g) + MnO_4^-(aq) \rightarrow Mn^{2+}(aq) + SO_4^{2-}(aq)$ (in acidic solution)
   d  $ClO^-(aq) + I^-(aq) \rightarrow Cl^-(aq) + I_2(aq)$

9. Nickel reacts with hydrochloric acid to evolve hydrogen gas and produce a green solution containing the $Ni^{2+}$ ion.
   a  Write a half-equation for the oxidation reaction.
   b  Write a half-equation for the reduction reaction.
   c  Write an ionic equation for the overall reaction.
   d  Identify the oxidizing agent in this reaction.

10. In each of the following equations, identify the oxidizing agent and the reducing agent.
    a  $Zn(s) + Pb^{2+}(aq) \rightarrow Zn^{2+}(aq) + Pb(s)$
    b  $2Mg(s) + O_2(g) \rightarrow 2MgO(s)$
    c  $CuO(s) + H_2(g) \rightarrow Cu(s) + H_2O(g)$
    d  $Cl_2(g) + 2HI(aq) \rightarrow 2HCl(aq) + I_2(s)$

11. State which species is acting as oxidizing agent and which is the reducing agent in each of the following equations.
    a  $Br_2(aq) + Mg(s) \rightarrow 2Br^-(aq) + Mg^{2+}(aq)$
    b  $2Ag^+(aq) + Sn^{2+}(aq) \rightarrow 2Ag(s) + Sn^{4+}(aq)$
    c  $Cu^{2+}(aq) + Pb(s) \rightarrow Cu(s) + Pb^{2+}(aq)$
    d  $MnO_4^-(aq) + 8H^+(aq) + 5Fe^{2+}(aq) \rightarrow Mn^{2+}(aq) + 5Fe^{3+}(aq) + 4H_2O(l)$
    e  $2Na(s) + 2H_2O(l) \rightarrow 2NaOH(aq) + H_2(g)$

12. When a piece of zinc was left to stand overnight in an aqueous solution of tin(II) nitrate, the mass of the zinc decreased by 0.50 g. Write a balanced equation to account for the loss in mass of the zinc.

13. For each of the equations given below, state whether the bolded chemical is being oxidized, reduced, neither oxidized nor reduced, or both oxidized and reduced.
    a  **$Pb^{2+}$**(aq) + Fe(s) → Pb(s) + $Fe^{2+}$(aq)
    b  Mg(s) + 2**$H_2O$**(l) → $Mg(OH)_2$(aq) + $H_2$(g)
    c  **$Ba(NO_3)_2$**(aq) + $H_2SO_4$(aq) → $BaSO_4$(s) + $2HNO_3$(aq)

14. Using the activity series shown in table 10.4.1 (p. 320), write ionic equations for any reactions that would occur when:
    a  cadmium (Cd) is added to a solution of $Cu(NO_3)_2$
    b  lead (Pb) is added to a solution of $Zn(NO_3)_2$.

| Reaction at anode (−) | Reaction at cathode (+) | Overall reaction equation | Oxidizing agent | Reducing agent |
|---|---|---|---|---|
| Ni(s) → Ni$^{2+}$(aq) + 2e$^-$ | Cu$^{2+}$(aq) + 2e$^-$ → Cu(s) | | | |
| | 2H$_2$O(l) + 2e$^-$ → H$_2$(g) + 2OH$^-$(aq) | | | K(s) |
| | | MnO$_4^-$(aq) + 8H$^+$(aq) + 5Fe$^{2+}$(aq) → Mn$^{2+}$(aq) + 4H$_2$O(l) + 5Fe$^{3+}$(aq) | | |
| Sn(s) → Sn$^{2+}$(aq) + 2e$^-$ | | | Sn$^{4+}$(aq) | |

**15** Copy and complete the table above.

**16** Complete the labelling of the voltaic cell shown, which uses the reaction between Fe(s) and Sn$^{2+}$(aq).

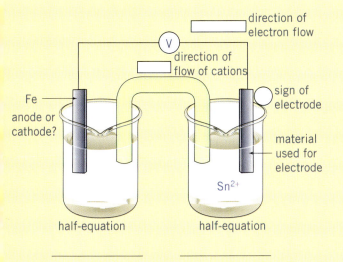

**17** A voltaic cell is constructed using the half-cells Ni$^{2+}$/Ni and Zn$^{2+}$/Zn. For this cell:

  **a** For this cell, which electrode (Ni or Zn):

  **i** is the anode?

  **ii** has a negative charge?

  **iii** will lose mass?

  **b** Name a chemical suitable for use in the salt bridge.

**18** Explain why it would be unwise to store silver nitrate solution (AgNO$_3$(aq)) in a copper container.

**19** A solution of hydrogen peroxide reacts with itself over time to produce water and oxygen gas, and so solutions of hydrogen peroxide are generally kept in the refrigerator.

  **a** By first writing the relevant half-equations, deduce an overall equation for this reaction.

  **b** Why are solutions of hydrogen peroxide refrigerated?

  **c** The addition of a small quantity of manganese dioxide to a hydrogen peroxide solution results in the vigorous evolution of oxygen gas. What is the role of MnO$_2$ in this reaction?

**20** Use the electrochemical series to deduce whether or not a reaction of any significant extent would occur in each of the following cases. Where a reaction would be expected, write the relevant partial equations and the overall equation.

  **a** Copper filings are sprinkled into a solution of silver nitrate.

  **b** A strip of magnesium is placed into a solution of hydrochloric acid.

  **c** Solutions of potassium bromide and zinc nitrate are mixed.

  **d** Chlorine gas is bubbled through a solution of tin(II) chloride.

  **e** Hydrogen sulfide gas is bubbled through a solution of copper(II) sulfate.

**21** A student is given a beaker containing an unknown solution of Q(NO$_3$)$_2$ and is asked to displace metal Q from the solution. The relevant half-equation is:

$$Q^{2+}(aq) + 2e^- \rightarrow Q(s) \qquad E^\ominus = -0.55 \text{ V}$$

By consulting the electrochemical series, deduce which of the following metals would be suitable to perform this function: iron, copper, zinc or lead.

**22** A chemist made the following observations using clean metal surfaces.

  • Metal B dissolved in 1 mol dm$^{-3}$ C(NO$_3$)$_2$ solution, forming a deposit of metal C.

  • Metal C would not dissolve in 1 mol dm$^{-3}$ A(NO$_3$)$_2$ solution.

  • Metal A would not dissolve in 1 mol dm$^{-3}$ B(NO$_3$)$_2$ solution.

List metals A, B and C in order from the strongest reducing agent to the weakest.

**23** As part of an investigative experiment, a student is provided with a quantity of a colourless solution of a metal nitrate and is asked to determine its identity. The solution is divided into several 100 cm$^3$ beakers and a number of experiments are performed, the results of which are shown below.

| Experiment | Observations |
| --- | --- |
| Turnings of copper metal added | No reaction observed |
| Iron nail placed in beaker | Crystals form over surface of iron |
| Hydrogen gas bubbled through solution | No reaction observed, but pH of solution decreases |

Use the electrochemical series to determine the possible identity of the metal cation, and so the chemical formula of the metal nitrate.

**24** An electrolytic cell is different from a voltaic cell in a number of important ways, and yet they also share a number of similarities.

Copy and complete the following table.

| Voltaic cells | Electrolytic cells |
| --- | --- |
| Spontaneous reactions produce energy. | Non-spontaneous reactions _____ energy. |
| Comprises two half-cells, connecting wires and a _____ _____ of KNO$_3$. | Comprises one cell only, connecting wires and a _____. |
| Oxidation occurs at the _____, reduction at the _____. | Oxidation occurs at the _____, reduction at the _____. |
| The anode has a _____ charge, the cathode has a _____ charge. | The anode has a _____ charge, the cathode has a _____ charge. |
| The polarity of electrodes is determined by the reactions occurring there. | The polarity of electrodes is determined by the _____ _____. |

**25** An electrolytic cell is used to pass an electric current through molten calcium chloride at sufficient voltage to ensure a reaction.

**a** Draw a diagram of the cell, showing the polarity of the electrodes and the direction of the flow of ions through the electrolyte and electrons through the conducting wires.

**b** State the products formed at each of the electrodes.

**c** Write half-equations for the reactions occurring at the anode and cathode.

**d** Write an overall equation to represent the reaction in the electrolytic cell.

www.pearsoned.com.au/schools

Weblinks are available on the Chemistry: For use with the IB Diploma Programme Standard Level Companion Website to support learning and research related to this chapter.

# Chapter 10 Test

## Part A: Multiple-choice questions

1. $MnO_2 + 4HCl \rightarrow Mn^{2+} + 2Cl^- + Cl_2 + 2H_2O$

   Which substance is produced by oxidation in the equation above?

   A $Mn^{2+}$      B $Cl^-$
   C $Cl_2$      D $H_2O$
   © IBO 2000, May P1 Q25

2. In the electrolysis of molten sodium chloride, the sodium ion goes to the:

   A positive electrode where it undergoes oxidation.
   B negative electrode where it undergoes oxidation.
   C positive electrode where it undergoes reduction.
   D negative electrode where it undergoes reduction.
   © IBO 2000, May P1 Q26

3. The oxidation number of sulfur in the $HS_2O_5^-$ ion is:

   A −1      B +3
   C +4      D +5
   © IBO 2000, Nov P1 Q25

4. $2AgNO_3(aq) + Zn(s) \rightarrow 2Ag(s) + Zn(NO_3)_2(aq)$
   $Zn(NO_3)_2(aq) + Co(s)$ No reaction
   $2AgNO_3(aq) + Co(s) \rightarrow Co(NO_3)_2(aq) + 2Ag(s)$

   Using the above information, the order of **increasing** activity of the metals is:

   A Ag < Zn < Co      B Co < Ag < Zn
   C Co < Zn < Ag      D Ag < Co < Zn
   © IBO 2000, Nov P1 Q26

5. Consider the following reaction.
   $H_2SO_3(aq) + Sn^{4+}(aq) + H_2O(l)$
   $\rightarrow Sn^{2+}(aq) + HSO_4^-(aq) + 3H^+(aq)$

   Which statement is correct?

   A $H_2SO_3$ is the reducing agent because it undergoes reduction.
   B $H_2SO_3$ is the reducing agent because it undergoes oxidation.
   C $Sn^{4+}$ is the oxidizing agent because it undergoes oxidation.
   D $Sn^{4+}$ is the reducing agent because it undergoes oxidation.
   © IBO 2004, Nov P1 Q25

6. During the electrolysis of a molten salt, which statement is **not** correct?

   A The ions only move when a current flows.
   B Positive ions are attracted to the negative electrode.
   C Positive ions gain electrons at the negative electrode.
   D Negative ions lose electrons at the positive electrode.
   © IBO 2001, May P1 Q24

7. What is the oxidation number of phosphorus in $NaH_2PO_4$?

   A +3      B −3
   C +5      D −5
   © IBO 2001, Nov P1 Q25

8. Which product is formed at the cathode (negative electrode) when molten $MgCl_2$ is electrolysed?

   A $Mg^{2+}$      B $Cl^-$
   C $Mg$      D $Cl_2$
   © IBO 2001, Nov P1 Q26

9. Which of the following changes represents a reduction reaction?

   A $Mn^{2+}(aq) \rightarrow MnO_4^-(aq)$
   B $2CrO_4^{2-}(aq) \rightarrow Cr_2O_7^{2-}(aq)$
   C $SO_4^{2-}(aq) \rightarrow SO_3^{2-}(aq)$
   D $Zn(s) \rightarrow Zn^{2+}(aq)$
   © IBO 2002, May P1 Q25

10. When the following equation is balanced, what is the coefficient for $Ce^{4+}$?

    $\_SO_3^{2-} + \_H_2O + \_Ce^{4+} \rightarrow \_SO_4^{2-} + \_H^+ + \_Ce^{3+}$

    A 1      B 2
    C 3      D 4
    © IBO 2003, Nov P1 HL Q31

(10 marks)

www.pearsoned.com.au/schools

For more multiple-choice test questions, connect to the Chemistry: For use with the IB Diploma Programme Standard Level Companion Website and select Chapter 10 Review Questions.

## Part B: Short-answer questions

1. For the following reaction:
   $2Cu^+ \rightarrow Cu + Cu^{2+}$

   a state the oxidation number of each species.
   (1 mark)

   b write a balanced half-reaction for the oxidation process.
   (1 mark)

**c** write a balanced half-reaction for the reduction process.

(1 mark)

© IBO 2000, Nov P2 Q2

**2** Chlorine can be prepared in the laboratory by reacting chloride ions with an acidified solution of manganate(VII) (permanganate) ions. The unbalanced equation for the reaction is:

$Cl^-(aq) + MnO_4^-(aq) + H^+(aq)$
$\rightarrow Mn^{2+}(aq) + Cl_2(g) + H_2O(l)$

Give the oxidation numbers of chlorine and manganese in the reactants and products. Write the balanced equation and identify the reducing agent in the reaction.

(5 marks)

© IBO 1999, Nov P2 Q5a

**3 a** Draw a diagram of apparatus that could be used to electrolyse molten potassium bromide. Label the diagram to show the polarity of each electrode and the product formed.

(3 marks)

**b** Describe the **two** different ways in which electricity is conducted in the apparatus.

(2 marks)

**c** Write an equation to show the formation of the product at each electrode.

(2 marks)

© IBO 2006, Nov P2 Q4d

## Part C: Data-based question

A part of the reactivity series of metals, in order of decreasing reactivity, is shown below.

Magnesium

Zinc

Iron

Lead

Copper

Silver

If a piece of copper metal were placed in separate solutions of silver nitrate and zinc nitrate:

**a** determine which solution would undergo reaction.

(1 mark)

**b** identify the type of chemical change taking place in the copper and write the half-equation for this change.

(2 marks)

**c** state, giving a reason, what visible change would take place in the solutions.

(2 marks)

© IBO 2004, May P2 Q6b

## Part D: Extended-response question

**a** Use these equations, which refer to aqueous solutions, to answer the questions that follow:

$Fe(s) + Cu^{2+}(aq) \rightarrow Fe^{2+}(aq) + Cu(s)$
$Cu(s) + 2Ag^+(aq) \rightarrow Cu^{2+}(aq) + 2Ag(s)$
$Mg(s) + Fe^{2+}(aq) \rightarrow Mg^{2+}(aq) + Fe(s)$

**i** List the four metals above in order of **decreasing** reactivity.

(1 mark)

**ii** Define *oxidation*, in electronic terms, using **one** example from above.

(2 marks)

**iii** Define *reduction*, in terms of oxidation number, using **one** example from above.

(2 marks)

**iv** State and explain which is the **strongest reducing agent** in the examples above.

(2 marks)

**v** State and explain which is the **strongest oxidizing agent** in the examples above.

(2 marks)

**vi** Deduce whether a silver coin will react with aqueous magnesium chloride.

(2 marks)

**b** Sketch a diagram of a voltaic cell made up of a $Mg/Mg^{2+}$ half-cell and a $Cu/Cu^{2+}$ half-cell. Label the essential components of the voltaic cell and state equations for the reactions that occur at the electrodes.

(4 marks)

**c** Describe how electrode reactions occur in an electrolytic cell and state the products at each electrode when molten copper(II) chloride is electrolysed.

(5 marks)

adapted from IBO 2000, May P2 Q4
Total marks 50

# 11 ORGANIC CHEMISTRY

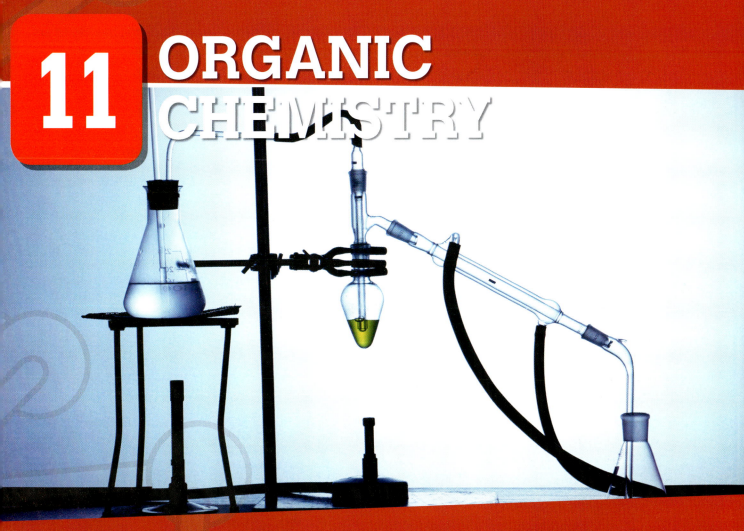

## Chapter overview

This chapter covers the IB Chemistry syllabus Topic 10: Organic Chemistry.

**By the end of this chapter, you should be able to:**

- describe the features of a homologous series
- predict and explain the trends in boiling points of members of a homologous series
- distinguish between empirical, molecular and structural formulas
- deduce structural formulas for the isomers of non-cyclic alkanes and straight-chain alkenes up to $C_6$
- recognize, draw structural formulas for and name (using the IUPAC system) alkanes, alkenes, alcohols, aldehydes, ketones, halogenoalkanes and carboxylic acids up to $C_6$
- identify the amino, benzene and ester functional groups in structural formulas
- identify primary, secondary and tertiary carbon atoms in alcohols and halogenoalkanes
- compare the volatility and solubility in water of the organic compounds studied
- explain the low reactivity of alkanes in terms of bond enthalpies and bond polarity
- write combustion reactions for alkanes, alkenes and alcohols
- recognize and write appropriate equations for addition, substitution and oxidation reactions involving alkanes, alkenes, halogenoalkanes and alcohols
- describe the polymerization of alkenes and describe the economic importance of the reactions of alkenes
- write mechanisms for nucleophilic substitution of halogenoalkanes and free radical substitution of methane and ethane with chlorine and bromine
- construct organic reaction pathways for the production of a range of organic compounds.

IBO Assessment statements 10.1.1 to 10.6.1

The diversity of life owes much to the diverse chemistry of one element: carbon. Over 90% of known compounds contain carbon. The ability of carbon to form this vast range of compounds has led to an equally vast range of living things composed of carbon-based molecules. Because the major source of carbon compounds is living or once-living material, it was originally thought that only living things had the 'vital force' needed to produce carbon-based compounds. The synthesis of urea (($NH_2$)$_2$CO) in 1828 by Friedrich Wöhler (1800–1882) showed this to be untrue, but the name 'organic' is still applied to the branch of chemistry dealing with the study of carbon-based compounds (excluding such compounds as CO, $CO_2$ and the carbonates).

The study of organic chemistry is central to understanding the chemistry (and biology) of living systems. Knowledge of organic chemistry is used in the manufacture of drugs, foods, pesticides, fertilizers and other chemicals used by humans in medicines, nutrition and agriculture. Production of other materials of huge importance to society, such as fuels, solvents and polymers, also requires a sound knowledge of organic chemistry. There can be little doubt as to the importance of this branch of chemistry, but with millions of known organic compounds, and with more being isolated and synthesized daily, where do we begin our study of carbon-based compounds?

organic = derived from living things

organic = grown without the use of synthetic fertilizers or pesticides

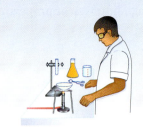
organic = the chemistry of carbon compounds

Figure 11.0.1 The term *organic* is used in a variety of ways.

## 11.1 INTRODUCTION TO ORGANIC CHEMISTRY

There are millions of compounds containing carbon. Carbon is present in all living things and in a wide range of non-living things. Several factors contribute to this vast range of carbon compounds:

- Carbon has four valence electrons, which allows each carbon atom to form up to four other bonds.
- Carbon can form single, double or triple bonds.
- Carbon can bond to itself, forming long chains and cyclic molecules.
- Carbon can bond to a range of other elements, including hydrogen, oxygen, nitrogen, sulfur and chlorine.

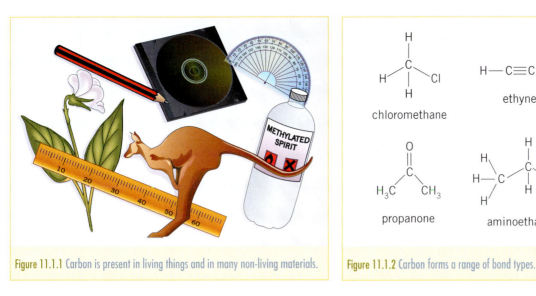

Figure 11.1.1 Carbon is present in living things and in many non-living materials.

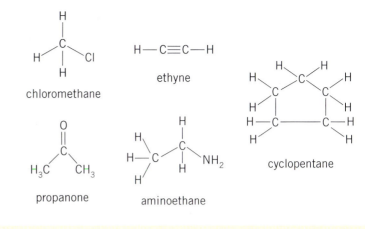

Figure 11.1.2 Carbon forms a range of bond types.

To deal with such a vast range of compounds, chemists look at various series or 'families' of carbon compounds that have common characteristics. The usual starting point is **hydrocarbons**, as they contain only two elements: hydrogen and carbon. A family of carbon compounds with common characteristics is called a **homologous series**. A homologous series is a group of carbon compounds with the same general formula. Each member of a homologous series differs from the previous member by one $CH_2$ group. Members of a homologous series have similar chemical properties, and there is a gradual change in physical properties as more carbon atoms are added.

**AS 10.1.1
Describe the features of a homologous series.** © IBO 2007

### CHEM COMPLEMENT

#### Is there life out there?

Absorption spectroscopy is being used to identify molecules present in interstellar space. More than 40 kinds of molecules have been detected. Among these are the organic compounds methanal (HCHO), methanoic acid (HCOOH), methanol ($CH_3OH$), ethanol ($CH_3CH_2OH$) and propyne ($CH_3CCH$). Does the presence of these relatively simple organic molecules in outer space lend support to the idea that living organisms (composed of more complex organic compounds) might exist elsewhere in the universe?

Figure 11.1.3 Organic compounds have been detected in space. Is this evidence of life in space?

## Alkanes

The most basic hydrocarbons are those in which only single bonds occur. The simplest of these is methane ($CH_4$), which is the first in a series of compounds called the **alkanes**. The alkanes are an example of a **homologous series**. The term *saturated* is also applied to alkanes. **Saturated hydrocarbons** have only single carbon–carbon bonds and therefore have the maximum number of hydrogen atoms; that is, they are saturated with hydrogen.

**General formula for the homologous series of alkanes is $C_nH_{2n+2}$**

For example, if a molecule has five carbon atoms ($n = 5$), the alkane formed is $C_5H_{12}$ ($2 \times 5 + 2 = 12$).

Table 11.1.1 shows the first six straight-chain alkanes. The naming of carbon compounds is very systematic. Both the prefix and suffix provide information. The prefix indicates the number of carbon atoms (see table 11.1.1) and the suffix '-ane' indicates an alkane.

**TABLE 11.1.1 THE FIRST SIX MEMBERS OF THE ALKANE SERIES**

| Name | Molecular formula | Structural formula | Properties | Uses |
|---|---|---|---|---|
| Methane | $CH_4$ | | Non-polar, gas, BP −164°C | Cooking, Bunsen burners, gas heating |
| Ethane | $C_2H_6$ | | Non-polar, gas, BP −87°C | Producing ethene |
| Propane | $C_3H_8$ | | Non-polar, gas, BP −42°C | Liquid petroleum gas (LPG) |
| Butane | $C_4H_{10}$ | | Non-polar, gas, BP −0.5°C | LPG, cigarette lighters, camp stoves |
| Pentane | $C_5H_{12}$ | | Non-polar, liquid, BP 36.1°C | Present in natural gas, cigarette lighters and aerosol propellants |
| Hexane | $C_6H_{14}$ | | Non-polar, liquid, BP 68.7°C | Present in petrol and many solvents |

Molecules can be represented by molecular, empirical or structural formulas. Notice in table 11.1.1 that the **molecular formula** shows the actual numbers of all the atoms in a molecule of the compound, but gives no information about the way the atoms are arranged, whereas the **structural formula** shows the way in which the atoms are bonded to each other and the shape of the molecule. A structural formula is much more useful when you need to visualize the molecule. Empirical formulas can also be written for these hydrocarbons, but, as discussed in chapter 2, the **empirical formula** is the lowest whole number ratio of elements in a compound, so the empirical formula of ethane would be $CH_3$ and for butane $C_2H_5$. Such a formula does little to help us determine the structure and bonding in a molecule.

The standard way of naming compounds (nomenclature) was developed by the International Union of Pure and Applied Chemistry, and hence it is referred to as the **IUPAC system**. The advantage of this system is that the name gives clues as to the structure of the molecule. Using this IUPAC system, it is possible to write the structural formula of any compound from its name. The IUPAC system is gradually replacing the older, less systematic way of naming organic compounds. For example, the chemical in which preserved animal samples are kept is called '**meth**anal'. Like **meth**ane, this compound has one carbon atom. It is an **aldehyde** (p. 360) and so has the ending '**-al**'. The name methanal is a more useful and informative name than the older term, 'formalin'.

Alkane nomenclature
Representations of methane

 **10.1.3**
**Distinguish between *empirical*, *molecular* and *structural* formulas.** © IBO 2007

### TABLE 11.1.2 PREFIXES USED FOR CARBON COMPOUNDS

| Number of carbon atoms | Prefix |
|---|---|
| 1 | Meth- |
| 2 | Eth- |
| 3 | Pro- |
| 4 | But- |
| 5 | Pent- |
| 6 | Hex- |
| 7 | Hept- |
| 8 | Oct- |
| 9 | Non- |
| 10 | Dec- |

### THEORY OF KNOWLEDGE

The IUPAC organization was formed in 1919 with the goal of fostering a common language or classification system for communicating Chemistry using clear, concise and unambiguous terminology.

- Outline some of the ways of communicating Chemistry you have met so far in IB Chemistry that use IUPAC rules.

- What do we gain and what do we lose by naming organic compounds using the IUPAC system?

- What are the advantages of having an international method of communicating the language of Chemistry?

- Give an example from one of your other subjects of knowledge that is communicated using a common international classification system.

### CHEM COMPLEMENT

#### Which name to use?

In the early days of organic chemistry, when relatively few pure compounds were known, new organic compounds were named by their discoverer. Thus, barbituric acid, a tranquilizer, was named by its discoverer, Adolf von Baeyer, in honour of his friend Barbara. Morphine, an analgesic isolated from the opium poppy, was named by German pharmacist Friedrich Sertürner after Morpheus, the Greek god of dreams. In some cases, we still use the common name of a substance in favour of its systematic name, because for complicated molecules the systematic name may be cumbersome. Thus we use the name 'glucose' in preference to the systematic name 2,3,4,5,6-pentahydroxyhexanal!

**10.1.2 Predict and explain the trends in boiling points of members of a homologous series.** © IBO 2007

The chemical properties of alkanes are similar. However, their physical properties vary systematically with the length of the molecule. You will recall from chapter 2 that the size of a non-polar molecule, such as an alkane, influences the strength of its intermolecular forces due to the increasing van der Waals' forces between the molecules. As the length of the molecule increases, melting and boiling points increase, viscosity and density increase, and volatility and solubility in water decrease.

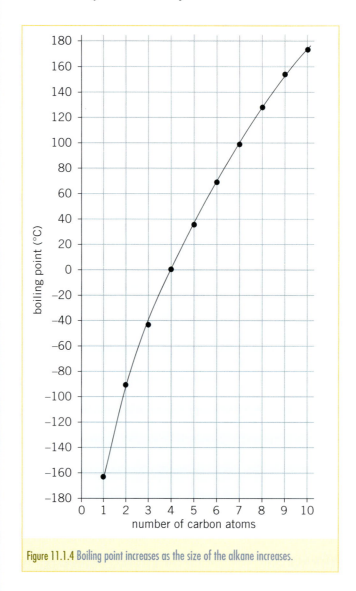

Figure 11.1.4 Boiling point increases as the size of the alkane increases.

Boiling points of organic compounds

PRAC 11.1
Properties of alkanes

## Structural isomers

The structural formulas shown in table 11.1.1 show only one possible arrangement of the atoms in the molecular formula, but some molecules can have several possible arrangements. For example, there are three different structures possible for the molecule $C_5H_{10}$. Molecules with the same molecular formula and different structural formulas are called **structural isomers**.

As the length and complexity of molecules increase, drawing full structural formulas becomes laborious. In those cases, more useful representations are semi-structural formulas, also known as **condensed structural formulas**, in which each carbon atom is listed alongside its attached hydrogen atoms.

**10.1.4 Describe isomers as compounds with the same molecular formula but with different arrangements of atoms.** © IBO 2007

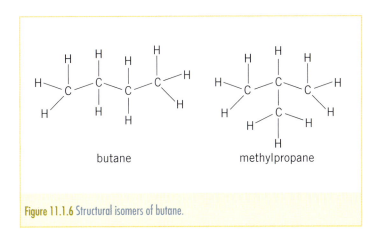

Figure 11.1.5 Molecular, semi-structural and structural formulas may be used to represent pentane.

Figure 11.1.6 Structural isomers of butane.

It is difficult to draw molecules to accurately represent their three-dimensional structure. In pentane, for example, each carbon atom forms bonds with four other atoms. The expected tetrahedral arrangement exists around each of these carbon atoms. The best representation of this is shown in figure 11.1.5 as the structural formula. However, for simplicity, as the size and complexity of the molecules increases, many molecules are simply drawn in two dimensions. Three-dimensional structures are drawn for simple inorganic and organic molecules wherever possible.

The first member of the alkane homologous series to exhibit structural isomerism is butane. Methane, ethane and propane each have only one possible structure. Butane has two isomers called butane and methylpropane. The structure of these isomers can be seen in figure 11.1.6.

In figure 11.1.7 you can see that there are three possible structures for a pentane molecule. As molecules became larger, the number of isomers increases. Isomers have different physical and chemical properties and so must be named differently. A systematic method of naming is therefore used; however, many compounds are also still known by their 'common' names, used before the introduction of the IUPAC system. For example, you may find the isomer of butane systematically called methylpropane referred to as 'isobutane'.

Using the IUPAC system, branched alkanes are named by considering them as straight-chain alkanes with side groups attached. Side groups are named using the prefix for the appropriate number of carbon atoms, and the suffix '-yl'. Thus a $CH_3$– side group is a methyl group; a $CH_3CH_2$– side group is an ethyl group and so on. Table 11.1.3 outlines the steps used in naming any alkane.

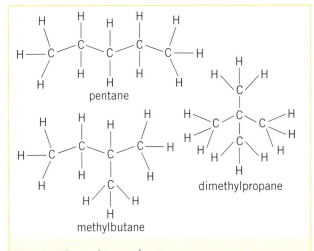

Figure 11.1.7 Structural isomers of pentane.

 10.1.5
**Deduce structural formulas for the isomers of the non-cyclic alkanes up to $C_6$.** © IBO 2007

 10.1.6
**Apply IUPAC rules for naming the isomers of the non-cyclic alkanes up to $C_6$.** © IBO 2007

Isomerism

### TABLE 11.1.3 NAMING HYDROCARBONS

| Steps | Example—use the steps to name the molecule |
|---|---|
| | 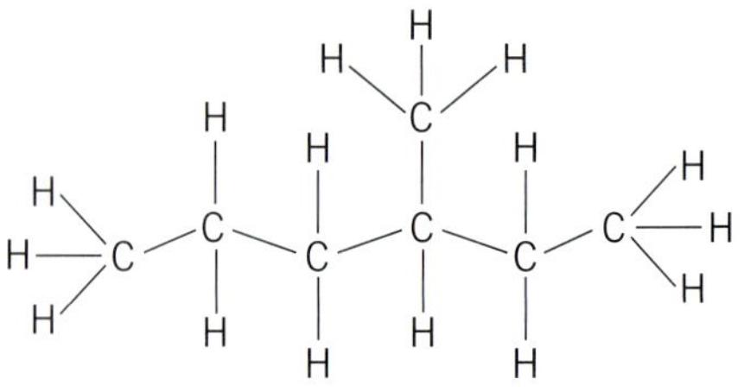 (molecular structure shown) |
| 1 Identify the longest carbon chain. | 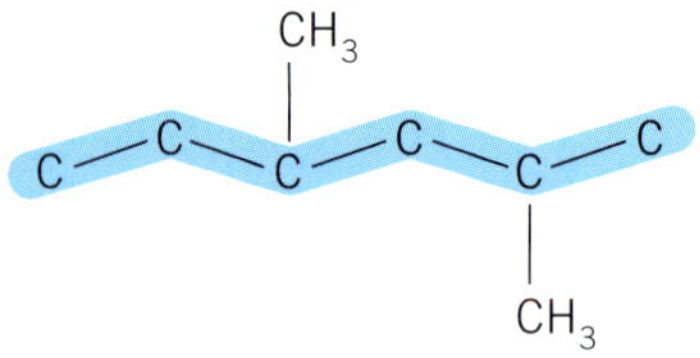 (structure with longest chain highlighted) |
| 2 Number the carbons, starting from the end closest to the branch. | 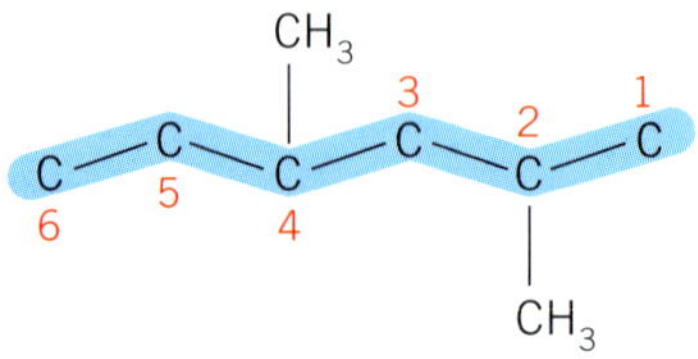 (structure with carbons numbered 1–6) |
| 3 Name the side branches and main chain. | 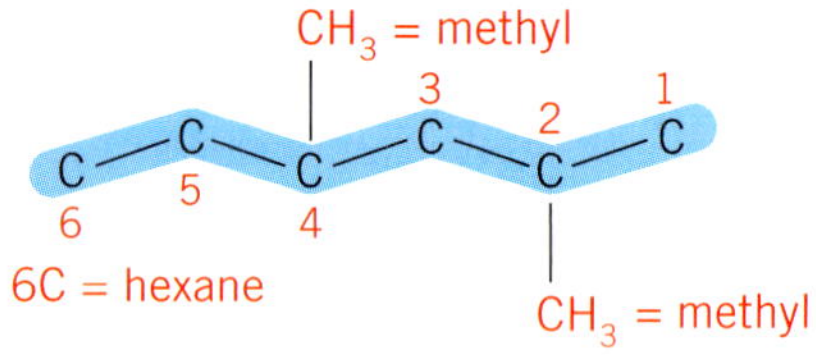 $CH_3$ = methyl<br>6C = hexane<br>$CH_3$ = methyl |
| 4 Combine to write the full name. | 2,4-dimethylhexane |

The carbon chain is numbered so that the side groups have the lowest possible numbers. For example, the molecule shown at right is named 3-methylhexane, not 4-methylhexane.

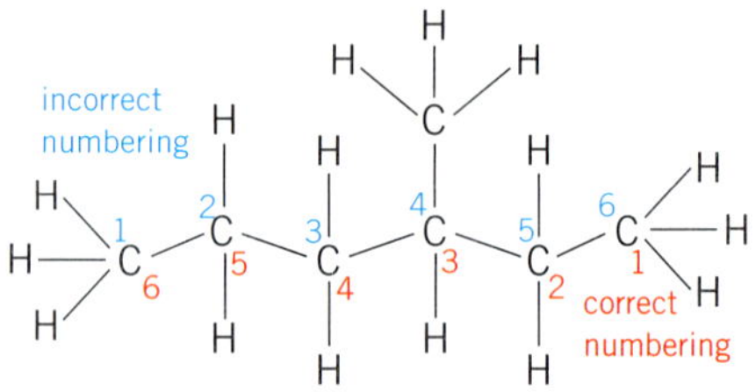

It is important to inspect a formula carefully, because the most obvious carbon chain is not always the longest one. For example, the longest chain in the molecule shown at right has five carbon atoms, not four.

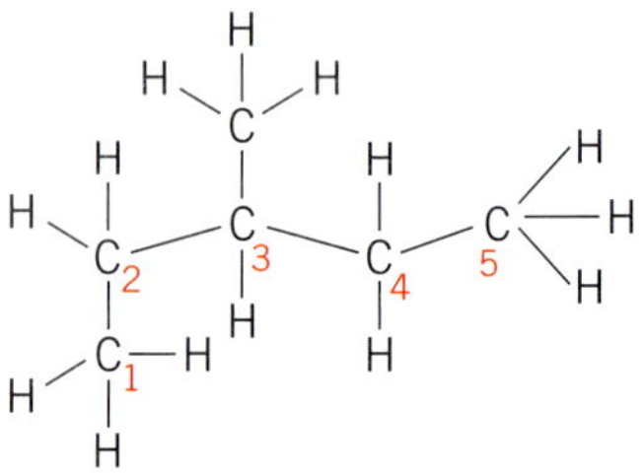

In cases where identical side groups are present on the hydrocarbon stem, they are given numbers, and the prefix 'di-', 'tri-', 'tetra-' and so on, is added to the hydrocarbon stem name. For example, the molecule shown at right is 2,3,4-trimethylhexane.

Alkane nomenclature

The name is written as a single word, using hyphens to separate the different prefixes and commas to separate numbers. If two or more different substituents are present, they are cited in alphabetical order (thus ethyl before methyl). If two or more identical substituents are present, a prefix ('di-', 'tri-' etc.) is used, but these prefixes are not used for alphabetizing purposes (thus ethyl before dimethyl).

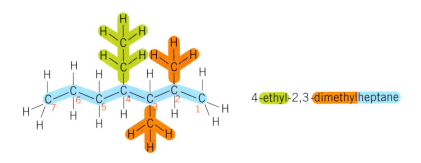

4-ethyl-2,3-dimethylheptane

**WORKSHEET 11.1**
Modelling and naming carbon compounds

Using these rules, we can now consider the five structural isomers of hexane.

[Structural diagrams: hexane, 2-methylpentane, 3-methylpentane, 2,3-dimethylbutane, 2,2-dimethylbutane]

Figure 11.1.8 Structural isomers of hexane.

### Alkenes

**Alkenes** are another homologous series of hydrocarbons. They are formed when the molecule contains one carbon–carbon double bond. Because at least two carbon atoms are required for a carbon–carbon double bond to be present, the simplest alkene is ethene, $C_2H_4$.

**Alkene series has the general formula $C_nH_{2n}$**

Alkenes are an example of an *unsaturated* hydrocarbon. **Unsaturated hydrocarbons** contain at least one carbon–carbon double bond, and so contain less than the maximum number of hydrogen atoms.

**TABLE 11.1.4 THE FIRST SIX ALKENES**

| Number of carbon atoms | Molecular formula | Name | Structural formula | Boiling point (°C) |
|---|---|---|---|---|
| 2 | $C_2H_4$ | Ethene | | −104 |
| 3 | $C_3H_6$ | Propene | | −47 |
| 4 | $C_4H_8$ | Butene (But-1-ene) | | −6 |
| 5 | $C_5H_{10}$ | Pentene (Pent-1-ene) | | 30 |
| 6 | $C_6H_{12}$ | Hexene (Hex-1-ene) | | 64 |

Alkenes exhibit the same trends in properties as the alkanes: as the molecular length increases, the boiling point and melting point increase, because the strength of the van der Waals' forces increases.

Alkenes can also exhibit structural isomerism. The double bond may occur in different places along the chain, or branching may occur.

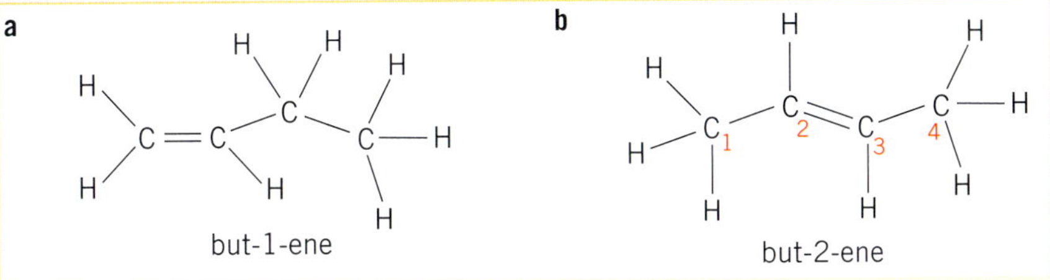

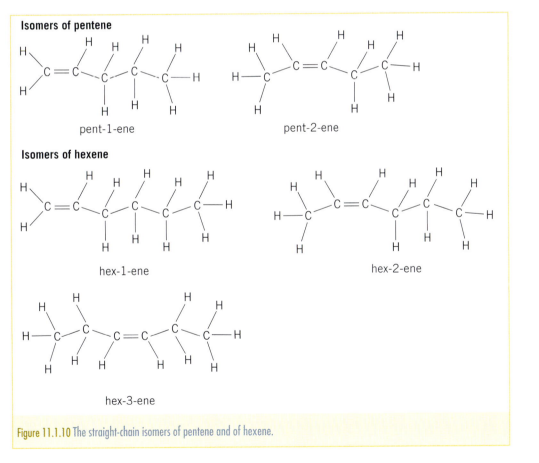

Figure 11.1.9 The two straight-chain isomers of butene.

There are two straight-chain isomers of butene: but-1-ene and but-2-ene.

The number indicates the position of the double bond in the chain. But-1-ene has its double bond in the first position along the chain, between carbons 1 and 2; but-2-ene has its double bond in the second position, between carbons 2 and 3.

The straight-chain isomers of pentene and of hexene follow the same pattern. There are only two structural isomers of pentene and there are three of hexene (see figure 11.1.10).

Figure 11.1.10 The straight-chain isomers of pentene and of hexene.

**AS 10.1.7**
**Deduce structural formulas for the isomers of the straight-chain alkenes up to $C_6$.**
© IBO 2007

**AS 10.1.8**
**Apply IUPAC rules for naming the isomers of the straight-chain alkenes up to $C_6$.**
© IBO 2007

## CHEM COMPLEMENT

### Blackening bananas

Ripening fruits give off ethene, which triggers further ripening. Damaged fruit produces extra ethene, causing any nearby fruit to ripen more quickly, hence the phrase 'one rotten apple can spoil the whole barrel'. Banana growers recommend that to keep newly purchased bananas fresh they should be stored more than one metre away from any ripe, blackening bananas due to the amount of ethene that this fruit is releasing.

Artificially introducing ethene increases the rate of the normal ripening process. For example, exposing 1 kg of tomatoes to as little as 0.1 mg of ethene for 24 hours will ripen them—but they do not taste as nice as those slowly ripened on the vine! The use of ethene in this way allows fruit growers to harvest fruit while it is still hard and green, then transport and cool-store it until it can be ripened shortly before sale.

Figure 11.1.11 Ripening bananas produces ethene which will make other bananas ripen more quickly.

PRAC 11.2
Models of hydrocarbons

Branched isomers of alkenes can also be formed. Methylpropene is an isomer of butene, and there are a number of branched isomers of pentene.

Figure 11.1.12 Methylpropene is a branched isomer of butene.

### Calculating the molecular formula of an unknown organic compound

When a sample of an unknown compound ($C_xH_yO_z$) is burnt in excess oxygen, carbon dioxide and water are produced. (See section 11.3, p. 372.) If the masses of the sample, carbon dioxide and water are measured, the empirical formula of the compound can be determined. Recall that the empirical formula shows the simplest whole-number ratio of the atoms present in the compound. The molar mass of the compound can also be determined by measuring the mass of a given volume of the gaseous compound at a specified temperature and pressure. Combining the results of these two experiments allows the molecular formula of the unknown compound to be determined. Recall that the molecular formula shows the actual numbers of atoms present in one molecule of the compound. The procedure used is shown in figure 11.1.13 and described in table 11.1.5.

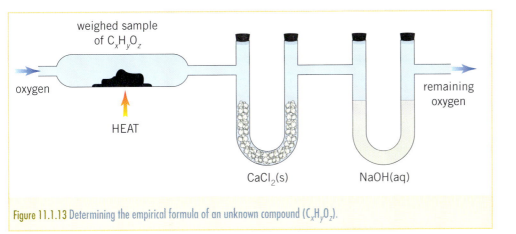

Figure 11.1.13 Determining the empirical formula of an unknown compound ($C_xH_yO_z$).

### TABLE 11.1.5 DETERMINING THE MOLECULAR FORMULA OF AN UNKNOWN COMPOUND

| | Procedure | Explanation |
|---|---|---|
| 1 | Weigh a sample of the compound. | Measures the mass of the sample. |
| 2 | Heat the sample in excess oxygen. | Combustion reaction occurs to produce $CO_2(g)$ and $H_2O(g)$. |
| 3 | Pass the unreacted $O_2(g)$ and products through $CaCl_2$. | $CaCl_2$ absorbs $H_2O$. The mass of $H_2O$ can therefore be determined. |
| 4 | Pass the remaining gases through NaOH(aq). | NaOH(aq) absorbs $CO_2$. The mass of $CO_2$ can therefore be determined. |
| 5 | Weigh a second sample of the compound. Heat this sample to its boiling point. | A known mass of gaseous sample is obtained. |
| 6 | Measure the volume, temperature and pressure of the gaseous sample. | Using the known $V$, $T$, $P$ and $m$ values, and $PV = \frac{m}{M}RT$, the molar mass of the gas can be determined. |

## Worked example

In order to determine the molecular formula of a compound known to contain only carbon, hydrogen and oxygen, two experiments were carried out.

### Experiment 1

A 0.60 g sample of the compound was burnt in excess oxygen. The gases evolved were passed through anhydrous $CaCl_2$, and its mass increased by 0.36 g. The remaining gases were bubbled through a NaOH solution, which increased in mass by 0.88 g.

### Experiment 2

A 1.21 g sample of the compound was vaporized. The vapour occupied 0.403 dm³ at 150°C and $1.17 \times 10^5$ Pa.

Calculate the molecular formula of the compound.

**Solution**

Step 1: Determine the mass of the products.

$m(H_2O) = 0.36$ g and $m(CO_2) = 0.88$ g

Step 2: Determine the mass of carbon in the sample.

We assume all the carbon in the sample is converted to $CO_2$.

$\therefore m(C)$ in sample $= m(C)$ in the $CO_2$ produced
$$= \frac{12.01}{44.01} \times 0.88 = 0.24 \text{ g}$$

Step 3: Determine the mass of hydrogen in the sample.

We assume all the hydrogen in the sample is converted to $H_2O$.

$\therefore m(H)$ in sample $= m(H)$ in $H_2O$ produced
$$= \frac{2.02}{18.02} \times 0.36 = 0.040 \text{ g}$$

Step 4: Determine the mass of oxygen in the sample.

$m(O)$ in sample $= m(\text{sample}) - m(C) - m(H)$
$= 0.60 - 0.24 - 0.040 = 0.32$ g

Step 5: Determine the empirical formula of $C_xH_yO_z$.

| C | : | H | : | O | |
|---|---|---|---|---|---|
| 0.24 | : | 0.040 | : | 0.32 | Elements in the compound / Mass ratio |
| $\frac{0.24}{12.01}$ | : | $\frac{0.040}{1.01}$ | : | $\frac{0.32}{16.00}$ | Divide by molar mass |
| 0.020 | : | 0.040 | : | 0.020 | Determine ratio |
| 1 | : | 2 | : | 1 | Simplify ratio |

The empirical formula (EF) of the compound is $CH_2O$.

Step 6: Determine the relative molecular mass of $C_xH_yO_z$.

$m(\text{sample}) = 1.21$ g         $V(\text{sample}) = 0.403$ dm$^3$
$T(\text{sample}) = 150°C = 423$ K         $P(\text{sample}) = 1.17 \times 10^5$ Pa $= 117$ kPa

$$PV = nRT = \frac{mRT}{M}$$

$$\therefore M = \frac{mRT}{PV} = \frac{1.21 \times 8.31 \times 423}{117 \times 0.403} = 90.2 \text{ g mol}^{-1}$$

The relative molecular mass (RMM) of the compound is 90.2.

Step 7: Determine the molecular formula (MF) of $C_xH_yO_z$.

EF is $CH_2O$

EFM is $12.0 + 2.0 + 16.0 = 30.0$

MF is $(CH_2O)_a$ where $a = \frac{RMM}{EFM} = \frac{90.2}{30.0} = 3$

The molecular formula of the compound is $C_3H_6O_3$.

## Section 11.1 Exercises

1. State two characteristics of a homologous series.

2. Give the molecular formula and draw structural formulas of each of the following molecules.
   a  Butane
   b  Pent-2-ene
   c  The alkane with 6 carbon atoms
   d  The alkene with 8 hydrogen atoms

3. Classify each of the following as alkanes or alkenes.
   a  $C_{12}H_{24}$
   b  $C_{16}H_{32}$
   c  $C_{15}H_{32}$
   d  $C_{72}H_{146}$

4. Draw structural formulas for each of the following hydrocarbon molecules.
   a  2-Methylbutane
   b  3-Methylpentane
   c  Pent-2-ene
   d  But-2-ene

5. Name the hydrocarbon molecules whose semi-structural (condensed structural) formulas are given below.
   a  $CH_3CH(CH_3)CH_2CH_2CH_3$
   b  $CH_2CHCH_2CH_3$
   c  $CH_3CHCHCH_3$
   d  $CH_3CHCHCH_2CH_3$

6. Name the hydrocarbons shown below.

   a
   $$CH_3CH_2\underset{\underset{CH_3}{|}}{CH}CH_3$$

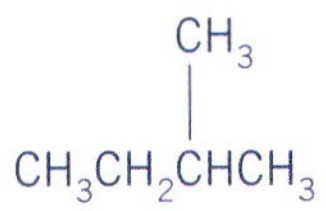

   b  $CH_3CH_2CH=CHCH_3$

   c
   $$CH_3CH_2\underset{\underset{\underset{\underset{CH_3}{|}}{CH_2}}{|}}{CH}CH_3$$

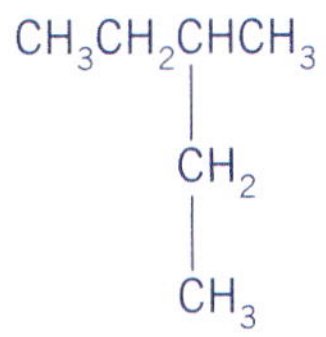

   d  $CH_3CH=CH_2$

7. Pentane boils at 36°C, while 2-methylbutane boils at 28°C. Explain why pentane has the higher boiling point.

8 Draw a Lewis structure for each of the following molecules.
  a  Methane, $CH_4$
  b  Ethene, $C_2H_4$
  c  Propane, $C_3H_8$

9 In order to determine the molecular formula of a compound known to contain only carbon and hydrogen, two experiments were carried out.

In the first experiment, a 1.122 g sample of the compound was burnt in excess oxygen. When the gases evolved were passed through anhydrous $CaCl_2$, its mass increased by 1.442 g. The remaining gases, when bubbled through a NaOH solution, increased its mass by 3.521 g.

In the second experiment a 2.000 g sample of the compound was vaporized. The vapour occupied a volume of 797 $cm^3$ at STP.

Calculate the empirical formula and hence the molecular formula of the compound.

10 An organic compound containing only carbon, hydrogen and oxygen was analysed gravimetrically. When completely oxidized in air, 0.900 g of the compound produced 1.80 g of carbon dioxide and 0.736 g of water. A separate 2.279 g hydrocarbon sample, when vaporized in a 1.00 $dm^3$ vessel at 100°C, had a pressure of 80.3 kPa. Determine the molecular formula of the compound.

## 11.2 INTRODUCING FUNCTIONAL GROUPS

The molecules considered so far contain only carbon and hydrogen. Introducing other types of atoms into a hydrocarbon results in a vast range of compounds with greater chemical reactivity than the alkanes. Atoms or groups of atoms attached to a hydrocarbon chain are called **functional groups**. These functional groups are significant because they change the reactivity of the molecule and frequently dictate the chemical properties of the compound.

When we are focusing on the presence of a functional group in a hydrocarbon, it is common to write the general formula with the symbols of the functional group written as they appear in the structure of the molecule and a single letter R to represent the alkyl group (the rest of the hydrocarbon). If there is more than one alkyl group, then R' represents the second alkyl group and R″ may represent a third alkyl group.

### Alcohols

The **alcohol** functional group is made up of an oxygen atom bonded to a hydrogen atom. This –OH group is known as a **hydroxyl group** and replaces one hydrogen in the structure of an alkane. The prefixes that are used to indicate the number of carbon atoms still apply to alcohols, however the suffix (ending) of the name of an alcohol is always -*ol*. The general formula of the alcohol homologous series is $C_nH_{2n+1}OH$ or an alcohol may generally be represented as R–OH.

---

**AS 10.1.9**
Deduce structural formulas for compounds containing up to six carbon atoms with one of the following functional groups: alcohol, aldehyde, ketone, carboxylic acid and halide.
© IBO 2007

**AS 10.1.10**
Apply IUPAC rules for naming compounds up to six carbon atoms with one of the following functional groups: alcohol, aldehyde, ketone, carboxylic acid and halogenoalkane.
© IBO 2007

## TABLE 11.2.1 THE FIRST SIX MEMBERS OF THE ALCOHOL SERIES

| Name | Condensed structural formula | Structural formula | Boiling point (°C) | Solubility in water |
|---|---|---|---|---|
| Methanol | $CH_3OH$ | | 64.7 | Soluble |
| Ethanol | $C_2H_5OH$ | | 78.4 | Soluble |
| Propanol (Propan-1-ol) | $C_3H_7OH$ | | 97.1 | Slightly soluble |
| Butanol (Butan-1-ol) | $C_4H_9OH$ | | 117.7 | Insoluble |
| Pentanol (Pentan-1-ol) | $C_5H_{11}OH$ | | 138 | Insoluble |
| Hexanol (Hexan-1-ol) | $C_6H_{13}OH$ | | 156 | Insoluble |

CHEMISTRY: FOR USE WITH THE IB DIPLOMA PROGRAMME STANDARD LEVEL

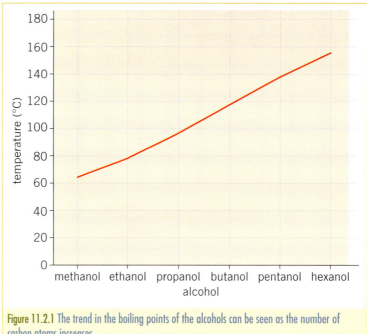

Figure 11.2.1 The trend in the boiling points of the alcohols can be seen as the number of carbon atoms increases.

**AS 10.1.13**
Discuss the volatility and solubility in water of compounds containing the functional groups listed in 10.1.9. © IBO 2007

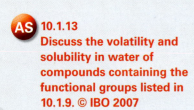

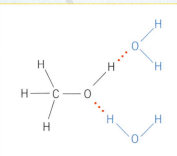

Figure 11.2.2 Hydrogen bonds form between methanol molecules and water.

Note that like the alkanes, the boiling points of the alcohols increase as the size of the alcohol increases; however, the boiling points of the alcohols are all higher than those of the alkanes of similar size. In contrast to the alkanes and alkenes, many of which are gases, all the alcohols are liquids at room temperature. This higher boiling point is due to the presence of the –OH group, which allows hydrogen bonding to occur between molecules and strengthens the intermolecular bonding. Hydrogen bonding and other types of intermolecular bonding which influence boiling and melting points were discussed in chapter 2, section 2.5. **Volatility** can be defined as how easily a substance evaporates, so the lower the boiling point of a substance, the greater its volatility; thus the volatility of the alcohols is *lower* than that of the alkanes.

Hydrogen bonding also influences the solubility of the alcohols. The presence of the hydroxyl (–OH) group allows hydrogen bonds to form between water molecules and alcohol molecules. For the smaller alcohols such as methanol and ethanol, this ensures that the alcohol readily dissolves in water, but as the size of the carbon chain in the alcohol increases, the solubility in water decreases. The increasing length of the hydrocarbon chain means that the influence of the –OH group diminishes in comparison to that of the rest of the molecule. A large proportion of the molecule is non-polar and unable to form hydrogen bonds, so its attraction to the water molecules is low. Propanol, $C_3H_7OH$, is slightly soluble in water, but alcohols with more than three carbons are insoluble in water.

The position of the hydroxyl group is indicated by a number corresponding to the position in the carbon chain of the carbon atom to which the hydroxyl group is bonded. As with all isomers, the carbon atoms are counted from the end closest to the hydroxyl group (see figure 11.2.3).

The position of the hydroxyl group also influences the ability of the molecule to form intermolecular bonds. This can be seen when we examine the boiling points of the structural isomers of butanol. These isomers are formed when the –OH group is attached to different carbon atoms in the four-carbon chain, or when the chain is branched.

DEMO 11.1
Ethanol and its properties

butan-1-ol    butan-2-ol    2-methylpropan-2-ol

Figure 11.2.3 Three structural isomers of butanol, $C_4H_9OH$.

In figure 11.2.4 the boiling point can be seen to decrease due to decreasing strength of the intermolecular forces between the molecules as the branching of the hydrocarbon chain increases.

10.1.12
Identify primary, secondary and tertiary carbon atoms in alcohols and halogenoalkanes.
© IBO 2007

Alcohols can be classified according to the carbon atom to which the hydroxyl group is bonded, and where that carbon atom is in the molecule. When the hydroxyl group is bonded to a carbon atom that is bonded to only one other carbon atom, that carbon atom is called a *primary* carbon atom and the alcohol may be termed a **primary alcohol**. When the hydroxyl group is bonded to a carbon atom that is bonded to two other carbon atoms, that carbon atom is called a *secondary* carbon atom and the alcohol may be termed a **secondary alcohol**. When the hydroxyl group is bonded to a carbon atom that is bonded to three other carbon atoms, that carbon atom is called a *tertiary* carbon atom and the alcohol is termed a **tertiary alcohol**.

Of the alcohols in figure 11.2.3, butan-1-ol is a primary alcohol, butan-2-ol is a secondary alcohol and 2-methylpropan-2-ol is a tertiary alcohol.

## Aldehydes and ketones

Aldehydes and ketones are both homologous series containing the **carbonyl functional group** (C=O). The carbonyl group is strongly polar and influences the physical and chemical properties of the compounds it is found in. The general formula of aldehydes is $C_nH_{2n}O$ or an aldehyde may generally be represented as RCHO. To avoid confusion with alcohols, the aldehyde functional group is written as –CHO, indicating that the H is not directly bonded to the O; in alcohols the functional group is written as –OH.

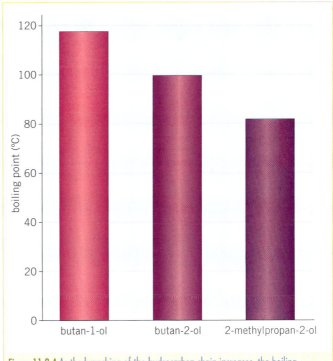

Figure 11.2.4 As the branching of the hydrocarbon chain increases, the boiling point decreases.

Ketones and aldehydes both have a single carbonyl group (C=O). In **aldehydes** this carbonyl group is bonded to a carbon that is on the end of the carbon chain; in **ketones** the carbonyl group is bonded to a carbon atom that is not on either end of the carbon chain. The general formula of ketones is the same as that of aldehydes: $C_nH_{2n}O$; however, they may generally be represented as RCOR′, where R and R′ may or may not be the same alkyl group. The consequence of this structure is that the smallest ketone, propanone, has three carbon atoms in its chain.

The names of aldehydes and ketones follow the same pattern as for other hydrocarbons. The number of carbon atoms in the chain including that in the carbonyl group is indicated by the prefix methan-, ethan-, propan- etc. and the presence of the aldehyde or ketone functional group is indicated by the suffix. The names of aldehydes finish with the suffix -*al* and those of ketones finish with the suffix -*one*. The position of the carbonyl group in a ketone is indicated by a number just before the suffix (-one). This number indicates which carbon atom is part of the carbonyl group.

| Name | Condensed structural formula | Structural formula | Boiling point (°C) | Solubility in water |
|---|---|---|---|---|
| Methanal | HCHO | | −21 | Soluble |
| Ethanal | $CH_3CHO$ | | 21 | Soluble |
| Propanal | $C_2H_5CHO$ | | 46 | Slightly soluble |
| Butanal | $C_3H_7CHO$ | | 75 | Very slightly soluble |
| Pentanal | $C_4H_9CHO$ | | 103 | Insoluble |
| Hexanal | $C_5H_{11}CHO$ | | 119 | Insoluble |

TABLE 11.2.2 ALDEHYDES WITH UP TO SIX CARBON ATOMS

## TABLE 11.2.3 KETONES WITH UP TO SIX CARBON ATOMS

| Name | Condensed structural formula | Structural formula | Boiling point (°C) | Solubility in water |
|---|---|---|---|---|
| Propanone | $CH_3COCH_3$ | | 56 | Soluble |
| Butanone | $C_2H_5COCH_3$ | | 80 | Slightly soluble |
| Pentan-2-one | $C_3H_7COCH_3$ | | 102 | Insoluble |
| Pentan-3-one | $C_2H_5COC_2H_5$ | | 100–102 | Insoluble |
| Hexan-2-one | $C_4H_9COCH_3$ | | 127 | Insoluble |
| Hexan-3-one | $C_3H_7COC_2H_5$ | | 123 | Insoluble |

The carbonyl group in ketones and aldehydes allows permanent dipole attraction to occur between the molecules and makes the intermolecular bonding stronger than that in alkanes of similar size but weaker than that in alcohols, in which the intermolecular bonding is hydrogen bonding. This results in the boiling points of aldehydes and ketones being higher than those of the alkanes, but lower than those of the alcohols. As the length of the carbon chain increases, the strength of the van der Waals' forces increases, thus

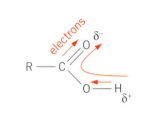

**Figure 11.2.5** Hydrogen bonding can occur between the polar carbonyl group and water molecules.

**Figure 11.2.6** Electrons are drawn away from the hydrogen of the carboxyl functional group.

increasing the boiling point of the compound. With the exception of methanal, all the aldehydes and ketones listed in tables 11.2.2 and 11.2.3 are liquids at room temperature, so their volatility could be described as low.

The carbonyl group of aldehydes and ketones is polar, and this enables hydrogen bonding between the carbonyl group and water molecules. The solubility of the small aldehydes and ketones in water is due to this hydrogen bonding. However, hydrogen bonding is not possible between aldehyde or ketone molecules.

### Carboxylic acids

**Carboxylic acids** are organic acids. The carboxylic acid functional group, the **carboxyl group**, has the formula –COOH. This functional group is made up of a carbonyl group (C=O) and a hydroxyl (–OH) group. Both of these groups are polar and the electrons are drawn away from the hydrogen atom, enabling it to be donated in an acid–base reaction.

These acids are not as strong as inorganic acids and are found in nature, for example, in citrus fruit (citric acid, $C_6H_8O_7$), stings of insects (methanoic acid, HCOOH) and vinegar (ethanoic acid, $CH_3COOH$).

As was discussed in chapter 9, section 9.3, weak acids such as carboxylic acids only partially dissociate in aqueous solution.

For example: $CH_3COOH(aq) + H_2O(l) \rightleftharpoons CH_3COO^-(aq) + H_3O^+(aq)$

**Figure 11.2.7** Some sources of carboxylic acids in nature.

**Figure 11.2.8** Sniffer dogs detect the characteristic combination of carboxylic acids in a person's sweat.

### CHEM COMPLEMENT

#### Tracking carboxylic acids

Larger carboxylic acids—$C_4$ to $C_8$ (butanoic to octanoic)—have strong, unpleasant odours. Butanoic acid is responsible for the smell of rancid butter and is present in human sweat. It can be detected by dogs at concentrations of $10^{-17}$ mol dm$^{-3}$. The usefulness of tracker dogs comes from this ability to detect the carboxylic acids in a person's sweat. Each person produces a characteristic blend of carboxylic acids, which the sensitive nose of a tracker dog can detect and recognize.

Naming of carboxylic acids follows the same pattern as for other hydrocarbons. The number of carbons in the carbon chain, including the carbon in the functional group, is indicated by the prefix methan-, ethan-, propan- etc. and the presence of the carboxyl functional group is indicated by the suffix –oic acid.

### TABLE 11.2.4 CARBOXYLIC ACIDS WITH UP TO SIX CARBON ATOMS

| Name | Condensed structural formula | Structural formula | Boiling point (°C) | Solubility in water |
|---|---|---|---|---|
| Methanoic acid | HCOOH | | 101 | Soluble |
| Ethanoic acid | CH$_3$COOH | | 118 | Soluble |
| Propanoic acid | C$_2$H$_5$COOH | | 141 | Soluble |
| Butanoic acid | C$_3$H$_7$COOH | | 164 | Soluble |
| Pentanoic acid | C$_4$H$_9$COOH | | 186 | Slightly soluble |
| Hexanoic acid | C$_5$H$_{11}$COOH | | 202 | Slightly soluble |

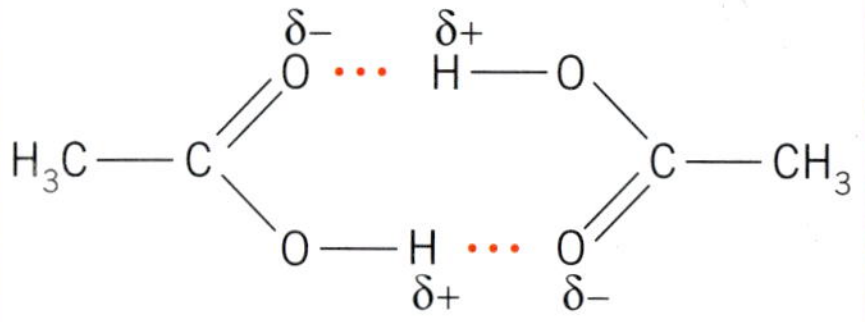

**Figure 11.2.9** Hydrogen bonding between two ethanoic acid molecules results in the formation of a dimer.

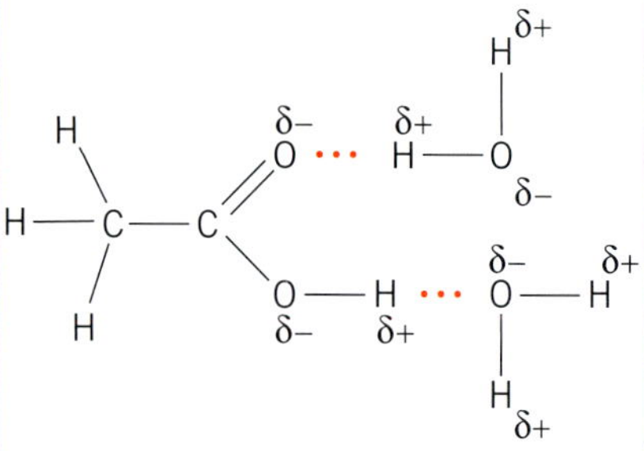

**Figure 11.2.10** Ethanoic acid forms hydrogen bonds with water molecules when it dissolves in water.

**DEMO 11.2**
Properties of vinegar

**AS 10.1.12**
Identify primary, secondary and tertiary carbon atoms in alcohols and halogenoalkanes.
© IBO 2007

**WORKSHEET 11.2**
Naming organic compounds

The high boiling points of the carboxylic acids is a direct result of the presence of hydrogen bonding between carboxylic acid molecules. They pair up, forming **dimers** whose increased molecular mass results in stronger van der Waals' forces between the molecules. All of the carboxylic acids shown in table 11.2.4 are liquids at room temperature. They have low volatility.

When dissolved in water, dimers do not form; instead, hydrogen bonding occurs between the carboxyl group of carboxylic acids and water molecules, making these compounds more soluble in water than those we have studied so far. The carbonyl group of the carboxylic acid forms hydrogen bonds with the slightly positive hydrogen atom in the water molecule, and the hydrogen in the hydroxyl group of the carboxylic acid forms hydrogen bonds with the non-bonding pair of electrons on the oxygen atom in the water molecule.

## Halogenoalkanes

The **halogenoalkanes** are molecules in which one or more of the hydrogen atoms within an alkane molecule has been replaced by a halogen atom, or **halide functional group**. The halogen may be any member of group 7, but is most commonly chlorine, bromine or iodine. The molecule is named according to the halogen found in the halogenoalkane: chloro-, bromo-, iodo-. The remainder of the name corresponds exactly to the alkane with the same number of carbons in the carbon chain. Hence, bromoethane has the formula $C_2H_5Br$. The position of the halogen is indicated by a number at the beginning of the name which corresponds to the number of the carbon to which the halogen is bonded (see figure 11.2.11(a) and (b)). Note that in this figure, molecules (a) and (b) are isomers of chloropropane, but molecule (c) is not. It is an isomer of chlorobutane.

Like alcohols, the halogenoalkanes can be classified into primary, secondary and tertiary compounds. When the halogen is bonded to a carbon atom which is only bonded to one other carbon atom, that carbon atom is called a primary carbon atom and the halogenoalkane may be termed a primary halogenoalkane. When the halogen is bonded to a carbon atom which is bonded to two other carbon atoms, that carbon atom is called a secondary carbon atom and the halogenoalkane may be termed a secondary halogenoalkane. When the halogen is bonded to a carbon atom which is bonded to three other carbon atoms, that carbon atom is called a tertiary carbon atom and the halogenoalkane is termed a tertiary halogenoalkane.

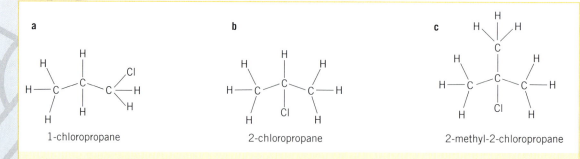

**Figure 11.2.11** (a) Primary, (b) secondary and (c) tertiary halogenoalkanes.

There is a large range of halogenoalkanes; however, they are almost all gases or liquids at room temperature, indicating a generally higher volatility than many of the hydrocarbons discussed in this section.

As for other hydrocarbons, the boiling points of the halogenoalkanes are influenced by the length of the carbon chain. As the number of carbons in the compound increases, the boiling point increases. The presence of the halide functional group makes halogenoalkanes slightly polar, so dipole–dipole attraction may have some influence on the boiling point. The nature of the halogen also influences the boiling point. The larger the halogen atom, the stronger the van der Waals' forces between the molecules, so the higher will be the boiling point and the less volatile the halogenoalkane.

Organic functional group names

### TABLE 11.2.5 SOME HALOGENOALKANES

| Name | Condensed structural formula | Structural formula | Boiling point (°C) | Solubility in water |
|---|---|---|---|---|
| Chloromethane | $CH_3Cl$ | | −24 | Slightly soluble |
| Bromomethane | $CH_3Br$ | | 4 | Slightly soluble |
| Iodomethane | $CH_3I$ | | 33 | Slightly soluble |
| Chloroethane | $C_2H_5Cl$ | | 12 | Slightly soluble |
| 1-Chloropropane | $C_3H_7Cl$ | | 46 | Insoluble |
| 1-Chlorobutane | $C_4H_9Cl$ | | 79 | Insoluble |

## Other functional groups

**AS 10.1.11**
Identify the following functional groups when present in structural formulas: amino (NH$_2$), benzene ring, esters (RCOOR). © IBO 2007

While there are many other homologous series that you may encounter in your future studies of organic chemistry, there are only a few others that you need to be familiar with in this course. One of these is the **amino group** –NH$_2$, a functional group that is present in organic bases called amines. Like ammonia, the amino group has a non-bonding pair of electrons that can accept a hydrogen ion in Brønsted-Lowry acid–base reactions or can be donated in Lewis acid–base reactions. This means that amines are organic bases. The amino group has particular importance in proteins where it occurs in (and gives its name to) amino acids. Amino acids will be discussed further in Option B Human Biochemistry.

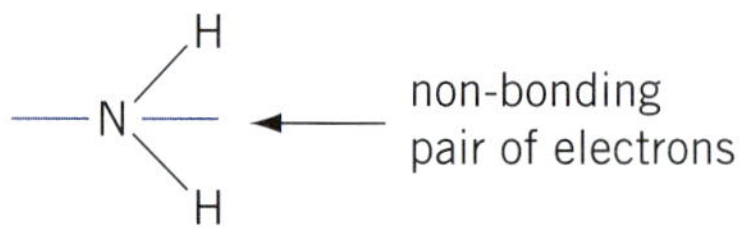

**Figure 11.2.12** The amino functional group.

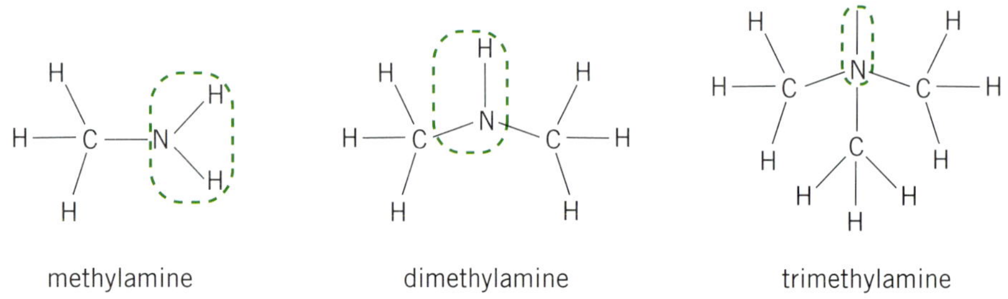

**Figure 11.2.13** A selection of amines showing the amino group.

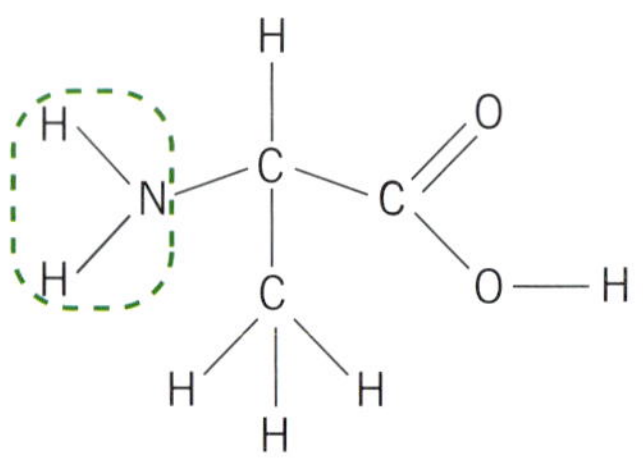

**Figure 11.2.14** The amino acid, alanine, with the amino group circled.

**Esters** are compounds that often have a characteristic sweet, fruity odour, and many of the esters occur naturally in fruits and flowers. The ester functional group is found in the middle of the molecule. It consists of a carbonyl group with a second oxygen bonded to the carbon atom.

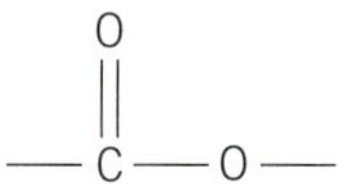

**Figure 11.2.16** The ester functional group.

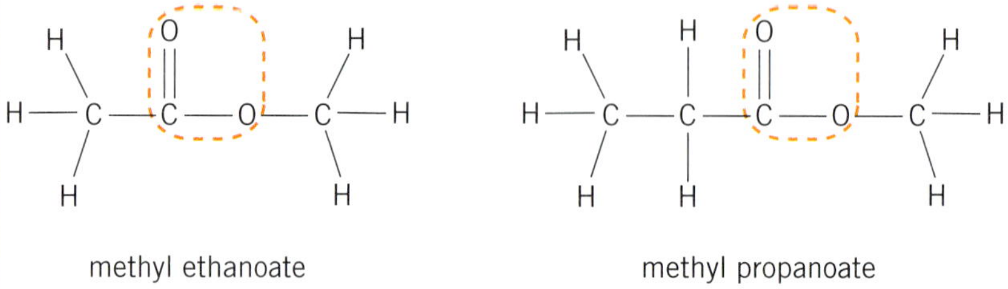

**Figure 11.2.17** Some ester molecules showing the ester functional group.

**Figure 11.2.15** Amines are partly responsible for the distinctive odour of fish.

### CHEM COMPLEMENT

#### Something smells fishy!

The NH$_2$ group occurs widely in biological molecules, especially proteins, but free amines are relatively rare in nature. They do, however, occur in decomposing material, formed by the action of bacteria on protein, and have an unpleasant odour. They are associated with the smell of decaying animal and human tissues—as the names putrescine (H$_2$N(CH$_2$)$_3$NH$_2$) and cadaverine (H$_2$N(CH$_2$)$_5$NH$_2$) suggest! Other amines—dimethylamine and trimethylamine—are partly responsible for the peculiar smell of rotting fish.

## CHEM COMPLEMENT

### What is that taste?

The pleasant, sweet odours and tastes that flavour many foods are due to complex mixtures of compounds, particularly esters. Manufacturers of such products as soft drinks, toppings and ice-cream like their products to smell and taste as natural as possible. Using extracts from natural foods may achieve this, but the extracts are often expensive and so the use of artificial flavourings is common. However, to reproduce synthetically the exact mixture found in an apple would be even more costly than using natural extract! Fortunately, manufacturers have found that the addition of only one or a few compounds, usually esters, achieves the desired taste and smell economically.

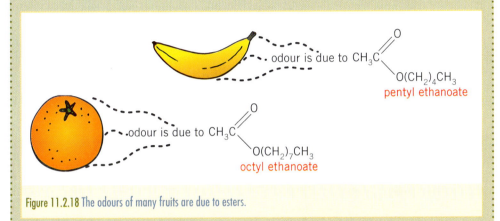

Figure 11.2.18 The odours of many fruits are due to esters.

Up to this point, all of the organic compounds we have considered have been **aliphatic** molecules, which are made up of a straight or branched carbon chain. If the carbon atoms make a closed ring, with carbon–carbon single or double bonds, the molecule is described as **alicyclic**. You may use the alicyclic compound cyclohexane in the laboratory. A third group of hydrocarbons is the group known as **aromatic** molecules. These molecules are related to **benzene**, $C_6H_6$.

Benzene can exist as a molecule on its own, or as a functional group. The carbons form a six-membered ring and there is a hydrogen atom bonded to each carbon, hence the formula $C_6H_6$. The bonding within benzene is not straightforward and is discussed in detail in the higher level course. At this stage of your studies it is sufficient that you are able to recognize the benzene ring in a chemical formula (see figures 11.2.19 and 11.2.20).

**WORKSHEET 11.3**
Functional groups

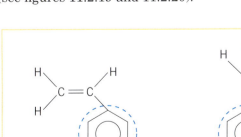

Figure 11.2.19 The symbol for benzene is drawn as a 'skeleton structure' showing only the bonding within the ring.

Figure 11.2.20 In these molecules, the benzene functional group is circled.

## CHEM COMPLEMENT

### Dangerously aromatic?

Benzene is a member of a group of compounds called 'aromatics'. This name came about because the first benzene compounds to be isolated from vegetables had pleasant aromas. This is not, however, the case for many aromatics. Naphthalene and toluene are also aromatics. Naphthalene is used to make the mothballs we keep in cupboards to repel moths that eat holes in our clothes. The moths certainly don't like the aroma of naphthalene and most humans find the smell of naphthalene overwhelming. Toluene smells like paint thinners—also a less than attractive smell!

The smell of benzene is often described as 'sweet'; however, this useful organic solvent has been shown to be a carcinogen (cancer-causing). Short-term exposure to high levels of benzene can cause drowsiness, dizziness, unconsciousness and death. This is not a substance to be using without the protection of gloves and a fume cupboard.

Figure 11.2.21 Benzene is a solvent that requires very careful treatment.

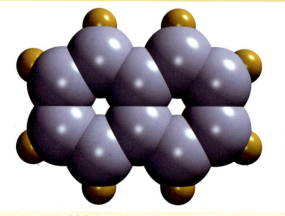

Figure 11.2.22 Naphthalene is made up of two benzene rings joined together.

## Section 11.2 Exercises

1. State the name of the homologous series to which each of the following compounds belongs.
   a. CH$_3$CH$_2$CH$_2$OH
   b. CH$_3$CH$_2$COOCH$_3$
   c. CH$_3$CH$_2$Cl
   d. CH$_3$CH$_2$COOH
   e. H$_2$NCH$_2$CH$_3$

2. Draw structural formulas to represent the following compounds.
   a. 2-methylbutane
   b. 1,1,1-trichloroethane
   c. 1-chlorohex-3-ene
   d. Pentanoic acid

3. State the name of the following organic molecules, given their condensed (semi-structural) formulas.
   a. CH$_3$CH$_2$CHO
   b. CH$_3$COCH$_2$CH$_3$
   c. CH$_2$ClCH$_2$Cl
   d. CH$_3$CHCHCH$_2$CH$_3$
   e. HCOOH
   f. CH$_3$CH$_2$CHOHCH$_2$CH$_3$

4. State the name of the molecules shown below.

   a, b, 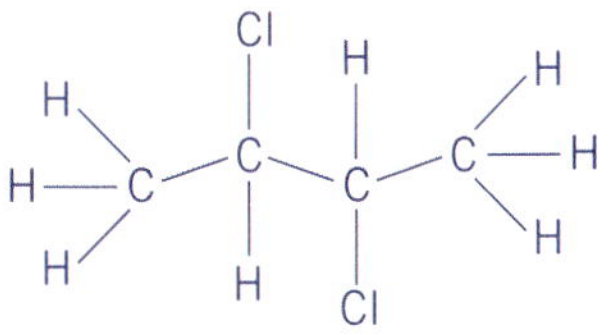 c, 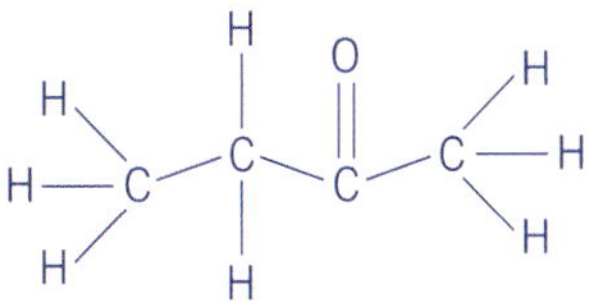 d, 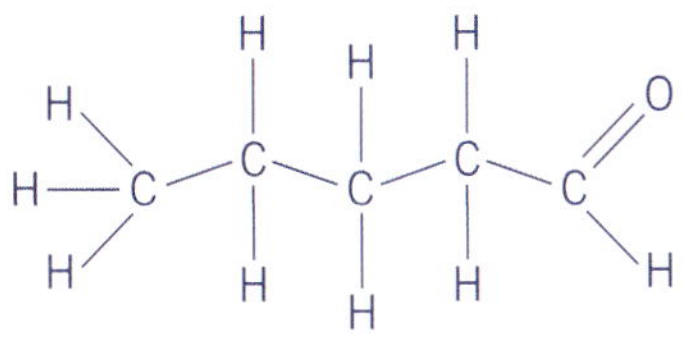 e [structural diagrams]

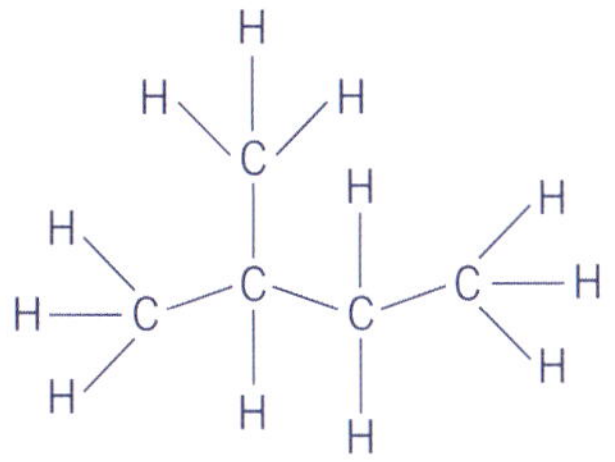

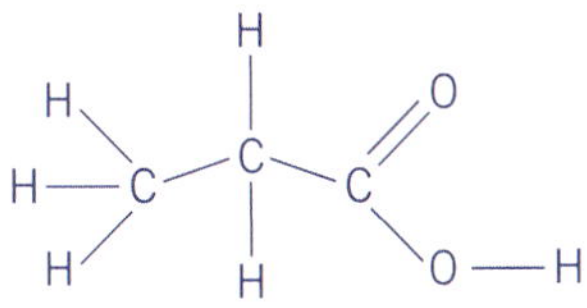

5. The formulas of a group of related molecules are: C$_2$H$_5$NH$_2$, C$_3$H$_7$NH$_2$, C$_4$H$_9$NH$_2$, C$_5$H$_{11}$NH$_2$, C$_6$H$_{13}$NH$_2$.
   a. State the name that is given to such a group of compounds.
   b. How would you expect the boiling points of these compounds to change with increasing numbers of carbon atoms? Explain.
   c. Compared to alkanes of a similar molar mass, state whether you would expect these compounds to be:
      i. more or less soluble in water
      ii. more or less volatile.

6 Draw structural formulas and state the names of all the straight (unbranched) chain alcohols with the molecular formula $C_5H_{12}O$.

7 The graph below shows the boiling points of the first six straight-chain alkanes and straight-chain alcohols.

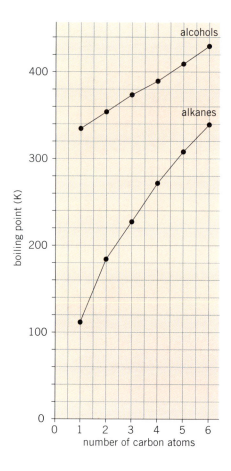

a Explain why the boiling points of alcohols are higher than those of the corresponding alkanes.

b Explain why the differences in boiling points between corresponding alcohols and alkanes decrease as the number of carbon atoms increases.

8 A compound has the formula $C_3H_8O$. Draw the structural formulas of three possible molecules with this formula.

9 a Draw the structural formulas for ethanol and hexan-1-ol.

b Explain why it is necessary to write the '1' in hexan-1-ol, but no number is given with ethanol.

10 In terms of structure and bonding, explain why:
a the boiling point of octane is higher than that of propane
b the volatility of ethane is greater than that of butane.

11 For each of the molecules below, state whether they are primary, secondary or tertiary alcohols.
a $CH_3CH_2C(CH_3)OHCH_3$
b $CH_3CH_2CHOHCH_2CH_3$
c $CH_3CH_2OH$
d $CH_3C(CH_3)OHCH_3$

12 For each of the molecules below, state whether they are primary, secondary or tertiary halogenoalkanes.
   a  $CH_3CHClCH_2CH_3$
   b  $CH_3CH_2CH_2Cl$
   c  $CH_3C(CH_3)ClCH_3$
   d  $CH_3CH_2CHClCH_3$

13 For each of the molecules below, circle the functional group/s and state the name of the functional group/s you have circled.

a

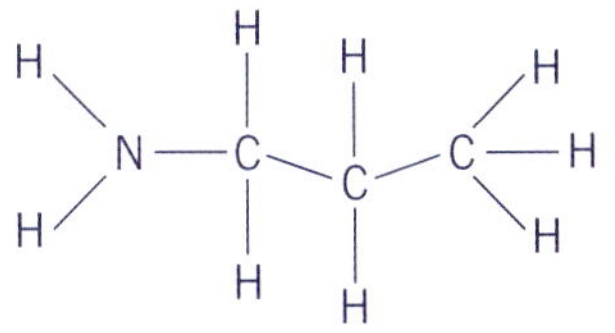

b

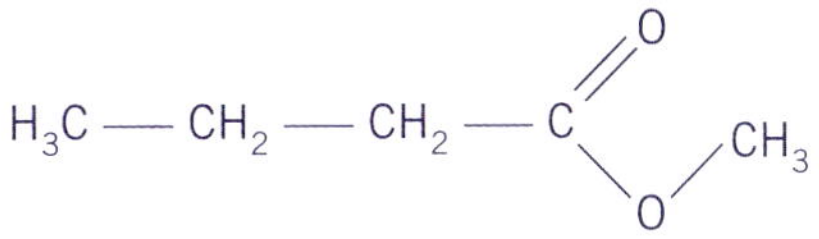

c

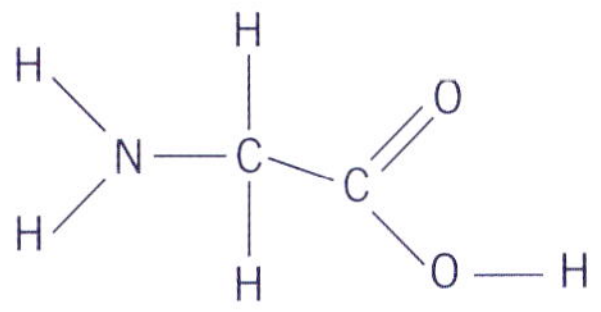

d

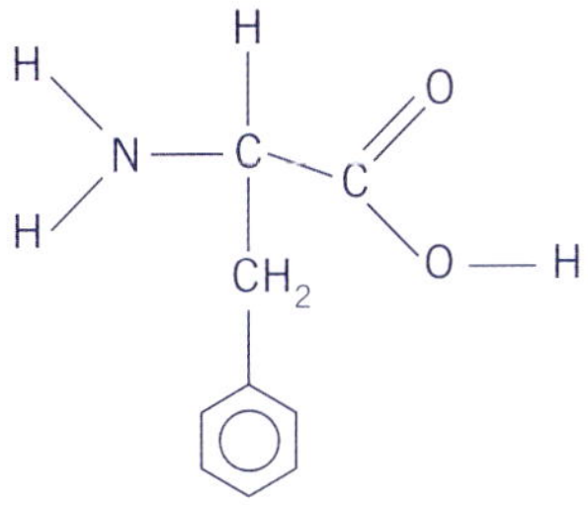

14 Name each of the molecules below.
   a  $CH_3CH_2CH_2COOH$
   b  $C_4H_9COCH_3$
   c  $CH_3CH_2C(CH_3)OHCH_3$
   d  $CH_3CH_2CH_2CHO$

## 11.3 REACTIONS OF ALKANES

Let us return to alkanes and their characteristic reactions. Alkanes are non-polar and so are insoluble in water. They are all less dense than water and therefore float on it. They are colourless when pure. Because C–C and C–H bonds are relatively strong, (their bond enthalpies are 348 kJ mol$^{-1}$ and 412 kJ mol$^{-1}$ respectively), the alkanes are fairly unreactive. The C–H and C–C bonds in alkanes are also non-polar. Even though carbon and hydrogen are different atoms, their electronegativities are very close (C has an electronegativity of 2.5 and that of H is 2.1). This means that the sharing of electrons in the bonds of an alkane is even. There are no charges to attract other polar or ionic species, explaining the lack of reactivity of the alkanes. For example, at room temperature alkanes do not react with acids, bases or strong oxidizing agents. This chemical inertness makes them valuable as lubricating materials and as the backbones for structural materials such as polymers.

The high bond enthalpy and non-polarity of the C–H and C–C bonds in an alkane results in any reaction that does occur having a high activation energy. Once enough energy has been absorbed to reach this activation energy, the reaction is rapid.

**AS** 10.2.1
Explain the low reactivity of alkanes in terms of bond enthalpies and bond polarity.
© IBO 2007

## Combustion reactions of alkanes

Figure 11.3.1 Alkanes are used as fuels.

At sufficiently high temperatures, alkanes react exothermically with oxygen. When there is a plentiful supply of oxygen, complete combustion occurs. This is the basis for the widespread use of alkanes as fuels. For example, the complete combustion of butane with oxygen can be written as:

$$2C_4H_{10}(g) + 13O_2(g) \rightarrow 8CO_2(g) + 10H_2O(g) \quad \Delta H = -5754 \text{ kJ mol}^{-1}$$

In a complete **combustion reaction**, the main products are water and carbon dioxide. For example, when a Bunsen burner is operated on a blue flame, the following reaction occurs:

$$CH_4(g) + 2O_2(g) \rightarrow CO_2(g) + 2H_2O(g) \quad \Delta H = -890 \text{ kJ mol}^{-1}$$

The complete combustion of petrol (octane) in a car engine can be represented by the equation:

$$2C_8H_{18}(l) + 25O_2(g) \rightarrow 16CO_2(g) + 18H_2O(g) \quad \Delta H = -11\,024 \text{ kJ mol}^{-1}$$

A general equation for the complete combustion of any hydrocarbon is:

$$C_xH_y(g) + \left(x + \frac{y}{4}\right)O_2(g) \rightarrow xCO_2(g) + \frac{y}{2}H_2O(g)$$

**AS 10.2.2**
Describe, using equations, the complete and incomplete combustion of alkanes.
© IBO 2007

**Incomplete combustion** of alkanes occurs when the supply of oxygen is not sufficient to fully oxidize the carbon in the alkane. The result is that carbon monoxide or elemental carbon can be one of the products of combustion. This can be a particular problem if the combustion is occurring in a confined space, such as the living room of a house or a busy traffic intersection, as carbon monoxide is poisonous even at very small concentrations.

Incomplete combustion often occurs in car engines when the engine has not been carefully maintained. This results in the car exhaust containing toxic carbon monoxide gas, as well as other polluting gases such as $NO_2$. Carbon

Figure 11.3.2 Toxic carbon monoxide adds to the hazards of traffic duty.

monoxide interferes with oxygen transport in the human body and in sufficiently high concentrations can even cause death. Particles of solid carbon (soot) also emitted can aggravate respiratory illnesses such as asthma, and increase the chance of developing bronchitis and other respiratory illnesses. This certainly does not make traffic duty at a busy intersection a pleasant experience for a policeman.

The incomplete combustion of petrol (octane) in a car engine can be represented by the equation:

$2C_8H_{18}(l) + 17O_2(g) \rightarrow 16CO(g) + 18H_2O(g)$

Notice that when incomplete combustion occurs, only 17 mole of oxygen is required to oxidize 2 mole of octane, but when complete combustion occurs, 25 mole of oxygen is required to oxidize the same amount of octane.

When a Bunsen burner is operated on a yellow flame, the supply of oxygen is being limited by closing of the holes of the burner. A test tube being heated by that burner will blacken, due to the production of carbon in the incomplete combustion of the methane gas:

$CH_4(g) + O_2(g) \rightarrow C(s) + 2H_2O(g)$

## CHEM COMPLEMENT

### Problems in the atmosphere

The use of hydrocarbons as fuels causes considerable environmental problems. The carbon dioxide emitted is a greenhouse gas and so contributes to global warming. The high temperature inside the internal combustion engine is sufficient for nitrogen and oxygen to combine, leading to the formation of oxides of nitrogen and, ultimately, acid rain. If there is incomplete combustion, as occurs in the internal combustion engine, then carbon monoxide and carbon itself are produced. Carbon monoxide is poisonous and the carbon can form particles that may interfere with the respiratory system.

Figure 11.3.3 Soot being given off in incomplete combustion will blacken a test tube that is heated in the flame.

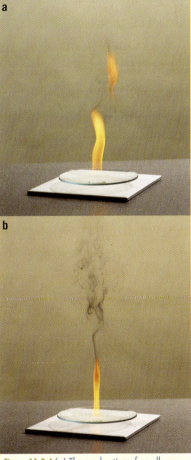

Figure 11.3.4 (a) The combustion of an alkane produces little smoke in comparison to (b) the combustion of an alkene.

## Substitution reactions of alkanes

A substitution reaction involves the replacement of one atom with another atom. The alkanes can undergo **substitution reactions**, primarily reactions in which halogen atoms replace hydrogen atoms. In order for this to occur, the covalent bond between the carbon atom and the hydrogen atom that is going to be replaced must be broken, and the bond between the reacting atoms (in this case halogen atoms) must also be broken. As the sharing of electrons in a halogen molecule such as Cl–Cl is even, when the bond splits **homolytic fission** occurs. In homolytic fission each atom receives one electron, and two species of the same type are produced. The presence of an unpaired electron is represented by a single dot:

$$Cl–Cl \rightarrow Cl\cdot + Cl\cdot$$

An atom or molecule with an unpaired electron is called a radical (or a **free radical**). The presence of the unpaired electron makes the radical very reactive, because it has a strong tendency to pair up with an electron from another molecule. The reaction of a free radical with another molecule usually results in the formation of another free radical:

$$C–H + Cl\cdot \rightarrow C–Cl + H\cdot$$

Methane can react with bromine in a series of substitution reactions that occur very rapidly by a free-radical mechanism.

The Br–Br bond is weaker than the C–H bond and will be easier to split:

$$Br_2 \rightarrow Br\cdot + Br\cdot \qquad \Delta H = +193 \text{ kJ mol}^{-1}$$

$$CH_4 \rightarrow CH_3\cdot + H\cdot \qquad \Delta H = +435 \text{ kJ mol}^{-1}$$

In the presence of ultraviolet (UV) light, sufficient energy is available to break the Br–Br bond to produce two bromine radicals. This is called *initiation*.

$$Br–Br \xrightarrow{UV \text{ light}} Br\cdot + Br\cdot$$

The bromine radicals then react with methane molecules forming hydrogen bromide and a methyl radical. This step is a *propagation* step, as it produces another radical.

$$CH_4 + Br\cdot \rightarrow CH_3\cdot + HBr$$

The methyl radical then reacts with another bromine molecule in another *propagation* step to make bromomethane and a bromine radical:

$$CH_3\cdot + Br–Br \rightarrow CH_3Br + Br\cdot$$

The propagation steps of this reaction continue until there is a *termination* step in which two radicals react. Such a step is very exothermic Three termination steps are possible:

$$Br\cdot + Br\cdot \rightarrow Br_2$$

$$CH_3\cdot + Br\cdot \rightarrow CH_3Br$$

$$CH_3\cdot + CH_3\cdot \rightarrow C_2H_6$$

If the supply of bromine is limited, then the main products of this reaction will be bromomethane and hydrogen bromide, with a little ethane also produced. If, however, there is an ample supply of bromine, further substitution reactions will occur, and dibromomethane, tribromomethane and tetrabromomethane will be made by processes similar to those above.

---

**AS 10.2.3**
Describe, using equations, the reactions of methane and ethane with chlorine and bromine. © IBO 2007

**AS 10.2.4**
Explain the reactions of methane and ethane with chlorine and bromine in terms of a free-radical mechanism. © IBO 2007

$$CH_3Br + Br_2 \xrightarrow{light} CH_2Br_2 + HBr$$
$$\text{dibromomethane}$$

$$CH_2Br_2 + Br_2 \xrightarrow{light} CHBr_3 + HBr$$
$$\text{tribromomethane}$$

$$CHBr_3 + Br_2 \xrightarrow{light} CBr_4 + HBr$$
$$\text{tetrabromomethane}$$

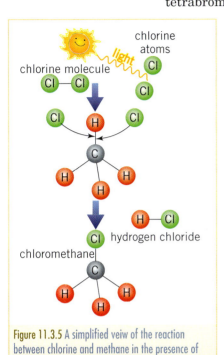

Figure 11.3.5 A simplified veiw of the reaction between chlorine and methane in the presence of UV light.

Other alkanes undergo similar substitution reactions with chlorine and bromine. Thus, methane and ethane react with chlorine to produce a range of chloromethanes and chloroethanes and bromine reacts with ethane to produce a range of bromoethanes.

As ethane has six hydrogen atoms, there are more steps that occur before its chlorination or bromination is complete.

Bromine is a red-brown colour but bromoalkanes are colourless, so it is easy to see whether this reaction has occurred. We say that the bromine is *decolourized* (it loses its colour) when it reacts with an alkane in the presence of UV light.

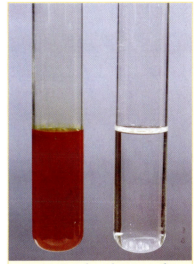

Figure 11.3.6 Test tubes with a mixture of hexane and bromine before (left) and after (right) being exposed to sunlight.

## CHEM COMPLEMENT

### More problems in the atmosphere

Substituted methanes (general formula $CF_xCl_{4-x}$) that contain both chlorine and fluorine are called chlorofluorocarbons (CFCs) or freons. These substances are very unreactive and have been extensively used as coolant fluids in refrigerators and air conditioners. Their chemical inertness allows freons to remain in the atmosphere for so long that eventually they reach high altitudes. There they react with ozone molecules, reducing the concentration of this essential UV-absorbing substance. The use of CFCs is now mostly prohibited, but as they have atmospheric lifetimes of 50–100 years, they are still to be found in significant concentrations in the stratosphere.

WORKSHEET 11.4
Reactions of alkanes

## Section 11.3 Exercises

1. Write balanced chemical equations to represent the following reactions.
   a. Complete combustion of pentane
   b. Complete combustion of butane
   c. Incomplete combustion of hexane to produce carbon monoxide and water
   d. Incomplete combustion of hexane to produce carbon and water

2. State why incomplete combustion of hydrocarbons is harmful to humans.

3. **a** State the type of reaction that occurs when chlorine reacts with an alkane.
   **b** State the type of mechanism that occurs during this reaction.
   **c** Explain what happens in the initiation step of the reaction between chlorine and ethane.

4. **a** Assuming there is a limited supply of bromine, write an equation to show the reaction that occurs between ethane and bromine.
   **b** Assuming there is a plentiful supply of chlorine, name the products that can be formed during the reaction of chlorine and ethane.

5. An organic liquid X has the following percentage composition by mass: C, 29.3%; H, 5.7%; and Br, 65.0%. Its molecular mass is approximately 120.
   **a** Determine the empirical and molecular formulas of X.
   **b** X can have two possible structures. Draw the structural formulas for these two compounds.
   **c** Name the two compounds.
   **d** Using any non-organic reagents necessary, show how each of the compounds named in part **c** could be prepared.

## 11.4 REACTIONS OF ALKENES

Alkenes are of great economic importance to society. Although alkenes occur naturally in only small quantities, when crude oil is put through the process of cracking, large amounts of alkenes can be obtained. In cracking (thermal cracking or catalytic cracking) alkanes with large molecular masses are broken into small alkenes and relatively small alkanes. This provides a source of alkenes, which have a wide variety of uses, and, at the same time, uses up the large, less useful alkanes.

Alkenes provide the starting materials for a wide range of synthetic polymers (plastics) that we depend on for so many uses. One of the most widely used alkenes is ethene. It can be changed into ethanol by the process of hydration, and it forms the basis of many polymerization reactions in which plastics are formed. In the manufacture of margarine, polyunsaturated vegetable oils (which have many carbon–carbon double bonds) are converted into a soft, low melting point solid by the addition of extra hydrogen. This creates a spread that closely resembles butter. The processes of addition polymerization (to make polymers), hydrogenation (to make margarine) and hydration (to make ethanol) will be considered further in this section.

### Addition reactions of alkenes

Alkanes undergo substitution reactions, but alkenes undergo **addition reactions**. In an addition reaction, two substances react together to form a single product. (The reactants *add* together.) You will recall from chapter 6 that a C=C double bond has a bond enthalpy of 612 kJ mol$^{-1}$ and the C–C single bond has a bond enthalpy of 348 kJ mol$^{-1}$. This means that less energy is required to break one half of a double bond than is needed to break a single bond in an alkane, making alkenes much more reactive than alkanes. The double bond can be a converted to a single bond, allowing extra atoms or groups of atoms to be added to the molecule. For example, ethene reacts readily with chlorine to produce 1,2-dichloroethane.

$$H_2C=CH_2(g) + Cl_2(g) \rightarrow CH_2ClCH_2Cl(l)$$

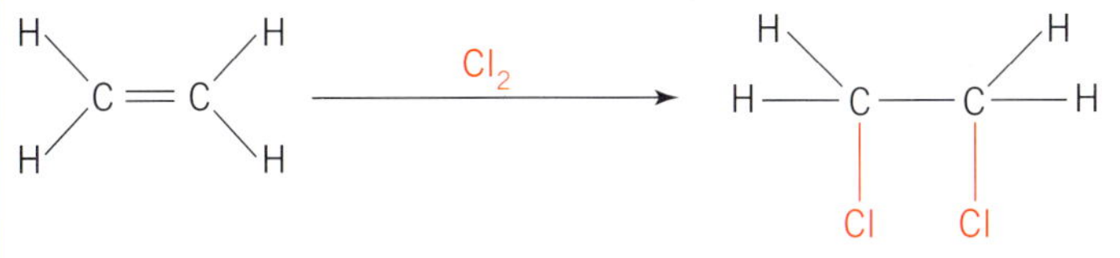

**Figure 11.4.1** Ethene readily reacts with chlorine at room temperature.

Although ethane reacts with chlorine only in sunlight or at high temperatures, the high reactivity of ethene's double bond means that chlorine readily reacts with ethene at room temperature. Note that the addition of chlorine occurs 'across the double bond', so that one chlorine atom attaches to each carbon atom. A similar addition reaction occurs with other halogens such as bromine and iodine.

## Worked example

State the products of the reaction with chlorine of:

**a** but-1-ene

**b** but-2-ene.

### Solution

**a**

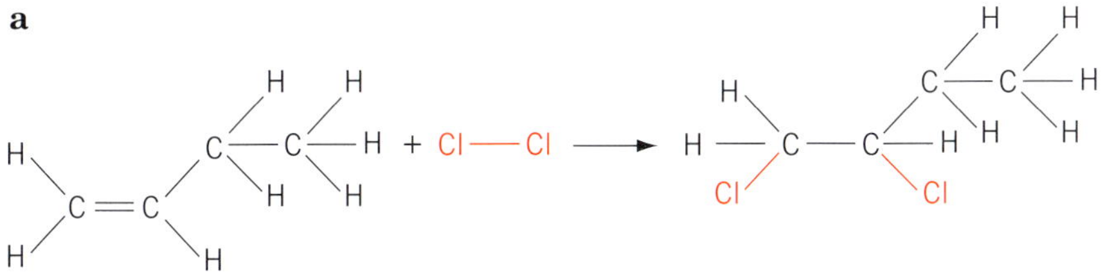

**b**

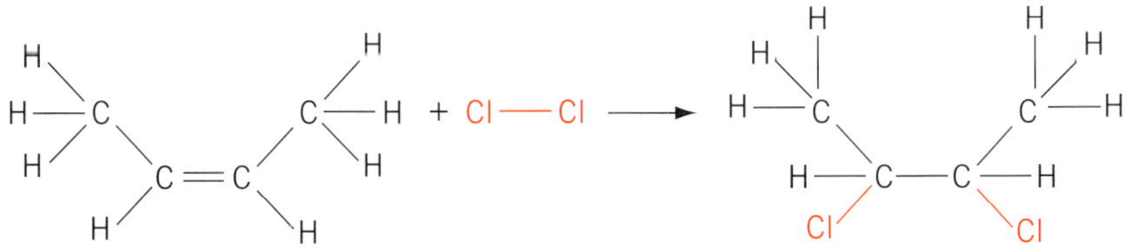

Alkenes also undergo an addition reaction with hydrogen. This reaction, sometimes called **hydrogenation**, results in the formation of a saturated hydrocarbon, an alkane. The reaction of ethene with hydrogen using a catalyst is shown below:

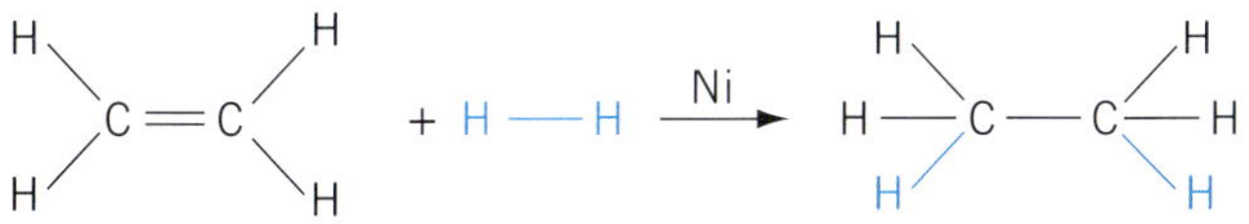

**Addition reactions
Catalysing hydrogenation**

Nickel is shown as the catalyst in this reaction, but platinum and palladium can also catalyse the hydrogenation of alkenes. This hydrogenation reaction is of particular use in the food industry. Margarine is a butter substitute that is made from vegetable oils. Vegetable oils are made up of a mixture of polyunsaturated fats which are recognized these days to be a healthy substitute for saturated fats like butter; however, it is difficult to 'spread' an oil on your bread! By adding hydrogen across some of the double bonds in the polyunsaturated fats the oils are converted to a soft, low melting point solid that is more like butter.

Figure 11.4.2 Addition reactions are used in the production of margarine from vegetable oils.

Alkenes can also react with hydrogen halides and with water. Both these reactions are addition reactions. The reaction between an alkene and a hydrogen halide provides a more controlled way to produce a particular halogenoalkane than the substitution reaction between an alkane and a halogen. The double bond controls the position of the halogen atom in that it will be on one of the two carbon atoms that were involved in the double bond.

**AS 10.3.2
Describe, using equations, the reactions of symmetrical alkenes with hydrogen halides and water.** © IBO 2007

but-2-ene + H—Cl ⟶ 2-chlorobutane

The added atoms will always bond to the two carbon atoms that were involved in the double bond. In the above equation, the double bond was in position 2, so the H and the Cl bonded to carbon atoms 2 and 3. The reaction of but-1-ene with HCl would result in the production of 2-chlorobutane, or 1-chlorobutane. Note that but-1-ene is described as an asymmetrical alkene.

but-1-ene + H—Cl ⟶ 2-chlorobutane

but-1-ene + Cl—H ⟶ 1-chlorobutane

Figure 11.4.3 When HCl reacts with but-1-ene, the product may be 2-chlorobutane or 1-chlorobutane.

An important industrial addition reaction is the reaction between ethene and water. This process, called **hydration**, is very important in the manufacture of ethanol. Ethene and steam are passed over a catalyst (phosphoric acid on silica) at 300°C and 70 atm pressure. The purpose of the phosphoric acid is to provide a lower activation energy pathway for the reaction. Hydrogen ions, $H^+$, are supplied by the phosphoric acid. These ions take part in the reaction but are regenerated at the end.

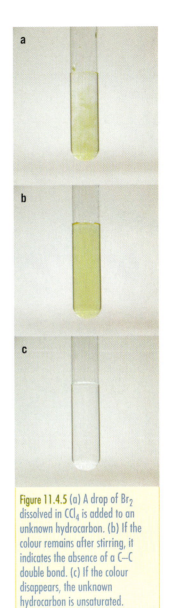

ethene + steam $\xrightarrow[\text{70 atm}]{\substack{H_3PO_4 \\ 300°C}}$ ethanol

This reaction is another example of an industrial process in which a compromise must be reached between rate of reaction and yield. The reaction to make ethanol is an exothermic reaction and would be favoured by low temperatures, at which the rate of reaction would be unacceptably low, so a catalyst is used in conjunction with moderately high temperatures and moderately high pressures (see chapter 8, section 8.3).

Figure 11.4.5 (a) A drop of $Br_2$ dissolved in $CCl_4$ is added to an unknown hydrocarbon. (b) If the colour remains after stirring, it indicates the absence of a C–C double bond. (c) If the colour disappears, the unknown hydrocarbon is unsaturated.

Figure 11.4.4 Ethene undergoes addition reactions.

The reaction between alkenes and bromine is used to distinguish between an alkene and an alkane. Both hydrocarbons are colourless and so cannot be distinguished by sight. Pure bromine is a red liquid that becomes yellow/orange in solution. When bromine water is added to an alkene, the colour of the bromine disappears immediately (that is, **decolourization** occurs) as the addition reaction occurs. The reaction between an alkane and bromine will not proceed unless a strong source of UV light is available. Similarly, purple acidified potassium permanganate, $KMnO_4$, is decolourized by reaction with alkenes.

**10.3.3**
**Distinguish between *alkanes* and *alkenes* using bromine water.** © IBO 2007

Testing hydrocarbons with bromine

PRAC 11.3
Reactions of alkanes and alkenes

CHEMISTRY: FOR USE WITH THE IB DIPLOMA PROGRAMME **STANDARD LEVEL**

**AS** 10.3.4
Outline the polymerization of alkenes. © IBO 2007

### Addition polymerization

A polymer (*poly* means 'many') is made up of very large chain-like molecules. These have been generated by the reaction of thousands of monomers (*mono* means 'one') that have joined together as repeating units in the chain.

The production of poly(ethene) is an example of an **addition polymerization reaction**. This refers to reactions in which the monomer contains a double bond. When the monomers join to each other, the double bond converts to a single bond and the second pair of electrons previously in the double bonds are used to form covalent bonds between the monomers. There are no by-products in this process—only one product forms. As the product is directly related to just one reactant, the name of the product is always the same as that of the reactant with 'poly' added as a prefix. So ethene polymerizes to make poly(ethene), propene polymerizes to make poly(propene) etc.

The equation can be written as follows, where $n$ is a large number and represents the number of repeating units in the polymer chain.

$$n\text{CH}_2\text{CH}_2 \rightarrow (\text{CH}_2\text{CH}_2)_n$$
ethene      poly(ethene)

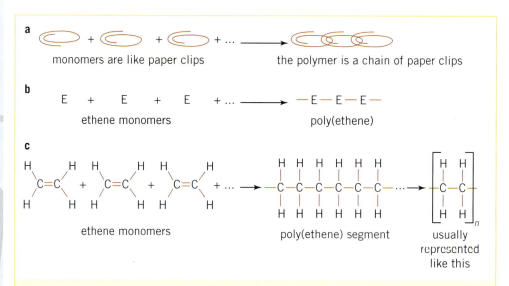

Figure 11.4.6 Ethene undergoes addition polymerization to form poly(ethene).

Ethene is a hydrocarbon, and so is non-polar. This leads to weak intermolecular forces between the poly(ethene) chains. Chemists have tried varying the atoms on the ethene monomer to produce polymers with different properties.

The chloroethene monomer used in the production of poly(chloroethene), also known as polyvinyl chloride (PVC), differs only slightly from ethene in that it has a chlorine atom replacing one of the hydrogen atoms. The effect of the chlorine atom is to introduce a significant dipole into the molecule. This increases the attraction between molecules and leads to a higher melting point. The presence of chlorine also lowers the flammability of the polymer. A poly(chloroethene) item that is burning in a Bunsen flame will extinguish itself when it is removed from the flame. This property leads to poly(chloroethene) being used as insulation on electrical wiring. It is also a popular choice for conveyor belts in coal mines, where fires can lead to serious explosions. Other uses include soft-drink bottles, fabric coatings and water pipes.

An addition polymer

Another addition polymer that has been successfully produced on a commercial scale is (poly)propene. The propene monomer has one double bond and can be regarded as an ethene molecule in which a hydrogen atom has been replaced by a methyl group.

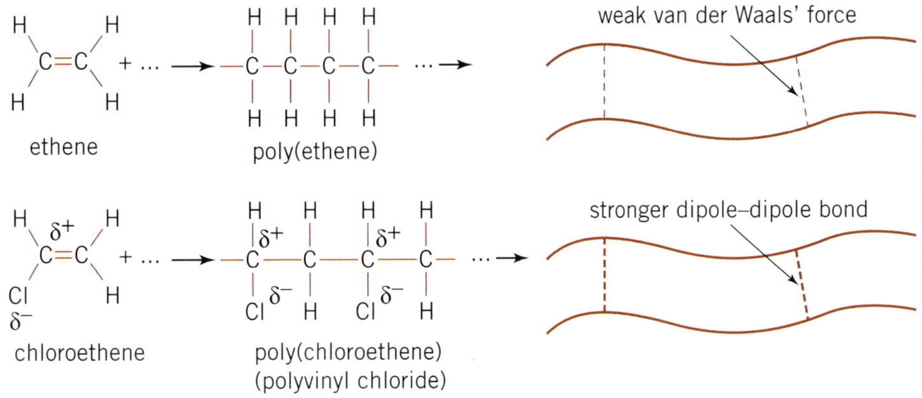

Figure 11.4.7 Dipoles lead to poly(chloroethene) having a higher melting point than poly(ethene).

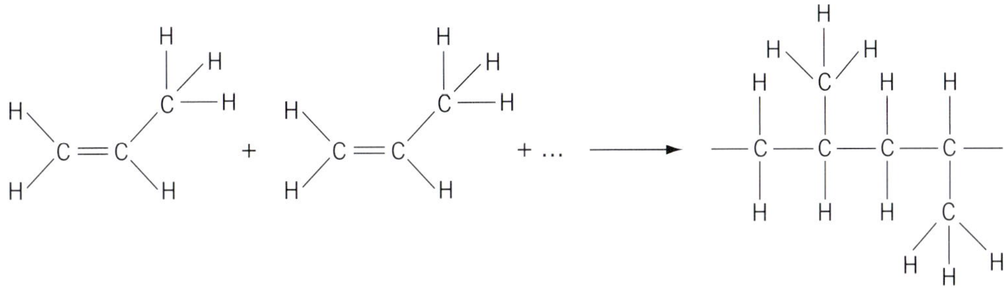

Figure 11.4.8 Poly(propene) is formed by addition polymerization.

### TABLE 11.4.1 A SELECTION OF ADDITION POLYMERS MADE USING MONOMERS DERIVED FROM ETHENE

| Monomer | Polymer | Polymer name | Some uses |
|---|---|---|---|
| Ethene 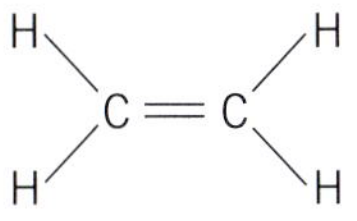 | 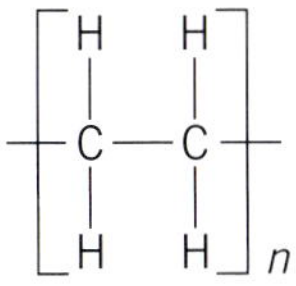 | Poly(ethene) | Plastic bags, bottles, toys |
| Propene 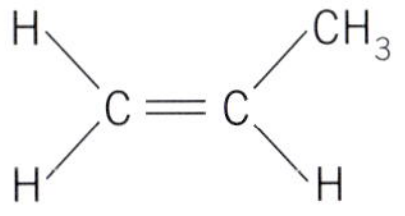 | 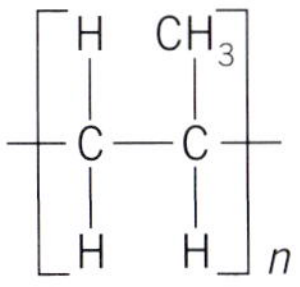 | Poly(propene) | Indoor–outdoor carpeting, bottles, luggage |
| Styrene 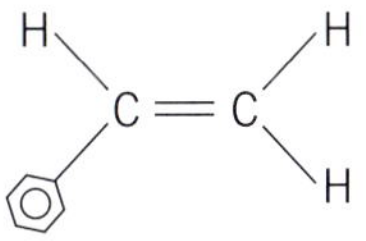 | 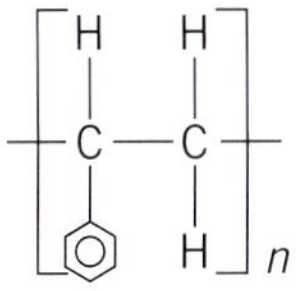 | Poly(styrene) | Styrofoam insulation, cups, packing materials |
| Chloroethene 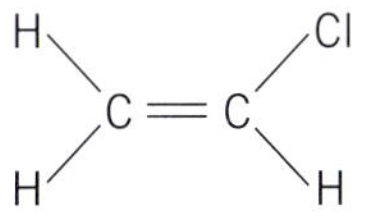 | 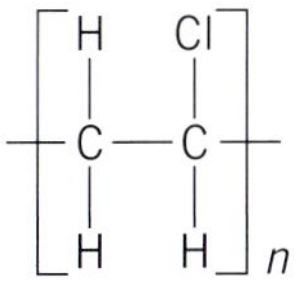 | Poly(chloroethene) or polyvinyl chloride (PVC) | Plastic wrap, plumbing, garden hoses |

## CHEM COMPLEMENT

### Teflon—wonder material

American polymer company DuPont first manufactured Teflon in 1938. It was used during World War II as part of the process of isolating uranium for the first atomic bomb. After the war, its uses spread to plumbing tape and non-stick frypans, as manufacturers looked to take advantage of its heat resistance and low-friction surface. It sounds like a simple process to apply a layer of Teflon to a metal frying pan, but the non-stick nature of Teflon has the disadvantage of making it a very difficult material to apply. The metal has to be grit-blasted and the Teflon applied in several layers, starting with a primer layer. Another innovative application of Teflon is in the fabric Goretex. Goretex is a Teflon-based material that 'breathes': rain does not penetrate Goretex, but moisture from the user can escape. High-quality camping gear is often made from Goretex.

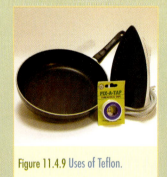

Figure 11.4.9 Uses of Teflon.

The table shows only a small sample of commercial addition polymers. Why manufacture so many polymers? The reason is that the changes to the monomer often lead to a unique or enhanced property. For example, the use of $CH_3$ in polypropene leads to a melting point that is about 60°C higher than that of polyethene. This allows polypropene to be used in microwave ovens and dishwashers. Teflon (poly(tetrafluorothene)) is a polymer that combines a non-stick surface with an extremely high melting point. Its formula is $(C_2F_4)_n$. The great length (and molecular mass) of polymer chains means that polymers have much higher melting temperatures than their small molecular mass monomers.

### Section 11.4 Exercises

1. **a** Give balanced equations for each of the following reactions.
   - **i** Reaction between but-2-ene and hydrogen
   - **ii** Addition of bromine to ethene
   - **iii** Complete combustion of ethanol
   - **iv** Reaction between ethane and chlorine
   
   **b** State any conditions necessary for the reaction in part **a iv**.
   
   **c** What would you observe during the reaction in part **a ii**?

2. Using any non-organic reagents necessary, show how you would prepare the following chlorinated hydrocarbons.
   - **a** Chloroethane from ethane
   - **b** Chloroethane from ethene

3. Describe the test that indicates whether a hydrocarbon is saturated or unsaturated.

4. 5.91 g of chlorine gas reacts with exactly 2.33 g of a pure alkene according to the equation:
   
   $C_nH_{2n} + Cl_2 \rightarrow C_nH_{2n}Cl_2$
   
   **a** What is the name given to this type of reaction?
   
   **b** Determine the molecular formula of the alkene.

5. Give the structural formulas of possible 'dibromo-' compounds that could be formed from the reaction of bromine with:
   - **a** ethane
   - **b** ethene.

6. **a** State an equation for the hydration of ethene.
   
   **b** State the conditions required for the reaction in part **a** to occur.
   
   **c** Write the equation for the reaction between ethene and HCl.

7. Explain what is meant by the term *polymer*.

8. State the balanced chemical equations for the formation of the following polymers:
   - **a** polyethene
   - **b** polypropene
   - **c** polychloroethene

9 Draw the polymer that forms from each of the monomers shown below.

**a** CH₂CH₂

**b** CH₂CHCl

**c**

**d** (structure with NC and H on C=C with H's)

10 A section of a polymer is drawn below. Draw the monomer from which it was made.

(polymer structure showing –C(H)(H)–C(CH₃)(H)–C(H)(H)–C(CH₃)(H)–)

## 11.5 REACTIONS OF ALCOHOLS

Alcohols contain an –OH group covalently bonded to a carbon atom. However, this –OH group does not behave in the same way as the hydroxide ion OH⁻ because NaOH is a base and CH₃OH is not. Alcohols, when dissolved in water, do not alter the pH of the water. Although the hydrogen atom is connected to an oxygen atom, alcohols do not readily donate the proton (they are weaker acids than water).

Like the alkanes and alkenes, alcohols undergo complete combustion in a plentiful supply of oxygen gas, producing only carbon dioxide and water as products. You may have used alcohol burners in an investigation in chapter 6. The complete combustion of ethanol is as follows:

$$C_2H_5OH(l) + 3O_2(g) \rightarrow 2CO_2(g) + 3H_2O(g)$$

When balancing an equation for the combustion of an alcohol it is important to remember that there is an oxygen atom in the alcohol, unlike alkanes and alkenes.

Figure 11.5.1 Flames from a sample of burning ethanol in a crucible. The flame is relatively clean (not sooty).

**AS 10.4.1**
Describe, using equations, the complete combustion of alcohols. © IBO 2007

Figure 11.5.2 An alcohol burner containing ethanol. The wick is soaked in ethanol, which moves up the wick by capillary action as the ethanol is burnt at the top of the wick.

## CHEM COMPLEMENT

### Ethanol as a biofuel

Rising oil prices and concern about the contribution to climate change of carbon dioxide from fossil fuel burning, as well as the danger of carbon monoxide and nitrogen oxides as air pollutants, are leading to the development of new 'cleaner' alternative fuels. Ethanol can be used as a fuel in different ways:

- Ethanol can be burnt in air, releasing energy according to the equation:
  $CH_3CH_2OH(l) + 3O_2(g) \rightarrow 2CO_2(g) + 3H_2O(g)$
  Small ethanol camping stoves are an example of the combustion of ethanol.
- It can be reacted with oxygen in a fuel cell, in which the chemical potential energy is converted directly to electrical energy.
- It can be blended with petrol to help extend supplies of petrol. There is some controversy as to the ideal ethanol concentration to use in these blends; however, cars are now being manufactured to specifically use bioethanol blends with as much as 85% bioethanol. These blends are also known as biofuels or 'gasohol'. Brazil has been one of the leading users of ethanol–petrol blends due to its abundance of high-sugar content crops. (Sugar is fermented to make the ethanol.) Many countries have taken up the challenge of providing bioethanol for those who can use it in their cars. It is estimated that a 5% use of biofuels in England would reduce carbon dioxide emissions so much that it would be equivalent to taking one million cars off the road.

Figure 11.5.3 Bioethanol E85 contains a mixture of 85% bioethanol with 15% petrol.

### The oxidation of alcohols

**AS 10.4.2**
Describe, using equations, the oxidation reactions of alcohols. © IBO 2007

The process of oxidation was defined in chapter 10 as the loss of electrons. In organic chemistry oxidation is easily recognized as the gain of oxygen or the loss of hydrogen from a compound. The **oxidation reactions** of alcohols vary, depending upon the type of alcohol involved. Primary, secondary and tertiary alcohols all give different reactions with strong oxidizing agents such as acidified potassium dichromate(VI) solution or acidified potassium manganate(VII) solution.

**AS 10.4.3**
Determine the products formed by the oxidation of primary and secondary alcohols. © IBO 2007

You will recall from section 11.2 that primary alcohols are those in which the hydroxyl functional group is bonded to a carbon atom that is bonded to only one other alkyl group. Butan-1-ol is an example of a primary alcohol.

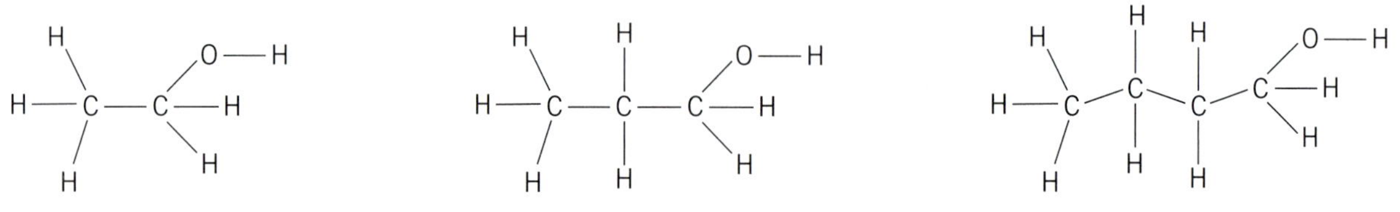

ethanol      propan-1-ol      butan-1-ol

Figure 11.5.4 Ethanol, propan-1-ol and butan-1-ol are all primary alcohols.

In the laboratory, when an aqueous solution of a primary alcohol such as ethanol is mixed with potassium dichromate(VI) and sulfuric acid, and the mixture heated under reflux, the alcohol is fully oxidized to a carboxylic acid.

During the process, the alcohol is initially oxidized to an aldehyde; however, by heating under reflux the aldehyde is further oxidized to a carboxylic acid. (See Chem Complement: Organic reactions in the laboratory.) When the reaction is 'complete', the condenser is turned around and the reaction mixture is distilled to collect an aqueous solution of the carboxylic acid. If the aldehyde is the desired product during this reaction, then the reaction can be carried out at room temperature and the aldehyde can be distilled off from the mixture. Some distillation apparatus is shown at the beginning of this chapter (p. 341).

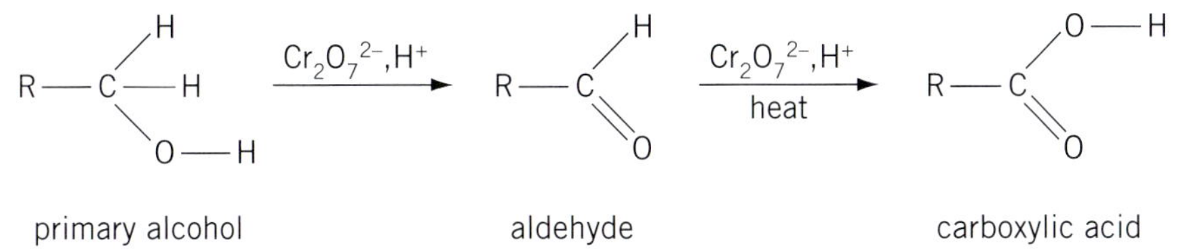

primary alcohol → aldehyde → carboxylic acid (with $Cr_2O_7^{2-}, H^+$ and $Cr_2O_7^{2-}, H^+$, heat)

Figure 11.5.5 Primary alcohols are oxidized at room temperature to aldehydes and to carboxylic acids, when heated under reflux.

## CHEM COMPLEMENT

### Organic reactions in the laboratory

When carrying out organic reactions in the laboratory, there are several important factors to consider: the rate of the reaction, the conditions for the reaction, and the separation of the product from the reaction mixture.

Most organic reactions occur slowly. Heat, and often a catalyst, is used to increase the rate of the reaction, but even then some reactions may take several hours. Heat may also cause a problem. Many organic compounds are volatile and have low boiling points. When an open flask is heated, these reactants will turn to a gas and be lost. However, if a condenser is joined to the flask, the gases condense and drop back into the flask. Condensing of compounds that evaporate during a reaction is called *refluxing*. The condenser is called a reflux condenser.

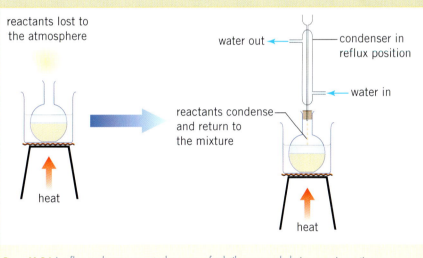

Figure 11.5.6 A reflux condenser prevents the escape of volatile compounds during organic reactions.

The conditions needed for many reactions include the correct temperature, appropriate quantities of reactants and the presence of a catalyst. Even a slight difference in conditions can affect the products obtained. During most organic preparations, side reactions occur. In addition, many organic reactions are reversible and so reactants may not completely change into the expected products. The result is that the reaction flask will contain a mixture of compounds.

The required product must be separated from the mixture in the flask. A solid is purified by filtration and washing, a liquid by distillation, and a gas by passing it through substances to absorb impurities. For example, sulfuric acid is used to remove water vapour.

The oxidation reaction of the primary alcohol (e.g. ethanol) to a carboxylic acid may be represented simply by an equation in which the symbol [O] represents the oxygen supplied by the oxidizing agent:

$$CH_3CH_2OH + [O] \xrightarrow{Cr_2O_7^{2-},\ H^+} CH_3COOH$$

Alternatively, we may write half-equations and a complete equation to represent the redox nature of the reaction:

$CH_3CH_2OH + H_2O \rightarrow CH_3COOH + 4H^+ + 4e^-$ × 3
$Cr_2O_7^{2-} + 14H^+ + 6e^- \rightarrow 2Cr^{3+} + 7H_2O$ × 2

$3CH_3CH_2OH(aq) + 2Cr_2O_7^{2-}(aq) + 16H^+(aq) \rightarrow 3CH_3COOH(aq) + 4Cr^{3+}(aq) + 11H_2O(l)$

A secondary alcohol has the hydroxyl group on a carbon that is bonded to two other carbons. Propan-2-ol and butan-2-ol are examples of secondary alcohols.

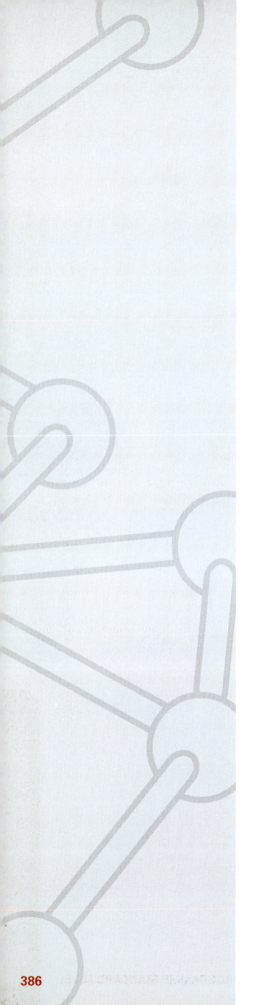

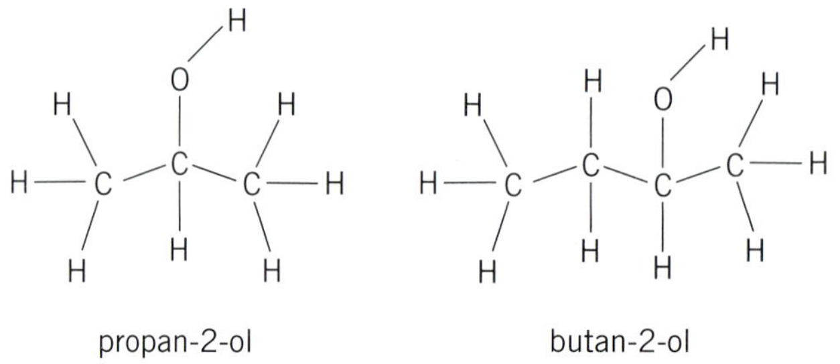

**Figure 11.5.7** Propan-2-ol and butan-2-ol are secondary alcohols.

When secondary alcohols are oxidized, ketones are formed. This reaction is very similar to the one in which aldehydes are produced, but the placement of the hydroxyl group results in the production of a ketone rather than an aldehyde and, ultimately, a carboxylic acid.

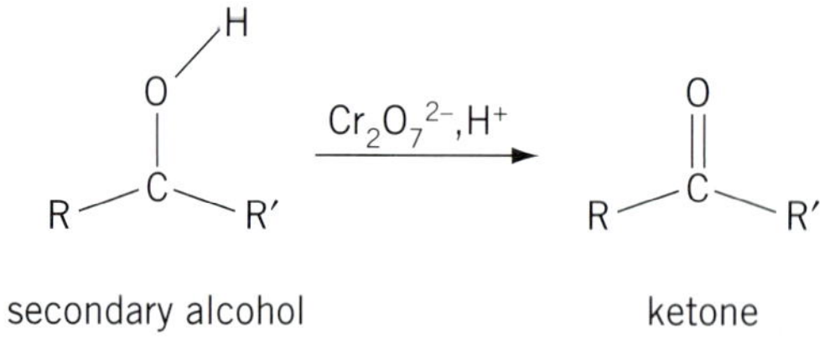

**Figure 11.5.8** Secondary alcohols are oxidized to ketones.

The oxidation reaction of a secondary alcohol such as propan-2-ol to a ketone may be represented simply as:

$$CH_3CH(OH)CH_3 + [O] \xrightarrow[H^+]{Cr_2O_7^{2-}} CH_3COCH_3$$

Alternatively, we may write half-equations and a complete equation to represent the redox nature of the reaction:

$CH_3CH(OH)CH_3 + H_2O \rightarrow CH_3COCH_3 + 4H^+ + 4e^-$ × 3
$Cr_2O_7^{2-} + 14H^+ + 6e^- \rightarrow 2Cr^{3+} + 7H_2O$ × 2

$3CH_3CH(OH)CH_3(aq) + 2Cr_2O_7^{2-}(aq) + 16H^+(aq) \rightarrow 3CH_3COCH_3(aq) + 4Cr^{3+}(aq) + 11H_2O(l)$

When these oxidation reactions are performed in a laboratory investigation, the change in colour of the oxidizing agent indicates that the reaction has proceeded. Potassium dichromate, $K_2Cr_2O_7$, changes colour from orange ($Cr_2O_7^{2-}$) to green ($Cr^{3+}$) during this reaction. If potassium manganate(VII), $KMnO_4$, is used instead, it changes colour from purple to colourless.

Tertiary alcohols, those with the hydroxyl group bonded to a carbon atom that is bonded to three other carbon atoms, are not easily oxidized. An example of a tertiary alcohol is 2-methylpropan-2-ol.

PRAC 11.4
Reactions of alcohols and carboxylic acids

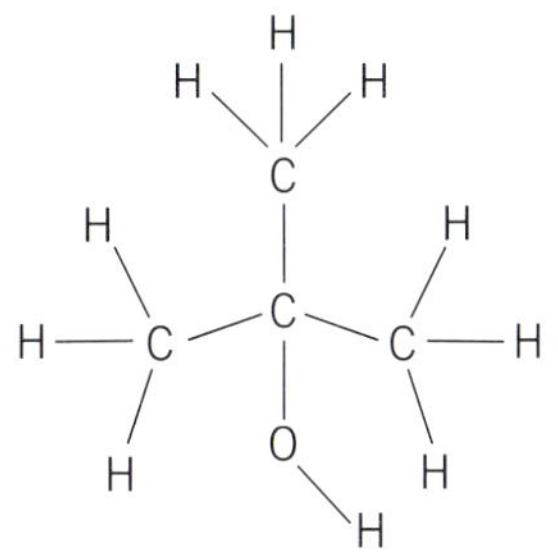

**Figure 11.5.9** 2-Methylpropan-2-ol is an example of a tertiary alcohol; however, these resist oxidation.

## Section 11.5 Exercises

1  Write the formula and the name of the main organic product in each of the following reactions.

   **a**  $CH_3CH_2OH + [O] \xrightarrow[H^+]{K_2Cr_2O_7}$

   **b**  $CH_2=CH_2 \xrightarrow[H^+]{H_2O}$

   **c**  $CH_3CH_2CHOHCH_3 + [O] \xrightarrow[H^+]{K_2Cr_2O_7}$

   **d**  $(CH_3)_3COH + [O] \xrightarrow[H^+]{K_2Cr_2O_7}$

2  A carboxylic acid is formed from the oxidation of a primary alcohol.
   **a**  State what other product can also be formed.
   **b**  State the process used to obtain this product only.
   **c**  State the process that is used to ensure the production of the carboxylic acid.

## 11.6 REACTIONS OF HALOGENOALKANES

Halogenoalkanes have a number of important uses such as the use of tetrachloromethane as a dry-cleaning solvent, but one of their most important roles is as the starting material for the synthesis of other organic compounds. Because the halogen atom is more electronegative than the carbon atom, the bond in halogenoalkanes is polar, with the carbon end having a partial positive charge. **Nucleophiles**, reactants with a non-bonding electron pair, are attracted to the slightly positive carbon atom in the halogenoalkane and a substitution reaction occurs. This type of substitution reaction is called a **nucleophilic substitution** reaction.

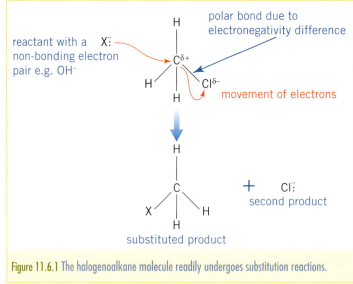

Figure 11.6.1 The halogenoalkane molecule readily undergoes substitution reactions.

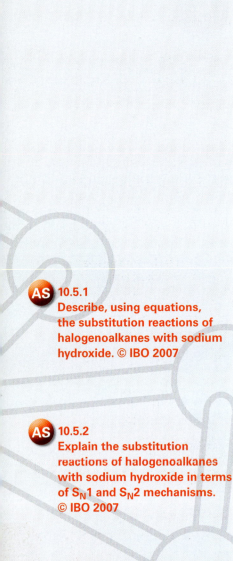

**10.5.1**
Describe, using equations, the substitution reactions of halogenoalkanes with sodium hydroxide. © IBO 2007

**10.5.2**
Explain the substitution reactions of halogenoalkanes with sodium hydroxide in terms of $S_N1$ and $S_N2$ mechanisms. © IBO 2007

**WORKSHEET 11.5**
Reations of alkenes, alcohols and halogenoalkanes

In the hydrolysis of a halogenoalkane a typical reacting species is the hydroxide ion. For example, ethanol may be produced when chloroethane reacts with hydroxide ions.

$$CH_3CH_2Cl + OH^- \rightarrow CH_3CH_2OH + Cl^-$$

The rate of this reaction is slow, so the chloroethane must be heated under reflux with aqueous sodium hydroxide for about an hour. The apparatus used would be similar to that in figure 11.5.6.

The ease with which a reaction between a halogenoalkane and sodium hydroxide occurs depends on the strength of the carbon–halogen bond. The standard bond enthalpies are C–Cl = 338 kJ mol$^{-1}$, C–Br = 276 kJ mol$^{-1}$ and C–I = 238 kJ mol$^{-1}$. Therefore the order of reactivity of these three halogenoalkanes towards the OH$^-$ nucleophile is RI > RBr > RCl.

The details of exactly how an organic reaction proceeds are known as its **reaction mechanism**, a proposed sequence of steps that shows how the reactants are converted into products.

There are two types of nucleophilic substitution reaction mechanisms: $S_N1$ and $S_N2$. They differ in the number of steps by which the reaction proceeds and, in particular, by the number of species that are needed to form the transition state: a reactive species that is formed in one step and then used in the next step. This intermediate may also be called an activated complex. It is the number of species involved in the first step that gives its number to the type of reaction. In an $S_N2$ reaction, two species are needed, whereas an $S_N1$ reaction needs only one species.

The reaction between a primary halogenoalkane and sodium hydroxide is an $S_N2$ reaction in which an OH group (the nucleophile) is substituted for the halide The proposed mechanism is shown in figure 11.6.2 and has one step.

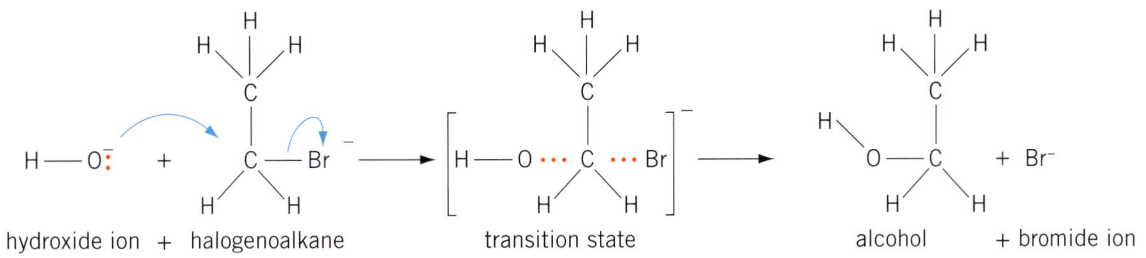

**Figure 11.6.2** The $S_N2$ mechanism for the reaction between the primary halogenoalkane bromoethane and the hydroxide ion. Note that two species are needed to form the transition state, hence the name $S_N2$.

In this reaction initially the nucleophilic OH⁻ is attracted to and attacks the slightly positive carbon atom that is bonded to the halogen (bromine). A reactive intermediate is formed in which the OH and the Br are both partially bonded to the carbon atom. This quickly results in the Br being displaced as a bromide ion, and the production of ethanol. This mechanism is supported by the dependence of the rate of reaction on the concentration of both the halogenoalkane and the hydroxide ions.

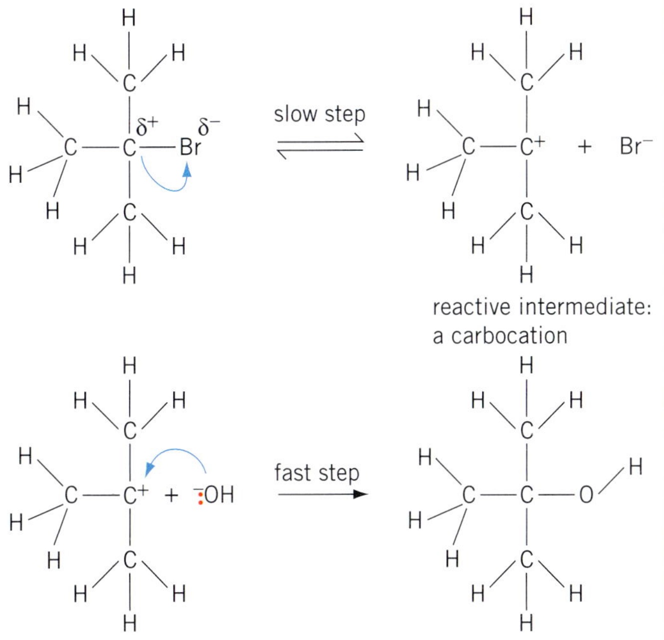

The reaction between a tertiary halogenoalkane, such as $(CH_3)_3CBr$ and sodium hydroxide has been proposed to take place in two steps. This proposal is based upon experimental evidence that the rate of the hydrolysis reaction is proportional to the concentration of the halogenoalkane, but does not depend upon the concentration of the hydroxide ions. This is therefore an $S_N1$ mechanism, as only one molecule is involved in forming the reactive intermediate, $(CH_3)_3C^+$.

In this reaction the first step is a slow one in which the hydroxide ion plays no part. The carbon–bromide bond undergoes **heterolytic fission** to form a bromide ion and a positive ion called a carbocation. In heterolytic fission, the two bonding electrons stay with the more electronegative of the bonding species, in this case the bromine atom. The reaction then proceeds rapidly in the second step in which the carbocation and the hydroxide ion bond together.

**Figure 11.6.3** The $S_N1$ mechanism for the reaction between the tertiary halogenoalkane 2-bromo-2-methylpropane and the hydroxide ion. Note that only one species is needed to form the reactive intermediate, hence the name $S_N1$.

Secondary halogenoalkanes can show evidence of both mechanisms, depending on the nature of the halogenoalkane and the solvent.

## Section 11.6 Exercises

1. Define the term *nucleophile*.

2. State the type of reactions that halogenoalkane molecules undergo.

3. Halogenoalkanes undergo reactions with sodium hydroxide. If both an iodoalkane and a chloroalkane undergo reactions with sodium hydroxide, state and explain which reaction would occur more rapidly and explain why this is the case.

4 State what type of halogenoalkanes undergo $S_N1$ mechanisms.

5 Write equations showing the mechanism involved in the reaction between chloroethane and the hydroxide ion, and state the type of mechanism involved.

6 Write equations showing the mechanism involved in the reaction between 2-chloro-2-methylbutane and the hydroxide ion, and state the type of mechanism involved.

## 11.7 REACTION PATHWAYS

10.6.1
**Deduce reaction pathways given the starting materials and the product.** © IBO 2007

In considering all of the reactions that have been discussed in sections 11.3 to 11.6, you may have noticed that there is often more than one way to make a particular type of compound. For example, a halogenoalkane may be the product of the addition reaction between an alkene and a hydrogen halide or the substitution reaction between an alkane and the halogen.

In this section we will see how all of the reactions you have studied can be interlinked. They are easier to learn when you can see them as part of a big picture rather than a series of unrelated events. The linking together of a series of reactions in order to make a particular product is known as a **reaction pathway**.

### Making an alcohol from an alkane

This is a two-step pathway. The starting material is an alkane, say ethane, the final product is ethanol. What is the common link between these two compounds? It is easy to change an alkane into a halogenoalkane (section 11.3) and we have just seen how to make an alcohol from a halogenoalkane (section 11.6), so it seems that the halogenoalkane, perhaps chloroethane, is the common link. The reaction pathway is shown in figure 11.7.1.

Figure 11.7.1 Reaction pathway for the formation of ethanol from ethane.

In the first reaction, the mechanism is by way of free radicals in the presence of UV light. In the initiation step the chlorine molecule forms two Cl· free radicals. In the propagation steps one of the Cl· free radicals attacks a C–H bond in ethane to form a ·$C_2H_5$ free radical, which then attacks a $Cl_2$ molecule, releasing another Cl· free radical and produces $C_2H_5Cl$.

In the second reaction the mechanism is an $S_N2$ mechanism. The hydroxide ion attacks the slightly positive C bonded to the Cl and a reactive intermediate is produced. This reactive intermediate reacts to expel a $Cl^-$ ion, leaving the product, $C_2H_5OH$.

## Alkene pathways

There are many compounds that can be made from alkenes. Table 11.7.1 shows a range of products that can be made from propene as the starting material, and the further products that can be made.

**TABLE 11.7.1 REACTION PATHWAYS STARTING WITH PROPENE**

| Reactant (in addition to propene) | Type and conditions of reaction | Product | Further reaction conditions and type of reaction | Product |
|---|---|---|---|---|
| Hydrogen | Addition reaction using nickel catalyst → | Propane | Substitution reaction (free radical mechanism) $Cl_2$, UV light → | 1-Chloropropane |
| Hydrogen chloride* | Addition reaction → | 1-Chloropropane | Substitution reaction ($S_N2$ mechanism) NaOH → | Propan-1-ol |
| Chlorine* | Addition reaction → | 1, 2-Dichloropropane | Substitution reaction (free radical mechanism) $Cl_2$, UV light → | 1,1,2-Trichloropropane |
| Steam | Addition reaction using $H_3PO_4$ catalyst → | Propan-1-ol | Oxidation by heating under reflux with $Cr_2O_7^{2-}$, $H^+$ → | Propanoic acid |
| Steam | Addition reaction using $H_3PO_4$ catalyst → | Propan-1-ol | Oxidation by gentle heating with $Cr_2O_7^{2-}$, $H^+$ → | Propanal |
| Propene | Addition polymerization using a catalyst → | Poly(propene) | | |

*Hydrogen chloride and chlorine are being used as specific examples of the range of reactions that could be performed with halogens ($Cl_2$, $Br_2$, $I_2$) and hydrogen halides (HCl, HBr, HI).

## Alcohol pathways

While the reactions that have been discussed for alcohols are not as varied as those for alkenes, there are two different types of alcohols—primary and secondary—which can be oxidized and so influence the pathways available.

We will consider the alcohol butanol. It has structural isomers that are a primary (butan-1-ol), secondary (butan-2-ol) or tertiary (2-methyl propan-2-ol) alcohols. As tertiary alcohols cannot easily be oxidized, we will not consider 2-methylpropan-2-ol.

**TABLE 11.7.2 OXIDATION REACTION PATHWAYS STARTING WITH BUTANOL**

| Alcohol | Oxidizing agent | Conditions | Product |
|---|---|---|---|
| Butan-1-ol | $K_2Cr_2O_7$ and $H^+$ | Gentle heat | Butanal |
| Butan-1-ol | $K_2Cr_2O_7$ and $H^+$ | Heating under reflux | Butanoic acid |
| Butan-2-ol | $K_2Cr_2O_7$ and $H^+$ | Gentle heat | Butanone |

### A 'web' of reaction pathways

In the following 'web', the reactions that have been studied in sections 11.3 to 11.6 are summarized. You should be able to apply these reactions to any of the hydrocarbons with up to six carbon atoms in their chain.

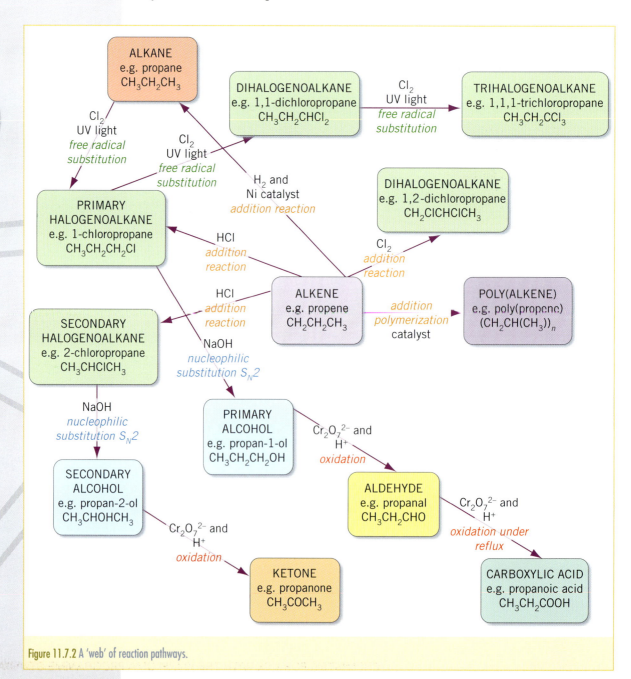

Figure 11.7.2 A 'web' of reaction pathways.

## Worked example

Show the reaction pathway that demonstrates how butanone can be made from but-2-ene.

### Solution

It can be seen from figure 11.7.2 that an alkene can be reacted with a hydrogen halide to produce a secondary halogenoalkane, which can be reacted with NaOH in a nucleophilic substitution reaction to make a secondary alcohol. The secondary alcohol can finally be oxidized to make a ketone. In the particular case of but-2-ene, the reaction pathway could appear as follows:

**WORKSHEET 11.6**
Organic reation pathways

But-2-ene $\xrightarrow{\text{HCl}}$ 2-chlorobutane $\xrightarrow{\text{NaOH}}$ butan-2-ol $\xrightarrow{\text{K}_2\text{Cr}_2\text{O}_7 \text{ and } \text{H}^+}$ butanone

$CH_3CH=CHCH_3$   $CH_3CHClCH_2CH_3$       $CH_3CHOHCH_2CH_3$       $CH_3COCH_2CH_3$

## Section 11.7 Exercises

1. State the type of reaction occurring in each of the following:
   a. alkene → primary halogenoalkane
   b. alkene → poly(alkene)
   c. alkene → alkane
   d. secondary alcohol → ketone

2. Write equations for the reaction pathway you would perform to convert each of the reactants to the stated product below. Indicate the type of reaction occurring and the reagents and conditions required for each part of the pathway.
   a. Ethane to ethanoic acid
   b. But-2-ene to butanone
   c. Ethene to ethanal

# Chapter 11 Summary

## Terms and definitions

**Addition reaction** A reaction in which a carbon–carbon double bond is replaced by a single bond and a new molecule is added into the original double-bonded molecule.

**Addition polymerization** A reaction in which many monomers with a carbon–carbon double bond react together in addition reactions to form a polymer chain.

**Alcohol** A homologous series with the general formula $C_nH_{2n+1}OH$.

**Aldehyde** A homologous series with the general formula $C_nH_{2n}O$, in which the functional group CHO is found.

**Alicyclic** Hydrocarbon compounds in which the carbon atoms are arranged in a ring.

**Aliphatic** Straight-chain and branched hydrocarbon compounds.

**Alkane** A homologous series with the general formula $C_nH_{2n+2}$.

**Alkene** A homologous series with the general formula $C_nH_{2n}$.

**Amino group** The functional group $NH_2$.

**Aromatic** Hydrocarbon compounds that are related to benzene.

**Benzene** A compound with the formula $C_6H_6$, consisting of six carbon atoms in a ring structure with six delocalized electrons. It can also be found as a functional group, $C_6H_5$.

**Carbonyl group** The functional group –C=O.

**Carboxyl group** The functional group –COOH.

**Carboxylic acid** A homologous series with the general formula $C_{n-1}H_{2n-1}COOH$ in which the functional group COOH is found.

**Combustion reaction** The reaction of a hydrocarbon with oxygen at high temperatures in which the products are carbon dioxide and water.

**Condensed structural formula** A formula that shows the arrangement of atoms of each element without showing all the bonds in one molecule of a compound.

**Decolourization** Removal of the colour of a bromine solution by the reaction with an alkene.

**Dimer** A molecule composed of two identical subunits linked together.

**Empirical formula** A formula that shows the simple whole-number ratio of elements in a compound.

**Ester** The functional group –CO-O- made up of a carbonyl group with another oxygen atom bonded to it. This functional group is only found in the middle of molecules, not at the end.

**Free radical** A reactive species containing an unpaired electron that is the product of homolytic fission.

**Functional group** An atom or group of atoms that influences the chemical properties of a compound.

**Halide group** A halogen atom covalently bonded to a hydrocarbon as a functional group.

**Halogenoalkane** A homologous series with the general formula $C_nH_{2n+1}X$, where X is a member of group 7 in the periodic table.

**Heterolytic fission** The breaking of a covalent bond in which one of the atoms involved in the bond retains both electrons from the bond, forming an anion, and the other atom (part of the molecule) forms a carbocation.

**Homologous series** A family of compounds that share the same general formula.

**Homolytic fission** The breaking of a covalent bond in which each atom involved in the bond retains one electron from the bond.

**Hydration** An addition reaction between an alkene and steam.

**Hydrocarbon** A compound predominantly made up of carbon and hydrogen.

**Hydroxyl group** The functional group –OH.

**Hydrogenation** An addition reaction between an alkene and hydrogen gas.

**Incomplete combustion** The reaction of a hydrocarbon with oxygen at high temperatures in which the products are carbon monoxide, and/or carbon and water.

**IUPAC system** The system that determines the names given to compounds, devised by the International Union of Pure and Applied Chemists.

**Ketone** A homologous series with the general formula $C_nH_{2n}O$, where $n \geq 3$, in which the functional group C=O is found.

**Molecular formula** A formula that shows the actual number of atoms of each element present in one molecule of a compound.

**Nucleophile** A reactant with a non-bonding electron pair or a negative charge, which is attracted to a centre of positive charge.

**Nucleophilic substitution** A substitution reaction in which a nucleophile is attracted to a positive charge centre.

**Oxidation reaction** A reaction in which oxygen is gained and/or hydrogen is lost.

**Primary alcohol or halogenoalkane** An alcohol or halogenoalkane in which the functional group is bonded to a carbon atom that is bonded to only one other carbon atom.

**Reaction mechanism** A sequence of reaction steps that shows in detail how a reaction possibly occurs.

**Reaction pathway** A series of steps tracing the reaction of a reactant to form a product it cannot form directly.

**Saturated hydrocarbon** A compound in which all carbon–carbon bonds are single bonds.

**Secondary alcohol/halogenoalkane** An alcohol or halogenoalkane in which the functional group is bonded to a carbon atom that is bonded to two other carbon atoms.

**$S_N1$** A nucleophilic substitution reaction in which only one species is needed to form the reactive intermediate.

**$S_N2$** A nucleophilic substitution reaction in which two species are needed to form the reactive intermediate.

**Structural formula** A formula that shows the arrangement of all the atoms and bonds in one molecule of a compound.

**Structural isomers** Molecules which have the same molecular formula but different structural formulas.

**Substitution reaction** A reaction in which one atom in a hydrocarbon is replaced by a different atom.

**Tertiary alcohol/halogenoalkane** An alcohol or halogenoalkane in which the functional group is bonded to a carbon atom that is bonded to three other carbon atoms.

**Unsaturated hydrocarbon** A compound that contains at least one carbon–carbon double or triple bond.

**Volatility** The ease with which a liquid turns into a vapour at low temperatures.

## Concepts

- Organic chemistry is the study of carbon-based compounds (excluding CO, $CO_2$ and the carbonates).

- A structural formula shows the exact arrangement and type of bonding of atoms in a molecule, whereas the molecular formula only shows how many and what type of atoms are in a molecule; an empirical formula gives the lowest whole number ratio of elements in a compound.

- A homologous series is a group of compounds whose successive members differ by $CH_2$. Examples of homologous series are shown in the table below, where R represents an alkyl group.

| Homologous series | General formula | Suffix or prefix used in naming |
|---|---|---|
| Alkanes | $C_nH_{2n+2}$ | -ane |
| Alkanes | $C_nH_{2n}$ | -ene |
| Alcohols | R–OH | -ol |
| Halogenoalkanes | R–X | 'halo'- |
| Aldehydes | R-CHO | -al |
| Ketones | R-CO-R' | -one |
| Carboxylic acids | R–COOH | -oic acid |

- Functional groups are atoms or groups of atoms attached to a hydrocarbon stem. Such groups often determine the chemical properties of the compound.

- The simplest organic molecules are hydrocarbons. Those with only single carbon–carbon bonds are saturated. Those with double or triple carbon–carbon bonds are unsaturated.

- The chemical properties of members of a homologous series are similar. Physical properties, such as boiling point, show a gradual trend that can usually be explained by the increasing strength of van der Waals' forces between molecules as molecular size increases.

- Hydrocarbons (and many other organic compounds) undergo complete combustion reactions with oxygen to form carbon dioxide and water and incomplete combustion reactions which may form carbon or carbon monoxide and water. These exothermic reactions are the basis for the use of hydrocarbons as fuels.

- Examples of reactions of organic compounds are shown in the table below, where R represents a hydrocarbon.

| Compound | Reaction type | Equations and examples | Product |
|---|---|---|---|
| Alkanes | Free-radical substitution | $RH + Cl_2 \xrightarrow{UV} RCl + HCl$ | Chloroalkane |
| Alkenes | Addition | $CH_2CH_2 \xrightarrow{A-B} CH_2ACH_2B$ | Range of products |
| Chloroalkanes | Nucleophilic substitution | $RCl + OH^- \rightarrow ROH + Cl^-$ | Alcohol |
| Primary alcohols | Oxidation | $RCH_2OH \xrightarrow{Cr_2O_7^{2-}, H^+} RCHO$ | Aldehyde |
| | | $RCHO \xrightarrow{Cr_2O_7^{2-}, H^+} RCOOH$ | Carboxylic acid |
| Secondary alcohols | Oxidation | $RCHOHR' \xrightarrow{Cr_2O_7^{2-}, H^+} RCOR'$ | Ketone |

- Polymerization reactions involve the joining together of a large number of small molecules called monomers. In addition polymerization, the monomer contains a double carbon–carbon bond. For example:

$$nCH_2CH_2 \xrightarrow{\text{high } T \text{ and } P} (CH_2CH_2)_n$$

- Various chemical reactions involving organic compounds may be combined to produce reaction pathways for the synthesis of particular organic compounds.

- Organic reactions are often slow and therefore require a catalyst and heat. Heating may cause evaporation of volatile reactants/products, so reflux condensing is often used. Reactions are often incomplete and may involve side reactions. The desired product may need to be separated from the reaction–product mixture, often by distillation.

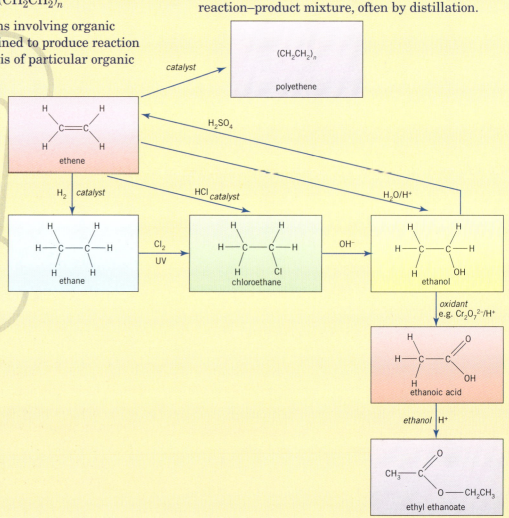

# Chapter 11 Review questions

1. Explain the meaning of each of the following terms.
   a. Homologous series
   b. Saturated molecule
   c. Structural isomer
   d. Functional group

2. Sketch and name all possible isomers of:
   a. hexane
   b. butene.

3. Correct the mistake in each of the following structures.

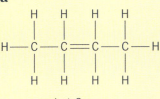
but-2-ene

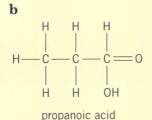
propanoic acid

4. Draw a structural formula for each of the following.
   a. Butan-1-ol
   b. Butanoic acid
   c. 2-Methylbutane
   d. Propanoic acid
   e. 3-Methylpentane
   f. Pent-2-ene

5. Select from the list below to match the descriptions that follow.

   $C_6H_{12}$
   $C_5H_9COOH$
   $C_7H_{16}$
   $C_2H_5OH$
   $C_2H_5COOH$
   $CH_3CH_2COOCH_3$

   a. A member of the alcohol series
   b. An alkene
   c. An unsaturated molecule
   d. An isomer of butanoic acid

6. Name two functional groups in each of the following molecules.
   a. [structure with O-CH₂-C(=O)-OH, Cl substituents on benzene ring]
   b. [structure: HO-CH(H)-C(=O)-C(H)(Br)-H]
   c. [structure: H₃C-C(=O)-O-C(Cl)=CH-H]

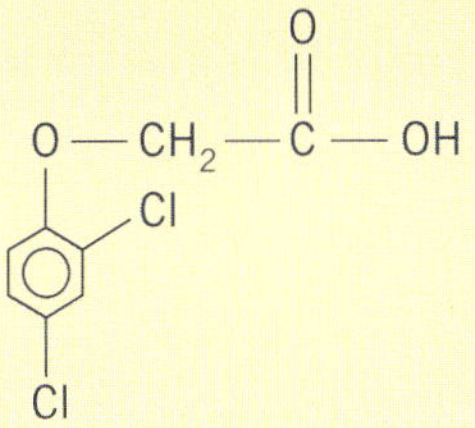

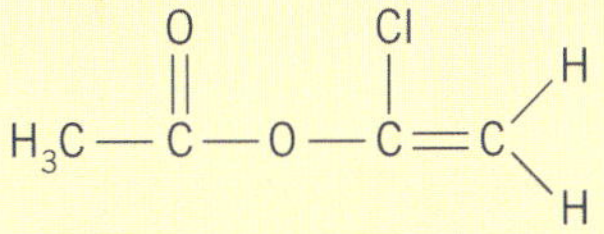

7. Which compound in each of the following pairs has the higher melting point? Explain your choices.
   a. $CH_3(CH_2)_4CH_3$ and $CH_3CH_3$
   b. NaCl and $CH_3Cl$
   c. $CH_3CH_2OH$ and $CH_3OCH_3$

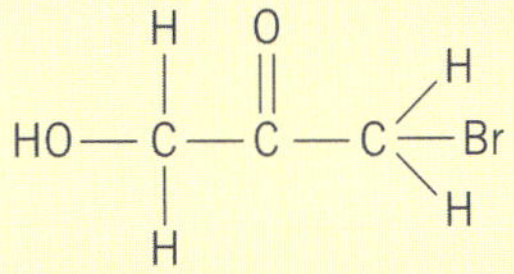

8. Name the following compounds.
   a. $CH_3—CH(CH_3)—CH_2—CH_2—CH_3$
   b. $CH_3—C(CH_3)_2—CH_2—CH_3$ (with structure shown)
   c. $CH_3—CH(CH_3)—CH(CH_3)—CH_3$ (with structure shown)

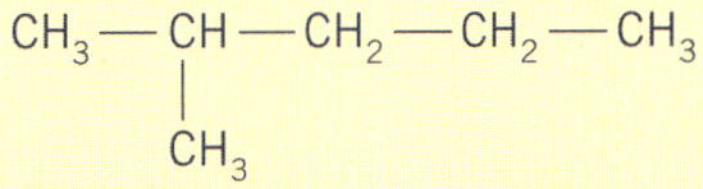

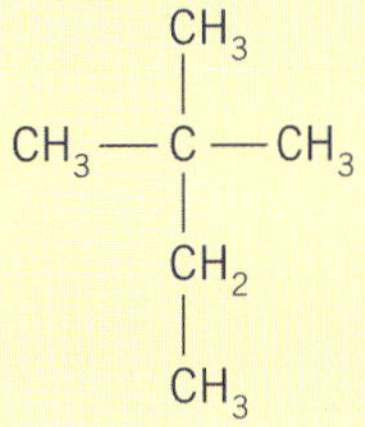

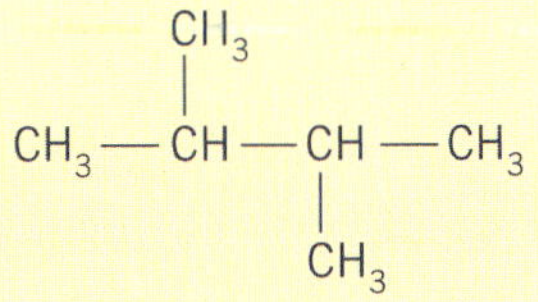

9. This question concerns pentane and pentan-1-ol.
   a. Both compounds are liquids at room temperature, but the boiling point of pentane (36°C) is significantly lower than that of pentan-1-ol (138°C). Suggest one reason for the difference in boiling point.
   b. Pentan-1-ol will react with acidified potassium dichromate. Write the structural formula of the organic product of this reaction.

10 The flowchart below depicts some of the possible reactions involving ethene. Name the substances A to D.

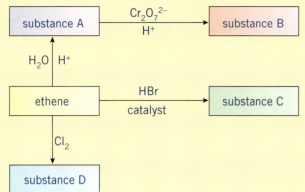

11 Two alcohols, X and Y, have the same composition by mass: carbon 60.0% and hydrogen 13.3%. Their relative molecular mass is 60.
   a Determine the empirical formula and the molecular formula of compounds X and Y.
   b Draw two structural formulas for these isomers.
   c State the names of the two isomers.

12 Write the full structural formula of the organic product of the following reactions.
   a Adding excess bromine to ethene
   b Heating ethanol with acidified potassium dichromate solution

13 Draw the polymer obtained from the following monomer.

14 Write a balanced chemical equation for the formation of polypropene.

15 State the strongest type of intermolecular force present in the compounds butan-1-ol, butanal and butanoic acid, and list the compounds in order of increasing boiling point.

16 a Sketch the intermediate formed when 1-bromopropane reacts with sodium hydroxide.
   b State how this reaction would differ from the one between 1-chloropropane and sodium hydroxide.

17 Describe the difference between an $S_N1$ reaction and an $S_N2$ reaction, and state an example of a compound which would undergo each type of reaction.

18 Write balanced chemical equations to represent the following reactions.
   a Complete combustion of propane
   b Incomplete combustion of butene
   c Complete combustion of propanol

19 Write equations showing the mechanism involved in the reaction between:
   a 1-bromopropane and the hydroxide ion, and state the type of mechanism involved
   b 2-chloro-2-methylpropane and the hydroxide ion, and state the type of mechanism involved.

20 The flowchart below shows a series of reactions involving the formation of organic compounds.

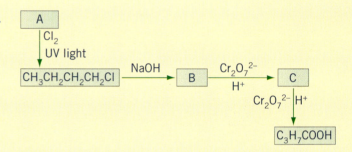

   a Name substances A to C.
   b State the type of reaction involved in the conversion of:
      i A to $C_4H_9Cl$
      ii $C_4H_9Cl$ to B
      iii B to C
   c State the type of mechanism involved in part b ii above.
   d State the conditions and experimental technique required to obtain the product C.

www.pearsoned.com.au/schools

Weblinks are available on the Chemistry: For use with the IB Diploma Programme Standard Level Companion Website to support learning and research related to this chapter.

# Chapter 11 Test

## Part A: Multiple-choice questions

1 Which names are correct for the following isomers of $C_6H_{14}$?

**I**

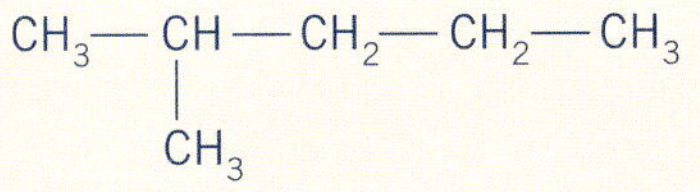

$CH_3-CH-CH_2-CH_2-CH_3$    2-methylpentane
    |
    $CH_3$

**II**

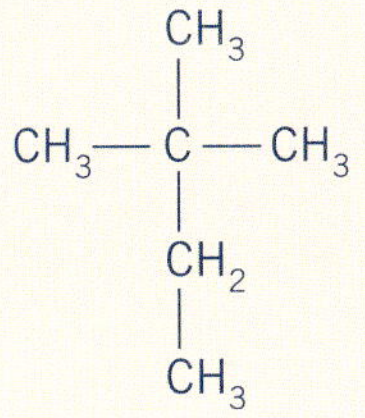

         $CH_3$
         |
$CH_3-C-CH_3$    2-ethyl-2-methylpropane
         |
         $CH_2$
         |
         $CH_3$

**III**

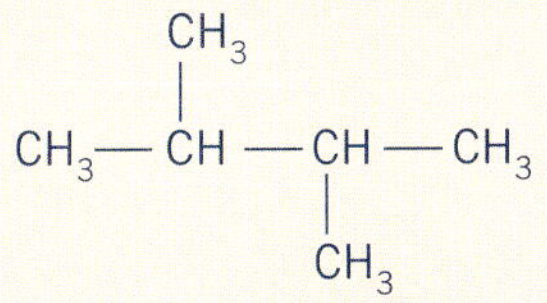

     $CH_3$
     |
$CH_3-CH-CH-CH_3$    2,3-dimethylbutane
         |
         $CH_3$

A  I only
B  I and II only
C  I and III only
D  I, II and III

© IBO 2000, May P1 Q29

2 Which of the following statements about single and double bonds between two carbon atoms is (are) correct?

 I  Double bonds are stronger than single bonds.
 II Double bonds are more reactive than single bonds.

A  I only
B  II only
C  Both I and II
D  Neither I nor II

© IBO 2001, Nov P1 Q28

3 When one mole of ethene reacts with two moles of oxygen gas:

A  $\Delta H$ is positive.
B  the oxidation number of carbon is unchanged.
C  an alcohol is formed.
D  carbon monoxide is produced.

© IBO 2002, May P1 Q27

4 What is the product of the reaction between bromine and ethene?

A  $CH_2=CHBr$
B  $CHBr=CHBr$
C  $CH_3CH_2Br$
D  $CH_2BrCH_2Br$

© IBO 2002, May P1 Q30

5 A compound with the empirical formula $C_2H_4O$ has a relative molecular mass of 88. What is the formula of the compound?

A  $CH_3CH_2COCH_3$
B  $CH_3COOH$
C  $HCOOCH_3$
D  $CH_3CH_2CH_2COOH$

© IBO 2002, Nov P1 Q27

6 Statement (S): Solubility of alcohols in water decreases with increase in $M_r$.

Explanation (E): The relative proportion of the hydrocarbon part in the alcohol increases with increasing $M_r$.

A  Both S and E are true.
B  Both S and E are false.
C  S is true but E is false.
D  S is false but E is true.

© IBO 2002, Nov P1 Q29

7 How many structural isomers are possible with the molecular formula $C_6H_{14}$?

A  4
B  5
C  6
D  7

© IBO 2005, Nov P1 Q28

8 Which compound is a member of the aldehyde homologous series?

A  $CH_3COCH_3$
B  $CH_3CH_2CH_2OH$
C  $CH_3CH_2COOH$
D  $CH_3CH_2CHO$

© IBO 2005, Nov P1 Q30

9 Propane, $C_3H_8$, undergoes incomplete combustion in a limited amount of air. Which products are most likely to be formed during this reaction?

A Carbon monoxide and water
B Carbon monoxide and hydrogen
C Carbon dioxide and hydrogen
D Carbon dioxide and water

© IBO 2006, May P1 Q29

10 Which compound will undergo oxidation when treated with acidified potassium dichromate(VI)?

A $CH_3CH_2CHO$
B $CH_3COCH_3$
C $CH_3COOH$
D $(CH_3)_3COH$

© IBO 2002, May P1 Q38

(10 marks)

www.pearsoned.com.au/schools

For more multiple-choice test questions, connect to the Chemistry: For use with the IB Diploma Programme Standard Level Companion Website and select Chapter 11 Review Questions.

## Part B: Short-answer questions

1 a State two characteristics of a homologous series.
(2 marks)

b Describe a chemical test to distinguish between alkanes and alkenes, giving the result in each case.
(3 marks)

© IBO 2006, May P2 Q5

2 a State and explain the trend in boiling points of the first 10 members of the alkene series.
(3 marks)

b Explain why methanol has a much higher boiling point than ethane ($M_r = 30$).
(2 marks)

© IBO 1999, May P2 Q3

3 Ethanol may be converted to ethanoic acid.

a Identify the reagent needed and state the type of reaction.
(2 marks)

b State the colour change observed during the reaction.
(1 mark)

© IBO 2002, May P2 Q6e

4 Butane and but-2-ene react with bromine in different ways.

a i Write an equation for the reaction between butane and bromine, showing the structure of a possible organic product.
(2 marks)

ii Identify the type of bond fission that occurs in bromine and the species that reacts with butane.
(2 marks)

b i Write an equation for the reaction between but-2-ene and bromine, showing the structure of the organic product.
(2 marks)

ii State the type of reaction occurring.
(1 mark)

© IBO 2006, Nov P2 Q4

## Part C: Data-based questions

1 Discuss the factors which affect the boiling points of covalently bonded compounds by reference to the following pairs of organic substances, whose boiling points are given.

a Ethane (184 K) and butane (273 K)
b Ethane (184 K) and bromoethane (311 K)
c Bromoethane (311 K) and ethanol (352 K)
(8 marks)

© IBO 2000, May P2 Q6a

2 Ethanol is used as a fuel because it undergoes combustion.

a Write a balanced chemical equation for the combustion of ethanol.
(2 marks)

b The standard enthalpy of combustion, $\Delta H_c^\circ$, and the relative molecular masses, $M_r$, of a series of alcohols are given below.

| Alcohol | $\Delta H_c^\circ$/kJ mol$^{-1}$ | $M_r$ |
|---|---|---|
| $CH_3OH$ methanol | −715 | 32.0 |
| $CH_3CH_2OH$ ethanol | −1371 | 46.0 |
| $CH_3CH_2CH_2OH$ propan-1-ol | −2010 | 60.0 |
| $CH_3CH_2CH_2CH_2OH$ butan-1-ol | −2673 | 74.0 |

**i** Calculate the relative molecular mass of pentan-1-ol and thus estimate $\Delta H_c^\ominus$ for pentan-1-ol using the graph below.

(2 marks)

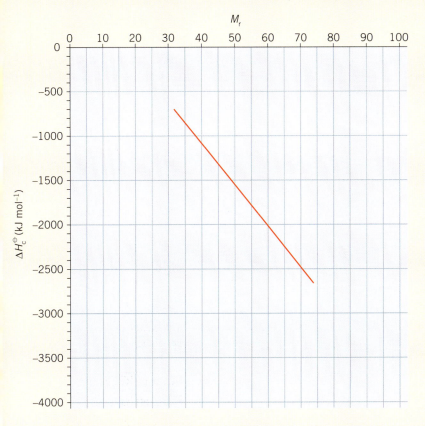

**ii** How would the value of the standard enthalpy of combustion of pentan-2-ol compare with that of pentan-1-ol? Explain your answer.

(2 marks)

© IBO 2001, Nov P2 Q2

## Part D: Extended-response question

**a** Three organic compounds have the same $M_r$ values.

$CH_3CH_2COOH$
$CH_3CH_2OCH_2CH_3$
$CH_3CH_2CH_2CH_2OH$

  **i** State and explain which compound has the lowest boiling point.

(2 marks)

  **ii** Calculate the volume that 0.0200 mol of the gaseous compound in part **a i** would occupy at 70°C and $1.10 \times 10^5$ Pa.

(3 marks)

**b** $CH_3COCH_3$ can be prepared in the laboratory from an alcohol. State the name of this alcohol, the type of reaction occurring and the reagents and conditions needed for the reaction.

(5 marks)

**c** 2-Bromobutane can be converted into butan-2-ol by a nucleophilic substitution reaction. This reaction occurs by two different mechanisms.

  **i** Give the structure of the transition state formed in the $S_N2$ mechanism.

(2 marks)

  **ii** Write equations for the $S_N1$ mechanism.

(2 marks)

**d** State and explain whether the rate of the $S_N1$ mechanism will decrease, increase or remain the same, if 2-chlorobutane is used instead of 2-bromobutane.

(2 marks)

© IBO 2006, Nov P2 Q7ai, ii; b, e, f
Total marks = 60

# Periodic Table

| 1 | 2 | | | | | | | | | | | | 3 | 4 | 5 | 6 | 7 | 0 |
|---|---|---|---|---|---|---|---|---|---|---|---|---|---|---|---|---|---|---|
| 1<br>**H**<br>1.01 | | | | | | | | | | | | | | | | | | 2<br>**He**<br>4.00 |
| 3<br>**Li**<br>6.94 | 4<br>**Be**<br>9.01 | | | | | | | | | | | | 5<br>**B**<br>10.81 | 6<br>**C**<br>12.01 | 7<br>**N**<br>14.01 | 8<br>**O**<br>16.00 | 9<br>**F**<br>19.00 | 10<br>**Ne**<br>20.18 |
| 11<br>**Na**<br>22.99 | 12<br>**Mg**<br>24.31 | | | | | | | | | | | | 13<br>**Al**<br>26.98 | 14<br>**Si**<br>28.09 | 15<br>**P**<br>30.97 | 16<br>**S**<br>32.06 | 17<br>**Cl**<br>35.45 | 18<br>**Ar**<br>39.95 |
| 19<br>**K**<br>39.10 | 20<br>**Ca**<br>40.08 | 21<br>**Sc**<br>44.96 | 22<br>**Ti**<br>47.90 | 23<br>**V**<br>50.94 | 24<br>**Cr**<br>52.00 | 25<br>**Mn**<br>54.94 | 26<br>**Fe**<br>55.85 | 27<br>**Co**<br>58.93 | 28<br>**Ni**<br>58.71 | 29<br>**Cu**<br>63.55 | 30<br>**Zn**<br>65.37 | | 31<br>**Ga**<br>69.72 | 32<br>**Ge**<br>72.59 | 33<br>**As**<br>74.92 | 34<br>**Se**<br>78.96 | 35<br>**Br**<br>79.90 | 36<br>**Kr**<br>83.80 |
| 37<br>**Rb**<br>85.47 | 38<br>**Sr**<br>87.62 | 39<br>**Y**<br>88.91 | 40<br>**Zr**<br>91.22 | 41<br>**Nb**<br>92.91 | 42<br>**Mo**<br>95.94 | 43<br>**Tc**<br>98.91 | 44<br>**Ru**<br>101.07 | 45<br>**Rh**<br>102.91 | 46<br>**Pd**<br>106.42 | 47<br>**Ag**<br>107.87 | 48<br>**Cd**<br>112.40 | | 49<br>**In**<br>114.82 | 50<br>**Sn**<br>118.69 | 51<br>**Sb**<br>121.75 | 52<br>**Te**<br>127.60 | 53<br>**I**<br>126.90 | 54<br>**Xe**<br>131.30 |
| 55<br>**Cs**<br>132.91 | 56<br>**Ba**<br>137.34 | 57<br>***La**<br>138.91 | 72<br>**Hf**<br>178.49 | 73<br>**Ta**<br>180.95 | 74<br>**W**<br>183.85 | 75<br>**Re**<br>186.21 | 76<br>**Os**<br>190.21 | 77<br>**Ir**<br>192.22 | 78<br>**Pt**<br>195.09 | 79<br>**Au**<br>196.97 | 80<br>**Hg**<br>200.59 | | 81<br>**Tl**<br>204.37 | 82<br>**Pb**<br>207.19 | 83<br>**Bi**<br>208.98 | 84<br>**Po**<br>(210) | 85<br>**At**<br>(210) | 86<br>**Rn**<br>(222) |
| 87<br>**Fr**<br>(223) | 88<br>**Ra**<br>(226) | 89<br>§**Ac**<br>(227) | | | | | | | | | | | | | | | | |

Atomic number<br>**Element**<br>Atomic mass

| * | 58<br>**Ce**<br>140.12 | 59<br>**Pr**<br>140.91 | 60<br>**Nd**<br>144.24 | 61<br>**Pm**<br>146.92 | 62<br>**Sm**<br>150.35 | 63<br>**Eu**<br>151.96 | 64<br>**Gd**<br>157.25 | 65<br>**Tb**<br>158.92 | 66<br>**Dy**<br>162.50 | 67<br>**Ho**<br>164.93 | 68<br>**Er**<br>167.26 | 69<br>**Tm**<br>168.93 | 70<br>**Yb**<br>173.04 | 71<br>**Lu**<br>174.97 |
|---|---|---|---|---|---|---|---|---|---|---|---|---|---|---|
| § | 90<br>**Th**<br>232.04 | 91<br>**Pa**<br>231.04 | 92<br>**U**<br>238.03 | 93<br>**Np**<br>(237) | 94<br>**Pu**<br>(244) | 95<br>**Am**<br>(243) | 96<br>**Cm**<br>(247) | 97<br>**Bk**<br>(247) | 98<br>**Cf**<br>(251) | 99<br>**Es**<br>(254) | 100<br>**Fm**<br>(257) | 101<br>**Md**<br>(258) | 102<br>**No**<br>(259) | 103<br>**Lr**<br>(260) |

© *International Baccalaureate Organization 2007*

# APPENDIX 1

## Table of relative atomic masses

| Element | Symbol | Atomic number | Atomic mass |
|---|---|---|---|
| Actinium | Ac | 89 | (227) |
| Aluminium | Al | 13 | 26.98 |
| Americium | Am | 95 | (243) |
| Antimony | Sb | 51 | 121.75 |
| Argon | Ar | 18 | 39.95 |
| Arsenic | As | 33 | 74.92 |
| Astatine | At | 85 | (210) |
| Barium | Ba | 56 | 137.34 |
| Berkelium | Bk | 97 | (247) |
| Beryllium | Be | 4 | 9.01 |
| Bismuth | Bi | 83 | 208.98 |
| Bohrium | Bh | 107 | (264) |
| Boron | B | 5 | 10.81 |
| Bromine | Br | 35 | 79.90 |
| Cadmium | Cd | 48 | 112.40 |
| Caesium | Cs | 55 | 132.91 |
| Calcium | Ca | 20 | 40.08 |
| Californium | Cf | 98 | (251) |
| Carbon | C | 6 | 12.01 |
| Cerium | Ce | 58 | 140.12 |
| Chlorine | Cl | 17 | 35.45 |
| Chromium | Cr | 24 | 52.00 |
| Cobalt | Co | 27 | 58.93 |
| Copper | Cu | 29 | 63.55 |
| Curium | Cm | 96 | (247) |
| Darmstadtium | Ds | 110 | (271) |
| Dubnium | Db | 105 | (262) |
| Dysprosium | Dy | 66 | 162.50 |
| Einsteinium | Es | 99 | (254) |
| Erbium | Er | 68 | 167.26 |
| Europium | Eu | 63 | 151.96 |
| Fermium | Fm | 100 | (257) |
| Fluorine | F | 9 | 19.00 |
| Francium | Fr | 87 | (223) |
| Gadolinium | Gd | 64 | 157.25 |
| Gallium | Ga | 31 | 69.72 |
| Germanium | Ge | 32 | 72.59 |
| Gold | Au | 79 | 196.97 |
| Hafnium | Hf | 72 | 178.49 |
| Hassium | Hs | 108 | (277) |
| Helium | He | 2 | 4.00 |
| Holmium | Ho | 67 | 164.93 |
| Hydrogen | H | 1 | 1.01 |
| Indium | In | 49 | 114.82 |
| Iodine | I | 53 | 126.90 |
| Iridium | Ir | 77 | 192.22 |
| Iron | Fe | 26 | 55.85 |
| Krypton | Kr | 36 | 83.80 |
| Lanthanum | La | 57 | 138.91 |
| Lawrencium | Lr | 103 | (260) |
| Lead | Pb | 82 | 207.19 |
| Lithium | Li | 3 | 6.94 |
| Lutetium | Lu | 71 | 174.97 |
| Magnesium | Mg | 12 | 24.31 |
| Manganese | Mn | 25 | 54.94 |
| Meitnerium | Mt | 109 | (268) |
| Mendelevium | Md | 101 | (258) |
| Mercury | Hg | 80 | 200.59 |
| Molybdenum | Mo | 42 | 95.94 |
| Neodymium | Nd | 60 | 144.24 |
| Neon | Ne | 10 | 20.18 |
| Neptunium | Np | 93 | (237) |
| Nickel | Ni | 28 | 58.71 |
| Niobium | Nb | 41 | 92.91 |
| Nitrogen | N | 7 | 14.01 |
| Nobelium | No | 102 | (259) |
| Osmium | Os | 76 | 190.21 |
| Oxygen | O | 8 | 16.00 |
| Palladium | Pd | 46 | 106.42 |
| Phosphorus | P | 15 | 30.97 |
| Platinum | Pt | 78 | 195.09 |
| Plutonium | Pu | 94 | (244) |
| Polonium | Po | 84 | (210) |
| Potassium | K | 19 | 39.10 |
| Praseodymium | Pr | 59 | 140.91 |
| Promethium | Pm | 61 | 146.92 |
| Protactinium | Pa | 91 | 231.04 |
| Radium | Ra | 88 | (226) |
| Radon | Rn | 86 | (222) |
| Rhenium | Re | 75 | 186.21 |
| Rhodium | Rh | 45 | 102.91 |
| Roentgenium | Rg | 111 | (272) |
| Rubidium | Rb | 37 | 85.47 |
| Ruthenium | Ru | 44 | 101.07 |
| Rutherfordium | Rf | 104 | (261) |
| Samarium | Sm | 62 | 150.35 |
| Scandium | Sc | 21 | 44.96 |
| Seaborgium | Sg | 106 | (266) |
| Selenium | Se | 34 | 78.96 |
| Silicon | Si | 14 | 28.09 |
| Silver | Ag | 47 | 107.87 |
| Sodium | Na | 11 | 22.99 |
| Strontium | Sr | 38 | 87.62 |
| Sulfur | S | 16 | 32.06 |
| Tantalum | Ta | 73 | 180.95 |
| Technetium | Tc | 43 | 98.91 |
| Tellurium | Te | 52 | 127.60 |
| Terbium | Tb | 65 | 158.92 |
| Thallium | Tl | 81 | 204.37 |
| Thorium | Th | 90 | 232.04 |
| Thulium | Tm | 69 | 168.93 |
| Tin | Sn | 50 | 118.69 |
| Titanium | Ti | 22 | 47.90 |
| Tungsten | W | 74 | 183.85 |
| Uranium | U | 92 | 238.03 |
| Vanadium | V | 23 | 50.94 |
| Xenon | Xe | 54 | 131.30 |
| Ytterbium | Yb | 70 | 173.04 |
| Yttrium | Y | 39 | 88.91 |
| Zinc | Zn | 30 | 65.37 |
| Zirconium | Zr | 40 | 91.22 |

Note: Numbers shown in brackets indicate the relative isotopic mass of the most stable isotope when the atomic mass is not known.

# APPENDIX 2

## Physical constants, symbols and units

In completing calculations during your chemistry course, you will encounter a number of physical constants. Table A2.1 lists a number of these constants, together with their symbol and value.

### TABLE A2.1 PHYSICAL CONSTANTS

| Name of constant | Symbol | Value (as frequently used) |
|---|---|---|
| Absolute zero | 0 K | $-273°C$ |
| Avogadro's constant | $L$ | $6.02 \times 10^{23}$ $mol^{-1}$ |
| Gas constant | $R$ | $8.31$ $J\ K^{-1}\ mol^{-1}$ |
| Molar volume of a gas at 273 K and $1.01 \times 10^5$ Pa | $V_m$ | $22.4$ $dm^3\ mol^{-1}$ ($2.24 \times 10^{-2}$ $m^3\ mol^{-1}$) |
| Planck's constant | $h$ | $6.63 \times 10^{-34}$ J s |
| Specific heat capacity of water | $c$ | $4.18$ $J\ g^{-1}\ K^{-1}$ ($4.18$ $kJ\ kg^{-1}\ K^{-1}$) |
| Ionic product constant for water | $K_w$ | $1.00 \times 10^{-14}$ $mol^2\ dm^{-6}$ at 298 K |
| 1 atm | | $1.01 \times 10^5$ Pa |
| 1 $dm^3$ | | 1 litre = $1 \times 10^{-3}$ $m^3$ = $1 \times 10^3$ $cm^3$ |

Accurate measurements are an essential part of Chemistry. Measurements can be made in a variety of units, but most chemists use the International System of Units, known as SI. This system, adopted by international agreement in 1960, is based on the metric system and units derived from the metric system. The fundamental units and some derived units are shown in table A2.2, along with relevant symbols.

### TABLE A2.2 FUNDAMENTAL AND DERIVED SI UNITS

| Quantity | Symbol | SI unit | Symbol of SI unit | Definition of derived SI unit |
|---|---|---|---|---|
| **Fundamental** | | | | |
| Mass | $m$ | kilogram | kg | |
| Length | $l$ | metre | m | |
| Time | $t$ | second | s | |
| Temperature | $T$ | kelvin | K | |
| Electric current | $I$ | ampere | A | |
| Amount of substance | $n$ | mole | mol | |
| **Derived** | | | | |
| Area | $A$ | square metre | $m^2$ | $m^2$ |
| Volume | $V$ | cubic metre | $m^3$ | $m^3$ |
| Force | $F$ | newton | N | $kg\ m\ s^{-2}$ |
| Energy | $E$ | joule | J | N m |
| Pressure | $P$ | pascal | Pa | $N\ m^{-2}$ |
| Electric charge | $Q$ | coulomb | C | A s |
| Electric potential | $V$ | volt | V | $J\ A^{-1}\ s^{-1}$ |
| Density | $\rho$ | | $kg\ m^{-3}$ | $kg\ m^{-3}$ |

The fundamental units are not always convenient. For example, expressing the mass of a short length of magnesium ribbon in kilograms is awkward, so the SI system uses a series of prefixes to alter the size of the unit. Commonly encountered prefixes are shown in table A2.3.

**TABLE A2.3 PREFIXES FOR UNITS**

| Prefix | Symbol | Fraction/multiple |
|---|---|---|
| pico | p | $10^{-12}$ |
| nano | n | $10^{-9}$ |
| micro | μ | $10^{-6}$ |
| milli | m | $10^{-3}$ |
| centi | c | $10^{-2}$ |
| deci | d | $10^{-1}$ |
| deca | da | $10^{1}$ |
| hecto | h | $10^{2}$ |
| kilo | k | $10^{3}$ |
| mega | M | $10^{6}$ |

Data recorded during experiments may often have to be converted from one unit to another, and otherwise mathematically manipulated. When conducting experiments and solving problems, the units for the quantities used must always be included. Table A2.4 shows a number of conversion factors for non-SI units. The following worked example shows a technique that may be used to convert from one unit to another.

## Worked example

Convert a pressure reading of $3.516 \times 10^3$ Pa to a reading in atmosphere (atm).

## Solution

Step 1: Write an equivalence statement. From table A2.4:

1 atm = 101.325 kPa = 101 325 Pa

Step 2: Write a statement for the required quantity ($x$).

$x$ atm = $3.516 \times 10^3$ Pa

Step 3: Solve the ratio for the unknown ($x$).

$$\frac{x}{1} = \frac{3.516 \times 10^3}{101\,325}$$

∴ $x$ = 0.034 70 atm

**TABLE A2.4 CONVERSION FACTORS FOR NON-SI UNITS**

| Quantity | Symbol | Unit | Symbol | SI equivalent |
|---|---|---|---|---|
| Length | $l$ | angstrom | Å | $10^{-10}$ m |
| Area | $A$ | hectare | ha | $10^4$ m$^2$ |
| Volume | $V$ | litre | L | $10^3$ cm$^3$ = $10^{-3}$ m$^3$ = 1 dm$^3$ |
| Mass | $m$ | tonne | t | $10^3$ kg = $10^6$ g |
| Pressure | $P$ | millimetre of mercury | mm Hg | 760 mm Hg = 101.325 kPa |
|  |  | atmosphere | atm | 1 atm = 101.325 kPa |
|  |  | bar | bar | $10^5$ Pa = $10^2$ kPa |
| Temperature | $T$ | degrees Celsius | °C | 0°C = 273 K |
| Energy | $E$ | calorie | cal | 4.184 J |

# SOLUTIONS

These solutions have been simplified and not all solutions are included here. Fully explained answers, including fully worked solutions to calculations and answers requiring a diagram, may be found on the *Chemistry: for use with the IB Diploma Programme Standard Level* student CD packaged with this book.

## Chapter 1 Atomic structure

### Section 1.1

1. protons, neutrons, electrons; protons, neutrons, nucleus; electrons, nucleus
2. Both have relative mass of 1.
3. Relative mass: electron $5 \times 10^{-4}$, proton 1
   Relative charge: electron −1, proton +1
4. Isotopes have the same atomic number but different mass numbers.
5. **a** $7 = Z$, $15 = A$    **b** 7 p, 8 n, 7 e: 2,5
6. **a** 30 p, 32 n, 30 e    **b** 36 p, 45 n, 36 e
   **c** 12 p, 12 n, 10 e    **d** 35 p, 46 n, 36 e
7. $^{90}_{45}X$ and $^{95}_{45}X$   $^{92}_{43}X$ and $^{90}_{43}X$
8. radioactivity, mass, density, melting point, boiling point
9. as medical tracers

### Section 1.2

3. It is the weighted mean of the isotopic masses of the isotopes of that element.
4. 6
5. 12.01
6. 30%, 70%; 204.4
7. 28.1
8. $^{10}B = 20\%$, $^{11}B = 80\%$
9. $A_r(Cu) = 63.6$
10. $^{72}_{32}Ge$ and $^{74}_{32}Ge$ are same element, $^{74}_{32}Ge$ and $^{74}_{34}Se$ are different elements.
11. Small peak at $m/z = 20$: $^{40}Ca^{2+}$ ions
    Larger peak at $m/z = 40$: $^{40}Ca^+$ ions
12. **a** $^{39}K^+$    **b** $^{10}B^+$    **c** $^{37}Cl^+$

### Section 1.3

1. Electrons move around the nucleus in energy levels, called shells.
2. An electron in shell 2 will have a higher energy.
4. **a** **i** 2,8,5   **ii** 2,7   **iii** 2,8,8   **iv** 2,8,8,1
   **b** **i** 2,8   **ii** 2,8   **iii** 2,8,8   **iv** 2,8,8
5. **a** **i** 4 e   **ii** 3 e   **iii** 7 e
   **b** **i** 8 e   **ii** 8 e   **iii** 8 e
6. **a** UV at lower wavelength than visible
   **b** IR at higher wavelength than visible
7. **a** Balmer: to 2nd shell
   **b** Lyman: to 1st shell
   **c** Paschen: to 3rd shell
9. red: shell 3 → 2; green: shell 4 → 2; purple: shell 5 → 2; violet: shell 6 → 2
10. shell 4 → 3 > shell 6 → 5

### Chapter 1 Review questions

1. **a** **i** 25 p   **ii** 29 n   **iii** manganese
   **b** **i** 36 p   **ii** 47 n   **iii** krypton
   **c** **i** 3 p   **ii** 3 n   **iii** lithium
   **d** **i** 98 p   **ii** 158 n   **iii** californium
   **e** **i** 37 p   **ii** 49 n   **iii** rubidium
3. **a** same number of protons, different number of neutrons
   **b** in radiocarbon dating
   **c** different number of protons
4. **a** aluminium   **b** 14 n   **c** 2,8,3
5. **a** **i** 2,8,8,2   **ii** 2,8,8
   **b** **i** 40   **ii** 20
   **c** 18 e, 20 p, 19 n
6. **a** fluorine   **b** carbon   **c** magnesium
   **d** nitrogen   **e** sulfur
7. **a** Magnitude of charge is the same.
   **b** They have similar mass and are found in the nucleus.
   **c** Size, charge and location in the atom are different.
9. **a** heavier   **b** 55   **c** not new
10. $^{70}Ga < {}^{56}Fe^+ < {}^{40}Ca < {}^{56}Fe^{2+} < {}^{16}O^+ < {}^{16}O^{2+}$
12. $A_r = 121.8$; $I_r({}^{121}Sb) = 120.90$, $A = 121$;
    $I_r({}^{123}Sb) = 122.90$, $A = 123$
13. 6.95
14. 192
15. $^{35}Cl = 76.0\%$, $^{37}Cl = 24.0\%$
16. 71.0
17. Continuous spectrum: like a rainbow; line spectrum: lines of colour on a black background

### Chapter 1 Test

#### Part A: Multiple-choice questions

| 1 | B | 4 | B | 7 | C | 10 | D |
| --- | --- | --- | --- | --- | --- | --- | --- |
| 2 | B | 5 | B | 8 | D | 11 | A |
| 3 | D | 6 | D | 9 | A | 12 | D |

## Chapter 2 Bonding

### Section 2.1

1. **a** metallic   **b** covalent   **c** ionic
   **d** ionic   **e** covalent   **f** metallic
   **g** ionic
2. **a** NaF   **b** CaS   **c** $CaCl_2$   **d** $Al_2O_3$
3. **a** $KNO_3$   **b** $CaCl_2$   **c** NaOH
   **d** $CuSO_4$   **e** $(NH_4)_2S$   **f** $Al(NO_3)_3$
4. **a** potassium chloride   **b** barium sulfate
   **c** hydrogen nitrate (nitric acid)   **d** aluminium oxide

| | | e | tin(II) iodide | | f | copper(II) phosphate |
|---|---|---|---|---|---|---|
| 5 | a | $Mn^{4+}$ | | b | $Co^{3+}$ | c | $Ni^{2+}$ |
| | d | $Pt^{2+}$ | | e | $Au^{3+}$ | f | $Ga^+$ |
| 6 | a | $BrO_3^-$ | | b | $SiO_3^{2-}$ | c | $AsO_2^-$ |
| | d | $SeO_3^{2-}$ | | e | $TeO_4^-$ | f | $IrCl_6^{3-}$ |

## Section 2.2

1. **a** mobile charged particles
   **b** strong bonding
   **c** particles that can be moved relative to each other
2. Layers of the metallic lattice slide over each other.
3. **a** **i** Electrons are not fixed in one place.
      **ii** Cations are in an ordered, 3D arrangement.
   **b** There is attraction between positive ions and electrons.
4. electrons
6. Increased vibration of cations slows flow of electrons.

## Section 2.3

1. The group number is the same as the number of valence electrons.
2. between the nuclei of the bonding atoms and the bonding electrons
3. I, carbon dioxide; II, ice; V, hydrogen chloride
4. **a** 2, 0   **b** 14, 12   **c** 12, 8
   **d** 8, 6   **e** 10, 4
6. The atomic radius of each element increases, as there are more electron shells.
7. **a** C≡C   **b** C–C
9. **a** bent   **b** linear
   **c** tetrahedral   **d** linear
   **e** trigonal pyramidal   **f** tetrahedral
   **g** bent
10. **a** $NH_3$   **b** $Cl_2$   **c** $H_2S$
    **d** $CO_2$   **e** $CH_3Cl$   **f** HF
11. **a** trigonal pyramidal, 107°
    **b** linear, 180°
    **c** tetrahedral around each C atom, 109.5°
    **d** tetrahedral, 109.5°
    **e** trigonal planar, 120°
12. II, hydrogen and oxygen
13. **a** δ– C, δ+ H   **b** δ– O, δ+ B   **c** δ– Cl, δ+ P
    **d** δ– S, δ+ H   **e** δ– O, δ+ Si

## Section 2.4

1. **a** different structural forms of the same element
2. **a** giant covalent network lattice
   **b** very hard, non-conductive
4. **a** network lattice structure
   **b** Si–Si: lower bond enthalpy, so less energy required to break bond and hence more reactive
5. **a** silica, sandstone, sand, quartz
   **b** network lattice structure
   **c** diamond, silicon, carbide

## Section 2.5

1. **a** dipole–dipole   **b** van der Waals'
   **c** hydrogen   **d** dipole–dipole
   **e** dipole–dipole   **f** van der Waals'
   **g** van der Waals'   **h** hydrogen
   **i** dipole–dipole
2. Octane exhibits stronger van der Waals' forces so will have a higher boiling point.
3. $He < CO_2 < SO_2 < HF < CH_4 < N_2$
4. **a** ethanol   **b** dichloromethane
   **c** water   **d** hydrogen fluoride
   **e** ammonia

## Section 2.6

1. **a** electrons   **b** $Na^+$, $Cl^-$ ions
   **c** $Fe^{2+}$ ions and electrons   **d** $Ag^+$, $I^-$ ions
2. K has weaker bonds between cations and delocalized electrons.
3. In the solid state, ions cannot move; when melted, the ions can move and conduct electricity.
4. Sufficient force can push ions out of alignment and generate electrostatic repulsions.
5. MgO has a stronger ionic bond than NaCl, resulting in a higher melting point.
6. Melting points decrease as strength of electrostatic attraction decreases.
7. calcium chloride, copper nitrate, methanol
8. butane

## Chapter 2 Review questions

1. **a** covalent   **b** metallic   **c** ionic
   **d** ionic   **e** covalent   **f** metallic
   **g** covalent   **h** ionic   **i** covalent
   **j** ionic
2. **a** Mg: 2,8,2; N: 2,5   **c** $Mg_3N_2$
4. MgO has a stronger ionic bond between its ions than in NaCl.
5. **a** $NiCl_2$   **b** $CaCO_3$   **c** $CuSO_4$
   **d** $CuNO_3$   **e** $PbI_4$   **f** KOH
   **g** $Al_2S_3$   **h** $Zn(HSO_4)_2$   **i** $Fe_2O_3$
   **j** $MgF_2$
6. **a** magnesium hydroxide   **b** potassium permanganate
   **c** silver oxide   **d** lead(II) oxide
   **e** potassium dichromate   **f** tin(II) nitrate
   **g** hydrogen sulfide   **h** iron(II) nitrate
   **i** zinc nitride   **j** calcium sulfate
7. Positive metal ions are surrounded by a 'sea' of mobile electrons.
9. **a** trigonal pyramidal   **b** linear
   **c** bent   **d** tetrahedral
   **e** bent   **f** tetrahedral
   **g** tetrahedral   **h** trigonal planar
   **i** linear   **j** trigonal planar
   **k** linear
10. **a** C=O   **b** C–O
11. Dative covalent bond: 2 bonding electrons from the same atom
12. **a** $BF_3$   **b** $CO_2$   **c** HF
    **d** HCN   **e** $NH_4^+$   **f** $SO_2$
15. **a** dipole–dipole   **b** van der Waals'
    **c** hydrogen   **d** dipole–dipole
    **e** dipole–dipole   **f** van der Waals'

| | | | | | | | | |
|---|---|---|---|---|---|---|---|---|
| | **g** | van der Waals' | | **h** | hydrogen | | | |
| | **i** | dipole–dipole | | | | | | |

**18** Water forms hydrogen bonds with ammonia. Octane is non-polar and cannot form intermolecular bonds with water.

## Chapter 2 Test

### Part A: Multiple-choice questions

| 1 | A | 3 | A | 5 | D | 7 | B | 9 | C |
|---|---|---|---|---|---|---|---|---|---|
| 2 | C | 4 | A | 6 | B | 8 | B | 10 | B |

## Chapter 3 Periodicity

### Section 3.1 Exercises

**1** Examples include
  **a** He    **b** Mg    **c** P    **d** Cu
**2 a** gallium    **b** magnesium    **c** tellurium
**3 a** A    **b** E    **c** C
**4 a** group 5, period 3    **b** group 2, period 4
**5 a** 2,5    **b** 2,8
**6 a** 2,8,1    **b** 2,5    **c** 2
**7** 2

### Section 3.2 Exercises

**1 a** strontium < magnesium
  **b** selenium < oxygen
**2 a** It is the effective nuclear charge experienced by a valence electron.
  **b** Core charge increases, so atomic radius decreases.
**3** Electrons are more strongly attracted to fluorine.
**4 a** magnesium > chlorine
**5 a** chlorine > phosphorus
**6 a** decrease from $Na^+$ to $Al^{3+}$
  **b** decrease from $Si^{4-}$ to $Cl^-$
  **c** Cl has greater nuclear charge than $Si^{4-}$
**7 a** <    **b** <    **c** >
  **d** <    **e** <    **f** >

### Section 3.3 Exercises

**1** Reactivity increases down the group with distance from the nucleus.
**2** Halogens become less reactive down the group.
**3 a** $2Li(s) + H_2O(l) \rightarrow Li_2O(aq) + H_2(g)$
  **b** $Li_2O(aq) + H_2O(l) \rightarrow 2LiOH(aq)$
  **c** alkaline    **d** use pH indicator
**4 a** $Br^-$ and $I^-$    **b** none
**5** $Cl_2(g) + 2Br^-(aq) \rightarrow 2Cl^-(aq) + Br_2(aq)$;
  $Cl_2(g) + 2I^-(aq) \rightarrow 2Cl^-(aq) + I_2(aq)$
**6 a** $P_4O_{10}(s) + 6H_2O(l) \rightarrow 4H_3PO_4(aq)$
  **b** acidic
**7 a** Na > Li    **b** Cl > Br    **c** F > Cl    **d** K > Na

### Chapter 3 Review questions

**1** Non-metals attract electrons more strongly than metals due to greater core charge.

**2 a** Mg: 2,8,2    **b** F: 2,7    **c** Ne: 2,8
  **d** Al: 2,8,3 (boron is a non-metal)
**3 a** Ar    **b** Ne or Kr    **c** if −2 ion
  **d** if +2 ion    **e** No
**4** It is the attraction of an atom for a shared pair of electrons when covalently bonded to another atom.
**5 a** It increases across a period, due to nuclear charge increasing and size of atom decreasing.
  **b** It decreases down a group.
**6 a** Na, Mg, Al: metals; Si: metalloid; others: non-metals
  **b** Na, Mg, Al, Si, P, S, Cl    **c** Cl, S, P, Si, Al, Mg, Na
**7 a** $Na^+ < Na$    **b** $Cl^- > Cl$    **c** $Cl^- > Na$
**8 a** $Cl^+$, Cl, $Cl^-$
  **b** More electrons increases electrostatic repulsion and hence the size.
**10 a** Both are group 2 metals with two valence electrons.
  **b** Ba is larger than Mg.
**11 a** O    **b** P    **c** B    **d** N
**12 a** 8 electrons in the 2nd shell, so only 8 elements in period 2.
  **b** Iodine is a larger atom.
  **c** Sulfer has greater core charge.
**13** fluorine
**14 a** It is the energy required to remove one mole of electrons from one mole of gaseous atoms.
  **b** $K(g) \rightarrow K^+(g) + e^-$
**15 a** It increases across a period.
  **b** It decreases down a group.
**16 a** D and F    **b** B    **c** I
  **d** C    **e** H    **f** G
**17 a** V    **b** He    **c** Te
  **d** O    **e** S    **f** Al
**19 a** Al > Na    **b** Mg > Cl    **c** Br > F
**20 a** Reactivity increases up the group
  **b** $I^-$, $Br^-$
  **c** $Cl_2(g) + 2I^-(aq) \rightarrow 2Cl^-(aq) + I_2(aq)$
  $Cl_2(g) + 2Br^-(aq) \rightarrow 2Cl^-(aq) + Br_2(aq)$
  **d** Add silver nitrate to the solution.

## Chapter 3 Test

### Part A: Multiple-choice questions

| 1 | B | 3 | B | 5 | B | 7 | A | 9 | D |
|---|---|---|---|---|---|---|---|---|---|
| 2 | A | 4 | C | 6 | C | 8 | C | 10 | D |

## Chapter 4 Quantitative chemistry

### Section 4.1 Exercises

**1 a** 1.00 mol    **b** 2.00 mol    **c** 4.00 mol    **d** 7.00 mol
**2 a** 1.25 mol    **b** 2.50 mol    **c** 6.50 mol    **d** 6.25 mol
**3 a** 1.50 mol    **b** 6.00 mol    **c** 0.25 mol    **d** 21.25 mol
**4 a** $2.11 \times 10^{24}$    **b** $5.66 \times 10^{23}$
  **c** $1.94 \times 10^{21}$    **d** $3.41 \times 10^{19}$
**5 a** 0.0100 mol    **b** 2.39 mol
  **c** $1.48 \times 10^{-6}$ mol    **d** $1.66 \times 10^{-21}$ mol

## Section 4.2 Exercises

1. **a** 63.02   **b** 123.88   **c** 114.14
   **d** 342.34   **e** 92.02
2. **a** 159.70   **b** 154.77   **c** 158.04
   **d** 294.14   **e** 244.24
4. **a** 63.55 g   **b** 30.01 g   **c** 84.32 g
   **d** 48.00 g   **e** 81.37 g   **f** 46.08 g
5. **a** 63.55 g mol$^{-1}$   **b** 30.01 g mol$^{-1}$
   **c** 84.32 g mol$^{-1}$   **d** 48.00 g mol$^{-1}$
   **e** 81.37 g mol$^{-1}$   **f** 46.08 g mol$^{-1}$
6. **a** 2.00 mol   **b** 0.100 mol   **c** 0.20 mol
   **d** 0.0483 mol   **e** 26.9 mol   **f** 939 mol
7. $3.72 \times 10^{-4}$ mol
8. **a** 14.4 g   **b** 8.20 g   **c** 564.1 g
   **d** 89.1 kg   **e** 0.164 g   **f** 0.578 g
9. 28.01 g mol$^{-1}$, CO
10. 165.28 g mol$^{-1}$, cobalt(III) chloride
11. 72.17 g mol$^{-1}$
12. **a** $2.51 \times 10^{24}$   **b** $6.68 \times 10^{25}$
    **c** $4.93 \times 10^{19}$   **d** $5.39 \times 10^{29}$
13. **a** $1.79 \times 10^{-19}$ g   **b** $6.49 \times 10^{-22}$ g
    **c** $5.33 \times 10^{-23}$ g   **d** $2.99 \times 10^{-23}$ g
    **e** $2.99 \times 10^{-22}$ g
14. **a** $8.4 \times 10^{-6}$ mol
    **b** $5.0 \times 10^{18}$ retinol molecules
    **c** $1.0 \times 10^{20}$ C atoms

## Section 4.3 Exercises

1. **a** 50.04%   **b** 40.04%   **c** 40.49%
   **d** 43.38%   **e** 15.77%
2. **a** HO   **b** $C_4H_9$   **c** $K_2S_2O_3$
   **d** $CH_2O$   **e** $Fe_2O_3$   **f** $C_2H_4O$
3. $Sn_2FeC_6N_6$ or $Sn_2Fe(CN)_6$
4. $TiO_2$
5. $C_3H_6O$
6. $Cr_2S_3O_{12}$ or $Cr_2(SO_4)_3$
7. Empirical formula: $C_2H_4O$; molecular formula: $C_4H_8O_2$
8. Empirical formula: $C_5H_7N$; molecular formula: $C_{10}H_{14}N_2$

## Section 4.4 Exercises

1. **a** $2Na(s) + Cl_2(g) \to 2NaCl(s)$
   **b** $4P(s) + 5O_2(g) \to 2P_2O_5(s)$
   **c** $2SO_2(g) + O_2(g) \to 2SO_3(g)$
   **d** $4P(s) + 3O_2(g) \to 2P_2O_3(s)$
   **e** $2Fe(s) + 3Cl_2(g) \to 2FeCl_3(s)$
   **f** $C_2H_4(l) + 2O_2(g) \to 2CO(g) + 2H_2O(g)$
   **g** $2CO(g) + O_2(g) \to 2CO_2(g)$
   **h** $N_2(g) + 3H_2(g) \to 2NH_3(g)$
2. **a** $C_6H_{12}O_6(aq) \to 2C_2H_5OH(aq) + 2CO_2(g)$
   **b** $4NH_3(g) + 5O_2(g) \to 4NO(g) + 6H_2O(l)$
   **c** $2AlI_3(s) + 3Cl_2(g) \to 2AlCl_3(s) + 3I_2(s)$
   **d** $2NaCl(aq) + 2H_2O(l) \to 2NaOH(aq) + H_2(g) + Cl_2(g)$
3. **a** $4Fe(s) + 3O_2(g) \to 2Fe_2O_3(s)$
   **b** $2AgNO_3(aq) + ZnI_2(aq) \to Zn(NO_3)_2(aq) + 2AgI(s)$
   **c** $2Fe_2O_3(s) + 3H_2(g) \to 4Fe(s) + 6H_2O(g)$
   **d** $Mg(NO_3)_2(aq) + Na_2CO_3(aq) \to 2NaNO_3(aq) + MgCO_3(s)$
   **e** $2Fe(OH)_3(s) \to Fe_2O_3(s) + 3H_2O(g)$
4. **a** $2C_8H_{18}(l) + 25O_2(g) \to 16CO_2(g) + 18H_2O(g)$
   **b** $2K(s) + 2H_2O(l) \to 2KOH(aq) + H_2(g)$
   **c** $8HF(aq) + Na_2SiO_3(s) \to H_2SiF_6(aq) + 2NaF(aq) + 3H_2O(l)$
5. 10 mol
6. 0.335 mol
7. $n(N_2) = 250$ mol; $n(H_2) = 750$ mol

## Section 4.5 Exercises

1. $n(HI) = 1.23$ mol; $n(KMnO_4) = 0.246$ mol; $n(H_2SO_4) = 0.368$ mol
2. 12.3 mol
3. 0.202 g
4. 0.101 g
5. $3.66 \times 10^6$ g
6. 16.6 g
7. $3.04 \times 10^5$ g
8. 6.58 g
9. 12.8 g
10. $5.62 \times 10^5$ g
11. $SO_2$, by 0.55 mol
12. **a** $Cl_2$, by 3.15 g   **b** 5.31 g
13. **a** C is in excess   **b** 32 g
14. **a** $AgNO_3(aq) + NaCl(aq) \to AgCl(s) + NaNO_3(aq)$
    **b** 0.844 g
15. **a** $Pb(NO_3)_2(aq) + 2NaCl(aq) \to PbCl_2(s) + 2NaNO_3(aq)$
    **b** 0.42 g
16. **a** 86.7%   **b** 98%   **c** Student 2 (from question 15)
17. **a** 5.31 g   **b** 83.4%
    **c** Not all of the MgO had been collected.

## Section 4.6 Exercises

1. **a** 200.600 kPa   **b** 107 kPa
   **c** $1.20 \times 10^5$ Pa   **d** $3.6 \times 10^3$ kPa
2. **a** 393 K   **b** 310 K   **c** 227°C
   **d** 37°C   **e** 298 K
3. **a** $4.1 \times 10^3$ dm$^3$   **b** 0.200 dm$^3$   **c** 0.050 dm$^3$
   **d** 0.520 m$^3$
4. **a** 105 kPa   **b** 316°C   **c** $5.63 \times 10^3$ cm$^3$
   **d** $6.29 \times 10^{-4}$ m$^3$   **e** 127 kPa   **f** 2.00 atm
   **g** 312 K   **h** $2.24 \times 10^4$ cm$^3$

## Section 4.7 Exercises

1. 200 cm$^3$
2. 6 dm$^3$
3. **a** amount of gas and temperature
   **b** reading 2
4. **a** 3.05 dm$^3$   **b** Ammonia is not an ideal gas.
5. 790 cm$^3$. The balloon will not burst.
6. 140 cm$^3$
8. reading 3

9  37.7 dm$^3$
10  44.0 g mol$^{-1}$
11  29.1 dm$^3$
12  $9.46 \times 10^2$ kPa
13  $-239$°C
14  sample I
15  30 dm$^3$

## Section 4.8 Exercises

1  **a** 0.080 mol dm$^{-3}$ **b** $1.58 \times 10^{-3}$ mol dm$^{-3}$
   **c** 0.458 mol dm$^{-3}$ **d** $4.3 \times 10^{-9}$ mol dm$^{-3}$
2  **a** 36.8 g **b** 4.99 g
   **c** 5.8 g **d** 0.228 g
3  **a** 1.57 mol dm$^{-3}$ **b** 3.31 mol dm$^{-3}$
4  0.531 mol dm$^{-3}$
5  0.993 mol dm$^{-3}$
6  71 cm$^3$
7  3.79 mol dm$^{-3}$
8  217 cm$^3$
9  2.09 mol dm$^{-3}$
10  1.91 g
11  5.67 cm$^3$
12  3.35 dm$^3$
13  6.89 g
14  21.3 cm$^3$
15  7.25 g
16  0.438 mol dm$^{-3}$
17  87.6 cm$^3$
18  NH$_3$ by 1.1 mol
19  0.695 mol dm$^{-3}$
20  5.58 g

## Chapter 4 Review questions

1  **b** $6.02 \times 10^{23}$ **c** Avogadro's constant
2  **a** $1.13 \times 10^{24}$ **b** $7.32 \times 10^{24}$
   **c** $1.51 \times 10^{23}$ **d** $1.79 \times 10^{20}$
3  **a** 1.5 mol **b** 0.100 mol **c** 0.285 mol
   **d** 5.55 mol **e** 124.0 mol **f** 0.0050 mol
4  118.68 g mol$^{-1}$, Sn
5  32.06 g mol$^{-1}$; element X is S
6  **a** $4.2 \times 10^{-22}$ g **b** $1.25 \times 10^{-20}$ g
   **c** $5.32 \times 10^{-23}$ g
7  $1.82 \times 10^{21}$
8  **a** Na = 22.34%; Br = 77.66%
   **b** H = 5.94%; O = 94.06%
   **c** Li = 30.21%; S = 69.79%
   **d** Cu = 39.82%; S = 20.09%; O = 40.09%
9  Empirical formula: CH$_3$; molecular formula: C$_2$H$_6$
10  PbO$_2$
11  C$_{17}$H$_{21}$NO$_4$, yes
12  **a** 2NaCN(s) + H$_2$SO$_4$ → Na$_2$SO$_4$(aq) + 2HCN(g)
    **b** 2C$_6$H$_{14}$(l) + 19O$_2$(g) → 12CO$_2$(g) + 14H$_2$O(l)
    **c** 3NaHCO$_3$(aq) + H$_3$C$_6$H$_5$O$_7$(aq) → 3CO$_2$(g) + 3H$_2$O(l) + Na$_3$C$_6$H$_5$O$_7$(aq)
13  **a** 2H$_2$O$_2$(aq) → 2H$_2$O(l) + O$_2$(g)
    **b** 2ZnS(s) + 3O$_2$(g) → 2ZnO(s) + 2SO$_2$(g)
14  14.75 mol
15  0.233 g
16  1.222 g
17  **a** H$_2$ by 50 mol **b** 300 mol
18  **a** Na$_2$SO$_4$(aq) + Pb(NO$_3$)$_2$(aq) → 2NaNO$_3$(aq) + PbSO$_4$(s)
    **b** Pb(NO$_3$)$_2$ **c** 0.303 g **d** 98.3%
19  0.13 dm$^3$
20  147 K
21  71.0 g mol$^{-1}$ chlorine
22  6.89 g
23  65.3 cm$^3$
24  $1.44 \times 10^4$ g

## Chapter 4 Test

### Part A: Multiple-choice questions

| 1 | C | 3 | D | 5 | A | 7 | C | 9 | A |
| --- | --- | --- | --- | --- | --- | --- | --- | --- | --- |
| 2 | B | 4 | A | 6 | C | 8 | C | 10 | A |

# Chapter 5 Measurement and data processing

## Section 5.1 Exercises

1  Not very accurate, but precise.
2  Results are quite accurate, since % difference = 2.24%
3  Systematic errors cannot be reduced by repetition.
5  **a** 21.43; 21.37; 27.07; 20.02; 22.38; 21.42; 21.38
   **b** 27.07, 20.02 & 22.38 should be ignored
   **c** 21.40 ± 0.2 cm$^3$
6  **a** 4 **b** 4 **c** 5 **d** 2
   **e** 5 **f** 1 **g** 4 **h** 3
7  **a** 0.96 **b** 34.7 **c** 22.22
   **d** 0.003 06 **e** 0.0031 **f** 1.00
   **g** 15.28 **h** 0.000 585 **i** 2.44
   **j** 0.83
8  **a** $5 \times 10^8$ **b** $4.8 \times 10^8$
   **c** $4.78 \times 10^8$ **d** $4.780 \times 10^8$
9  **a** 485.7 **b** 33.79
   **c** $1.7 \times 10^2$ **d** 0.18

## Section 5.2 Exercises

1  **a** 25.82 to 25.86 cm$^3$ **b** 14.1 to 14.7 cm$^3$
   **c** 249.5 to 250.5 cm$^3$
2  **a** 0.4% **b** 0.2% **c** 6%
   **d** 2% **e** 0.2%
3  **a** 25.0 ± 0.5°C **b** 10.00 ± 0.5 cm$^3$
   **c** 2.507 ± 0.025 g **d** 25.50 ± 1.28 cm$^3$
4  **a** 23.0; ±1.0 **b** 5; ±2 **c** 25.00; ±0.1
5  **a** 40; ±1.5; 4% **b** 14.02; ±0.04; 0.3%
   **c** 175; ±5.5; 3%

6 Student B's measurement has less uncertainty than student A's.
7 $2.00 \text{ mol dm}^{-3} \pm 1.7\%$
8 $0.227 \pm 0.002 \text{ mol dm}^{-3}$

## Section 5.3 Exercises
1 Volume is inversely proportional to pressure.
3   **a** C     **b** A     **c** B

## Chapter 5 Review questions
1 Yes
2 The percentage difference is 5%, ∴ not very accurate.
4 Repeat the experiment many times and average concordant results.
5   **a** 0.266     **b** 15.13     **c** 1348
6   **a** 0.6%     **b** 14%     **c** 2%
7   **a** ±0.1     **b** ±2     **c** ±0.01
8   **a** $3.694 \pm 0.004$ g     **b** $21 \pm 1°C$
  **c** $115.27 \pm 0.06$ g
9   **a** 0.1%     **b** 5%     **c** 0.05%
10   **a** $0.00889 \text{ mol} \pm 0.6\%$     **b** $0.0499 \pm 8\%$
  **c** $0.0400 \text{ mol dm}^{-3} \pm 0.9\%$
11   **a** $\pm 5 \times 10^{-5}$ mol     **b** $\pm 4 \times 10^{-3}$ mol
  **c** $\pm 4 \times 10^{-4} \text{ mol dm}^{-3}$
12 Volume is proportional to temperature at constant pressure.
14   **a** $Ba(OH)_2(aq) + H_2SO_4(aq) \rightarrow BaSO_4(s) + 2H_2O(l)$
  **b** It decreased then increased.
  **c** Ions are being removed from solution.
16   **a** $3.6 \text{ g min}^{-1}$; $1.1 \text{ g min}^{-1}$     **b** greater value at $t = 0$

## Chapter 5 Test
### Part A: Multiple-choice questions

| 1 | B | 3 | C | 5 | B | 7 | C | 9 | D |
|---|---|---|---|---|---|---|---|---|---|
| 2 | C | 4 | B | 6 | D | 8 | D | 10 | A |

# Chapter 6 Energetics

## Section 6.1 Exercises
1 **a, b, f** endothermic     **c, d, e** exothermic
2 reaction e
4   **a** products > reactants     **b** endothermic
  **c** ionic bonds
5   **a** $1100 \text{ kJ mol}^{-1}$     **b** $-900 \text{ kJ mol}^{-1}$.
6 endothermic
7   **a** gases     **b** 101.3 kPa (1 atm)
8   **b** exothermic     **c** $-170 \text{ kJ mol}^{-1}$
  **d** $290 \text{ kJ mol}^{-1}$     **e** $290 \text{ kJ mol}^{-1}$
9   **a** enthalpy of NaOH(s) > NaOH(aq)
  **b** exothermic     **c** ionic bonds, H-bonds
10 $E_a = +150 \text{ kJ mol}^{-1}$; $\Delta H = -100 \text{ kJ mol}^{-1}$

## Section 6.2 Exercises
1 25.9°C
2 236 MJ
3 2.65 g
4 17.6°C
5 $5.03 \times 10^3$ kJ
6 $2.91 \times 10^4$ kJ
7 10.2 g
8 $\Delta H = -2.04 \times 10^3 \text{ kJ mol}^{-1}$
9   **a**   **i** $725 \text{ kJ mol}^{-1}$
      **ii** $22.6 \text{ kJ g}^{-1}$
      **iii** $17.8 \text{ MJ dm}^{-3}$
  **b** $9.49 \times 10^4 \text{ dm}^3$

## Section 6.3 Exercises
1 $-302 \text{ kJ mol}^{-1}$
2 $-296.1 \text{ kJ mol}^{-1}$
3 $-58.0 \text{ kJ mol}^{-1}$
4 $-1114.8 \text{ kJ mol}^{-1}$
5 $+226 \text{ kJ mol}^{-1}$
6 $+130 \text{ kJ mol}^{-1}$
7 $-426.5 \text{ kJ mol}^{-1}$

## Section 6.4 Exercises
3 $-184 \text{ kJ mol}^{-1}$
4 $-535 \text{ kJ mol}^{-1}$
5   **a** $-1194 \text{ kJ mol}^{-1}$     **b** $-1076 \text{ kJ mol}^{-1}$
  **c** C=C has a higher bond enthalpy than C–C.
6 ethane
7   **a** $-1145 \text{ kJ mol}^{-1}$     **b** $-1875 \text{ kJ mol}^{-1}$

## Chapter 6 Review questions
2 **a**, **c** and **e** are endothermic; **b** and **d** are exothermic
3 $+2.0 \text{ kJ mol}^{-1}$
4   **a** $-544 \text{ kJ mol}^{-1}$     **b** $+1648 \text{ kJ mol}^{-1}$
  **c** $-302 \text{ kJ mol}^{-1}$
5   **b** exothermic     **c** 250 kJ
  **d** $330 \text{ kJ mol}^{-1}$
7 6.13 kJ
8 $2.44 \times 10^3$ kJ
9   **a** 6.19 kJ     **b** 2.87 kJ     **c** $4.37 \times 10^8$ J
10 53.2°C
11 328 kJ
12 +41.0 kJ
13 12.8 kJ
14 $-134 \text{ kJ mol}^{-1}$
15 $-8316 \text{ kJ mol}^{-1}$
16 $-6356$ kJ
17 $2677 \text{ kJ mol}^{-1}$
18 13.9 g
19 $-124$ kJ
20   **a** 1.56 mol     **b** 1115 kJ     **c** $3.14 \text{ dm}^3$ (L)
  **d** Less due to heat lost to the surroundings.

## Chapter 6 Test

### Part A: Multiple-choice questions

| 1 | C | 3 | B | 5 | A | 7 | A | 9  | C |
|---|---|---|---|---|---|---|---|----|---|
| 2 | C | 4 | D | 6 | D | 8 | D | 10 | D |

# Chapter 7 Kinetics

## Section 7.1 Exercises

1. **b**
   - **i** $2.05 \times 10^{-3}$ mol dm$^{-3}$ s$^{-1}$
   - **ii** $1.38 \times 10^{-3}$ mol dm$^{-3}$ s$^{-1}$
   - **iii** $2.25 \times 10^{-4}$ mol dm$^{-3}$ s$^{-1}$
   
   **c** $t = 1600$ s to $t = 2000$ s

2. **b** NO$_2$ is decomposing,
   **c** The rate of increase of [O$_2$] = ½ rate of increase of [NO].

3. **a** 1.0 cm$^3$ min$^{-1}$    **b** 0.60 cm$^3$ min$^{-1}$
   **c** 0 cm$^3$ min$^{-1}$

## Section 7.2 Exercises

2. **a** increase    **b** increase
   **c** stay the same    **d** increase

3. **b**
   - **i** stays the same
   - **ii** increases
   - **iii** increases

4. They collide with energy $\geq E_a$ and have correct orientation.

7. **a** Concentration of N$_2$O$_5$ was greatest at start of experiment. Rate decreases after that.
   **b** It would take longer for N$_2$O$_5$ to decompose.
   **c** Use a higher temperature or catalyst.

## Chapter 7 Review questions

2. Particles have a range of speeds, and hence energies, at any given temperature.
3. Particles have energy $\geq E_a$ and collision must have correct orientation.
5. Rate is decreased if the temperature is low.
6. The energy required to break the bonds of the reactants, hence endothermic.
7. Use powdered Zn, increase temperature, increase concentration of HCl
8. **a** Reactant bonds must be broken.
   **b** $E_a > 280$ kJ mol$^{-1}$
   **c** It provides an alternative reaction pathway.
10. **a** **i** increase    **ii** increase
    **b** **i** no change   **ii** decrease
    **c** **i** increase    **ii** no change
    **d** **i** decrease    **ii** no change

## Chapter 7 Test

### Part A: Multiple-choice questions

| 1 | B | 3 | A | 5 | C | 7 | B | 9  | D |
|---|---|---|---|---|---|---|---|----|---|
| 2 | D | 4 | D | 6 | C | 8 | C | 10 | A |

# Chapter 8 Equilibrium

## Section 8.1 Exercises

2. Open system is open to the surroundings, whereas a closed system is sealed off.
3. Constant colour intensity, gas pressure, temperature, electrical conductivity, or pH
4. The rate of the forward reaction is exactly equal to the rate of the reverse reaction.

## Section 8.2 Exercises

1. **a** $K_c = \dfrac{[CuO]^2[NO_2]^4[O_2]}{[Cu(NO_3)_2]^2}$    **b** $K_c = \dfrac{[CH_3OH]}{[CO][H_2]^2}$

   **c** $K_c = \dfrac{[HI]^2}{[H_2][I_2]}$    **d** $K_c = \dfrac{[I_2][I^-]}{[I_3^-]}$

2. **a** $K_c^2$    **b** $\dfrac{1}{\sqrt{K_c}}$

3. **b** to the left    **c** decrease
4. **a** **i** moves to right    **ii** increases
   **b** **i** moves to left     **ii** decreases
   **c** **i** moves to right    **ii** increases
   **d** **i** moves to left     **ii** decreases
6. **a** right    **b** left    **c** right    **d** left
7. **a** right    **b** right   **c** right
8. **a** right    **b** right   **c** right    **d** right
9. **a** Addition of excess oxygen will ensure greater yield as excess reactant used.
   **b** Using a catalyst will improve only rate of reaction, not yield.
   **c** Supplying extra heat will not increase the yield, as the reaction is exothermic.

## Section 8.3 Exercises

1. porous iron
2. Increase concentration of reactant, remove product, increase gas pressure or decrease temperature.
3. **a** There is no effect.
   **b** Increases the rate of the reaction.
4. vanadium(V) oxide, V$_2$O$_5$, in pellet form
5. see question 1

## Chapter 8 Review questions

1. **a** $K_c = \dfrac{[SO_3]^2}{[SO_2]^2[O_2]}$ mol$^{-1}$ dm$^3$

   **b** $K_c = \dfrac{[PH_3]^4}{[P_4][H_2]^6}$ mol$^{-3}$ dm$^9$

   **c** $K_c = \dfrac{[Li_2CO_3][H_2O]}{[LiOH]^2[CO_2]}$ mol$^{-1}$ dm$^3$

   **d** $K_c = \dfrac{[NO_2]^4[H_2O]^6}{[NH_3]^4[O_2]^7}$ mol$^{-1}$ dm$^3$

2. **a** reactants    **b** products    **c** reactants    **d** products

3   a   $K_c = \dfrac{[SO_3]^2}{[SO_2]^2[O_2]}$ mol$^{-1}$ dm$^3$   b   experiment 2

4   a   $K' = \dfrac{1}{x}$   b   $K' = x^2$   c   $K' = \dfrac{1}{x^3}$

5   $K_c = \dfrac{[NH_3]^2}{[N_2][H_2]^3}$ mol$^{-2}$ dm$^6$; smaller equilibrium constant

6   a   i   right          ii   increases
    b   i   left           ii   decreases
    c   i   left           ii   decreases
    d   i   right          ii   increases

7   a   right   b   the right   c   left   d   left
9   a   left    b   left        c   no effect
    d   right   e   no effect
11  a   right   b   left        c   left   d   left
12  a   four times
    b   COCl$_2$ was added.
    c   Reaction was proceeding to the left.
    d   $K_c = \dfrac{[COCl_2]}{[CO][Cl_2]}$

13  a   low temperatures and high pressure
    b   catalyst used
14  High pressure is very expensive; high temperatures favour faster reactions.
15  $N_2 + 3H_2 \rightleftharpoons 2NH_3$ Ammonia; $K_c = \dfrac{[NH_3]^2}{[N_2][H_2]^3}$ mol$^{-2}$ dm$^6$

## Chapter 8 Test

### Part A: Multiple-choice questions

| 1 | B | 3 | D | 5 | C | 7 | B | 9 | B |
|---|---|---|---|---|---|---|---|---|---|
| 2 | A | 4 | D | 6 | A | 8 | C | 10 | C |

# Chapter 9 Acids and bases

## Section 9.1 Exercises

1   a   HClO$_4$(aq) + H$_2$O(l) → ClO$_4^-$(aq) + H$_3$O$^+$(aq)
    b   CH$_3$NH$_2$(aq) + H$_2$O(l) → CH$_3$NH$_3^+$(aq) + OH$^-$(aq)
2   a   HBr, HSO$_4^-$, NH$_4^+$, HClO$_4$, H$_2$CO$_3$
    b   I$^-$, PO$_4^{3-}$, NH$_3$, SO$_4^{2-}$, H$_2$O
3   a   HSO$_4^-$(aq) + H$_2$O(l) → H$_2$SO$_4$(aq) + OH$^-$(aq)
    b   HSO$_4^-$(aq) + H$_2$O(l) → SO$_4^{2-}$(aq) + H$_3$O$^+$(aq)
4   Acid: HCO$_3^-$(aq) + OH$^-$(aq) → CO$_3^{2-}$(aq) + H$_2$O(l)
    Base: HCO$_3^-$(aq) + HCl(aq) → H$_2$CO$_3$(aq) + Cl$^-$(aq)
5   a   HCO$_3^-$/CO$_3^{2-}$ and HCl/Cl$^-$
    b   HSO$_4^-$/SO$_4^{2-}$ and NH$_4^+$/NH$_3$
    c   HCl/Cl$^-$ and H$_3$O$^+$/H$_2$O
6   a   base    b   base    c   neither
    d   base    e   acid
7   HSO$_4^-$, SO$_4^{2-}$ and H$_3$O$^+$ ions allow conduction of electricity.

8   Acids: HF, NH$_4^+$, H$_3$O$^+$; bases: Cl$^-$, CO$_3^{2-}$, NH$_3$, SO$_4^{2-}$; both acid and base: HSO$_4^-$, HPO$_4^{2-}$

## Section 9.2 Exercises

1   An alkali is a base that dissolves in water.
3   a   2Al(s) + 6HF(aq) → 2AlF$_3$(aq) + 3H$_2$(g)
    b   2KHCO$_3$(s) + H$_2$SO$_4$(aq) → K$_2$SO$_4$(aq) + 2H$_2$O(l) + 2CO$_2$(g)
    c   Fe$_2$O$_3$(s) + 6HNO$_3$(aq) → 2Fe(NO$_3$)$_3$(aq) + 3H$_2$O(l)
    d   Ca(OH)$_2$(aq) + 2HCl(aq) → CaCl$_2$(aq) + 2H$_2$O(l)
    e   NH$_3$(aq) + HNO$_3$(aq) → NH$_4$NO$_3$(s)

## Section 9.3 Exercises

1   [H$_3$O$^+$] in concentrated solution (e.g. 6 mol dm$^{-3}$) > [H$_3$O$^+$] in dilute solution (e.g. 0.01 mol dm$^{-3}$)
2   An acid such as HCl is a strong acid, whereas an acid such as CH$_3$COOH is a weak acid.
3   A strong acid dissociates in water to a greater extent than does a weak acid.
4   a   HCl has higher [H$_3$O$^+$] and is a better conductor.
    b   H$_2$SO$_4$ has higher [H$_3$O$^+$] and is a better conductor.
5   a   NaOH has better conductivity and higher [OH$^-$].
    b   Na$_2$CO$_3$ has higher conductivity, but KOH has higher [OH$^-$].
6   a   HNO$_3$(aq) + H$_2$O(l) → H$_3$O$^+$(aq) + NO$_3^-$(aq)
    b   HCN(g) + H$_2$O(l) → H$_3$O$^+$(aq) + CN$^-$(aq)
    c   KOH(s) + aq → K$^+$(aq) + OH$^-$(aq)
    d   KF(s) + aq → K$^+$(aq) + F$^-$(aq)
        F$^-$(aq) + H$_2$O(l) → HF(aq) + OH$^-$(aq)
7   a   good    b   good    c   poor    d   poor
8   It can act as either an acid or a base, i.e. it can donate or accept an H$^+$ ion.
9   Measure electrical conductivity; higher conductivity indicates stronger base.

## Section 9.4 Exercises

1   Neutral = 7; acidic pH < 7.
2   a   tomato juice    b   milk    c   vinegar
3   a   seawater        b   black coffee
    c   household ammonia
4   pH = 5
5   pH = 10
6   100 times more acidic
7   10$^{-2}$ mol dm$^{-3}$
8   pH = 5
9   NaOH(s) $\xrightarrow{H_2O}$ Na$^+$(aq) + OH$^-$(aq)
    NH$_3$(g) + H$_2$O(l) ⇌ NH$_4^+$(aq) + OH$^-$(aq)

## Chapter 9 Review questions

1   A Brønsted–Lowry acid is a H$^+$ donor (e.g. HCl), a Lewis acid is an electron pair acceptor (e.g. BF$_3$).
2   a   Brønsted–Lowry acid and base; Lewis base
    b   Brønsted–Lowry base; Lewis base
    c   Brønsted–Lowry acid
    d   Brønsted–Lowry base; Lewis base
    e   Lewis acid
    f   Brønsted–Lowry base; Lewis base

3   a   $ClO_3^-$, $S^{2-}$, $NH_3$, $OH^-$
    b   $H_2CO_3$, $H_2S$, $HClO_4$, $H_3PO_4$
4   a   $H_2C_2O_4/HC_2O_4^-$ & $H_3O^+/H_2O$
    b   $H_2O/OH^-$ and $HCN/CN^-$
    c   $CH_3COOH/CH_3COO^-$ and $HS^-/S^{2-}$
    d   $HSO_4^-/SO_4^{2-}$ and $HF/F^-$
5   a   A           b   B           c   B
    d   AB          e   A           f   AB
    g   A           h   A           i   N
8   a   $Zn(s) + 2HCl(aq) \rightarrow ZnCl_2(aq) + H_2(g)$
    b   There is no proton transfer.
9   a   i   Does not dissociate readily or completely in water.
        ii  Able to donate or accept $H^+$ ions.
11  Measure the conductivity of the two solutions.
12  a   $H_2SO_4$, $HSO_4^-$, $SO_4^{2-}$, $H_2O$, $H_3O^+$
    b   $H_2O$                      c   $SO_4^{2-}$
13  NaOH, $NH_3$, KCl, $H_2CO_3$, $HSO_4^-$, HCl
14  Different concentration of $H_3O^+$ ions results in different pH values.
15  a   $Ba(OH)_2$   b   $NH_3$     c   HCl, NaOH
16  a   tomato juice             b   milk
17  a   seawater                 b   household ammonia
18  pH = 4
19  a   pH = 5                   b   acidic
20  1000 times

# Chapter 9 Test

## Part A: Multiple-choice questions

| 1 | A | 3 | D | 5 | C | 7 | A | 9 | A |
| 2 | B | 4 | D | 6 | C | 8 | B | 10 | D |

# Chapter 10 Oxidation and reduction

## Section 10.1 Exercises

1   a   redox                b   Not redox
    c   redox                d   redox
2   a   Mg: +2; O: −2        b   H: +1; O: −2
    c   H: +1; P: +6; O: −2  d   C: −2; H: +1
    e   N: +4; O: −2         f   K: +1; Mn: +7; O: −2
3   a   +2 copper(II) oxide  b   +1 copper(I) oxide
    c   +4 manganese(IV) oxide
    d   +5 vanadium(V) oxide
4   nitrogen(I) oxide; nitrogen(II) oxide; nitrogen(IV) oxide; nitrogen(III) oxide; nitrogen(V) oxide
6   a   R     b   O     c   N     d   B
7   In $C_6H_{12}O_6$, ON(C) = 0; in $CO_2$, ON(C) = +4, ∴ carbon is oxidised. ON(O) = 0 in $O_2$ and = −2 in $CO_2$, ∴ oxygen is reduced.
8   a   $H_2$     b   $I^-$     c   S in $SO_2$

## Section 10.2 Exercises

1   a   Oxidation is the loss of electrons. Reduction is the gain of electrons.
    b   An oxidizing agent accepts electrons. A reducing agent donates electrons.
2   Oxidizing agent listed first:
    a   $Fe^{2+}$; Cu    b   $Ag^+$; $Sn^{2+}$    c   H in $H_2O$; Ca
    d   $O_2$; Mg        e   $Cu^{2+}$; Fe        f   $F_2$; Ni
    g   $Fe^{3+}$; $I^-$
4   Oxidizing agent listed first:
    a   $Cu^{2+}$; $H_2$   b   $Cr_2O_7^{2-}$; $I^-$   c   $MnO_2$; $SO_2$
5   Oxidizing agent listed first:
    a   $O_2$; $H_2$                 b   not redox
    c   not redox                    d   $KMnO_4$; $H_2S$
7   a   $2S_2O_3^{2-}(aq) + I_2(aq) \rightarrow S_4O_6^{2-}(aq) + 2I^-(aq)$
    b   $Zn(s) + 2NO_3^-(aq) + 4H^+(aq) \rightarrow Zn^{2+}(aq) + 2NO_2(g) + 2H_2O(l)$
    c   $3ClO^-(aq) \rightarrow 2Cl^-(aq) + ClO_3^-(aq)$
    d   $2Cr_2O_7^{2-}(aq) + 3CH_3CH_2OH(aq) + 16H^+(aq)$
         $\rightarrow 4Cr^{3+}(aq) + 3CH_3COOH(aq) + 11H_2O(l)$

## Section 10.3 Exercises

2   high solubility, non-reactive
4   Salt bridge may dry out or $Pb^{2+}$ ions may all be reacted.
5   $A^{2+}(aq) + B(s) \rightarrow A(s) + B^{2+}(aq)$
7   It is used to complete the circuit by allowing the flow of ions.
10  a   cathode              b   right to left
    c   right to left        d   negative
    e   Pt or C              f   $Fe^{3+}(aq) + e^- \rightarrow Fe^{2+}(aq)$
    g   $2I^-(aq) \rightarrow I_2(aq) + 2e^-$
    h   $2Fe^{3+}(aq) + 2I^-(aq) \rightarrow I_2(aq) + 2Fe^{2+}(aq)$

## Section 10.4 Exercises

1   a   They donate electrons easily.
    b   They gain electrons readily.
2   Reactivity: Z > X > W > Y
3   a   $Fe(s) + Ni^{2+}(aq) \rightarrow Fe^{2+}(aq) + Ni(s)$
    b   No reaction
    c   $Mg(s) + Sn^{2+}(aq) \rightarrow Mg^{2+}(aq) + Sn(s)$
4   a   no reaction
    b   $Mg(s) + Zn^{2+}(aq) \rightarrow Mg^{2+}(aq) + Zn(s)$
    c   no reaction
    d   $2Na(s) + 2H_2O(l) \rightarrow 2Na^+(aq) + H_2(g) + 2OH^-(aq)$
5   a   $Ag^+$    b   $Na^+$    c   Na    d   Ag
7   It could be Co, Cd, Fe or Cr.
8   a   $Sn(s) \rightarrow Sn^{2+}(aq) + 2e^-$; $Ag^+(aq) + e^- \rightarrow Ag(s)$
    b   anode: Sn; cathode: Ag
    c   negative: Sn; positive: Ag
    d   $Sn(s) + Ag^+(aq) \rightarrow Sn^{2+}(aq) + Ag(s)$
    e   oxidizing agent: $Ag^+$; reducing agent: Sn
    f   from Sn to Ag
    g   Positive ions flow into the Ag half-cell; negative ions flow into the Sn half-cell.
11  $Cl_2(g) + 2NaBr(aq) \rightarrow Br_2(aq) + 2NaCl(aq)$
    $F_2(g) + 2NaBr(aq) \rightarrow Br_2(aq) + 2NaF(aq)$

## Section 10.5 Exercises

1. It drives the non-spontaneous endothermic redox reactions in the forward direction.
2. The negative electrode is the electron source.
3. **a** $2Cl^-(l) \rightarrow Cl_2(g) + 2e^-$; positive
   **b** $Cl_2(g)$, Li
   **c** $Li^+(l) + 2Cl^-(l) \rightarrow Li(l) + Cl_2(g)$
5. **b** $2F^-(l) \rightarrow F_2(g) + 2e^-$; $Al^{3+}(l) + 3e^- \rightarrow Al(l)$
   **e** $2Al_3^+(l) + 6F^-(l) \rightarrow 2Al(l) + 3F_2(g)$
6. Ions move through the electrolyte (cations to cathode, anions to anode). Electrons move through the wire.

## Chapter 10 Review questions

2. gain of oxygen, loss of electrons, loss of hydrogen
3. **a** N +4, O –2  **b** K +1, Cr +6, O –2
   **c** H +1, S +6, O –2  **d** H +1, N +3, O –2
   **e** H +1, P +5, O –2  **f** S +2, O –2
   **g** Ga +3, C +4, O –2  **h** Zn +2, Br +5, O –2
4. **a** $MnO_4^-$ +7, $MnO_2$ +4, $Mn^{2+}$ +2
   **b** $MnO_4^- + 4H^+ + 3e^- \rightarrow MnO_2 + 2H_2O$
   $Mn^{2+} + 2H_2O \rightarrow MnO_2 + 4H^+ + 2e^-$
5. $ClO^- + 2H^+ + 2e^- \rightarrow Cl^- + H_2O$
   $ClO^- + 3H_2O \rightarrow ClO_3^- + 6H^+ + 6e^-$
8. **a** $2Al(s) + 6H^+(aq) \rightarrow 2Al^{3+}(aq) + 3H_2(g)$
   **b** $3Cu(s) + 2NO_3^-(aq) + 8H^+(aq)$
   $\rightarrow 3Cu^{2+}(aq) + 2NO(g) + 4H_2O(l)$
   **c** $5SO_2(g) + 2H_2O(l) + 2MnO_4^-(aq)$
   $\rightarrow 5SO_4^{2-}(aq) + 4H^+(aq) + 2Mn^{2+}(aq)$
   **d** $ClO^-(aq) + 2I^-(aq) + 2H^+(aq) \rightarrow Cl^-(aq) + I_2(aq) + H_2O(l)$
9. **a** $Ni(s) \rightarrow Ni^{2+}(aq) + 2e^-$
   **b** $2H^+(aq) + 2e^- \rightarrow H_2(g)$
   **c** $Ni(s) + 2H^+(aq) \rightarrow Ni^{2+}(aq) + H_2(g)$
   **d** $H^+(aq)$
10. Oxidizing agent listed first:
    **a** $Pb^{2+}$; Zn  **b** $O_2$; Mg  **c** CuO; $H_2$  **d** $Cl_2$; $I^-$
11. **a** $Br_2$; Mg  **b** $Ag^+$; $Sn^{2+}$  **c** $Cu^{2+}$; Pb
    **d** $MnO_4^-$; $Fe^{2+}$  **e** $H_2O$; Na
12. $Zn(s) + Sn^{2+}(aq) \rightarrow Zn^{2+}(aq) + Sn(s)$
13. **a** reduced  **b** reduced  **c** neither oxidised nor reduced
14. **a** $Cu^{2+}(aq) + Cd(s) \rightarrow Cu(s) + Cd^{2+}(aq)$
    **b** no reaction
17. **a i** Zn  **ii** Zn  **iii** Zn
    **b** e.g. $KNO_3(aq)$
18. $2Ag^+(aq) + Cu(s) \rightarrow 2Ag(s) + Cu^{2+}(aq)$
19. **a** $2H_2O_2(aq) \rightarrow 2H_2O(l) + O_2(g)$
    **b** to slow the rate of reaction
    **c** a catalyst
21. zinc
22. B > A > C
23. $Sn^{4+}(aq)$; $Sn(NO_3)_4$.
25. **b** calcium and chlorine
    **c** $2Cl^-(l) \rightarrow Cl_2(g) + 2e^-$; $Ca^{2+}(l) + 2e^- \rightarrow Ca(l)$
    **d** $Ca^{2+}(l) + 2Cl^-(l) \rightarrow Ca(l) + Cl_2(g)$

## Chapter 10 Test

### Part A: Multiple-choice questions

| 1 | C | 3 | C | 5 | B | 7 | C | 9 | C |
|---|---|---|---|---|---|---|---|---|---|
| 2 | D | 4 | D | 6 | A | 8 | C | 10 | B |

# Chapter 11 Organic chemistry

## Section 11.1 Exercises

1. They have same general formula, and successive members differ by $CH_2$.
2. **a** $C_4H_{10}$  **b** $C_5H_{10}$
   **c** $C_6H_{14}$  **d** $C_4H_8$
3. **a** alkene  **b** alkene
   **c** alkane  **d** alkane
5. **a** 2-methylpentane  **b** but-1-ene
   **c** but-2-ene  **d** pent-2-ene
6. **a** 2-methylbutane  **b** pent-2-ene
   **c** 3-methylpentane  **d** propene
9. Empirical formula: $CH_2$; molecular formula: $C_4H_8$
10. Empirical formula: $C_3H_8$; molecular formula: $C_3H_8$

## Section 11.2 Exercises

1. **a** alcohol  **b** ester
   **c** chloroalkane  **d** carboxylic acid
   **e** amine
3. **a** propanal  **b** butanone
   **c** 1,2-dichloroethane  **d** hex-2-ene
   **e** methanoic acid  **f** pentan-3-ol
4. **a** 2-methylbutane  **b** 2,3-dichlorobutane
   **c** butanone  **d** pentanal
   **e** propanoic acid
6. pentan-1-ol; pentan-2-ol; pentan-3-ol
8. Condensed structural formulas: $CH_3CH_2CH_2OH$, $CH_3CH(OH)CH_3$, $CH_3OCH_2CH_3$
9. **a** Semi-structural formulas: $CH_3CH_2OH$ and $CH_3(CH_2)_4CH_2OH$
11. **a** tertiary  **b** secondary
    **c** primary  **d** tertiary
12. **a** secondary  **b** primary
    **c** tertiary  **d** secondary
13. **a** amine
    **b** ester
    **c** 1: amine; 2: carboxyl group
    **d** 1: amine; 2: carboxyl group; 3: benzene ring
14. **a** butanoic acid  **b** hexan-2-one
    **c** 2-methylbutan-2-ol  **d** butanal

## Section 11.3

1. **a** $C_5H_{12} + 8O_2 \rightarrow 5CO_2 + 6H_2O$
   **b** $2C_4H_{10} + 13O_2 \rightarrow 8CO_2 + 10H_2O$
   **c** $2C_6H_{14} + 13O_2 \rightarrow 12CO + 14H_2O$
   **d** $2C_6H_{14} + 7O_2 \rightarrow 12C + 14H_2O$

3  a  substitution
   b  free-radical mechanism
   c  homolytic fission
4  a  $C_2H_6 + Br_2 \rightarrow C_2H_5Br + HBr$
5  a  $C_3H_7Br$
   b  $CH_3CH_2CH_2Br$ and $CH_3CH(Br)CH_3$
   c  1-bromopropane and 2-bromopropane

## Section 11.4 Exercises

1  a  i   $CH_3CHCHCH_3 \xrightarrow{H_2/catalyst} CH_3CH_2CH_2CH_3$
      ii  $CH_2CH_2 \xrightarrow{Br_2} CH_2BrCH_2Br$
      iii $CH_3CH_2OH(l) + 3O_2(g) \rightarrow 2CO_2(g) + 3H_2O(l)$
      iv  $CH_3CH_3 \xrightarrow{Cl_2/UV} CH_3CH_2Cl + HCl$
   b  UV light
   c  decolourization of the bromine solution
2  a  $CH_3CH_3 \xrightarrow{Cl_2/UV} CH_3CH_2Cl + HCl$
   b  $CH_2CH_2 \xrightarrow{HCl/catalyst} CH_3CH_2Cl$
4  a  addition                b  $C_2H_4$
5  a  $CH_3CHBr_2$ or $CH_2BrCH_2Br$ (by substitution)
   b  $CH_2BrCH_2Br$ only (by addition)
6  a  $CH_2CH_2 + H_2O \rightarrow CH_3CH_2OH$
   c  $CH_2CH_2 \xrightarrow{HCl/catalyst} CH_3CH_2Cl$
8  a  $nCH_2CH_2 \rightarrow (CH_2CH_2)_n$
   b  $nCH_2CHCH_3 \rightarrow (CH_2CHCH_3)_n$
   c  $nCH_2CHCl \rightarrow (CH_2CHCl)_n$

## Section 11.5 Exercises

1  a  $CH_3COOH$, ethanoic acid    b  $CH_3CH_2OH$ ethanol
   c  $CH_3CH_2COCH_3$ butanone    d  no product
2  a  an aldehyde

## Section 11.6 Exercises

2  nucleophilic substitution reactions
3  The iodoalkane would react more rapidly than the chloroalkane.
4  tertiary halogenoalkanes
5  $S_N2$ mechanism
6  $S_N1$ mechanism

## Section 11.7 Exercises

1  a  addition reaction          b  addition polymerisation
   c  addition reaction          d  oxidation

## Chapter 11 Review questions

2  a  Five isomers: hexane, 2-methylpentane, 3-methylpentane, 2,2-dimethylbutane, 2,3-dimethylbutane
   b  Three isomers: but-1-ene, but-2-ene, 2-methylprop-1-ene
3  a  There are too many hydrogens on middle carbons.
   b  There should not be a hydrogen on the final carbon.
5  a  $C_2H_5OH$           b  $C_6H_{12}$
   c  $C_6H_{12}$          d  $CH_3CH_2COOCH_3$
6  a  chloro group, carboxyl group, benzene ring
   b  carbonyl group (C=O), bromo group, hydroxyl group
   c  ester group (COO), chloro group, C=C double bond
8  a  2-methylpentane
   b  2,2-dimethylbutane
   c  2,3-dimethylbutane
10 A: ethanol; B: ethanoic acid; C: bromoethane; D: 1,2-dichloroethane
11 a  Empirical and molecular formulas are both $C_3H_8O$.
   b  $CH_3CH_2CH_2OH$ and $CH_3CH(OH)CH_3$
   c  propan-1-ol and propan-2-ol
14 $nCH_2CHCH_3 \rightarrow (CH_2CHCH_3)_n$
15 butanal < butan-1-ol < butanoic acid
16 b  slower
17 $S_N1$: one species forms the reactive intermediate; $S_N2$: two species form intermediate
18 a  $C_3H_8 + 5O_2 \rightarrow 3CO_2 + 4H_2O$
   b  $C_4H_8 + 4O_2 \rightarrow 4CO + 4H_2O$ or $C_4H_8 + 2O_2 \rightarrow 4C + 4H_2O$
   c  $2C_3H_7OH + 9O_2 \rightarrow 6CO_2 + 8H_2O$
20 a  A: butane; B: butan-1-ol; C: butanal
   b  i   free-radical substitution
      ii  nucleophilic substitution
      iii oxidation
   c  $S_N2$
   d  Reacted at room temperature, then aldehyde is distilled off

## Chapter 11 Test

### Part A: Multiple-choice questions

| 1 | C | 3 | D | 5 | D | 7 | B | 9  | A |
|---|---|---|---|---|---|---|---|----|---|
| 2 | C | 4 | D | 6 | A | 8 | D | 10 | A |

# GLOSSARY

**Absolute uncertainty**
The size of an uncertainty, including its units.

**Accuracy**
An expression of how close the measured value is to the 'correct' or 'true' value.

**Acid rain**
Rain with pH < 5.

**Acidic solution**
A solution with pH < 7 at 25°C.

**Activation energy, $E_a$**
The energy required to break the bonds of the reactants, initiating a chemical reaction and hence allowing it to progress.

**Addition polymerization reaction**
A reaction in which many monomers with a carbon–carbon double bond react together in addition reactions to form a polymer chain.

**Addition reaction**
A reaction in which a carbon–carbon double bond is replaced by a single bond and a new molecule is added into the original double-bonded molecule.

**Alcohol**
A homologous series with the general formula $C_nH_{2n+1}OH$.

**Aldehyde**
A homologous series with the general formula $C_nH_{2n}O$, in which the functional group CHO is found at the end of the chain.

**Alicyclic**
Hydrocarbon compounds in which the carbon atoms are arranged in a ring.

**Aliphatic**
Straight-chain and branched hydrocarbon compounds.

**Alkali**
A base that dissolves in water.

**Alkali metals**
The name given to group 1 of the periodic table.

**Alkaline earth metals**
The name given to group 2 of the periodic table.

**Alkaline solution**
A solution with pH > 7 at 25°C.

**Alkane**
A homologous series with the general formula $C_nH_{2n+2}$.

**Alkene**
A homologous series with the general formula $C_nH_{2n}$.

**Allotropes**
Different structural forms of an element.

**Amino group**
The functional group $NH_2$.

**Amphiprotic**
A substance that can donate a proton or accept a proton.

**Amphoteric**
A substance that can behave as an acid or a base.

**Anion**
A negatively charged ion.

**Anode**
The electrode at which oxidation takes place.

**Aromatic**
Hydrocarbon compounds that are related to benzene.

**Atomic number**
The number of protons in the nucleus of an atom.

**Atomic radius**
The distance from the centre of the nucleus to the outermost electron shell.

**Average bond enthalpy**
The amount of energy required to break one mole of bonds in the gaseous state averaged across a range of compounds containing that bond.

**Average rate of reaction**
The change in the concentration, mass or volume of the reactants or products over a period of time during the reaction.

**Avogadro's constant**
The number of elementary particles in one mole of a substance. Equal to approximately $6.02 \times 10^{23}$ $mol^{-1}$. Symbol: $L$

**Avogadro's law**
Equal volumes of gases at the same temperature and pressure contain equal numbers of particles.

**Benzene**
Compound with the formula $C_6H_6$, consisting of six carbon atoms in a ring structure with six delocalized electrons. It can also be found as a functional group, $C_6H_5$.

**Best-fit line or curve**
A line, or curve, drawn on a graph that represents the general trend in the measurements but on which not all the data points fall.

**Brønsted–Lowry acid**
A hydrogen ion ($H^+$) donor.

**Brønsted–Lowry base**
A hydrogen ion ($H^+$) acceptor.

**Captive zeros**
Zeros between non-zero integers.

**Carbonyl group**
A functional group consisting of C=O.

**Carboxyl group**
The functional group –COOH.

**Carboxylic acid**
A homologous series with the general formula $C_{n-1}H_{2n-1}COOH$ in which the carboxyl functional group is found.

**Catalyst**
A substance that changes the rate of reaction without itself being used up or permanently changed.

**Cathode**
The electrode at which reduction takes place.

**Cation**
A positively charged ion.

**Chemical property**
A characteristic that is exhibited as one substance is chemically transformed into another.

**Closed system**
A reaction mixture in a container that is sealed and does not allow products to escape to the surroundings.

**Collision theory**
A theory that explains rates of reaction on a molecular level.

**Combustion reaction**
An exothermic reaction in which a fuel is oxidized by oxygen.

**Complete combustion**
The reaction of a hydrocarbon with oxygen at high temperatures in which the products are carbon dioxide and water.

**Concentration**
The amount of solute in a given volume of solution, expressed in $g\ dm^{-3}$ or $mol\ dm^{-3}$.

**Condensed structural formula**
A formula that shows the arrangement of atoms of each element without showing all the bonds in one molecule of a compound.

**Conjugate acid**
Acid formed when a Brønsted–Lowry base gains a hydrogen ion.

**Conjugate base**
Base formed when a Brønsted–Lowry acid donates a hydrogen ion.

**Contact process**
The industrial process in which sulfuric acid is made in a number of stages, starting with the reaction between sulfur and oxygen.

**Continuous spectrum**
Spectrum of light in which there are no gaps; each region blends directly into the next.

**Continuum**
A series of lines that become so close that they merge together.

**Convergence**
The decreasing of the distance between lines in an emission spectrum as the energy of a set of spectral lines increases.

**Core charge**
The effective nuclear charge experienced by the outer-shell electrons of an atom.

**Covalent bond**
The electrostatic attraction of one or more pairs of shared electrons to the two nuclei they are shared between.

**Covalent network lattice**
An arrangement of atoms in a lattice in which there are strong, covalent bonds between the atoms in three dimensions.

**Dative (coordinate) bond**
A covalent bond formed between two atoms, only one of which has provided electrons for the bond.

**Decolourization**
Removal of the colour of a bromine solution by the reaction with an alkene.

**Delocalized electrons**
Electrons which are not confined to a particular location, but are able to move throughout a structure.

**Dependent variable**
The variable that changes as a result of the independent variable changing during an experiment.

**Diatomic molecule**
A molecule which is made up of two atoms.

**Dimer**
A molecule composed of two identical subunits linked together.

**Directly proportional**
As the independent variable increases the dependent variable also increases.

**Displacement reaction**
A reaction in which one substance (a solid) replaces another (aqueous ions). Often involves the reaction of metal atoms with the ions of another metal.

**Dissociation (ionization)**
The process by which ions are formed and separate when a compound dissolves in and reacts with water.

**Double covalent bond**
A covalent bond made up of two pairs of shared electrons (4 electrons).

**Ductility**
The ability to be drawn into a wire.

**Dynamic equilibrium**
When the forward and the backward reactions are still progressing, but macroscopic properties are constant.

**Electrical conductivity**
The ability to allow electricity to pass through a substance.

**Electrochemical series**
A list of oxidizing agents and their conjugate reducing agents written as half-equations in order of increasing strength of the oxidizing agent.

**Electrolysis**
The decomposition of a compound into its constituent elements using electricity to drive the non-spontaneous reactions.

**Electrolytic cell**
A single chamber in which two electrodes are dipped into an electrolyte that is to be decomposed.

**Electromagnetic spectrum**
The range of all possible electromagnetic radiation.

**Electron**
Negatively charged subatomic particle which orbits the nucleus of the atom.

**Electron arrangement**
The pattern of electrons around a nucleus, written as numbers each of which represents the number of electrons in an electron shell, starting from the shell closest to the nucleus and proceeding outwards.

**Electron shell**
Region of space surrounding the nucleus in which electrons may be found.

**Electron transfer**
The movement of electrons from a reducing agent to an oxidizing agent.

**Electronegativity**
A measure of the attraction that an atom has for a shared pair of electrons when it is covalently bonded to another atom.

**Electrostatic attraction**
The attraction between positive and negative charges.

**Element**
A substance made up of only one type of atom.

**Emission spectrum**
Line spectrum generated when an element is excited and then releases energy as light.

**Empirical formula**
A formula that shows the simple whole number ratio of elements in a compound.

**Endothermic reaction**
A reaction that absorbs heat energy from the surroundings: $H_{products} > H_{reactants}$

**Enthalpy**
Heat energy.

**Enthalpy level diagram (energy profile diagram)**
A diagram which shows the relative stabilities of reactants and products, as well as the activation energy, $E_a$, and the change in enthalpy of the reaction.

**Equilibrium constant expression**
The fraction formed by the concentrations of products raised to the powers of their coefficients and divided by the concentrations of reactants raised to the powers of their coefficients.

**Equilibrium constant, $K_c$**
The temperature-dependant value of the equilibrium constant expression.

**Ester**
The functional group –CO-O– made up of a carbonyl group with another oxygen atom bonded to it. This functional group is found only in the middle of molecules, not at the end.

**Exothermic reaction**
A reaction in which heat energy is released to the surroundings: $H_{products} < H_{reactants}$

**Experimental yield**
The amount of product made during an experiment.

**Extent of reaction**
The degree to which products are made.

**First ionization energy**
The amount of energy required to remove one mole of electrons from one mole of atoms in the gaseous state.

**Free radical**
A reactive species containing an unpaired electron, which is the product of homolytic fission.

**Frequency**
The number of waves passing a given point each second.

**Functional group**
An atom or group of atoms that influences the chemical properties of a compound.

**Gradient**
The slope of a line. Calculated by dividing the change in data from the vertical axis (the 'rise') by the change in data from the horizontal axis (the 'run').

**Group**
Vertical column in the periodic table.

**Haber process**
The industrial process in which nitrogen and hydrogen react together to make ammonia.

**Half-cell**
Half a voltaic cell. The half-cell contains a conjugate redox pair: an electrode and an electrolyte.

**Half-equation**
An equation that shows the loss or gain of electrons by an individual reactant.

**Halide group**
A halogen atom that is covalently bonded to a hydrocarbon as a functional group.

**Halide ion**
A negative ion formed when a halogen atom gains one electron.

**Halogenoalkane**
A homologous series with the general formula $C_nH_{2n+1}X$ where X is a member of group 7 in the periodic table.

**Halogens**
The name given to group 7 of the periodic table.

**Heat energy change ($\Delta H$)**
The energy released or absorbed during a reaction.

**Heat of combustion**
The amount of energy produced by the combustion of one mole of a substance. Measured in $kJ\ mol^{-1}$ or $kJ\ g^{-1}$.

**Hess's law**
A law which states that the heat evolved or absorbed in a chemical process is the same, whether the process takes place in one or in several steps.

**Heterogeneous**
In different physical states.

**Heterolytic fission**
The breaking of a covalent bond in which one of the atoms involved in the bond retains both electrons from the bond, forming an anion, and the other atom (part of the molecule) forms a carbocation.

**Highest oxide**
The compound formed with oxygen in which the element is in its highest possible oxidation state.

**Homogeneous**
In the same physical state.

**Homologous series**
A family of compounds which share the same general formula.

**Homolytic fission**
The breaking of a covalent bond in which each atom involved in the bond retains one electron from the bond.

**Hydrocarbon**
A compound predominantly made up of carbon and hydrogen.

**Hydrogen bonding**
A type of strong, dipole–dipole bonding occurring between molecules that contain hydrogen bonded to fluorine, oxygen or nitrogen.

**Hydroxyl group**
The functional group –OH.

**Ideal gas**
A gas in which the gas particles are completely independent.

**Incomplete combustion**
The reaction of a hydrocarbon with oxygen at high temperatures in which the products are carbon monoxide, and/or carbon and water.

**Independent variable**
The variable that is deliberately changed during the course of an experiment.

**Initial rate of reaction**
The change in the concentration, mass, or volume of the reactants or products at the start of the reaction.

**Instantaneous rate of reaction**
The change in the concentration, mass, or volume of the reactants or products at a particular point in time during the reaction.

**Intermolecular forces**
Electrostatic attraction between molecules.

**Inversely proportional**
The dependent variable is proportional to $\dfrac{1}{\text{independent variable}}$.

**Ionic bond**
The electrostatic attraction between a positively charged ion and a negatively charged ion.

**Ionic lattice**
A regular arrangement of ions in which every positive ion is surrounded by negative ions and every negative ion is surrounded by positive ions.

**Isotopes**
Atoms that have the same atomic number but different mass numbers.

**IUPAC system**
The system that determines the names given to compounds, devised by the International Union of Pure and Applied Chemists.

**Ketone**
A homologous series with the general formula $C_nH_{2n}O$, where $n \geq 3$, in which the functional group C=O is found on one of the carbon atoms not at the end of the chain.

**Kinetic molecular theory**
A model or a set of proposals, used to explain the properties and behaviour of gases.

**Law of conservation of mass**
During a chemical reaction the total mass of the reactants is equal to the total mass of the products.

**Le Chatelier's principle**
If a change is made to a system at equilibrium, the reaction will proceed in such a direction as to partially compensate for this change.

**Leading zeros**
Those zeros that precede the first non-zero digit.

**Lewis acid**
An electron pair acceptor.

**Lewis base**
An electron pair donor.

**Lewis structure (electron dot structure)**
A diagram of a molecule or other covalent species in which the outer shell (valence) electrons of the atom are represented by dots or crosses and the sharing of electrons to form a covalent bond is shown.

**Limiting reactant**
The reactant that is used up completely in a chemical reaction.

**Line spectrum**
Discrete lines representing light of discrete energies on a black background.

**Macroscopic properties**
The properties of a reaction that can be measured, e.g. concentration, pressure, temperature, pH.

**Malleability**
Ability of a metal to be bent or beaten into shape without breaking.

**Mass number**
Sum of the numbers of protons and neutrons in the nucleus of an atom.

**Mass spectrometer**
Instrument which enables the relative masses of atoms to be determined.

**Maxwell–Boltzmann energy distribution curve**
A graph showing the distribution of velocities of particles in a gas.

**Metallic bond**
The electrostatic attraction between a positively charged metal ion and the sea of delocalized electrons surrounding it.

**Mistake**
A measurement that does not fit the general trend of the results due to some form of carelessness or lack of skill on the part of the operator.

**Molar mass**
The mass of one mole of a substance measured in grams. Symbol: $M$; unit: g mol$^{-1}$

**Molar ratio**
The ratio in which reactants and products in a chemical equation react. It is indicated by the coefficients written in front of each reactant and product in the equation.

**Molar volume**
The volume occupied by one mole of a gas under a given set of conditions of temperature and pressure.

**Mole**
The amount of substance containing the same number of elementary particles as there are atoms in 12 g exactly of carbon-12. Symbol: $n$; unit: mol

**Molecular formula**
A formula that shows the actual number of atoms of each element present in one molecule of a compound.

**Molecule**
A discrete group of non-metallic atoms, of known formula, covalently bonded together.

**Molten salt**
Ionic compound that has been heated until it is a liquid.

**Monoprotic acid**
An acid that contains only one acidic hydrogen atom per molecule that can be ionized in aqueous solution.

**Neutral solution**
A solution with pH = 7 at 25°C.

**Neutralization reaction**
An exothermic reaction in which an acid reacts with an alkali.

**Neutron**
Uncharged subatomic particle found in the nucleus of the atom.

**Noble gases**
The gaseous elements of group 0 of the periodic table, all of which have a full valence shell.

**Non-bonding pairs of electrons (Lone pairs)**
Pairs of valence electrons that are not involved in bonding.

**Nucleophile**
A reactant with a non-bonding electron pair, or a negative charge, which is attracted to a centre of positive charge.

**Nucleophilic substitution**
A substitution reaction in which a nucleophile is attracted to a positive charge centre.

**Nucleus**
The small dense central part of the atom.

**Open system**
A reaction mixture in a container that allows products to escape to the surroundings.

**Oxidation**
The loss of electrons (or gain of oxygen or loss of hydrogen).

**Oxidation number**
A number that is assigned to an element in a compound, indicating the degree to which the atom has lost electrons.

**Oxidation reaction**
A reaction in which oxygen is gained and/or hydrogen is lost.

**Oxidizing agent (oxidant)**
A reactant, often a non-metal, which causes another reactant to lose electrons (to be oxidized) by accepting electrons from it.

**Percentage composition**
The amount of each element in a compound expressed as a percentage.

**Percentage difference**
This is calculated as
$$\frac{\text{experimental value} - \text{accepted value}}{\text{accepted value}} \times \frac{100}{1}.$$

**Percentage uncertainty**
This is calculated as $\frac{\text{absolute uncertainty}}{\text{measurement}} \times \frac{100}{1}.$

**Percentage yield**
A calculation of the experimental yield as a percentage of the theoretical yield.

**Period**
Horizontal row within the periodic table.

**Periodicity**
The repetition of properties at regular intervals within the periodic table.

**Permanent dipole attraction**
The electrostatic attraction that occurs between polar molecules.

**Physical property**
A characteristic that can be determined without changing the chemical composition of the substance.

**Polar covalent bond**
A covalent bond in which the electrons are not equally shared between the two nuclei due to a difference in the electronegativities of the element involved.

**Polyatomic ion**
A group of two or more atoms covalently bonded together with an overall positive or negative charge.

**Precision**
An expression of how closely a group of measurements agree with one another.

### Primary alcohol/halogenoalkane
An alcohol or halogenoalkane in which the functional group is bonded to a carbon atom that is bonded to only one other carbon atom.

### Proton
Positively charged subatomic particle found in the nucleus of the atom.

### Qualitative data
Observations made during an investigation that would enhance the interpretation of the results. These include observations of changes of colour, production of a gas, liquid or solid, release or absorption of heat.

### Quantitative data
Numerical measurements of the variables associated with an investigation.

### Radioisotope
An isotope that is radioactive.

### Random uncertainty
A minor uncertainty that is inherent in any measurement, such as the error associated with estimating the last digit of a reading.

### Rate of reaction
The change in the concentration, mass or volume of the reactants or products with time.

### Reactant in excess
A reactant that may not be used up completely in a chemical reaction.

### Reaction mechanism
A sequence of reaction steps which shows in detail how a reaction possibly occurs.

### Reaction pathway
A series of steps which traces the reaction of a reactant to form a product which it cannot form directly.

### Reactivity series
A list of metals in order of decreasing reactivity.

### Reducing agent (reductant)
A reactant, usually a metal, which causes another reactant to gain electrons (to be reduced) by donating electrons to it.

### Reduction
The gain of electrons (or loss of oxygen or gain of hydrogen, or decrease in oxidation number).

### Relative atomic mass
The weighted mean of the masses of the naturally occurring isotopes of an element on the scale in which the mass of an atom of carbon-12 is taken to be 12 exactly. Symbol: $A_r$

### Relative formula mass
The sum of the relative atomic masses of the elements as given in the formula for any non-molecular compound.

### Relative isotopic mass
Mass of a particular isotope measured relative to carbon-12. Symbol: $I_r$

### Relative molecular mass
The sum of the relative atomic masses of the elements as given in the molecular formula of a compound. Symbol: $M_r$

### Saturated hydrocarbon
A compound in which all carbon–carbon bonds are single bonds.

### Scattergraph
A graph on which points are plotted individually (as would experimental data points).

### Secondary alcohol/halogenoalkan
An alcohol or halogenoalkane in which the functional group is bonded to a carbon atom that is bonded to two other carbon atoms.

### Self-ionization
The reaction between two molecules of the same type resulting in two ions, e.g. the self-ionization of water:
$H_2O(l) + H_2O(l) \rightleftharpoons H_3O^+(aq) + OH^-(aq)$

### Significant figures
All those digits that are certain plus one estimated (uncertain) digit.

### Single covalent bond
A covalent bond made up of one pair of shared electrons (2 electrons).

### $S_N1$
A nucleophilic substitution reaction in which one species is needed to form the reactive intermediate.

### $S_N2$
A nucleophilic substitution reaction in which two species are needed to form the reactive intermediate.

### Solute
The dissolved component of a solution.

### Solution
A homogeneous mixture of a solute in a solvent.

### Solvent
A substance, usually a liquid, that is able to dissolve another substance, the solute.

### Spectator ions
Ions that do not participate in an aqueous reaction, but are present in the solution.

### Stability
Internal energy of a reactant or product, low energy = high stability.

### Standard electrode potential, $E^\ominus$
The potenial difference generated by a voltaic cell in which a particular half-cell under standard conditions is connected to a standard hydrogen half-cell.

### Standard enthalpy change of reaction
The difference between the enthalpy of the products and the enthalpy of the reactants under standard conditions.

### Standard temperature and pressure (STP)
A set of conditions applied to gaseous calculations where the temperature is 0°C and the pressure is 1 atm (101.3 kPa).

### State symbols (s), (l), (g), (aq)
Symbols used in chemical equations to indicate the state of a reactant or product.

### Strong acid
An acid that dissociates almost completely in aqueous solution.

### Strong base
A base that readily accepts a hydrogen ion from water, and dissociates almost completely in an aqueous solution.

### Strong bonding
Ionic, metallic or covalent bonding.

### Structural formula
A formula that shows the arrangement of atoms and all the bonds in one molecule of a compound.

### Structural isomers
Molecules that have the same molecular formula but different structural formulas.

### Substitution reaction
A reaction in which one atom in a hydrocarbon is replaced by a different atom.

### Systematic error
An error in the system which is usually associated with poor accuracy in measurements.

### Tertiary alcohol/halogenoalkane
An alcohol or halogenoalkane in which the functional group is bonded to a carbon atom which is bonded to three other carbon atoms.

### Theoretical yield
The amount of product that is expected to be produced in a reaction, based on 100% reaction of the reactants.

### Thermochemical equation
A chemical equation that includes the enthalpy change $\Delta H$.

### Trailing zeros
Zeros to the right of a number.

### Triple covalent bond
A covalent bond made up of three pairs of shared electrons (6 electrons).

### Triprotic acid
An acid that contains three acidic hydrogen atoms per molecule which can be ionized in aqueous solution.

### Uncertainty
An expression of the range of values between which a measurement may fall.

### Unsaturated hydrocarbon
A compound that contains at least one carbon–carbon double or triple bond.

### Valence electrons
The electrons in the outermost shell of an atom.

### van der Waals' forces
The weak attraction between molecules that occurs due to the instantaneous dipoles formed as a result of the random movement of electrons within the molecules. Also called dispersion or London forces.

### Volatility
The ease with which a liquid turns into a vapour at low temperatures.

### Voltaic cell (also electrochemical cell or galvanic cell)
An arrangement, consisting of two half-cells connected internally by a salt bridge and externally by wires, that uses a redox reaction to generate electricity.

### VSEPR theory
The theory that accounts for the shape of molecules as a result of electrostatic repulsion between pairs of electrons (bonding and non-bonding) to a position as far from the others as possible in three-dimensional space. Also known as the valence shell electron pair repulsion theory.

### Wavelength
Distance between successive crests of a light wave.

### Weak acid
An acid that dissociates to a small extent in aqueous solution.

### Weak base
A base that accepts a hydrogen ion from water with difficulty, and dissociates to a small extent in an aqueous solution.

# INDEX

Page numbers in **bold** refer to key terms in **bold** type in the text. These terms are also defined in the 'terms and definitions' section for each chapter, and in the cumulative glossary for the coursebook.

## A

absolute uncertainty **171**–4
absolute zero 404
abundance fraction 10, 11–12
accuracy (measurement) 161
acid(s)
    Brønsted–Lowry **277**–83
    carboxylic **362**–4
    Lewis **277**–9
    see also acids and bases; pH
acidic solution **279**
acidity, Lewis 277–9
acid rain **290**–91
acids and bases
    Brønsted–Lowry 277–82
    conjugate pairs 281–2
    indicators 285–6
    Lewis 277–9
    properties 282–6
    strong/weak 286–92
    theories 275–82
    see also acid(s); base(s); carboxylic acids; pH
actinides (periodic table) 79
activation energy **191**–3, **228**–9
    catalysts 235, 258
activation energy barrier 228
activity series 320
addition polymerization reaction **380**–82
addition reactions **376**, 378, 391–2
alcohol (functional group) **356**
alcohols 356–9, 390–91
    oxidation 384–7
    pathways 391–2
    reactions of 383–7
aldehydes **359**–62
alicyclic (molecules) **367**
aliphatic (molecules) **367**
aliquot 165
alkali metals **78**, 87, 91
    periodic table trends 81
    physical properties 81–5
alkaline earth metals **78**–9, 91
alkaline solution **283**
alkalis **283**
alkanes 136–7, **343**–9, 371–6, 390–91
alkenes **350**–52, 376–82, 391
allotropes **51**, 66
aluminium 30, 31, 196

amino group **366**
ammonia
    Haber process 261–3
    Lewis structure 42
    production 261–3
ammonium ion
    formation 44
    structure 44
amount (as number of mole) 101, 104, 128, 404
    gases, factors affecting 128–31, 177
amphiprotic **279**
amphoteric **89**, 279
anion **6**, 20
anode **315**
    see also electrolytic cells; oxidation
area see surface area
Aristotle 2
aromatic (molecules) **367**–8
atom(s)
    Bohr's model 4, 15–18
    central 40, 46
    bonding, metal and metal 34–6
    bonding, metal and non-metal 28–34
    bonding, non-metal and non-metal 36–57
    calculating mass 106
    kinetic energy 226–7
    nuclear, summarized 4
    number, in mole 99, 100–102
    polar covalent bonds 47
    production of spectra 16–18
    properties 4–5
    radii measurement 37, 82
    Rutherford's model 2
    separating by mass 9–11
    Thomson's model 2
    trigonal shapes 41–6
    see also electron(s); electron shells; energy levels (electrons)
atomic number ($Z$) **5**–6, 20, 78
atomic radii
    measurement 37, 82
    periodic table 83–5
atomic radius **86**
atomic spectra 16–17
atomic spectrometry 9–11
atomic structure see atom(s)
attraction
    dipole-dipole 59–60
    permanent dipole 59–60, 66
    see also electrostatic attraction; force(s)
attractive forces (electrons) 34–40
average bond enthalpy **208**
average rate of reaction **222**

Avogadro, Amedeo 100, 134
Avogadro's constant ($L$) **99**–102, 404
    and mole concept 98–102
Avogadro's law **134**–6, 140, 177

## B

balanced equations 114–17, 309–11
barometer 129
base(s)
    Brønsted–Lowry **277**–83
    Lewis **277**–9
    see also acids and bases
battery 312
    see also cells (electric/energy); electrolytic cells
benzene **367**–8
best-fit curve **179**
best-fit line **179**
Bohr model of the atom 15–18
Bohr, Niels 4, 15–18
boiling point see melting/boiling points
Boltzmann, Ludwig 227
bond angle 40–46
bond dissociation enthalpy **27**, 40, 66
bond enthalpies **207**–9, 371
bonding 27
    and melting point 84
    and physical properties 62–4
    alkanes 343–9
    alkenes 350–52
    carbon 342–3
    covalent 36–57, **37**, 66
    dipole-dipole 59–60
    dissociation enthalpies **27**, 40, 66
    functional groups 356–68
    hydrogen **60**–61
    intermolecular forces 57–62
    ionic **28**–34, 66
    lone (non-bonding) pairs **37**, 40–46
    metallic 34–6, 66
    network lattice 51–7
    non-polar 49–50, 58–9
    single covalent **37**
    strong chemical **27**, 34–5, 66
    types, summarized 63
    see also chemical bonds; covalent bonds; ionic bonds; metallic bonds
bond lengths 40
bond polarity 46–50
Boyle, Robert 131
Boyle's law 131–2, 136, 140, 177
Brønsted–Lowry acids/bases **277**–83
buckyballs (fullerenes) 54–6
burette, correct use 165

## C

calculations
- atomic radii measurement 37, 82
- concentration 143–5
- enthalpy changes 196–204
- gas relationships 128–42
- mass 103–7
- mass, molecules 106
- molecular formula of gas 131
- mole numbers 103–7
- number of moles in substance 101, 104, 128
- percentage composition by mass 108–9
- percentage yield 125–6
- relative atomic mass 11–12
- thermochemical equations 202–4
- uncertainty in 161–5, 167, 171–4
- see also equations/formulas; formulas (theory of); redox equations

calorimetry, simple 198
captive zeros (measurement) **167**
carbon
- and relative scales 98
- allotropes 55
- as diamond 52
- as fullerenes 54–6
- Bohr's atomic model 16
- compounds 345
- isotopes 7, 8, 10
- organic chemistry 342–3

carbon dioxide
- formulas 111
- Lewis structure 43
- non-polar molecule 50

carbon monoxide structure 45
carbon nanotubes 56
carbonyl functional group **359**
carboxyl group **362**
carboxylic acids **362**–4
catalyst(s) **234**–5
- and position of equilibrium **249**
- equilibrium system 258
- rate of reaction 234

cathode **315**
- see also electrolytic cells; oxidation

cation **6**, 20, 34–5
Celsius scale 129–30
cells (electric/energy) 312
- dry 313
- electrolytic 327–32, **329**
- half-cells 315–16, 322
- voltaic **312**–19, 328
- see also electrochemical series

central atom 40, 46
certainty (measurement) 167
Chadwick, James 2, 4
change
- in enthalpy 189, 196–204
- of state 226–7

Charles, Jacques 132–3
Charles' law 132–4, 136, 140, 177
chemical bonds 27–64
- Linus Pauling 47
- metal and metal 34–6
- metal and non-metal 28–34
- non-metal and non-metal 36–57
- strong **27**, 34–5
- see also bonding; covalent bonds; ionic bonds

chemical energy see energy
chemical equations 114–18
- balancing 114–17
- thermochemical **190**, 202–4
- word equations 114
- see also equations/formulas; formulas (theory of); redox equations

chemical equilibrium see equilibrium; equilibrium constant; dynamic equilibrium
chemical kinetics 217
chemical properties **87**
- elements 87–9
- oxides 87–9
- periodic table group trends 87–8

chemical reactions see reaction(s)
chlorine 88
- as simple ion 30
- electron arrangement 15
- physical properties 82
- reaction pathways 391

closed system **245**
collision theory 225–36, **228**
colorimetry, rate of reaction 219–20
combustion reactions **189**–90, **372**
- complete **372**–3
- incomplete **372**–3
- measuring heat 198–209

compounds
- determining composition 108–9
- formation 36
- ionic 29–33
- ionic, writing formulas 30–32
- stability of 36
- see also organic chemistry; organic compounds

concentration **143**–7
- and physical equilibria 255–6
- and position of equilibrium 253–5
- and reaction rate 217–19, 231–2
- and time 253–5
- calculating 143–7
- determining mass via 146–7
- solutions 143–5

condensed structural formulas **346**–65
conductivity see electrical conductivity
conjugate acid-base **281**
- see also acids and bases

conjugate redox pair 324
- see also redox reactions

conservation of mass see law of conservation of mass
Contact process 264–7, **265**
continuous spectrum (light) 14–**15**, 20
continuum (electron shells) **18**, 20
convergence (electron shells) **18**, 20
conversion factors
- cubic units 130
- non-SI units 405
- pressure units 129, 405
- temperature 130, 405

coordinate bond **278**
copper
- as ion 30, 31
- mass spectra 10
- specific heat capacity 196

core charge **82**, 84, 86
covalent bonds 36–57, **37**, 66
- dative **45**, 66
- in hydrogen 60
- nature of 37–8
- network lattices 51–7
- polar 47, **49**, 66
- pure 47
- single **37**, 66

covalent molecular substances, properties of 63
covalent network lattices **51**, 63, 66
current see cells (electric/energy)
crystal lattice 33

## D

Dalton, John 2, 37
data 175–81
- see also measurement; qualitative data; quantitative

dative coordinate bond **278**
dative covalent bond **45**, 66
decolourization **379**
delocalized (electrons) **34**–5, 66
Democritus 2
density 404
dependent variable **175**
derived units (listed) 404
diagrams
- energy level **191**
- enthalpy level **191**
- Lewis 40–46
- see also graphical measurement; graphs

diamond
- as carbon allotrope 55
- structure 52

diatomic molecules **37**, 41, 66
diesel, heat of combustion 199

dilution of solutions 145
dimers **364**
dipole-dipole
    attraction 59–61
    bonding 59–60
    *see also* bonding
diprotic **289**
directly proportional **176**
dispersion forces *see* van der Waals' forces
displacement reactions **320**
dissociation (ionization) **277**
ductility **34**, 66
dynamic equilibrium **245**

# E

electrical conductivity **34**–5, 66
    of bonding products 63
    of fullerenes 54–5
    rate of reaction 220
electrochemical series **322**–4
electrode 315
electrolysis **328**
    molten salts 330–32
electrolytic cells 327–32, **329**
    *see also* cells (electric/energy)
electromagnetic spectrum **14**–15, 20
electron dot structure *see* Lewis structure
electron(s) **4**, 20
    arrangement 14–18, **15**, 20, 27, 36–7, 41
    bonding pair 37
    changing energy levels 14–18
    delocalized **34**–5, 66
    forces between 34–40
    history 2–3
    in shells 14–18, 38–9
    instantaneous dipole 58
    Lewis bases 277–9
    metal structure 34–5
    non-bonding pairs 40–46, 66
    outer shell 16, 27, 28, 36, 40–46, 82–4
    oxidation/reduction 303
    relative mass 4
    shell diagrams 38
    transfer (energy level) 29
    valence **15**, 20, **27**, 34, 40–46
    *see also* atom(s); electron shells; energy levels (electrons); Lewis structure
electronegativity **46**, 66, **81**–2
    and bond polarity 46–50
    alkali metals 82
    defined 46
    halogens 82
electron shells **5**, 15–18, 20
    continuum **18**, 20
    convergence **18**, 20
    hydrogen covalent bond 37
    *see also* atom(s); electron(s); energy levels (electrons)
electron transfer **303**
electrostatic attraction **27**, 58, 66
    and ion formation 29
    and ionization energy 82–4
    and metallic bonding 33, 34
    and stable molecules 37
    and temporary dipoles 59
element(s) **28**, 66, 91
    atomic number **5**
    electronegativities 48–9
    ionization energies 83
    isotopes **5**
    mass number **5**
    melting/boiling points 84
    metal 28, 78
    non-metal 28, 78
    Period 3 oxides 89
    physical properties 81–6, 91
    symbols 114
    transuranium 79
    *see also* ionic bonds; periodic table
emission spectroscopy 16–17
emission spectrum **16**–18, 20
    Balmer/Lyman/Paschen series 17, 18
    continuum **18**
    convergence **18**
empirical formula(s) **31**, 66, 108–14, **109**, **345**
    determining 109–10, 352–3
endothermic reactions 189–96, **190**
    and equilibrium constant 250–52
    as bond breaking 208
    summary of 194
energetics 188–209
    exothermic/endothermic reactions 189–96
energy 14–15, 404
    and equilibrium constant 250–52
    activation **191**–3, **228**–9, 235, 258
    as light 14–15
    bond enthalpies **207**–9
    changes via reaction 36
    chemical 36
    electromagnetic spectrum 14
    heat energy change **197**–209
    ionization 82
    kinetic 36–7, 225–8
    light 14–15
    molecule formation 37
    potential 36
    water characteristics 196–7
    *see also* atom(s); electron(s); endothermic reactions; exothermic reactions
energy levels (electrons) 14–18
    high vs low 14, 16–17
energy profile diagram 191

enthalpies, bond **207**–9, 371
enthalpy (*H*) **189**–209
    calculating changes 196–204
enthalpy level diagram **191**
enthalpy of reaction 208
equations, chemical *see* chemical equations; equations/formulas; formulas (theory of); redox equations
equations/formulas
    acids 275
    balanced 114–17
    bases 276
    common negative ions 31
    common positive ions 31
    determining empirical 109–10
    determining molecular 110
    gas relationships 140, 177
    ideal gas 136–8
    ionic 284
    ionization 287
    light energy/wavelength 14
    mass number 5
    number of moles 103–7
    percentage abundance 10
    percentage composition **108**
    percentage difference **161**
    percentage uncertainty **171**
    percentage yield **125**–6
    polyatomic ions 31
    rate of reaction **217**
    relative atomic mass 11
    simple ions 30
    stability (reactants/products) **191**
    unbalanced 114–17
    *see also* Boyle's law; Avogadro's law; calculations; Charles' law; chemical equations; formulas (theory of); redox equations
equations, writing *see* formulas (theory of)
equilibrium
    dynamic 244–6
    gaseous 256–7
    heterogeneous **246**
    homogenous **246**
    physical systems 255–6
    position of 246–60
    *see also* dynamic equilibrium
equilibrium constant **247**–9
    and catalyst 258
    and concentration 254
    and pressure/volume 261
    and temperature 250–52
    *see also* dynamic equilibrium
equilibrium constant expression **247**
equilibrium law 247–9
equilibrium system, dilution of aqueous 252
error 162–5, **163**, 200
    *see also* measurement
esters **366**

ethanoic acid, formulas 111
ethanol
    as biofuel 384
    burning experiment 200
    heat of combustion 199
    specific heat capacity 196
ethene, formulas 111
evaporation, and kinetic energy 226
exothermic reactions **189**–96
    and equilibrium constant 250–52
    as bond making 208
    summary of 194
experiment(s)
    data, graphing 178–81
    fair 175
    heat of combustion, ethanol 200
    measuring heat energy change 198–201
    measuring rate of reaction 219–25
    titration 165–6
experimental yield **125**
extent of reaction **246**

## F
Fahrenheit scale 130
fair experiment 175
first ionization energy **82**–5
fluorine, physical properties 82
force(s) 404
    dispersion 58–9
    electrostatic **27**, 29, 33, 34, 37, 58
    intermolecular **57**–61, 66
    intermolecular vs intramolecular 57
    van der Waals' **58**–60, 66, 85, 346, 350
formulas (theory of)
    comprising symbols 114
    naming, for ionic compounds 30–32
    structural **40**, 66
    writing, for ionic compounds 30–32
    see also equations, chemical; equations/formulas; redox equations
free radical mechanism **374**
fullerenes (buckyballs) 54–6
functional groups **356**–68
fundamental units (listed) 404

## G
gallium 77
gamma see radiation
gas(es)
    and particle motion 226–7
    amount (mole) 128
    carbon monoxide 45
    determining molecular formula 110
    ethene 111
    factors affecting amounts 128–31
    four properties of 128, 177

ideal gas equation 136–8, 140
inert 27
    pressure 128–9, 177
    quantitative behaviour 128–42
    rate of reaction 219
    temperature 129–30, 177
    volume 130–42, 177
    see also carbon dioxide; gas laws; gas relationships; hydrogen
gas constant 404
gaseous volume relationships 131–40
gas laws
    Avogadro's law 134–6, 140
    Boyle's law 131–2, 136, 140
    Charle's law 132–4, 136, 140
gas relationships 128–42
gas-solid-liquid
    change of state 226–7
    chemical equilibrium 245
gas systems, pressure 256–7
general gas equation 140
glossary 417–24
gradient **180**
graphical measurement 175–81
    best-fit line/curve **179**
graphite 53–5
graphs 175–81
    gradient **180**
    sketching 175–7
    see also diagrams
ground state (atom) **15**, 17, 20
group (periodic table) **78**–91

## H
Haber, Fritz 264
Haber process 261–4, **261**
half-cells **314**–16, 322
half-equations **303**, 309–10
halide functional group **364**
halide ion **88**
halogenoalkanes **364**–5
    reactions of 387–9
halogens **78**
    melting/boiling points 59
    periodic table trends 79, 81, 88
    physical properties 81–5
heat
    change of state 226–7
    equilibrium constant 250–52
    see also enthalpy; melting/boiling points; temperature
heat content see enthalpy
heat energy change **197**
    in ethanol experiment 200
    measuring reactions 198–201
heat of combustion 198–209
    ethanol experiment 200
    measuring 198–209

Hess's law **204**–7
heterogeneous equilibrium **246**
heterolytic fission **389**
highest oxides **89**
homogenous equilibrium **246**
homologous series **343**
homolytic fission **374**
hydration **379**
hydrocarbons **343**
    drawing and naming 346–9
    heats of combustion 199
    saturated **343**
    unsaturated **350**
    see also organic chemistry; organic compounds
hydrogen
    bond dissociation enthalpy 28
    bonding **60**–61, 66
    Brønsted–Lowry theory 277
    covalent bonding in 37, 60
    emission spectrum 17–18
    halides 60
    heat of combustion 199
    instantaneous dipole 58
    ions and hydrogenation 379
    isotopes 7
    molecule formation 37
hydrogenation **377**
hydronium ion
    formation 44, 277
    structure 45
hydroxyl group **356**

## I
ideal gas equation 136–8, 140
incomplete combustion **372**
independent variable **175**
industrial processes 261–7
inert gases see gas(es)
infrared
    radiation 14, 18
    rays 14
    spectrometers 11
    spectrum 14
initial rate of reaction **222**
instantaneous dipole 58–9
instantaneous rate of reaction **222**
intermolecular forces **57**–61, 66
    hydrogen bonding **60**–61
    permanent dipole attraction 59–60
    van der Waals' forces **58**–60
inversely proportional **176**
iodine 8, 82
ionic bonding **28**–34, 66
ionic bonds 28–34, 44–5
ionic compounds 29–33
    structure 32–3
    properties 63

ionic equations 284
ionic lattice **32**–33, 34–5, 66
ionic radii
    periodic table 83–5
    values 83
ionization 277
ions **6**, 20
    Brønsted–Lowry theory 276–9
    calculating concentration 144–5
    common, Lewis structures 44–5
    electrical conductivity 35
    formation 29
    kinetic energy 226–7
    mass spectrometer 10
    metal 34–5
    polyatomic 30–32, 66
    positive and negative 31–3
    spectator **284**
iron
    as ion 30, 31
    bond dissociation enthalpy 28
    specific heat capacity 196
isotopes **5**–7, 8–11, 20, 98
    properties of 6–7
IUPAC system **345**

## K

Kelvin scale 129–30
ketones **359**–62
kinetic energy 36–7, 225–8
    electrons 36
    evaporation 226
    melting 226
    particles 225–7
    see also energy
kinetic molecular theory 131, **225**–7
    changes in state 226
    modes of movement 226–7
kinetic particle theory 225–7, 256
kinetics, rates of reaction **217**–25
Kuhn, Thomas 4

## L

lanthanides (periodic table) **79**
lattice
    diamond 52
    graphite 53
    ionic 33, 34–5, 66
    network 51–9
    three-dimensional 34
Lavoisier, Antoine 115
law of conservation of energy 204
law of conservation of mass **114**–17
law of equilibrium 247–9
law of octaves 76
leading zeros (measurement) **167**
Le Chatelier, Henri-Louis 249

Le Chatelier's principle 249–52, **250**, 261
Lewis
    acids **277**–9
    bases **277**–9
Lewis diagram(s) 40–46
Lewis, Gilbert 27, 37, 277
Lewis structure 38–46, 66
    drawing 39
    common molecules 41–4
light
    absorption, rate of reaction 220
    calorimetry 219–20
    continuous spectrum 14–**15**
    emission spectrum 16–18
    line spectra 16–17
    visible 14–18
    wavelength 14–15
    white 15
light energy/wavelength, equation 14
limiting reactant **122**
limiting reagent(s) 122
    and solutions 149–50
linear molecules 41
line spectra **16**–17, 20
liquid-gas-solid
    change of state 226–7
    chemical equilibrium 245
London forces see van der Waals' forces
lone pairs (non-bonding pairs) **37**, 40, 66

## M

macroscopic properties **245**
magnesium 10, 30, 31, 89
malleability **34**, 66
mass 404
    calculations of 103–7, 118–28
    kinetic energy 227
    known and unknown 118–28
    law of conservation of 114–17
    molar (M) **104**, 118–28
    relative atomic (RAM) 10 12, **11**, 20
    relative isotopic (RIM) **11**
    separating atoms by 9–11
    uncertainty (measurement) 164
    via concentrations 146–7
mass/charge ratio ($m/z$) 9, 10, 11
mass number ($A$) **5**, 20
mass spectra 10, 11
mass spectrometer **5**, 9–11, 20, 98
    ion deflection 10
    stages in operation 9–10
mass spectrometry 9–11
matter, change in state 226–7
Maxwell–Boltzmann distributions 227–9
Maxwell–Boltzmann energy distribution curve **227**
Maxwell, James 227
measurement 161–80

    graphical technique 175–81
    liquid volumes 162
    mistake **163**
    of heat energy changes 198–201
    precision **161**
    systematic error 162–5
    tabulations 175
    theory (discussed) 161–4, 167–9, 171–2
    trailing zeros **167**
    uncertainty in 161–5, 167, 171–4
melting/boiling points
    and bonding 84
    alcohol series 357–9
    aldehydes 360
    alkali metals 82
    alkenes 350
    carboxylic acids 363–4
    change of state 226–7
    halogenoalkanes 365
    hydrides 61
    hydrogen bonding 61
    hydrogen halides 60
    ketones 361
    of bonding products 63
Mendeleev, Dmitri 76–7
meniscus, reading 163
mercury 64
metal(s)
    alkali **78**
    alkaline earth **78**–9
    bonding 28–36
    ions 34–5
    properties 35, 63
    reactivity 320–27
    see also atom(s); metallic bonds; transition metals
metallic bonding **34**
metallic bonds 34–6, 66
metallic lattice 34–5
methane
    heat of combustion 199
    Lewis structure 42
    non-polar molecule 49
microwaves 14
mistake (measurement) **163**
molar mass ($M$) **104**
molar volume $V_m$ **138**–43
mole **99**
    and bond enthalpies **207**–9
    as amount 101, 104, 128, 404
    calculating number 103–7
    see also mole concept
mole concept 98–102, 118–28
    see also mole
molecular arrangement see Lewis structure
molecular formula(s) 108–14, **109**, 345
    calculating 111, 352–3
    see also organic chemistry

molecules **37**, 66
  alkane 343–5
  calculating mass 106
  central atom 40, 46
  common arrangement 41
  diatomic 37, 41, 66
  dimer 250
  functional groups 356–68
  kinetic energy 226–7
  Lewis structures 38–46
  linear 41
  non-polar 58–9
  organic 342–92
  polar 59–60
  polyatomic 41
  shapes of 40–45
  symmetrical structure 49–50
  see also atom(s); bonding; force(s)
monomer(s) 380–82
monoprotic **289**
motion, particle 226–7
movement, in collision theory 226–7

## N

naming ionic compounds 30–31
negative charge centres, two, three, four 41, 44
network lattices 51–7
  covalent **51**, 66
neutralization reactions **190**
neutral (substance) **285**
neutron(s) **4**–5, 20
nitrogen
  as simple ion 30
  bond dissociation enthalpy 28
  electron arrangement 15
  STP molar volume 138
Nobel, Alfred 193
noble gases **78**–9
non-bonding pairs (lone pairs) **37**, 40–46, 66
non-polar bonding 49–50, 58–9, 371
normal distribution curve 164
nuclear atom, summarized 4
nucleophiles **388**
nucleophilic substitution **388**
nucleus **5**, 20
nuclide notation 6

## O

ON see oxidation numbers
open system **245**
organic chemistry 342–92
  alcohols 383–7
  alkanes 343–6, 371–6
  alkenes 376–82
  functional groups 356–67

halogenoalkanes 387–9
  reaction pathways 390–93
  see also organic compounds
organic compounds 342–92
  naming 344–5
  see also organic chemistry; hydrocarbons
Ostwald process 261
oxidant 88
oxidation **302**–7, 315
  alcohols 384–7
oxidation numbers (ON) **303**–7
oxidation reactions **384**
  pathways 392
oxides, elements 87–9, 302–6
oxidizing agents **87**, 302–9
oxygen
  as simple ion 30
  electron arrangement 15
  STP molar volume 138

## P

paradigms 4
particle(s)
  and Le Chatelier's principle 249–52
  as gas 226–7
  calculating numbers 100–102
  kinetic energy 226–7
  size, and surface area 233
  subatomic, properties 4–5
  see also energy
particle motion 226–7
Pauling, Linus 47
pentene, formulas 111
percentage abundance 10
percentage composition **108**
percentage difference **161**
percentage uncertainty **171**–4
percentage yield **125**
  calculating 125–6
periodicity 79–89, **81**
periodic table 27, 28, 77–89, 91
  actinides **78**–9
  atomic radii values 83
  chemical properties trends 87–8
  core charge **82**, 84, 86
  electronegativity trends 46, **81**–6
  elements, melting/boiling points 84
  first ionization energy 82–3
  groups **78**
  history of 76–7
  ionic radii values 83
  Mendeleev's 77
  metals/non-metal division 79
  modern 78–81
  period 3 trends 85
  see also element(s)
period (periodic table) **78**

permanent dipole **49**, 59–61
pH 285, 292–5
phosphorus 16, 90, 192
physical constants (listed) 404
physical equilibria 255–6
physical properties **81**
  elements 81–9
  see also properties/structures
pipette, correct use 165
polar covalent bonds 47, **49**, 66
pollution 290–91, 373
polyatomic ions **30**–32, 66
polymer(s) 380–82
potassium, physical properties 82
potential difference 322–4, 404
potential energy 34–6
  and compound formation 36
precision (measurement) **161**
prefixes, units (listed) 169, 404
pressure
  and reaction rate 219, 231–2
  ammonia production 262–3
  gaseous systems 256–7
  gases, factors affecting 128–31
  SI unit of 128, 404
primary alcohol **359**
primary halogenoalkane 364
products (chemical)
  and conservation of mass law 114–17
  and mole concept 118
  and rate of reaction 217–25
  as solutions 148–9
  calculating mass of 118–19
  concentration of 217–19
  dynamic equilibrium 245
  stabilities of **191**
  surface area 233
  see also rate(s) of reaction; reaction(s)
propane
  heat of combustion 199
  reaction pathways 391
propene
  formulas 111
  reaction pathways 391
properties
  acids 282–6
  alkali metals 82
  alkanes/alkenes 344, 350
  bases 282–6
  carbon allotropes 55
  chemical, of elements and their oxides 87–90, 91
  covalent molecular substances 63
  covalent network lattice 63
  elements, melting/boiling points 84
  halogens 59, 81–5
  ionic compounds 29, 63
  isotopes 6–7, 8
  metallic substances 63

metals 28, 35
period 3 elements 85
period 3 oxides 89
physical, of elements 81–6, 91
subatomic particles 4–5
temporary dipole 59

## Q

qualitative data **175**–81
    see also measurement
quantitative
    behaviour of gases 128–42
    data **175**–81
    see also measurement
quantities, getting values from graphs 180–81
quartz, structure 57

## R

radiation 7
    gamma 14
    infrared 14, 18
    poisoning 8
    ultraviolet (UV) 14, 18
radioisotopes **7**, 20
    uses of 7–8
radio waves 14
RAM see relative atomic mass
random uncertainty (measurement) **162**, **163**–5
rate(s) of reaction **217**–25
    and surface area 233
    catalysts 234–5, 258
    colorimetry 219–20
    electrical conductivity 220
    gas volume change 219
    measuring 219–25
    titrimetric analysis 219
    see also reaction rate
reactant(s) 210–12
    and conservation of mass law 114–17
    and mole concept 118
    and rate of reaction 217–25
    as solutions 148–9
    concentration of 217–19
    dynamic equilibrium 245
    excess 122
    limiting **122**
    stabilities of **191**
    surface area 233
    see also rate(s) of reaction; reaction rate
reaction(s)
    and change of state 226–7
    and enthalpy 189–209
    and equilibrium law 247–9
    and gas volume relationships 131–42, 177, 219

addition **376**
addition polymerization **380**
alcohols 383–7
alkanes 371–6
alkenes 376–82
bond breaking 208
bond making 208
calculating theoretical yield of 120
combustion **189**–90, 198–209, **372**
displacement **320**
endothermic 189–96, **190**, 208–9
exothermic **189**–96, 208–9
extent of **246**
halogenoalkanes 387–9
heat energy changes **197**–201
known/unknown mass in 118–28
Le Chatelier's principle 249–52
mass relationships in 118–28
organic, in laboratory 385
oxidation **384**
particle motion 226–7
rates of **217**–25
    see also reactant(s); reaction rate; redox reactions
reaction mechanism **388**
reaction pathways **390**–93
reaction rate **217**–25
    and catalyst 234–5
    and surface area 233
    and temperature 229–32
    concentration in 217–19, 231–2
    see also rate(s) of reaction; reaction(s)
reactivity 320–27, 371–2
reactivity series **320**
redox equations 307–12
redox reactions 307–12, 320–27
reducing agents **88**, 302–9, **307**
reductant **307**
reduction **302**–7, 315
relative atomic mass (RAM) 10–12, **11**, 20, **103**
    calculating 11–12
    list of values 403
relative charges 4
relative formula mass **103**
relative isotopic mass, $I_r$ 11
relative masses 4
relative molecular mass, $M_r$ **103**
relative scale(s) 98
repulsive forces (electrons) 34–40
Roentgen, Wilhelm 7
rounding off (measurement) 168
Rutherford 2–3, 4

## S

salt 29
salt bridge 316
saturated hydrocarbons 343

scattergraph **178**
scientific notation 167
secondary alcohol **359**
secondary halogenoalkane 364
self-ionization **279**
shapes of molecules 40–45
shells (electron) **5**
    see also electron(s); electron shells; energy levels (electrons)
significant figures (measurement) **167**–71
    rules for 168
silicon, structure and bonding 56–7
silicon dioxide, structure and bonding 56–7
simple calorimetry 198
single covalent bond **37**, 66
SI unit(s)
    of pressure 128
    listed 404
$S_N1$ mechanism **388**
$S_N2$ mechanism **388**
sodium 86–7
    as ion 30, 31
    bond dissociation enthalpy 28
    ionic bonding 29–30
    physical properties 82
    water reaction 89
sodium chloride formation 29–30, 33
solid–liquid–gas
    change of state 226–7
    chemical equilibrium 245
solids(s), in reaction 233
solubility
    alcohol series 357
    aldehydes 360
    alkane 136–7
    carboxylic acids 363
    ketones 361
solute(s) **142**–5
solution(s) **142**–51
    and limiting reagent questions 149–51
    alkali **283**
    calculating concentration 143–5
    dilution of aqueous 257–8
    preparing standard 166
    simple calorimeter 198
solvent **142**–5
specific heat capacity **196**, 197, 200
spectator ions **284**
spectra see atom(s); atomic spectra; electron(s); emission spectra; light; line spectra; mass spectra
spectrometers, IR (infrared) 11
spectrometry, mass 9–11
    see also mass spectrometer
spectroscopy 7
    emission 16–17
speed, collision theory 227
    see also particle motion
stabilities (of reactants/products) **191**

stability 191
    and compound formation 36
standard conditions 138, 189
standard electrode potential **322**
standard enthalpy change of reaction **189**–90
standard form (notation) 167
standard solutions, preparing 166
standard temperature and pressure (STP) **138**
standard (volumetric) flask 165–6
states of matter
    changes in 226–7
    chemical equilibrium 245
stoichiometry 119
    and thermochemical equations 202–4
STP (standard temperature and pressure) **138**
strong acid(s) **284**, 286–91
strong base(s) **286**–91
strong chemical bonds **27**, 34–5, 66
structural formula **40**, 66, **345**
structural isomers **346**–9, 351
structure
    Lewis 38–46
    symmetrical 49–50
subatomic particles 4–5
subshells (electron) 15–16, 18
substitution reactions **374**
sulfur 88, 89
sulfur dioxide structure 45
sulfuric acid 264–7
surface area 233, 404
symbols 4, 114, 404
systematic uncertainty (error, measurement) 162–5, **163**

# T

Teflon 382
temperature 404
    and activation energy barrier 228
    and changes of state 226
    and equilibrium constant 250–52
    and gases 129–30, 177
    and kinetic energy 227
    and particle speed 227
    and reaction rate 229–30
    ammonia production 262–3
    Celsius scale 129–30
    Fahrenheit 130
    Kelvin 129–30
    Maxwell–Boltzmann distributions 227
    uncertainty (measurement) 164
    *see also* heat; melting/boiling points
temporary dipole 58–9
tertiary alcohol **359**

tertiary halogenoalkane 364
tetrahedral arrangement 41
    diamond 52
theoretical yield **125**
theories, acids and bases 275–82
theory of knowledge
    data graphing 176
    deductive reasoning 201
    inductive reasoning 78, 201
    kinetic molecular theory 131
    Linus Pauling 47
    paradigms 4
    phlogiston theory 115
    role of serendipity 56
    scientific method 232
    Stephen Hawking's definition 3
    subatomic particles 5
    visible spectrum 18
Thermit reaction 321
thermochemical equation(s) **190**
    enthalpy changes 202–4
Thomson, J.J. 2
time 404
    and concentration 253–5
    in rates of reaction 217–25
    uncertainty (measurement) 164
titration experiments 165–6
titration curve(s) 178
trailing zeros (measurement) **167**
transition metals 30, **79**
    electronegativity trends 46
    periodic table 79
trigonal planar 41
trigonal pyramid 42
triprotic **289**

# U

ultraviolet light 14
ultraviolet (UV) radiation *see* radiation
unbalanced equations 114–17
uncertainty (measurement) **162**–5, 171–4
    absolute **171**–4
    calculating 172
    in results 171–5
    percentage **171**–4
    random uncertainty and systematic errors 162–4
universal gas constant ($R$) 136
units
    cubic, converting 130
    non-SI, conversion factors 405
    prefixes (listed) 169
    pressure, converting 129
    SI (listed) 128, 404
    temperature, converting 130
unsaturated hydrocarbons **350**

# V

valence electrons **15**, 20, **27**, 34, 42–6
    and electronegativity 46–50
    in diamond 52
van der Waals' forces **58**–60, 66, 346, 350
variables
    dependent **175**
    experimental 175–6
    independent **175**
velocity
    kinetic energy 227
    particles 227
visible light *see* light
volatility **358**
Volta, Alessandro 312–13
voltaic cells **312**–19
    vs electrolytic cells 328
volume(s) 404
    gas 130–42, 162, 177, 219
    uncertainty (measurement) 164–5
volumetric glassware, random uncertainties 165
VSEPR theory **40**, 66

# W

water
    as solvent 142
    bond dissociation enthalpy 28
    formulas 111
    heat absorption 196–7
    hydrogen bonding 60
    Lewis structure 43
    reactions, period 3 oxides 89
    self-ionization **279**
    specific heat capacity 196, 404
wavelength (light) **14**–18, 20
    and energy, equation 14
    electromagnetic spectrum 14
    *see also* light; radiation
weak acid(s) **286**–91
weak base(s) **287**–91
wood, heat of combustion 199
word equations 114

# X

X-rays 14

# Z

$Z$ (atomic number) **5**–6, 20